# Applied Research in Uncertainty Modeling and Analysis

## THE INTERNATIONAL SERIES IN INTELLIGENT TECHNOLOGIES

**Series Editor:**
Hans-Jürgen Zimmermann
Aachen Institute of Technology
Operations Research
Aachen, Germany

**Other books in the series:**

*Zimmermann, H.-J.; Tselentis, G.; van Someren, M.; Dounias, G. (Eds.):*
Advances in Computational Intelligence and Learning
*Angstenberger, Larisa:*
Dynamic Fuzzy Pattern Recognition with Applications to Finance and Engineering
*Jain, Lakhmi; De Wilde, Philippe (Eds.):*
Practical Applications of Computational Intelligence Techniques
*Teodorescu, H.-N.; Mlynek, D.; Kandel, A.; Zimmermann, H.-J. (Eds.):*
Intelligent Systems and Interfaces
*Verbruggen, H.B.; Zimmermann, Hans-Jürgen; Babuska, Robert (Eds.):*
Fuzzy Algorithms for Control
*Teodorovic, Dusan, Vukadinovic, Katarina:*
Traffic Control and Transport Planning:
*Babuska, Robert:*
Fuzzy Modeling for Control:
*Ayyub, Bilal M.; Gupta, Madan M. (Eds.):*
Uncertainty Analysis in Engineering and Sciences
*Kacprzyk, J.; Nurmi, H.; Fedrizzi, M. (Eds.):*
Consensus under Fuzziness
*Sebastian, Hans-Jürgen; Antonsson, Erik K. (Eds.):*
Fuzzy Sets in Engineering Design and Configuration
*Hua Harry Li; Gupta, Madan M. (Eds.):*
Fuzzy Logic and Intelligent Systems

# Applied Research in Uncertainty Modeling and Analysis

edited by

**Nii O. Attoh-Okine**
*University of Delaware, Newark, Delaware, USA*

*and*

**Bilal M. Ayyub**
*University of Maryland, College Park, Maryland, USA*

Springer

Library of Congress Cataloging-in-Publication Data

A C.I.P. Catalogue record for this book is available
from the Library of Congress.

ISBN 978-1-4419-3635-6    e-ISBN 978-0-387-23550-9    Printed on acid-free paper.



Printed in the United States of America.

9 8 7 6 5 4 3 2 1

springeronline.com

# DEDICATION

*To my wife Yaa and our children Nii Attoh and Naa Djama*

**Nii O. Attoh-Okine**

*To my wife Deena and our children, Omar, Rami, Samar, and Ziad*

**Bilal M. Ayyub**

# LIST OF CONTRIBUTORS


Achintya Haldar
University of Arizona
Department of Civil and Engineering Mechanics
Tucson, Arizona, USA

Ching-Chung Li
Laboratory for Comp. Neuroscience
Department of Electrical Engineering
University of Pittsburgh
Pittsburgh, Pennsylvania, USA

Daliana B. L.M. Santos
University of Brasilia
Brazil

Deidre E. Paris
Clark Atlanta University
Atlanta, Georgia, USA

Domonkos Tikk
Department of Telecom. and Media Informatics
Budapest University of Technology and Economics
Budapest, Hungary

Dusan Teodorovic
Department of Civil and Environmental Engineering
Virginia Polytechnic Institute and State University
Falls Church, Virgina, USA

Giulio Erberto Cantarella
University of Salerno
Italy

G.M. Ostrovsky
Karpov Institute of Physical Chemistry
Moscow, Russia

György Biró
2Textminer Ltd
Hungary

Hasan Katkhuda
Department of Civil and Engineering Mechanics
University of Arizona
Tucson, Arizona, USA

Hitoshi Furuta
Department of Informatics
Kansai University
Osaka, Japan

Iisakki Kosonen
Helsinki University of Technology
Helsinki, Finland

Jae Dong Yang
Dept. of Computer Science
Chonbuk National University
Chonju, Korea

James J. Buckley
University of Alabama at Birmingham
Birmingham,. Alabama, USA

Jungwon Huh
Yosu National University, Yeosu
Jellanam-dom, Korea

Kenichi Abe
Graduate School of Engineering
Tohoku University
Sendai, Japan

Kevin Reilly
University of Alabama at Birmingham
Birmingham, Alabama, USA

Luke E. K. Achenie
Department of Chemical Engineering
University of Connecticut
Stoors, Connecticut, USA

Madan M. Gupta
College of Engineering
University of Saskatchewan
Saskatoon, Canada

Makoto Yoshizawa
Information Synergy Center
Tohoku University
Sendai, Japan

Maria Alice P. Jacques
University of Brasilia
Brazil

Mark Jablonowski
United Technologies Corporation
Hartford, Connecticut, USA

Mark L. Scheuer
University of Pittsburgh
Pittsburgh, Pennsylvania, USA

Masahiro Yasui
Department of Informatics
Kansai University
Osaka, Japan

Masao Sakai
Graduate School of Engineering
Tohoku University
Sendai, Japan

Matti Pursula
Helsinki University of Technology
Helsinki, Finland

Mauro Dell'Orco
Department of Highways
and Transportation
Polytechnic University of Bari
Bari, Italy

Metin Senbil
Department of Urban Management
Graduate School of Engineering
Kyoto University
Kyoto, Japan

Mingui Sun
University of Pittsburgh
Pittsburgh, Pennsylvania, USA

Musharraf Zaman
University of Oklahoma
Norman, Oklahoma, USA

Neal D. Ryan
University of Pittsburgh
Pittsburgh, Pennsylvania, USA

Noriyasu Homma
College of Medical Sciences
Tohoku University
Sendai, Japan

Qiang Liu
Department of Electrical
Engineering
University of Pittsburgh
Pittsburgh, Pennsylvania, USA

Rafael E. Herreraa
University of Pittsburgh
Pittsburgh, Pennsylvania, USA

Rafiqul Alam Tarefder
Idaho State University
Pocatello, Idaho, USA

Robert J. Sclabassi
University of Pittsburgh
Pittsburgh, Pennsylvania, USA

Robyn R. Bates
University of Pittsburgh
Pittsburgh, Pennsylvania, USA

Ronald E. Dahl
University of Pittsburgh
Pittsburgh, Pennsylvania, USA

Ryuichi Kitamura
Department of Urban Management
Graduate School of Engineering
Kyoto University
Kyoto, Japan

Seung Y. Lee
Arizona Dept of Transportation
Arizona, USA

Shinya Kikuchi
Department of Civil and Environmental
Engineering
University of Delaware
Newark, Delaware, USA

Stefano de Luca
University of Salerno
Italy

Tarek N Kudsi
Senior Structural Engineer
Dar Al- Handasah Consultants
Shair & Partners

Xidong Zheng
University of Alabama at Birmingham
Birmingham Alabama, USA

Zoubir Lounis
Institute for Research in Construction
National Research Council Canada
Ottawa, Ontario, Canada

# FOREWORD

The origins of uncertainty theory and its application can be traced to the inception of philosophy. Now this theory has become an integral part of various disciplines such as engineering, medicine, finance, and computer science.

Uncertainty, which was considered synonymous with random, stochastic, and probabilistic processes has grown to incorporate many more uncertain tools and methodologies. Today the questions with which many practitioners are struggling are:

(a) What is uncertainty? Is it just a lack of knowledge and limited information?

(b) What are the correct approaches to addressing, analyzing, and modeling uncertainty?

(c) How does the quality and quantity of information affect uncertainty analysis and modeling?

(d) How robust are answers obtained from uncertainty analysis and modeling?

Some of these questions have been addressed philosophically in the literature. Since the late 1990's, computational intelligence, or soft computing, which consists of the areas of fuzzy set theory, neural networks and genetic algorithms, has been successfully applied as a tool in uncertainty analysis and modeling. This book on Applied Research in Uncertainty Modeling and Analysis which contains twenty-three invited chapters authored by world researchers focuses on the improved computational techniques to uncertainty modeling and analysis, and it presents both theoretical and practical applications of real world problems, thus exploiting the innovative uses of uncertainty theories.

**Part I** of the book concentrates on the philosophical and theoretical foundations of uncertainty. **Part II** provides biomedical and chemical engineering applications. The biomedical applications use hidden Markov models, Markov random field and mean field theory. These models have wide applications in the field of engineering and other disciplines. **Part III** touches on civil infrastructure systems, and **Part IV** deals with the management of risks and documents classification problems. In **Part V** of the book, a well-balanced mixture of fuzzy systems, neural networks, and agent-based approach and prospect theory is applied to soft transportation applications.

The last section of the book, **Part VI**, demonstrates how uncertainty modeling is important to structural engineering, as uncertainty modeling and analysis is needed in both design and decision making in structural systems. **Parts III, V** and **VI** of the book present concrete applications in civil engineering.

The invited chapters were carefully selected from papers presented at the ISUMA 2003. This book is designed to serve as both a reference guide and a partial textbook in the field of applied uncertainty modeling and analysis. This volume will be an important addition to the International Series in Intelligent Technology.

Madan M. Gupta
Intelligent Systems Research Laboratory
College of Engineering
University of Saskatchewan

# PREFACE

The application areas of uncertainty are numerous and diverse, including all fields of engineering, computer science, systems control and finance. Determining appropriate ways and methods of dealing with uncertainty has been a constant challenge.

The theme for this book is better understanding and the application of uncertainty theories. This book, with invited chapters, deals with the uncertainty phenomena in diverse fields. The book is an outgrowth of the Fourth International Symposium on Uncertainty Modeling and Analysis (ISUMA), which was held at the center of Adult Education, College Park, Maryland, in September 2003. All of the chapters have been carefully edited, following a review process in which the editorial committee scrutinized each chapter.

The contents of the book are reported in twenty-three chapters, covering more than ..... pages. This book is divided into six main sections.

**Part I (Chapters 1-4)** presents the philosophical and theoretical foundation of uncertainty, new computational directions in neural networks, and some theoretical foundation of fuzzy systems.

**Part II (Chapters 5-8)** reports on biomedical and chemical engineering applications. The sections looks at noise reduction techniques using hidden Markov models, evaluation of biomedical signals using neural networks, and changes in medical image detection using Markov Random Field and Mean Field theory. One of the chapters reports on optimization in chemical engineering processes.

**Part III (Chapters 9-11)** describes the application of neural networks and artificial life to civil infrastructure systems. One chapter focuses on pavement deterioration, while the second chapter uses neural networks in residential infrastructure management and the application of underground mall.

**Part IV (Chapters 12-13)** describes the management of risks and document classification problems. **Part V (Chapters 14-18)** presents the applications of uncertainty theory to a variety of transportation engineering problems. This section includes topics on neural networks, fuzzy systems, and agent-based approach and prospect theory in soft transportation applications.

**Part VI (Chapters 19-23)** describes various uncertainties in structural engineering. The section covers risk and reliability applications and chloride contamination in concrete and underground structures.

The studies were designed to accommodate the interests of a large segment of researchers and engineers from a wide variety of subject areas. This clearly indicates an increasing interest in uncertainty applications. Hopefully, the studies will stimulate the interest of other researchers in the theory of uncertainty and innovative applications. The editors are grateful to the authors and the researchers. Also we express our thanks to the editorial staff, especially Sean Lorre of Kluwer Academic Publishers for providing useful feedback and comments during the editorial phases.

Nii O. Attoh Okine
Newark, Delaware

Bilal M. Ayyub
College Park, MD

# CONTENTS

# Chapter 1

# PHILOSOPHICAL AND THEORETICAL BASES FOR ANALYZING AND MODELING UNCERTAINTY AND IGNORANCE

**Bilal M. Ayyub**

## 1. DATA ABUNDANCE AND UNCERTAINTY

The ability of a living system or a machine to make appropriate decisions can be taken as a measure of intelligence. This decision-making ability requires the processing of data and information, construction of knowledge, and assessment of associated uncertainties and risks. The analysis and modeling of uncertainty enhances this ability of making appropriate decisions, thereby increasing intelligence. A need to model and analyze uncertainties is also stemming from the awareness that data abundance does not necessarily give us certainty, and sometimes can lead to error in decision-making with undesirable outcomes due to either overwhelming-confusing situations, or a sense of overconfidence leading to an improper information use. The former situations can be an outcome of the limited capacity of a human mind in some situations to deal with complexity and data abundance; whereas the latter can be attributed to a higher order of ignorance, called the ignorance of self-ignorance.

As our society advances in many scientific dimensions and invents new technologies, human knowledge is being expanded through observation, discovery, information gathering, and logic. Also, the access to newly generated information is becoming easier than ever as a result of computers and the Internet. We have entered an exciting era where electronic libraries, online databases, and information on every aspect of our civilization such as patents, engineering products, literature, mathematics, physics, medicine, philosophy, and public opinions, are becoming a mouse-click or a few clicks away. In this era, computers can generate even more information from abundantly available online data. Society can act or react based on this information at the speed of its generation, creating sometimes non-desirable situations, for example, price and/or political volatilities. There is a great need to assess uncertainties associated with information, and quantify our state of knowledge and/or ignorance. The accuracy, quality, and incorrectness of such information, and knowledge incoherence are coming under focus by our philosophers, scientists, engineers, technologists, decision and policy makers, regulators and lawmakers, and our society as a whole. As a result uncertainty and ignorance analyses are receiving a lot of attention by our society. We are moving from emphasizing the state of knowledge expansion and creation of information to a state that includes knowledge and information assessment by critically evaluating them in terms of relevance, completeness, non-distortion, coherence, and other key measures.

Our society is becoming less forgiving and demanding from our knowledge base. The use of non-credible information, leading to questionable decisions, could place decision makers on the defensive. On the other hand, untimely processing and use of any available information, even if it might be inconclusive, can be treated worse than lack of knowledge and ignorance. In the January 2003 State of the Union address, the U. S. President George W. Bush stated "*The British government has learned that Saddam Hussein recently sought significant quantities of uranium from Africa.*" A few month later, after the conclusion of the 2003 U. S. war on Iraq, senior White House officials have conceded the information that former Iraqi President Saddam Hussein tried to buy uranium from Niger was inaccurate, but they said Bush's State of the Union speech was based on a broader range of intelligence. The assertion that Iraq was trying to reconstitute its nuclear weapons program was a key point in the administration's rationale for war. Although in July 2003, the White House spokesman said "*The issue of Iraq's attempts to acquire uranium from abroad was not an element underpinning the judgment reached by most intelligence agencies that Iraq was reconstituting its nuclear weapons program.*" These statements and decisions were made despite the March 2003 International Atomic Energy Agency dismissal as forgeries documents that alleged Iraq may have tried to buy 500 tons of uranium from Niger. The news elevated the problem to scandalous levels for this action on uncertain information, although inaction on uncertain intelligence, such as the "intelligence failure" in the case of the 2001 World Traded Center attacks, was treated as scandalous and was investigated due to its unacceptability. Any inaction due to non-credible information can be easily taken by our demanding society to be as erroneous as an action based on non-credible information; hence the need for uncertainty assessment, modeling and analysis.

Making appropriate decision commonly entails risk control and management. Although people have some control over the levels of technology-caused risks to which they are exposed, reduction of risk needs to pursued by governments and corporations as a result of increasing demands by our society, and generally entails a reduction of benefits, thus posing a serious dilemma. The public and policy-makers are required, with increasing frequency, to subjectively "weigh benefits against risks" and assess associated uncertainties when making decisions. Further, lacking a systems or a holistic approach, vulnerability exists for overpaying to reduce one set of risks that may introduce offsetting or larger risks of another kind. Such risk-based decisions require uncertainty modeling and analysis.

## 2. KNOWLEDGE

Philosophers defined knowledge, its nature, and methods of acquisitions that evolved over time producing various schools of thought. Table 1 provides a summary of key terminology related to knowledge. Figure 1 shows knowledge types, sources and objects. Figure 2 shows the relationships among information, opinions and knowledge.

*Table 1:* Selected Knowledge and Epistemology Terms [1]

| Term | Definition |
|---|---|
| Philosophy | The fundamental nature of the world, the grounds for human knowledge, and the evaluation of human conduct. |
| Epistemology | A branch of philosophy that investigates the possibility, origins, nature, and extent of human knowledge. |
| Metaphysics | The investigation of ultimate reality. A branch of philosophy concerned with providing a comprehensive account of the most general features of reality as a whole, and the study of being as such. Questions about the existence and nature of minds, bodies, God, space, time, causality, unity, identity, and the world are all metaphysical issues. |
| • Ontology | • A branch of metaphysics concerned with identifying, in the most general terms, the kinds of things that actually exist. |
| • Cosmology | • A branch of metaphysics concerned with the origin of the world. |
| • Cosmogony | • A branch of metaphysics concerned with the evolution of the universe. |
| Ethics | A branch of philosophy concerned with the evaluation of human conduct. |
| Aesthetics | A branch of philosophy that studies beauty and taste, including their specific manifestations in the tragic, the comic, and the sublime; where beauty is the characteristic feature of things that arouse pleasure or delight, especially to the senses of a human observer, and sublime is the aesthetic feeling aroused by experiences too overwhelming (i.e., awe) in scale to be appreciated as beautiful by the senses. |
| Knowledge | A body of propositions that meet the conditions of justified true belief. |
| Priori | Knowledge derived from reason alone. |
| Posteriori | Knowledge gained by reference to the facts of experience. |
| Rationalism | Inquiry based on priori principles, or knowledge based on reason. |
| Empiricism | Inquiry based on posteriori principles, or knowledge based on experience. |

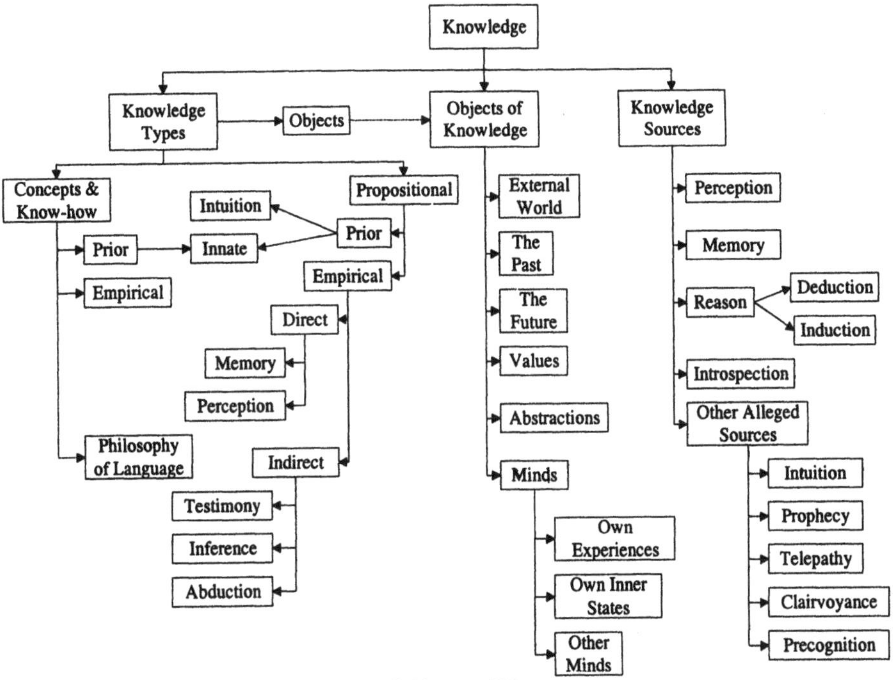


*Figure 1:* Knowledge Types, Sources and Objects [1]

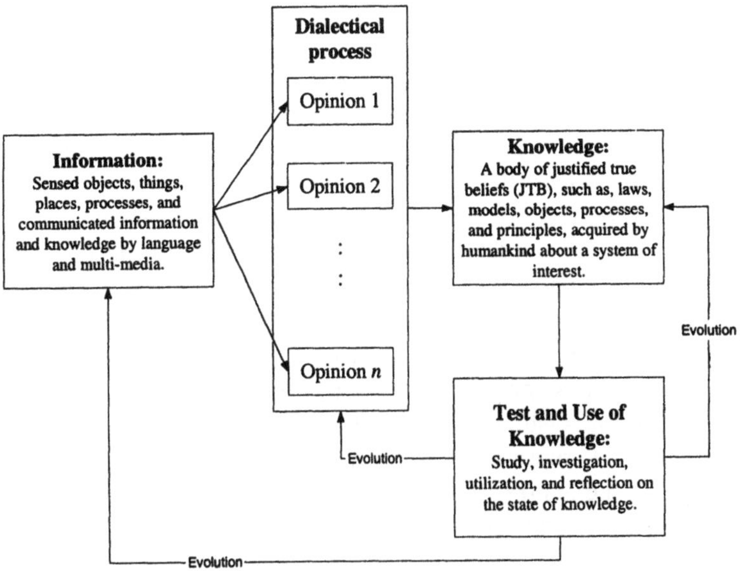


*Figure 2:* Knowledge, Information, Opinions, and Evolutionary Epistemology [1]

# 3. IGNORANCE

## 3.1. Ignorance and Knowledge

Generally, engineers and scientists, and even almost all humans, tend to focus on what is known and not on the unknowns. Even the English language lends itself for this emphasis. For example, we can easily state that Expert A *informed* Expert B, whereas we cannot directly state the contrary. We can only state it by using the negation of the earlier statement as "Expert A *did not inform* Expert B." Statements such as "Expert A *misinformed* Expert B," or "Expert A *ignored* Expert B" do not convey the same (intended) meaning. Another example is "John *knows* David," for which a meaningful direct contrary statement does not exist. The emphasis on knowledge and not on ignorance can also be noted in sociology by having a field of study called the *sociology of knowledge* and not having *sociology of ignorance*, although Weinstein and Weinstein [20] introduced the *sociology of non-knowledge*, and Smithson [14] introduced the *theory of ignorance*.

Engineers and scientists tend to emphasize knowledge and information, and sometimes intentionally or unintentionally brush aside ignorance. In addition, information (or knowledge) can be misleading in some situations because it does not have the truth content that was assigned to it leading potentially to overconfidence. In general, knowledge and ignorance can be classified as shown in Figure 3 using squares with crisp boundaries for the purpose of illustration. The shapes and boundaries can be made multi-dimensional, irregular and/or fuzzy. The evolutionary infallible knowledge (EIK) about a system is shown as the top-right square in the figure, and can be intrinsically unattainable due to the fallacy of humans and the evolutionary nature of knowledge. The state of reliable knowledge (RK) is shown using another square, i.e., the bottom left square, for illustration purpose. The reliable knowledge represents the present state of knowledge in an evolutionary process, i.e., a snapshot of knowledge as a set of know-how, object and prepositions that meet justifiable true beliefs within reasonable reliability levels. At any stage of human knowledge development, this knowledge base about the system is a mixture of truth and fallacy. The intersection of EIK and RK represents the knowledge base with the infallible knowledge components (i.e., know-how, objects and propositions). Therefore, the following relationship can be stated using the notations of set theory:

$$\text{Infallible Knowledge (IK)} = \text{EIK I RK} \qquad (1)$$

where I means intersection. Infallible knowledge is defined as knowledge that can survive the dialectic processes of humans and societies, and passes the test of time and use. This infallible knowledge can be schematically defined by the intersection of these two squares of EIK and RK. Based on this representation, two primary types of ignorance can be identified: (1) ignorance within the knowledge base RK due to factors such as irrelevance, and (2) ignorance outside the knowledge base due to unknown objects, interactions, laws, dynamics, and know-how.

Expert A of some knowledge about the system can be represented as shown in Figure 3 using ellipses for illustrative purposes. Three types of ellipses can be identified: (1) a subset of the evolutionary infallible knowledge (EIK) that the expert has learned, captured and/or created, (2) self-perceived knowledge by the expert, and (3) perception by others of the expert's knowledge. The EIK of the expert might be smaller than the self-perceived knowledge by the expert, and the difference between the two types is a measure of overconfidence that can be partially related to the expert's ego. Ideally, the three ellipses should be the same, but commonly they are not. They are greatly affected by communication skills of experts and their successes in dialectic processes that with time might lead to evolutionary knowledge marginal advances or quantum leaps. Also, their relative sizes and positions within the infallible knowledge (IK) base are unknown. It can be noted from Figure 3 that the expert's knowledge can extend beyond the reliable knowledge base into the EIK area as a result of creativity and imagination of the expert. Therefore, the intersection of the expert's knowledge with the ignorance space outside the knowledge base can be viewed as a measure of creativity and imagination. Another expert (i.e., Expert B) would have her/his own ellipses that might overlap with the ellipses of Expert A, and might overlap with other regions by varying magnitudes.

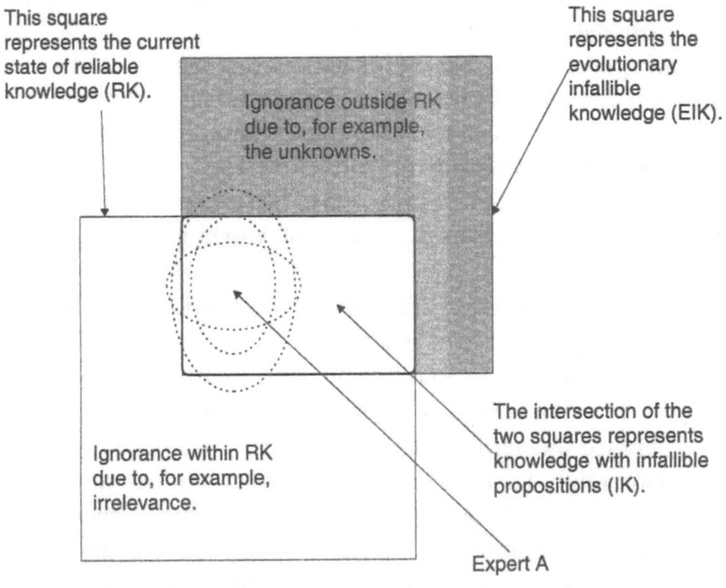


*Figure 3:* Human Knowledge and Ignorance

## 3.2. Classification of Ignorance

The state of *ignorance* for a person or society can be unintentional or deliberate due to an erroneous cognition state and not knowing relevant information, or ignoring information and deliberate inattention to something for various reasons such as limited resources or cultural opposition, respectively. The latter type is a state of *conscious ignorance* which is not intentional, and once recognized evolutionary species try to correct for that state for survival reasons with varying levels of success. The former ignorance type belongs to the *blind ignorance* category. Therefore, ignoring means that someone can either *unconsciously* or *deliberately* refuse to acknowledge or regard, or leave out an account or consideration for relevant information [7]. These two states should be treated in developing a hierarchal breakdown of ignorance.

Using the concepts and definitions from evolutionary knowledge and epistemology, ignorance can be classified based on the three knowledge sources as follows:

- *Know-how ignorance*: It can be related to the lack of, or having erroneous know-how knowledge. Know-how knowledge requires someone to know how to do a specific activity, function, procedure, etc., such as, riding a bicycle.
- *Object ignorance*: It can be related to the lack of, or having erroneous object knowledge. Object knowledge is based on a direct acquaintance with a person, place or thing, for example, Mr. Smith knows the President of the United States.
- *Propositional ignorance*: It can be related to the lack of, or having erroneous propositional knowledge. Propositional knowledge is based on propositions that can be either true or false, for example, Mr. Smith knows that the Rockies are in North America.

The above three ignorance types can be cross-classified against two possible states for a knowledge agent, such as a person, of knowing their state of ignorance. These two states are

- *Non-reflective* (or *blind*) *state*: The person does not know of self-ignorance, a case of ignorance of ignorance.
- *Reflective state*: The person knows and recognizes self-ignorance. Smithson [14] termed this type of ignorance *conscious ignorance*, and the blind ignorance was termed *meta-ignorance*. As a result, in some cases the person might formulate a proposition but still be ignorant of the existence of a proof or disproof, i.e., *ignoratio elenchi*. A knowledge agent's response to reflective ignorance can be either passive acceptance or a guided attempt to remedy one's ignorance that can lead four possible outcomes: (1) a successful remedy that is

recognized by the knowledge agent to be a success leading to fulfillment, (2) a successful remedy that is not recognized by the knowledge agent to be a success leading to searching for a new remedy, (3) a failed remedy that is recognized by the knowledge agent to be a failure leading to searching for a new remedy, and (4) a failed remedy that is recognized by the knowledge agent to be a success leading to blind ignorance, such as ignoratio elenchi or irrelevant conclusion.

The cross classification of ignorance is shown in Figure 4 in two possible forms that can be used interchangeably. Although the blind state does not feed directly into the evolutionary process for knowledge, but it represent a becoming knowledge reserve. The reflective state has a survival value to evolutionary species; otherwise it can be argued that it never would have flourished [4]. Ignorance emerges as a lack of knowledge relative to a particular perspective from which such gaps emerge. Accordingly, the accumulation of beliefs and the emergence of ignorance constitute a dynamic process resulting in old ideas perishing and new ones flourishing [3]. According to Bouissac [3], the process of scientific discovery can be metaphorically described as not only a cumulative sum (positivism) of beliefs, but also an activity geared towards relentless construction of ignorance (negativism), producing architecture of holes, gaps and lacunae so to speak.

Hallden [9] examined the concept of evolutionary ignorance in decision theoretic terms. He introduced the notion of gambling to deal with blind ignorance or lack of knowledge according to which there are times when, in lacking knowledge, gambles must to be taken. Sometimes gambles pay off with success, i.e., continued survival, and sometimes they do not leading to sickness or death.

According to evolutionary epistemology, ignorance has factitious, i.e., human-made, perspectives. Smithson [15] provided a working definition of ignorance based on "Expert A is ignorant from B's viewpoint if A fails to agree with or show awareness of ideas which B defines as actually or potentially valid." This definition allows for self-attributed ignorance, and either Expert A or B can be attributer or perpetrator of ignorance. Our ignorance and claimed knowledge depend on our current historical setting which is relative to various natural and cultural factors such as language, logical systems, technologies and standards which have developed and evolved over time. Therefore, humans evolved from blind ignorance through gambles to a state of incomplete knowledge with reflective ignorance recognized through factitious perspectives. In many scientific fields, the level of reflective ignorance becomes larger as the level of knowledge increases. Duncan and Weston-Smith [8] stated in the *Encyclopedia of Ignorance* that compared to our pond of knowledge, our ignorance remains atlantic. They invited scientists to state what they would like to know in their respective fields, and noted that the more eminent they were the more readily and generously they described their ignorance. Clearly, before solving a problem, it needs to be articulated.

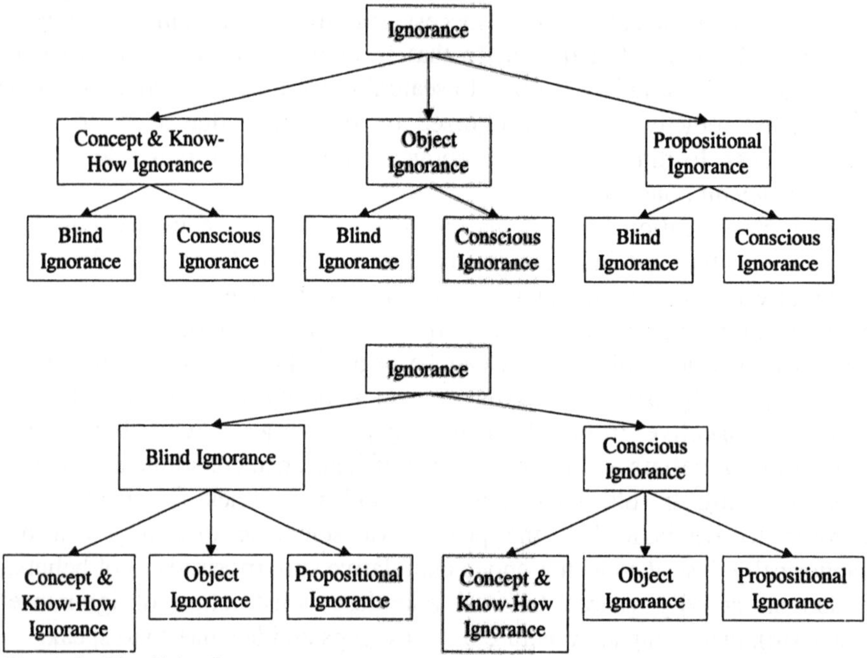

*Figure 4:* Classifying Ignorance

## 3.3. Ignorance Hierarchy

Figures 2 and 3 express knowledge and ignorance in evolutionary terms as they are socially or factitiously constructed and negotiated. Ignorance can be viewed to have a hierarchal classification based on its sources and nature as shown in Figure 5. Ignorance can be classified into two types, blind ignorance (also called meta-ignorance), and conscious ignorance (also called reflective ignorance).

Blind ignorance includes *not knowing* relevant know-how, objects-related information, and relevant propositions that can be justified. The unknowable knowledge can be defined as knowledge that cannot be attained by humans based on current evolutionary progressions, or cannot be attained at all due to human limitations, or can only be attained through quantum leaps by humans. Blind ignorance also includes irrelevant knowledge that can be of two types: (1) relevant knowledge that is dismissed as irrelevant or ignored, and (2) irrelevant knowledge that is believed to be relevant through non-reliable or weak justification or as a result of ignoratio elenchi. The irrelevance type can be due to untopicality, taboo, and undecidability. Untopicality can be attributed to intuitions of experts that could not be negotiated with others in terms of cognitive relevance. Taboo is due to socially reinforced irrelevance. Issues that people must not know, deal with, inquire about, or investigate define the domain of taboo. The undecidedness type deals with issues that cannot be designated true or false because they are considered insoluble, or solutions that are not verifiable, or as a result of ignoratio elenchi. A third component of blind ignorance is fallacy that can be defined as erroneous beliefs due to misleading notions.

Kurt Gödel (1906-1978) showed that a logical system could not be both consistent and complete; and could not prove itself complete without proving itself inconsistent and vise versa. Also, he showed that there are problems that cannot be solved by any set of rules or procedures; instead for these problems one must always extend the set of axioms. This philosophical view of logic can be used as a basis for classifying the conscious ignorance into *inconsistency* and *incompleteness*.

Inconsistency in knowledge can be attributed to distorted information as a result of inaccuracy, conflict, contradiction, and/or confusion as shown in Figure 5. Inconsistency can result from assignments and substitutions that are wrong, conflicting or biased producing confusion, conflict or inaccuracy, respectively. The confusion and conflict results from an in-kind inconsistent assignments and substitutions; whereas inaccuracy results from a level bias or error in these assignments and substitutions.

Incompleteness is defined as incomplete knowledge, and can be considered to consist of (1) absence and unknowns as incompleteness in kind, and (2) uncertainty. The unknowns or unknown knowledge can be viewed in evolutionary epistemology as the difference between the *becoming* knowledge state and *current* knowledge state. The knowledge absence component can lead to one of the scenarios: (1) no action and working without the knowledge, (2) unintentionally acquiring irrelevant knowledge leading to blind ignorance, (3) acquiring relevant knowledge that can be with various uncertainties and levels. The fourth possible scenario of deliberately acquiring irrelevant knowledge is not listed since it is not realistic.

Uncertainty can be defined as knowledge incompleteness due to inherent deficiencies with acquired knowledge. Uncertainty can be classified based on its sources into three types: ambiguity, approximations, and likelihood. The ambiguity comes from the possibility of having multi-outcomes for processes or systems. Recognition some of the possible outcomes creates uncertainty. The recognized outcomes might constitute only a partial list of all possible outcomes leading to unspecificity. In this context, unspecificity results from outcomes or assignments that are not completely defined. The incorrect definition of outcomes, i.e., error in defining outcomes, can be called nonspecificity. In this context, nonspecificity results from outcomes or assignments that are improperly defined. The unspecificity is a form of knowledge absence and can be treated similar to the absence category under incompleteness. The nonspecificity can be viewed as a state of blind ignorance.

The human mind has the ability to perform approximations through reduction and generalizations, i.e., induction and deduction, respectively, in developing knowledge. The process of approximation can involve the use of vague semantics in language, approximate reasoning, and dealing with complexity by emphasizing relevance. Approximations can be viewed to include vagueness, coarseness and simplification. Vagueness results from the non-crisp nature of belonging and non-belonging of elements to a set or a notion of interest; whereas coarseness results from approximating a crisp set by subsets of an underlying partition of the set's universe that would bound the crisp set of interest. Simplifications are assumptions made to make problems and solutions tractable.

The likelihood can be defined in the context of chance, odds and gambling. Likelihood has primary components of randomness and sampling. Randomness stems from the non-predictability of outcomes. Engineers and scientists commonly use samples to characterize populations, hence the last type.

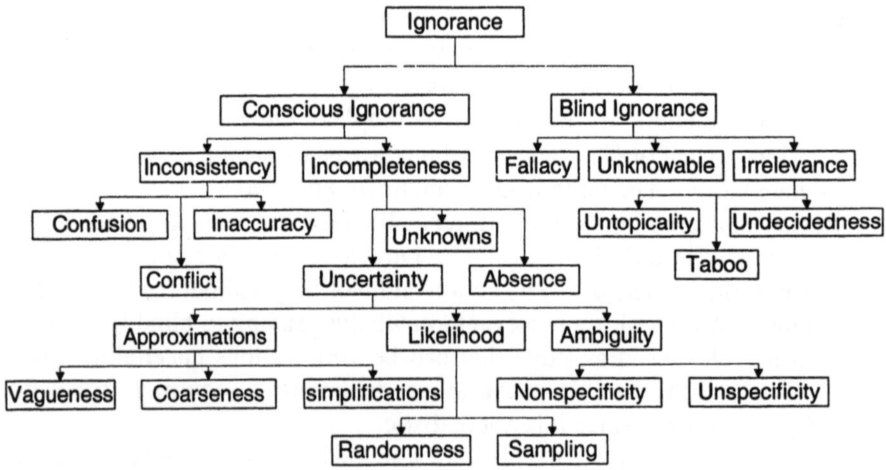


*Figure 5:* Ignorance Hierarchy [1]

# 4. MODELS FOR IGNORANCE AND UNCERTAINTY TYPES

Systems analysis provides a general framework for modeling and solving various problems and making appropriate decisions. For example, an engineering model of an engineering project starts by defining the system including a segment of the project's environment that interacts significantly with it. The limits of the system are drawn based on the nature of the project, class of performances (including failures) under consideration and the objectives of the analysis. The system definition can be based on observations at different system levels in the form of a hierarchy. Each level of knowledge that is obtained about an engineering problem can be said to define a system on the problem. As additional levels of knowledge are added to previous ones, higher epistemological levels of system definition and description are generated which, taken together, form a hierarchy of such system descriptions. An epistemological hierarchy of systems suited to the representation of engineering problems with a generalized treatment of uncertainty can provide realistic assessments of systems [10, 11].

The ignorance types as summarized in Figure 5 might require a mix of theories that most appropriately and effectively model its ignorance content [5, 6, 11, 12, 13, 16, 17, 18, 21, and 22). According to Table 2, classical sets theory can effectively deal with ambiguity by modeling nonspecificity; whereas fuzzy and rough sets can be used to model vagueness, coarseness and simplifications. The theories of probability and statistics are commonly used to model randomness and sampling uncertainty. Bayesian methods can be used to combine randomness or sampling uncertainty with subjective information that can be viewed as a form of simplification. Ambiguity, as an ignorance type, forms a basis for randomness and sampling; hence its cross-shading in the table with classical sets, probability, statistics, and Bayesian methods. Inaccuracy, as an ignorance type, that can be present in many problems, is cross-shaded in the table with probability, statistics, and Bayesian methods. The theories of evidence, possibility and monotone measure can be used to model confusion and conflict, and vagueness. Interval analysis can be used to model vagueness and simplification; whereas interval probabilities can be used to model randomness and simplification. Table 2 provides example application of various theories to address respective ignorance types.

System definition commonly involves data collection and encoding, and expressing information. The process of encoding data and information expression needs to be performed for each aspect of the system in the context of a universe or a universal set ($U$). In probability theory, the universal set is called the sample space ($S$). A universe can be defined as the totality of all the things that exist pertaining to the attribute of interest. Mathematically, a universal set is defined as the set of all objects or elements considered in a given problem or for modeling the attribute of the system. The universe is commonly treated as a "complete" set that is known with absolute certainty, termed in this case the "closed-world" assumption. This assumption can be relaxed to allow for case involving modeling based on an "open-world" assumption. In this section, all modeling cases are theories are based on the "closed-world" assumption as shown in the first column of Figure 6.

The elements of the universe ($U$ or $S$) are commonly assumed as precise objects without any uncertainty in defining such objects. The meaning of the term "precisely defined elements" might vary by application. It could mean that the elements are (1) strictly described; or (2) accurately stated; or (3) definite; or (4) distinctly defined with no variation; or (5) strictly conform to usage and/or rules. This case of precise elements defining *U* is shown in Figure 6 as the first branching of a tree representing several cases that are discussed in this section. The second branch in the first-level of branching in the second column is the case of imprecise or vague objects. Vaguely defined elements carry a contrary meaning to precisely defined elements. This term could mean, depending on the application, (1) not clearly, precisely, or definitely expressed or stated; or (2) indefinite in shape, form, and/or character; (3) hazily or indistinctly seen or sensed; (3) not sharp, certain, or precise in thought, feeling, or expression; or (4) imprecisely determined or known; or (5) uncertain in nature. In this case, the elements f the universe cannot be defined precisely, and are defined in vague terms that are nevertheless meaningful. Examples of the precise elements are integer numeric values, or letters of the alphabet. For the case of vague elements, an example is the illnesses of humans that are of varying imprecision levels.

The third column of Figure 6 addresses a notion of interest that can be expressed by a set or an event that is defined herein as a collection of elements from a universe of interest. Such a notion can be precisely or vaguely expressed. This second level of branching is added to the first-branching level to produce part of the tree in Figure 6. The next column addresses uncertainty in belonging (i.e., membership) of an element to a set or event. Two cases are considered, the case of certain or binary belonging (i.e., 0 for nonbelonging and 1 for belonging to a set), and the case of uncertain or gradual belonging (i.e., belonging is assigned a membership value in the continuous interval 0 to 1). Adding the belonging-uncertainty branching to the tree produces the eight cases as shown in Figure 6 with branches corresponding to various theories that are built on the assumptions enumerated along each branch. The top branch of precisely defined elements in a universe with precisely defined notions and certain belonging form the basis for classical set theory. In cases where a set is not fully known in terms of what elements belong to it, the set can be approximated by rough sets [12]. Cases involving vague notions with uncertain membership can be modeled using fuzzy sets [21]. The branch of precisely defined elements in a universe with vaguely defined notions and certain belonging is not logical and impractical, and is disregarded. Similarly, vaguely defined elements in a universe with precisely defined notions and uncertain belonging form the basis for fuzzy measure theory [19] as a generalization of measure theory. Making the notions in this case vague form the basis for the generalized fuzzy measure theory [19]. The remaining two cases under vaguely defined elements in a universe are illogical and impractical, and are disregarded. Ayyub [2] provides additional information on these theories and their applications.

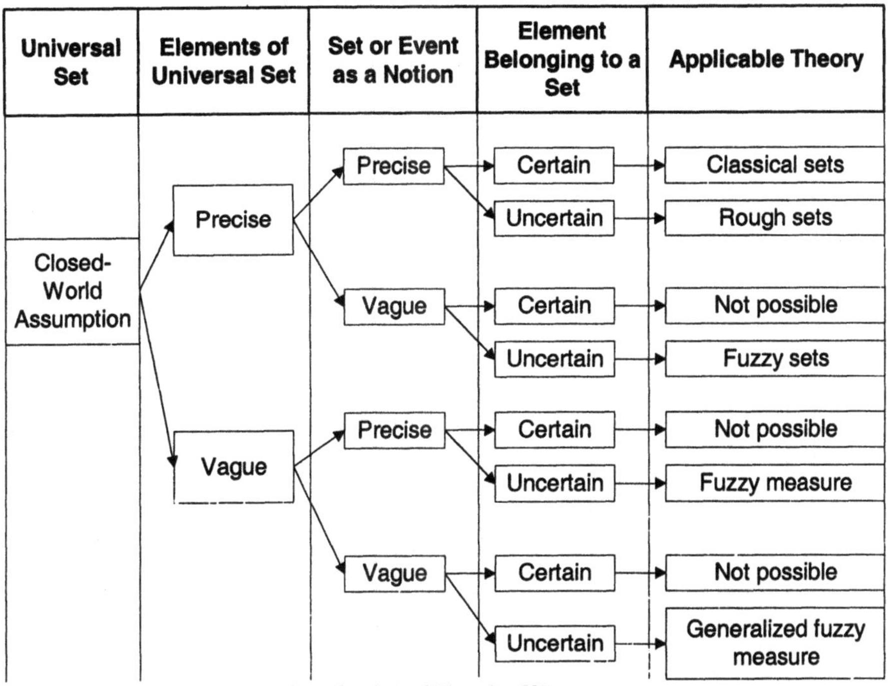


*Figure 6:* Identification and Classification of Theories [2]

| | Ignorance Type | | | | | | |
|---|---|---|---|---|---|---|---|
| Theory | Confusion & Conflict | Inaccuracy | Ambiguity | Randomness & Sampling | Vagueness | Coarseness | Simplification |
| Classical sets | | | Modeling | | | | |
| Probability | | Forecasting | | Quality control | | | Modeling |
| Statistics | | | | Sampling | | | |
| Bayesian | | | | Reliability analysis | | | |
| Fuzzy sets | | | | | Control | | |
| Rough sets | | | | | | Classification | Modeling |
| Evidence | Diagnostics | | | | | | |
| Possibility | | Forecasting | | | Control | | |
| Monotone measure | | | | | | | |
| Interval probabilities | | Risk analysis | | | | | |
| Interval analysis | | Validation | | | | | |

# REFERENCES


Ayyub, B. M., 2001, Elicitation of Expert Opinions for Uncertainty and Risks: Theory, Applications and Guidance, CRC Press, FL.

Ayyub, B. M., 2004. Uncertainty Modeling and Analysis in Engineering and the Sciences, Draft Book Manuscript, Department of Civil and Environmental Engineering, University of Maryland, College Park, MD 20742.

Bouissac, P. 1992. "The Construction of Ignorance and the Evolution of Knowledge," University of Toronto, Quarterly, Vol. 61, No. 4, Summer.

Campbell, D. T. 1974. "Evolutionary Epistemology," In P.A. Schilpp ed., The Philosophy of Karl Popper, LaSalle, Ill.: Open Court Publishing, 1,413-463.

Dempster, A. P., 1976a. "Upper and Lower Probabilities Induced by Multivalued Mapping," Annals of Mathematical Statistics, 38(2), 325-339.

Dempster, A. P., 1976b. "Upper and Lower Probabilities Inferences Based on a Sample From a Finite Univariate Population," Biometrika, 54(3 & 4), 515-528.

di Carlo, C. W., 1998. Evolutionary Epistemology and the Concept of Ignorance, PhD thesis, University of Waterloo, Ontario, Canada.

Duncan, R., and Weston-Smith, M., ed. 1977. The Encyclopedia of Ignorance, New York: Pergamon Press Ltd.

Hallden, S. 1986. The Strategy of Ignorance: From Decision Logic to Evolutionary Epistemology. Library of Theoria, Stockholm.

Klir, G. J., 1985, Architecture of Systems Problem Solving, Plenum Press, New York.

Klir, G. J., and Folger, T. A., 1988, Fuzzy Sets, Uncertainty, and Information, Prentice Hall, N.J.

Pawlak, Z., 1991. Rough Sets: Theoretical Aspects of Reasoning About Data. Kluwer, Boston.

Shafer, G., 1976, A Mathematical Theory of Evidence, Princeton University Press, Princeton, N.J.

Smithson, M., 1985. "Towards a Social Theory of Ignorance," J. of the Theory of Social Behavior, Vol. 15, 151-172.

Smithson, M., 1988, Ignorance and Uncertainty, Springer-Verlag, New York, NY.

Smithson, M., 1989, Ignorance and Uncertainty, Springer-Verlag, New York, NY.

Sugeno, M., 1974. Theory of Fuzzy Intervals and Its Applications. PhD Dissertation, Tokyo Institute of Technology, Tokyo, Japan.

Sugeno, M., 1977. "Fuzzy Measures and Fuzzy Integrals: A Survey," In Gupta, M. M., Saridis, G. N., and Gaines, B. R. (eds.), Fuzzy Automata and Decision Processes, North-Holland, Amsterdam and New York, 89-102.

Wang, Z., and Klir, G. J., 1992. Fuzzy Measure Theory, Plenum Press, New York.

Weinstein, D., and Weinstein, M. A., 1978. "The Sociology of Non-knowledge: a Paradigm." In R. A. Jones (Ed.) Research in the Sociology of Knowledge, Science and Art, Volume 1, JAI Press, NY.

Zadeh, L. A., 1965, "Fuzzy Sets," Information and Control, 8, 338-353.

Zadeh, L. A., 1987, "Fuzzy Sets as a Basis for a Theory of Possibility," Fuzzy Sets and Systems, 1, 3-28.

# Chapter 2

# A SELF-ORGANIZING NEURAL NETWORK BY DYNAMIC AND SPATIAL CHANGING WEIGHTS†

**Noriyasu Homma, Madan M. Gupta, Masao Sakai, Makoto Yoshizawa, and Kenichi Abe**

## 1. INTRODUCTION

In neural networks, synaptic weights are considered to store knowledge of the past experiences (Gupta et al., 2003). As an evidence of synaptic long-term memory structure in biological brain, some spines on which synaptic connections exist could persist throughout a mouse's lifetime (Grutzendler et al., 2002), while some spines in another cortical area had a limited lifetime in contrast (Trachtenberg et al., 2002). This discrepancy seems to imply the fact that several biological learning mechanisms exist and they are still not clear.

Generally speaking, learning and adaptation algorithms provide how to change the weights in order to accomplish given missions. Although the supervised backpropagation learning is one of the most successful neural applications, unsupervised learning or self-organization is an attractive function in biological neural networks. For example, Hebbian learning is a biologically based unsupervised learning (Hebb, 1949) and the self-organizing map (SOM) can learn a topological relation of the environment as in its network structure (Kohonen, 1982; Kohonen, 1989).

On the other hand, cognition using incomplete information or observation is another attractive and intrinsic function in brain. The SOM,

---

† This work is partially supported by The Ministry of Education, Culture, Sports, Science and Technology under Grant-in-Aid for Scientific Research #15700119, by Japan Society for the Promotion of Science under Grant-in-Aid for Scientific Research #15300152, and by The Okawa Foundation for Information and Telecommunications under Research Grant #03-18.

however, needs the complete observation of the input information. This implies the SOM cannot self-organize its connection weights by using incomplete information of the inputs.

In this paper we propose a self-organizing neural structure for storing feature space representation of logical concepts from incomplete observation by using a neuron model with dynamic and spatial changing weights (Homma and Gupta, 2002; Homma et al., 2002). To form the complete informational structure of concepts, (i) a necessary connection structure is created by an extended Hebbian rule and (ii) unnecessary connections are deleted by an unsupervised competitive learning. An ability of the proposed neural network for acquiring the informational structure is proven by using a concept formation problem.

# 2. CONCEPT FORMATION PROBLEM

## 2.1. Human Cognition using Incomplete Information

In general, to understand or express a logical concept we use or have to use an incomplete information or only a subset of the complete information. For example, a concept "apple" is described by its attributes such as "shape is round," "color is red or green," "taste is good," and so on. There are many other attributes of the concept "apple," but it is very difficult to explain all the attributes of the "apple." Indeed we do not recognize, at least consciously, all the attributes of the concept "apple." Thus we use an incomplete informational set of the attributes and our cognition is based on an imperfect informational structure of the concept. The imperfect structure can be formed by our past experience or knowledge.

Note that each of these attributes and their components such as "shape," "round," and "color" is also a *concept*. Then if we do not know about a concept of "shape," we might not be aware of the attribute "shape is round" even if we *receive* this information of the shape.

An interesting thing of our cognition is that even if we could not *recognize* some of these attributes of a concept we can still *understand* the concept in our way. That is, we can understand "this is an apple" even if we are not aware of its shape. In fact, we did not know a concept of "mass -- energy equivalence $E = mc^2$ " before Einstein had discovered. This equivalence is also an important attribute of the concepts "mass" and "energy." We, however, *did* know and understand "mass" and "energy" before the discovery in our way. An important difference between before and after the discovery is *deepness* of our understanding.

## 2.2. Problem Definition

We consider a case of that the incompleteness is only due to a lack of our awareness of some attributes of target concepts. That is, neural networks *receive* the complete information of a concept, but they can *recognize* only an incomplete information through their underdeveloped neural connection structure that represents their imperfect informational structure of the inner awareness. Neural networks are required to develop the imperfect neural connections in order to observe and acquire the complete information of the target.

We define a vector representation of a *concept*, $\mathbf{x}$, using a set of feature variables $\{x_1, x_2, \ldots, x_n\}$

$$\mathbf{x} = [x_1, x_2, \ldots, x_n]^T \in \Re^n \tag{1}$$

Let us consider a set of *N* vector representations of *N* concepts, *S*, given as

$$S = \{\mathbf{x}^1, \mathbf{x}^2, \ldots, \mathbf{x}^N\} \tag{2}$$

where $\mathbf{x}^i$, $i = 1, 2, \ldots, N$, are the vector representations of concepts *i*. A neuron *receives* the complete information of concept *i*, $\mathbf{x}^i$, but can *recognize* only an incomplete observation, $\hat{\mathbf{x}}^i = [\hat{x}_1^i, \hat{x}_2^i, \ldots, \hat{x}_n^i]^T$, through its synaptic connection vector $\mathbf{w} = [w_1, w_2, \ldots, w_n]^T$

$$\hat{x}_j^i = w_j x_j^i, \quad j = 1, 2, \ldots, n \tag{3}$$

Let the initial weight vector be $\mathbf{w}(0) = 0$ and it will be developed to a connection structure $\mathbf{w} \neq 0$. For example, at the initial stage a neuron cannot *recognize* any input information through the initial connections, $\mathbf{w}(0) = 0$.

Here the goal of the *i*th concept formation by the *i*th neuron is to satisfy the following condition

$$\hat{\mathbf{x}}^i \to \mathbf{x}^i \tag{4}$$

In other words, a concept formation defined in this paper is an acquisition process of *awareness* structure for observing the complete information by developing connection weights $\mathbf{w}^i$. In this sense, the connection weights imply strength of *awareness* for the received information.

A trivial solution of Eq. (4) is

$$\mathbf{w}^i \to [1, 1, \ldots, 1]^T \tag{5}$$

This solution means that neurons have all the possible connections and thus any informational structure of the concept vectors will not affect on this neural structure. A biological neuron, however, is connected with only of the order of $10^4$ neurons out of over $10^{10}$ neurons on average (Gupta et al., 2003).

That is, the neural network in the brain has a *sparse* or *local* connection structure that may associate with its local function. The local structure can be constructed by a signal-driven biological process such as Hebbian rule. Indeed, by using Hebbian rule, there is little possibility of making connections that carry few signals: if $x_i$ is always 0, then the possibility of $w_i = 0$ is very high. Furthermore, it is very difficult to discover all the attributes of a concept and the trivial solution is the easiest one only when someone knows all the attributes such as in a supervised learning scheme.

To seek another meaningful solution, let us consider the unipolar binary representation that is a simple cording convention of feature values. For example, features (that are also *concepts* themselves) "round," "cube," "red," and "green" can be corded as $\mathbf{x} = [x_1, x_2, x_3, x_4]^T$. If features of a target concept are "shape is round" and "color is green," then the unipolar binary feature vector can be represented as $\mathbf{x} = [1, 0, 0, 1]^T$.

If we use the unipolar binary representation for the feature values, that is,

$$x_i \in \{0, 1\}, \quad i = 1, 2, \ldots, n \tag{6}$$

then $x_i^2 = x_i$. This implies that the following weights vector is also the solution of Eq. (4)

$$\mathbf{w}^i \to \mathbf{x}^i \tag{7}$$

In this case, the neural structure will converge to the *informational structure* of the concept.

Finally, formation of a set of concepts $S$ using a neural network with $M$ neurons, ($M \geq N$), can be represented by

$$S \subset S_{NN} = \{\mathbf{w}^1, \mathbf{w}^2, \ldots, \mathbf{w}^M\} \tag{8}$$

In addition to this definition of the goal, as mentioned in Section 2.1, human cognition may be subjective and relative. Thus even if an incomplete information $\hat{\mathbf{x}}^i$ is observed through underdeveloped connections, the concept $i$ can be recognized using the underdeveloped observation. We assume that the neural cognition is a process of matching its inner awareness structure $\mathbf{w}$ with an observed informational structure $\hat{\mathbf{x}}$, not with the complete informational structure $\mathbf{x}$. The cognition *result* may be independent of the deepness of understanding, although the deepness is improved as new attributes of the concept are discovered ("mass" is mass even if we do not know about the mass-energy equivalence. see Section 2.1).

For simplicity let $0 \leq w_i \leq 1$ for all $i = 1, 2, \ldots, n$, then the deepness of understanding can be represented by the total amount of awareness given as the sum of the weights $W = \sum_{i=1}^{n} w_i$. Thus the neural output as a result of the cognition process for concept $i$ can be defined as

$$y = \begin{cases} \dfrac{1}{W}\sum_{j=1}^{n} \hat{x}_j^i = \dfrac{1}{W}\mathbf{w}^T\mathbf{x}, & (W \neq 0) \\ 0, & (W = 0) \end{cases} \tag{9}$$

The normalized output implies the ratio of a sum of the observed information to the total amount of awareness. If a neuron is aware of a subset of the complete informational structure, then the output is equal to 1, otherwise the output is less than 1, regardless of the total amount of awareness $W$.

# 3. SELF-ORGANIZING NEURAL STRUCTURE WITH DYNAMIC AND SPATIAL CHANGING WEIGHTS

## 3.2. General Model of Discrete-Time Neurons with Dynamic and Spatial Changing Weights

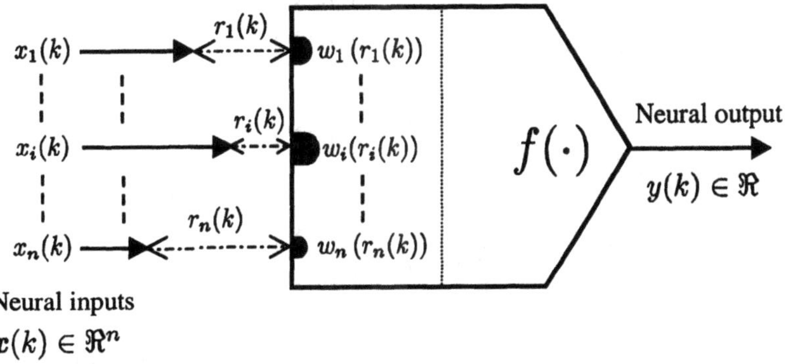


*Figure 1:* A discrete-time neuron with dynamic and spatial changing weights (DSCWs)

Fig. 1 shows the model of a discrete-time neuron with dynamic and spatial changing weights (DSCWs) (Homma and Gupta, 2002; Homma et al., 2002). In this neural model, the following variables are defined:

$\mathbf{x}(k) = [x_1(k), x_2(k), \ldots, x_n(k)]^T \in \Re^n$ : neural input vector;

$\mathbf{r}(k) = [r_1(k), r_2(k), \ldots, r_n(k)]^T \in \Re^n$ : spatial distance vector between the sensory devices or another neuron's axon branches and the corresponding target dendrites (Fig. 1);

$\mathbf{w}(\mathbf{r}(k)) = [w_1(r_1(k)), w_2(r_2(k)), \ldots, w_n(r_n(k))]^T \in \Re^n$ : weight vector as a function of the spatial distance vector $\mathbf{r}$; and

$y(k) \in \Re$ : neural output.

The input $\mathbf{x}(k)$ to the neuron may be external signals or outputs from other neurons, but no self-feedback connection is considered in this paper.

By the definition of the weight vector, the derivatives $dw_i / dr_j$ are 0 for $i \neq j$. Then the spatial changes in the weights are defined by the diagonal matrix as

$$\frac{d\mathbf{w}(\mathbf{r})}{d\mathbf{r}} = \mathrm{diag}\left(\frac{dw_1}{dr_1}, \frac{dw_2}{dr_2}, \ldots, \frac{dw_n}{dr_n}\right) \tag{10}$$

In general a necessary condition to form the *i*th synaptic weight is that the spatial distance $r_i$ is sufficiently short. As listed below various functions that satisfy this condition can be used for representing the relation between weights $w_i$ and distances $r_i \geq 0, \quad i = 1, 2, \ldots, n$.

**(i) linear function**

$$w_i(r_i) = \begin{cases} w_0\left(-\dfrac{r_i}{r_0} + 1\right), & (r_i \leq r_0) \\ 0, & (r_i > r_0) \end{cases} \tag{11}$$

**(ii) exponential function**

$$w_i(r_i) = w_0 \exp\left(-\frac{r_i}{r_0}\right) \tag{12}$$

**(iii) sigmoid function**

$$w_i(r_i) = \frac{w_0}{1 + \exp\left(\dfrac{r_i - r_0}{\alpha}\right)} \tag{13}$$

where $r_0$, $w_0$, and $\alpha$ are positive constants in this paper. Note that the following step-function can be represented by the sigmoid function with $\alpha \to 0$.

$$w_i(r_i) = \begin{cases} w_0, & (r_i < r_0) \\ 0, & (r_i > r_0) \end{cases} \tag{14}$$

A discrete-time *dynamic change* in the connecting weight is defined as

$$\mathbf{w}(\mathbf{r}(k+1)) = \mathbf{w}(\mathbf{r}(k)) + \Delta\mathbf{w}(\mathbf{w}(\mathbf{r}(k)), \mathbf{x}(k), y(k)) \tag{15}$$

where $\Delta\mathbf{w} = [\Delta w_1, \Delta w_2, \ldots, \Delta w_n]^T$ is a vector function describing discrete-time changes in the weight vector. Note that this discrete-time dynamic changes can be used for the discrete-time *learning* scheme that provides how to change the *existing* synaptic weights.

To create new synaptic connections, another mechanism such as *structural adaptation* is needed. A dynamic change in the distance is introduced as a structural adaptation mechanism given by

$$\mathbf{r}(k+1) = \mathbf{r}(k) + \Delta\mathbf{r}(\mathbf{r}(k), \mathbf{x}(k), y(k)) \tag{16}$$

where $\Delta\mathbf{r} = [\Delta r_1, \Delta r_2, \ldots, \Delta r_n]^T$ is a vector function describing discrete-time changes in the distance vector. Note that both the dynamic changes in Eqs. (15) and (16) imply the changes in the distance $\mathbf{r}(k)$. Eq. (15) implies a *learning* algorithm for only *existing* weights $w_i \neq 0$, that is, it only *changes* a strength of awareness. On the other hand, *discovering* a new attribute, that is a creation of a new synaptic connection, is not a subset of the *change*, but achieved by a structural adaptation given in Eq. (16). A specific combination of these two rules can be used for a specific task.

Substituting the dynamic and spatial changing weights into Eq. (9) for $W \neq 0$, the neural output, as a result of neural cognition of a concept, is defined as

$$\begin{aligned} y &= f(\mathbf{x}(k), \mathbf{w}(\mathbf{r}(k))) \in \Re \\ &\equiv \frac{1}{W}(\mathbf{w}(\mathbf{r}(k)))^T \mathbf{x}(k) \end{aligned} \tag{17}$$

where $f$ is an activation function (Fig. 1).

## 3.2. Unsupervised Competitive Learning

We select a neuron $c$ whose output is the largest among the neural network

$$c = \arg\max_j y_j(k) \tag{18}$$

If there are some neurons $c_j \geq 0, \quad j = 1, 2, \ldots$, that satisfy Eq. (18), then the neuron with the largest sum of weights is selected.

$$c = \arg\max_{c_j} W^{c_j}(k) \tag{19}$$

If there are still several candidates, a neuron that satisfies Eq. (19) is selected randomly. Then the neural outputs are re-defined as $y_j(k) = 0, (j \neq c)$, and $y_c(k) = y_c(k)$.

The learning rule in Eq. (15) is achieved by using observed information $\hat{\mathbf{x}}$ instead of the complete information $\mathbf{x}$

$$\begin{cases} w_j^c(k+1) = w_j^c(k) + \gamma[\hat{x}_j(k) - w_j^c(k)] \\ \qquad\qquad\;\; = w_j^c(k)[1 + \gamma(x_j(k) - 1)] \\ w_j^i(k+1) = w_j^i(k), \quad (i \neq c) \end{cases} \tag{20}$$

where $\gamma > 0$ is a learning constant. For simplicity, let $\gamma = 1$ then this competitive learning rule can be written as

$$\begin{cases} w_j^c(k+1) = w_j^c(k)x_j(k) = \hat{x}_j(k) \\ w_j^i(k+1) = w_j^i(k), \quad (i \neq c) \end{cases} \tag{21}$$

In the case of $w_j^c(k) = 1$, note that if $\hat{x}_j(k) = 1$, that implies $x_j(k) = 1$, then this feature is needed to be recognized and thus the connection structure will not be changed $w_j^c(k+1) = 1$. On the other hand, if $\hat{x}_j(k) = 0$, that is $x_j(k) = 0$, this feature is an unnecessary information and thus this connection should be disappear $w_j^c(k+1) = 0$. Thus this learning rule can delete unnecessary connections. However, if $w_j^c(k) = 0$, this learning rule cannot change this weight. The creation of a new connection can be achieved by the following structural adaptation.

## 3.3. Extended Hebbian Rule for Structural Adaptation

Using a spatial changing function as listed in Section 3.1, if a distance is sufficiently small the synaptic weight is relatively large. By a structural adaptation, any connection from sensory neuron *i* to cognitive neuron *j*, $w_{ji}(k)$, can be formed by growth of an axon branch of neuron *i* and a dendrite of neuron *j*. In this paper Eq. (16) is achieved by an extended Hebbian rule as follows:

$$\begin{aligned} r_{ji}(k+1) &= r_{ji}(k) - \eta r_{ji}(k)x_i(k)y_j(k) \\ &= r_{ji}(k)(1 - \eta x_i(k)y_j(k)) \end{aligned} \tag{22}$$

where $\eta > 0$ is a growing constant and $x_i$ and $y_j$ are outputs of sensory and cognitive neurons, respectively. Here, if the two neurons *i* and *j* fire simultaneously, the distance $r_{ji}(k+1) = r_{ji}(k)(1-\eta)$. On the other hand, the distance will not change if they do not fire simultaneously. For simplicity, let $\eta = 1$, $x_i \in \{0, 1\}$, $y_c = 1$, and $y_{j \neq c} = 0$ then

$$r_{ji}(k+1) = \begin{cases} 0, & (x_i(k)y_c(k) = 1) \\ r_{ji}(k), & (x_i(k)y_j(k) = 0) \end{cases} \tag{23}$$

The input information $x_i$, however, can be recognized only through the *existing* synaptic connection with $w_i^c = 1$ since if $w_i^c = 0$ then the recognizable input $\hat{x}$ is always equal to 0, $\hat{x}_i = w_i^c x_i = 0$, regardless of the input value of $x_i$. To avoid this dilemma, let us introduce a supplemental growing scheme that if the $i$th sensory neuron fires, $x_i = 1$, distances between its axon and the dendrites of some possible target neurons, $r_{ji}$, become sufficiently short to form synaptic connections *temporary*. Since the $i$th sensory neuron does not know whether a target cognitive neuron is the candidate neuron $c$ or not, target neurons with $w_i^j = 0$ are selected randomly in this paper. Then according to Eq. (23), the temporal synaptic connections will be *deleted* for the target neurons that do not fire simultaneously, $y_{j \neq c} = 0$, but if the target is the candidate neuron $c$, the connection will be formed permanently since these two neurons $i$ and $c$ fire simultaneously, $x_i y_c = 1$. Note that, thus only a necessary connection structure such that $x_i(k) y_c(k) = 1$ can be created by using Eq. (23).

Such random search of formation and deletion needs so-called *trial and error* mechanism. In general if the number of trials, $N_t$, is sufficiently large at each iteration, a potentially necessary connection with $w_i^c = 0$ for $x_i = 1$ can be selected correctly, and thus this necessary connection will be formed.

# 4. CONCEPT FORMATION FROM INCOMPLETE OBSERVATION

## 4.1. Self-Organizing Network for Concept Formation

Here a self-organizing algorithm to get the solutions given by Eqs. (7) and (8) is proposed. Let us consider the unipolar binary case for features of the concept vector $\mathbf{x}$.

1. Initialize iteration $k = 0$, the number of neurons $m(0) = 0$, and the connection vector $\| \mathbf{w}^1(0) = 0 \|$.

2. Select one concept vector $\mathbf{x}^{i(k)} \in S$, $i(k) \in \{1, 2, K, N\}$ and input it to the neural network.

3. Select a candidate neuron for the concept $\mathbf{x}^{i(k)}$ by the method described in Section 3.2.

4. Self-organizing the weights according to the following condition

Case 1: If the candidate output $y_c$ is equal to 1, try to create a new *awareness* weight $w_i^c$ so that $\hat{x}_j(k) = 1$ by the extended Hebbian rule given in Eq. (23).

Case 2: If $0 < y_c < 1$, create new neuron $m(k)+1$ with the same weight vector $\mathbf{w}^{m(k)+1} = \mathbf{w}^c$ and $m(k)+1 \to m(k)$. Then $c = m(k)$ and learn the weight to delete unnecessary connections by using Eq. (21).

Case 3: In the case of $y_c = 0$ if there is a neuron without any connections, then select this neuron for a candidate. Otherwise, create a new candidate neuron and $m(k)+1 \to m(k)$. Then try to create a randomly selected new *awareness* weight for this candidate by Eq. (23).

5. $m(k+1) = m(k)$, $k \to k+1$ and return to Step 2.

## 4.1. Some Remarks on Ability of Concept Formation Network

A concept formation ability of the proposed self-organizing network is summarized in the following theorem.

**Theorem 1** (Concept Formation Neural Structure Theorem) For a unipolar binary concept set $S = \{\mathbf{x}^1, \mathbf{x}^2, \ldots, \mathbf{x}^N\}$, there exists a self-organizing network with $M$ neurons, ($M \geq N$), such that

$$\mathbf{x}^i = \mathbf{w}^{j_i}, \quad i = 1, 2, \ldots, N, \quad j_i \in \{1, 2, \ldots, M\} \tag{24}$$

**Proof**: For a concept $\mathbf{x}(k) = \mathbf{x}^i$, a neuron $c$ satisfying the candidate condition in Section 3.2 is selected. Then even if the current candidate output as a function of the current weight vector $\mathbf{w}^c(k)$ is less than 1, $y_c(\mathbf{x}^i, \mathbf{w}^c(k)) < 1$, the output after the learning in Step 4 can be changed into 1

$$y_c(\mathbf{x}^i, \mathbf{w}^c(k+1)) = 1 \tag{25}$$

This is because if $y_c(k) < 1$, we can delete only the wrong connections for the target concept by using the self-learning rule given as Eq. (21). Also if $y_c(k) = 1$, then

$$\| \mathbf{w}^c(k+1) \| > \| \mathbf{w}^c(k) \| \tag{26}$$

except for $\mathbf{w}^c(k) = \mathbf{x}^i$ by using the Hebbian rule that can create only the correct connection if the number of trials is sufficiently large; there is no wrong connection weight for the candidate. That is,

$$\| \mathbf{x}^i - \mathbf{w}^c(k+1) \| < \| \mathbf{x}^i - \mathbf{w}^c(k) \| \tag{27}$$

Thus if this neuron is a candidate only for concept *i*

$$\mathbf{w}^c \to \mathbf{x}^i \tag{28}$$

Otherwise, if this neuron *c* will also be a candidate (i.e. $y_c(k) = 1$) for another concept $\mathbf{x}^j$, $j \neq i$ and $\| \mathbf{x}^j \| > \| \mathbf{w}^c \|$ then

$$\mathbf{w}^c \to \mathbf{x}^j \tag{29}$$

Once $\mathbf{w}^c$ converges to $\mathbf{x}^j$, another candidate neuron $c'(\neq c)$ for concept *i* will not be a candidate for concept *j* since neuron *c* is the only candidate for concept *j*. Thus after all concepts like concept *j* have their own candidates, the weight of a candidate neuron *c'* will converge to concept *i*

$$\mathbf{w}^{c'} \to \mathbf{x}^i \tag{30}$$

Therefore if *M* is sufficiently large

$$S \subset S_{NN} = \{\mathbf{w}^1, \mathbf{w}^2, \mathrm{K}\ , \mathbf{w}^M\} \tag{31}$$

This proves the theorem. ■

In addition to this ability, it is worth to compare the proposed self-organizing learning with Kohonen's self-organizing map (SOM) (Kohonen, 1982; Kohonen, 1989). A major difference between the SOM and the proposed method given in Eqs. (18) and (20) is completeness of the input information. SOM supposes that the complete information of the input $\mathbf{x}$ can be given, while only an incomplete observation of the input, $\hat{\mathbf{x}}$, is possible to use in the proposed method. In the sense of a hypothesis that human cognition is based on an incomplete information, the proposed network provides better model of human cognition.

Furthermore, the extended Hebbian rule proposed for structural adaptation is a possible informational formulation of a biological evidential perspective in which some signal-driven special chemical matters can lead to growing directions of synaptic connections (Kohara et al., 2001).

## 5. CONCLUSIONS

We have developed a new self-organizing network model of concept formation by using neurons with dynamic and spatial changing weights. The proposed network can recognize concepts using incomplete information of concepts at current degree of understanding and develop its synaptic connections of inner informational structure to discover new information of concepts and finally the complete information.

# REFERENCES


Grutzendler, J., Kasthuri, N., and Gan, W. B., 2002. "Long-term dendritic spine stability in the adult cortex," Nature, vol. 420, pp. 812-816.

Gupta, M. M., Jin, L., and Homma, N., 2003. Static and Dynamic Neural Networks - From Fundamentals to Advanced Theory -, IEEE Press & Wiley, Hoboken, NJ.

Hebb, D. O., 1949. The Organization of Behavior, Wiley, New York.

Homma, N., and Gupta, M. M., 2002. "Incremental Neural Learning by Dynamic and Spatial Changing Weights," Proc. 15th IFAC World Congress, vol. 3e, T-We-A04-2.

Homma, N., Gupta, M. M., Abe, K., and Takeda, H., 2002. "Superimposing Memory by Dynamic and Spatial Changing Synaptic Weights," Proc. 2002 SICE Annual Conf., pp. 2903-2908.

Kohara, K., Kitamura, A., Morishima M., and Tsumoto T., 2001. "Activity-Dependent Transfer of Brain-Derived Neurotrophic Factor to Postsynaptic Neurons," Science, vol. 291, pp. 2419-2423.

Kohonen, T., 1982. "Self-organized formation of topologically correct feature maps," Biol. Cybern., vol. 43, pp. 59-69.

Kohonen, T., 1989. Self-Organization and Associative Memory, 3rd ed., Springer-Verlag.

Trachtenberg, J. T., Chen, B. E., Knott, G. W., Feng, G., Sanes, J. R., Welker, E., and Svoboda, K., 2002. "Long-term *in vivo* imaging of experience-dependent synaptic plasticity in adult cortex," Nature, vol. 420, pp. 788-794.

## Chapter 3

# SIMULATION OF FUZZY SYSTEMS I

**James J. Buckley, Kevin D. Reilly and Xidong Zheng**

## 1. INTRODUCTION

We begin in the next section with a discussion of how we obtain fuzzy numbers for arrival and service rates in a basic queuing network. The fuzzy queuing model is presented in the third section. Since this model is discussed in [2] we only present an overview. This is followed by our simulation models and results. We go through the fuzzy calculations in detail so the reader can appreciate how the simulation method simplifies the process of going from initial data to the optimization models.

However, the simulations sometimes produce fuzzy results at variance with the fuzzy calculations. The matter will be discussed at appropriate points in succeeding sections and will be related to our belief that some modeling and simulation styles may be approximating the results that would be obtained using the extension principle.

The last section has a brief summary and our plans for future research on this topic. This chapter expands on previous work reported in conference papers [6],[23],[27],[28].

Let us now introduce the notation we will use in the paper. We place a "bar" over a symbol to denote a fuzzy set. So, $\overline{a}_i$, $\overline{A}$, $\overline{x}$, ... all represent fuzzy sets. If $\overline{A}$ is a fuzzy set, then $\overline{A}(x) \in [0,1]$ is the membership function for $\overline{A}$ evaluated a real number $x$. An $\alpha$-cut of $\overline{A}$, written $\overline{A}[\alpha]$, is defined as $\{x|\overline{A}(x) \geq \alpha\}$, for $0 < \alpha \leq 1$. $\overline{A}[0]$ is separately defined as the closure of the union of all the $\overline{A}[\alpha]$, $0 < \alpha \leq 1$. A fuzzy number $\overline{N}$ is a fuzzy subset of the real numbers satisfying: (1) $\overline{N}(x) = 1$ for some $x$ (normalized); and (2) $\overline{N}[\alpha]$ is a closed, bounded interval for $0 \leq \alpha \leq 1$. A triangular fuzzy number $\overline{T}$ is defined by three numbers $a_1 < a_2 < a_3$ where the graph of $y = \overline{T}(x)$ is a triangle with base on the interval $[a_1, a_3]$ and vertex at $x = a_2$ ( $\overline{T}(a_2) = 1$). We write $\overline{T} = (a_1/a_2/a_3)$ for triangular fuzzy numbers. A triangular shaped fuzzy number has curves, not straight line segments, for the sides of the triangle. For any fuzzy number $\overline{N}$ we have $\overline{N}[\alpha] = [n_1(\alpha), n_2(\alpha)]$ all $\alpha$, which describes the closed, bounded, intervals as functions of $\alpha$.

Probably the first paper on fuzzy simulation was [7], followed by [8]. Here the authors wanted to "randomly" generate values of a fuzzy variable $\overline{F}$. The fuzzy variable $\overline{F}$ has its values restricted by a possibility distribution $\overline{N}$, which is usually just a fuzzy number. This means that $Poss(\overline{F} = x) = \overline{N}(x)$. They randomly produce a value for $\overline{F}$ using a two step process: (1) first randomly generate $\alpha$ from a uniform distribution over $[0, 1]$; (2) then randomly generate $x$ from a uniform distribution over the interval $\overline{N}[\alpha]$. This procedure has subsequently been employed in the simulation models in [20] and [25]. A completely different approach is in [3] where the authors show how to generate a chaotic sequence of fuzzy numbers, all of the same type and in the same interval, to estimate the fuzzy solution to a fuzzy optimization problem. In [11] the author is also concerned with fuzzy simulation. In this paper the author selects specific values from the fuzzy numbers in the problem (called Fuzzy Method 3 in the paper), inputs these numbers and computes the result, and then constructs a fuzzy set for the final result from the individual outputs. This is similar to what we will be doing. Finally, the last paper involving fuzzy simulation is [26] where the author simulates verbal models. In this paper the author substitutes fuzzy sets for linguistic variables, uses hedges and the compositional rule of inference, to obtain a fuzzy set conclusion which is then translated back to a linguistic term.

## 2. FUZZY ARRIVAL/SERVICE RATES

We start our model from crisp arrival/service rates ($\lambda/\mu$). Recall that a complete specification of a (crisp, single station) queuing system involves specifying only these rates and a queue discipline; since we will be assuming FIFO disciplines throughout, we can focus on the arrival/service rates. For a fuzzy system, the fuzzy rates ($\overline{\lambda}/\overline{\mu}$) must replace the crisp values at some point. Refer to Figure 1, which shows the stages of computation in the (totally) crisp case starting from arrival/service rates.

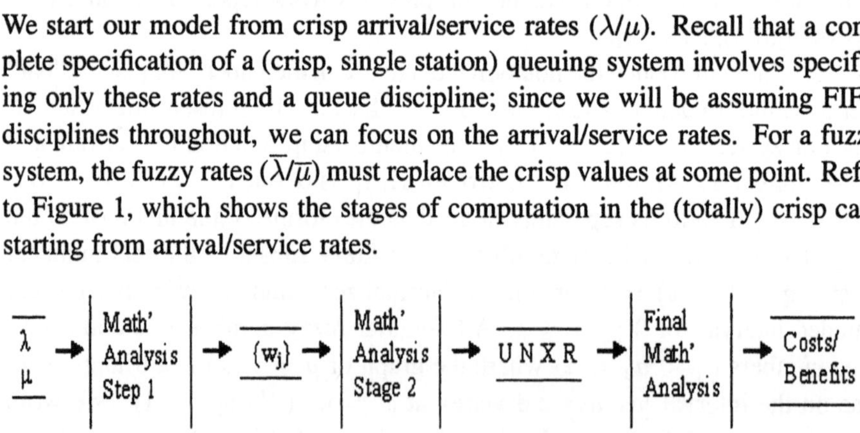


Figure 1: Steps in Crisp Calculation of System Performance Values Starting from Arrival/Service Rates, Culminating in Cost and Benefi ts.

The overall progression of the mathematical argument and/or simulation is as follows. We first construct fuzzy numbers for $\lambda$ = the arrival rate and for $\mu$ = the service rate. Using these we determine fuzzy steady state probabilities $(\overline{w}_j)$. From the fuzzy steady state probabilities we compute the fuzzy numbers for system performance $\overline{U}$ = server utilization, $\overline{N}$ = number of customers in the system, $\overline{X}$ = throughput, $\overline{R}$ = response time and $\overline{LC}$ = lost customers (see [19]). Finally, we input the fuzzy numbers for system performance into models to find the optimal mix of the variables (number of servers, capacity,...) to $min\overline{R}$, $max\overline{U}$, max fuzzy profit, etc. We now discuss each part of the model, except the final optimization, before we present our simulation results which will take us directly from $\alpha$-cuts of $\overline{\lambda}$ and $\overline{\mu}$ to the optimization procedure.

For the rest of this section we concentrate on deriving fuzzy numbers for the arrival rate, and the service rate in a queuing system. We consider the fuzzy arrival rate first.

## 2.1. Fuzzy Arrival Rate

We assume that we have Poisson arrivals [24] which means that there is a positive constant $\lambda$ so that the probability of $k$ arrivals per unit time is

$$\lambda^k \exp(-\lambda)/k!, \tag{1}$$

the Poisson probability function. We need to estimate $\lambda$, the arrival rate, so we take a random sample $X_1, ..., X_m$ of size $m$. In the random sample $X_i$ is the number of arrivals per unit time in the $i^{th}$ observation. Let $S$ be the sum of the $X_i$ and let $\overline{X}$ be $S/m$. Here, $\overline{X}$ is not a fuzzy set but the mean.

Now $S$ is Poisson with parameter $m\lambda$ ([14],p.298). Assuming that $m\lambda$ is sufficiently large (say, at least 30), we may use the normal approximation ([14],p.317), so the statistic

$$W = \frac{S - m\lambda}{\sqrt{m\lambda}}, \tag{2}$$

is approximately a standard normal. Then

$$P[-z_{\beta/2} < W < z_{\beta/2}] = 1 - \beta, \tag{3}$$

where the $z_{\beta/2}$ is defined as

$$\int_{-\infty}^{z_{\beta/2}} N(0,1)dx = 1 - \beta/2, \tag{4}$$

and $N(0,1)$ denotes the normal density with mean zero and unit variance. Now we divide numerator and denominator of $W$ by $m$ and get

$$P[-z_{\beta/2} < Z < z_{\beta/2}] = 1 - \beta, \tag{5}$$

where

$$Z = \frac{\overline{X} - \lambda}{\sqrt{\lambda/m}}. \tag{6}$$

From these last two equations we may derive an approximate $(1-\beta)100\%$ confidence interval for $\lambda$. Let us call this confidence interval $[l(\beta), r(\beta)]$.

We now show how to compute $l(\beta)$ and $r(\beta)$. Let

$$f(\lambda) = \sqrt{m}(\overline{X} - \lambda)/\sqrt{\lambda}. \tag{7}$$

Now $f(\lambda)$ has the following properties: (1) it is strictly decreasing for $\lambda > 0$; (2) it is zero for $\lambda > 0$ only at $\overline{X} = \lambda$; (3) the limit of $f$, as $\lambda$ goes to $\infty$ is $-\infty$; and (4) the limit of $f$ as $\lambda$ approaches zero from the right is $\infty$. Hence, (1) the equation $z_{\beta/2} = f(\lambda)$ has a unique solution $\lambda = l(\beta)$; and (2) the equation $-z_{\beta/2} = f(\lambda)$ also has a unique solution $\lambda = r(\beta)$.

We may find these unique solutions. Let

$$V = \sqrt{z^2_{\beta/2}/m + 4\overline{X}}, \tag{8}$$

$$z_1 = [-\frac{z_{\beta/2}}{\sqrt{m}} + V]/2, \tag{9}$$

and

$$z_2 = [\frac{z_{\beta/2}}{\sqrt{m}} + V]/2. \tag{10}$$

Then $l(\beta) = z_1^2$ and $r(\beta) = z_2^2$.

We now substitute $\alpha$ for $\beta$ to get the $\alpha$-cuts of fuzzy number $\overline{\lambda}$. Add the point estimate, when $\alpha = 1$, $\overline{X}$, for the 0% confidence interval. Now as $\alpha$ goes from 0.01 (99% confidence interval) to one (0% confidence interval) we get the fuzzy number for $\lambda$. We drop the graph straight down at the ends to obtain a complete fuzzy number.

**Example 1**

Suppose $m = 100$ and we obtained $\overline{X} = 25$. We evaluated equations (8) through (10) using Maple [18] and then the graph of $\overline{\lambda}$ is shown in Figure 2, without dropping the graph straight down to the $x-$axis at the end points. However, in the rest of the paper we will use a triangular fuzzy number for $\overline{\lambda}$.

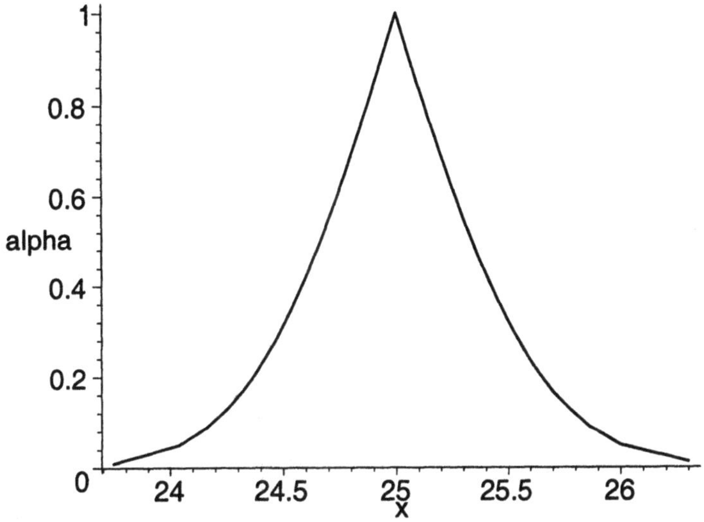


Figure 2: Fuzzy Arrival Rate $\overline{\lambda}$ in Example 1.

## 2.2. Fuzzy Service Rate

Let $\mu$ be the average (expected) service rate, in the number of service completions per unit time for a busy server. Then $1/\mu$ is the average (expected) service time. The probability density of the time interval between successive service completions is ([24], Chapter 15)

$$(1/\mu)\exp(-t/\mu), \tag{11}$$

for $t > 0$, the exponential probability density function. Let $X_1, ..., X_n$ be a random sample from this exponential density function. Then the maximum likelihood estimator for $\mu$ is $\overline{X}$ ([14],p.344), the mean of the random sample (not a fuzzy set). We know that the probability density for $\overline{X}$ is the gamma ([14],p.297) with mean $\mu$ and variance $\mu^2/n$ ([14],p.351). If $n$ is sufficiently large we may use the normal approximation to determine approximate confidence intervals for $\mu$. Let

$$Z = (\sqrt{n}[\overline{X} - \mu])/\mu, \tag{12}$$

which is approximately normally distributed with zero mean and unit variance, provided $n$ is sufficiently large. See Figure 6.4-2 in [14] for $n = 100$ which shows the approximation is quite good if $n = 100$. The graph in Figure 6.4-2 in [14] is for the chi-square distribution which is a special case of the gamma distribution. So we now assume that $n \geq 100$ and use the normal approximation to the gamma.

An approximate $(1-\beta)100\%$ confidence interval for $\mu$ is obtained from

$$P[-z_{\beta/2} < Z < z_{\beta/2}] = 1 - \beta, \tag{13}$$

where $\beta$ was defined in equation (4). After solving for $\mu$ we get

$$P[L(\beta) < \mu < R(\beta)] = 1 - \beta, \tag{14}$$

where

$$L(\beta) = [\sqrt{n}\,\overline{X}]/[z_{\beta/2} + \sqrt{n}], \tag{15}$$

and

$$R(\beta) = [\sqrt{n}\,\overline{X}]/[\sqrt{n} - z_{\beta/2}]. \tag{16}$$

An approximate $(1-\beta)100\%$ confidence interval for $\mu$ is

$$[\frac{\sqrt{n}\,\overline{X}}{z_{\beta/2} + \sqrt{n}}, \frac{\sqrt{n}\,\overline{X}}{\sqrt{n} - z_{\beta/2}}]. \tag{17}$$

**Example 2**

If $n = 400$ and $\overline{X} = 1.5$, then we get

$$[\frac{30}{z_{\beta/2} + 20}, \frac{30}{20 - z_{\beta/2}}], \tag{18}$$

for a $(1-\beta)100\%$ confidence interval for the service rate $\mu$. Now we can put these confidence intervals together, one on top of another, to obtain a fuzzy number $\overline{\mu}$ for the service rate. We evaluated equation (18) using Maple [18] for $0.01 \leq \beta \leq 1$ and the graph of the fuzzy service rate, without dropping the graph straight down to the $x$-axis at the end points, is in Figure 3. For simplicity we use triangular fuzzy numbers for $\overline{\mu}$ in the rest of the paper.

Having fuzzy numbers for $\lambda$ and $\mu$ we may proceed to the next computational stage.

## 3. FUZZY QUEUING MODEL

We go through the crisp and fuzzy calculations needed to get to the fuzzy numbers describing system performance.

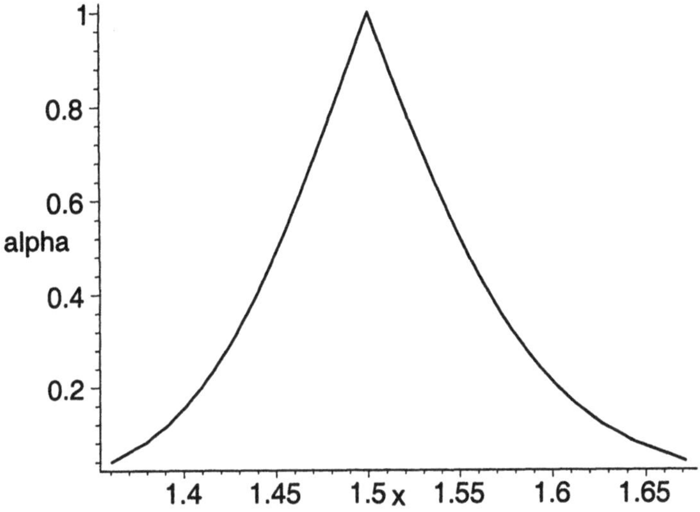


Figure 3: Fuzzy Service Rate $\overline{\mu}$ in Example 2.

## 3.1. Fuzzy Steady State Probabilities

We first discuss the crisp queuing model. The system has $c$ parallel and identical servers, system capacity $M$ (in the servers and in the queue) and an infinite calling source. A basic assumption is that we are in steady state, all transient behavior has died down and can be neglected, and the time interval $\delta$ is sufficiently small so that the probability of two or more events occurring during $\delta$ is zero.

The main objective at this point is to compute the steady state probabilities $w_i$ = the probability that there are $i$ customers in the system, $0 \leq i \leq M$, from which we may determine various measures of system performance. The result is, from standard queuing theory, that $w_i = F_i(\lambda, \mu, c, M)$, $0 \leq i \leq M$. Expressions for $F_i$ can be found, e.g., in [19] or [24]. To obtain fuzzy steady state probabilities, we directly fuzzify crisp expressions. Then

$$\overline{w}_i = F_i(\overline{\lambda}, \overline{\mu}, c, M), \tag{19}$$

to be evaluated using $\alpha$-cuts. We have

$$\overline{w}_i[\alpha] = \{F_i(\lambda, \mu, c, M) \mid \lambda \in \overline{\lambda}[\alpha], \mu \in \overline{\mu}[\alpha]\}, \tag{20}$$

for all $\alpha$ in $[0, 1]$. Then we find $\alpha$-cuts of the fuzzy steady state probabilities. Let $\overline{w}_i[\alpha] = [w_{i1}(\alpha), w_{i2}(\alpha)]$, then [4]

$$w_{i1} = min\{F_i(\lambda, \mu, c, M) \mid \lambda \in \overline{\lambda}[\alpha], \mu \in \overline{\mu}[\alpha]\}, \tag{21}$$

and

$$w_{i2} = max\{F_i(\lambda, \mu, c, M) \,|\, \lambda \in \overline{\lambda}[\alpha], \mu \in \overline{\mu}[\alpha]\}, \tag{22}$$

for $0 \leq i \leq M, 0 \leq \alpha \leq 1$. These are non-linear optimization problems which we solved for selected values of alpha through use of the Premium Solver Platform V5.0 from Frontline Systems [10]. In the future we simply call the optimization software Solver

Now assume we have all the needed fuzzy steady state probabilities.

## 3.2. Fuzzy System Performance Variables

We first discuss the computing of $\overline{U}$ =server utilization, $\overline{N}$ =expected number of customers in the system and $\overline{X}$ =average server throughput because the first two problems involve solving a linear programming problem. Then $\overline{R}$ = average response time is simply $\overline{N}/\overline{X}$. Finally we see how to get $\overline{LC}$ = the expected number of lost customers per unit time due to finite system capacity. To motivate the fuzzy calculations we will first present the crisp definition. The crisp steady state probabilities are $w_i$, $0 \leq i \leq M$. Let us assume that there are now two servers ($c = 2$) in the system.

The crisp definition of $U$ is

$$U = \sum_{i=2}^{M} w_i, \tag{23}$$

which is the probability that BOTH servers are busy. $U \times 100$ gives the percentage of time we expect both servers to be busy. In the fuzzy case

$$\overline{U} = \sum_{i=2}^{M} \overline{w}_i, \tag{24}$$

and is evaluated by $\alpha$-cuts

$$\overline{U}[\alpha] = \{\sum_{i=2}^{M} w_i | \quad \mathbf{S} \quad \}, \tag{25}$$

for all $\alpha$, where **S** is "$w_i \in \overline{w}_i[\alpha], 0 \leq i \leq M, w_0 + ...w_M = 1$". This is restricted fuzzy arithmetic, first presented in ([15]-[17],[21],[22]). Notice that we cannot simply add up the $\overline{w}_i$ for $i = 2, 3, ..., M$ because it may result in a fuzzy number not in the interval $[0, 1]$. The restriction is that the sum of the $w_i$ must equal one so that it is a discrete probability distribution. Also see ([1],[2],[5]) for more details on restricted fuzzy arithmetic applied to this type of calculation. We compute the alpha-cuts of $\overline{U}$ by solving a linear programming problem. Let $\overline{w}_i[\alpha] = [w_{i1}(\alpha), w_{i2}(\alpha)]$, $0 \leq i \leq M$. Let $\overline{U}[\alpha] = [u_1(\alpha), u_2(\alpha)]$. The objective functions are

$$max/min[w_2 + ... + w_M], \tag{26}$$

subject to constraints

$$w_{i1}(\alpha) \leq w_i \leq w_{i2}(\alpha), i = 0, ..., M, w_0 + ... + w_M = 1. \tag{27}$$

The solution to the min (max) problem gives $u_1(\alpha)$ ($u_2(\alpha)$). In general, we always have $0 \leq u_1(0)$, $u_2(0) \leq 1$.

$N$ is just the expected number of customers in the system

$$N = \sum_{k=0}^{M} kw_k, \tag{28}$$

and $\overline{N}$ is determined by its $\alpha$-cuts

$$\overline{N}[\alpha] = \{\sum_{k=0}^{M} kw_k | \quad \mathbf{S} \quad \}. \tag{29}$$

The end points of the interval $\overline{N}[\alpha] = [n_1(\alpha), n_2(\alpha)]$ may also be found by solving linear programming problems

$$max/min[w_1 + 2w_2 + 3w_3 + ... + Mw_M], \tag{30}$$

subject to the same constraints given above. We get $n_1(\alpha)$ ($n_2(\alpha)$) from the min (max) problem. In general, we always have $n_1(0) \geq 0$ and $n_2(0) \leq M =$ system capacity.

$X$ is the expected number of customers leaving the system per time period $\delta$. We first derive a crisp expression for $X$ and then fuzzify it. From [19]

$$X = \mu w_1 + 2\mu(w_2 + ... + w_M). \tag{31}$$

So we need to solve

$$max/min[\mu w_1 + 2\mu(w_2 + ... + w_M)], \tag{32}$$

subject to $\mathbf{S}$ and $\mu \in \overline{\mu}[\alpha]$. This is a non-linear programming problem. In general we always have $x_1(0) \geq 0$ where $\overline{X}[\alpha] = [x_1(\alpha), x_2(\alpha)]$. Also, since $\overline{R} = \overline{N}/\overline{X}$ and $\overline{R}[\alpha] = [r_1(\alpha), r_2(\alpha)]$, we will always have $r_1(0) \geq 0$.

Lastly, $LC = \lambda w_M$ so

$$\overline{LC} = \overline{\lambda}\overline{w}_M. \quad (33)$$

Here we simply multiply two fuzzy numbers. If $\overline{LC}[\alpha] = [lc_1(\alpha), lc_2(\alpha)]$, then we must get $lc_1(0) \geq 0$.

Having fuzzy numbers for system performance we may go on to final models for costs, benefits, and related quantities, but we shall not discuss this in any detail in this book.

# 4. FUZZY CALCULATIONS

The goal is: determine $\alpha$-cuts of fuzzy system performance values $\overline{U}, \overline{N}, \overline{X}, \overline{R}$ and $\overline{LC}$. We have two methods: (1) the one-step approach; and (2) the two-step procedure. For the computations in this section, and in the rest of this chapter, let us have $c = 2$ and $M = 4$, and the triangular fuzzy numbers $\overline{\lambda} = (3/4/5)$ for the fuzzy arrival rate and $\overline{\mu} = (5/6/7)$ for the fuzzy service rate..

## 4.1. One-Step Approach

Here we put the functions for the $w_i$ into the expressions for $U$, $N$, $X$, $R$ and $LC$ to get each a function of $\lambda$, $\mu$, $c$ and $M$. Then we have $U = f_u(\lambda, \mu)$, $N = f_n(\lambda, \mu)$, $X = f_x(\lambda, \mu)$, $R = f_r(\lambda, \mu)$ and $LC = f_{lc}(\lambda, \mu)$. For example, let $\rho = \lambda/\mu$, then

$$U = w_2 + w_3 + w_4, \quad (34)$$

where $w_2 = (\rho^2/2)w_0$, $w_3 = (\rho^3/4)w_0$, $w_4 = (\rho^4/8)w_0$ and

$$w_0 = [1 + \rho + \frac{\rho^2(1 - \{\rho^3/8\})}{2(1 - \rho/2)}]^{-1}. \quad (35)$$

Now go back to section 3.2 and substitute the above expressions for the $w_i$ ($w_1 = (\rho)w_0$) into the formulas for $N$, $X$, $R$ and $LC$ and we have the needed one-step expressions

Using the extension principle we now calculate $\overline{U} = f_u(\overline{\lambda}, \overline{\mu})$,...,$\overline{LC} = f_{lc}(\overline{\lambda}, \overline{\mu})$. The results, only for the $\alpha = 0, 1$ cuts are in Table 1. We employed Solver (or Matlab) for these optimizations.

As we shall see, crisp simulation can approximate these one-step results. For the rest of this chapter we will be using these triangular fuzzy numbers for $\overline{\lambda}$ and $\overline{\mu}$. Figures 2 and 3 show $\overline{\lambda}$ and $\overline{\mu}$ to be triangular shaped fuzzy numbers, but for simplicity we will now use triangular fuzzy numbers for $\overline{\lambda}$ and $\overline{\mu}$.

Now that we are employing $3 \leq \lambda \leq 5$ and $5 \leq \mu \leq 7$, there are other bounds on $\overline{X}$, $\overline{R}$ and $\overline{LC}$ in addition to those stated in the previous section. For

Table 1: One-Step Alpha Cuts of $\overline{U}$,$\overline{N}$,$\overline{X}$,$\overline{R}$,$\overline{LC}$. ($c = 2, M = 4$)

| | Alpha=Zero Cut | Alpha=One Cut |
|---|---|---|
| $\overline{U}$ | $[0.0749, 0.3043]$ | 0.1615 |
| $\overline{N}$ | $[0.4456, 1.1304]$ | 0.7205 |
| $\overline{X}$ | $[2.9737, 4.9223]$ | 3.9504 |
| $\overline{R}$ | $[0.1489, 0.2364]$ | 0.1824 |
| $\overline{LC}$ | $[0.0082, 0.2174]$ | 0.0497 |

Table 2: Alpha Cuts of the Fuzzy Probabilities. ($c = 2, M = 4$)

| | Alpha=Zero Cut | Alpha=One Cut |
|---|---|---|
| $\overline{w}_0$ | $[0.3478, 0.6475]$ | 0.5031 |
| $\overline{w}_1$ | $[0.2775, 0.3503]$ | 0.3354 |
| $\overline{w}_2$ | $[0.0595, 0.1739]$ | 0.1118 |
| $\overline{w}_3$ | $[0.0127, 0.0870]$ | 0.0373 |
| $\overline{w}_4$ | $[0.0027, 0.0435]$ | 0.0124 |

throughput $x_2(0) \leq 5$ since the maximum value of $\lambda$ is 5. Also, $lc_2(0) \leq 5$. Lastly, $r_2(0) \leq 0.6$ since the maximum time in the system would be 3 service completions ($M = 4$) at maximum service time of $0.2 = 1/5$ time units. We see that the values in Table 1 all satisfy these new constraints.

## 4.2. Two-Step Approach

We continue to assume that $c = 2$, $M = 4$, $\overline{\lambda} = (3/4/5)$ and $\overline{\mu} = (5/6/7)$. The two-step method first finds the fuzzy steady state probabilities $\overline{w}_i$, $0 \leq i \leq 4$ by solving equations (21) and (22). Using Solver (or Matlab) the $\alpha = 0, 1$ cut results in Table 2 are found. Next, using the fuzzy numbers for $\overline{w}_i$ solve for the fuzzy numbers $\overline{U}, ..., \overline{LC}$ given in section 3.2. Again, Solver was used and the $\alpha = 0, 1$ cuts are in Table 3. Notice that $x_2(0) = 5.0000$ in Table 3. The actual value computed was 6.6962, but from the previous subsection, we have the constraint $x_2(0) \leq 5.0000$. The two-step $\overline{X}$ gets "cut off" at 5.0000.

Now compare Table 1 (one-step results) to Table 3 (two-step results). We see that (one-step)$\overline{X}^{(1)} \leq \overline{X}^{(2)}$(two-step) but (one-step)$\overline{Z}^{(1)} \approx \overline{Z}^{(2)}$(two-step) for $\overline{Z} \in \{\overline{U}, \overline{N}, \overline{R}, \overline{LC}\}$. We note that the two-step procedure has potential for generating results with more fuzziness than the one-step method. We will see

Table 3: Two-Step Alpha Cuts of $\overline{U}$,$\overline{N}$,$\overline{X}$,$\overline{R}$,$\overline{LC}$.($c = 2, M = 4$)

| | Alpha=Zero Cut | Alpha=One Cut |
|---|---|---|
| $\overline{U}$ | $[0.0749, 0.3043]$ | 0.1615 |
| $\overline{N}$ | $[0.4455, 1.1306]$ | 0.7205 |
| $\overline{X}$ | $[2.1370, 5.0000]$ | 3.9504 |
| $\overline{R}$ | $[0.1464, 0.2529]$ | 0.1824 |
| $\overline{LC}$ | $[0.0081, 0.2175]$ | 0.0497 |

Table 4: Optimal Values of $\overline{U}$,$\overline{N}$,$\overline{X}$,$\overline{R}$,$\overline{LC}$.($c = 2, M = 4$)

| | Min | $\lambda$ | $\mu$ | Max | $\lambda$ | $\mu$ |
|---|---|---|---|---|---|---|
| $U$ | 0.0749 | 3 | 7 | 0.3043 | 5 | 5 |
| $N$ | 0.4456 | 3 | 7 | 1.1304 | 5 | 5 |
| $X$ | 2.9737 | 3 | 5 | 4.9223 | 5 | 7 |
| $R$ | 0.1489 | 3 | 7 | 0.2364 | 5 | 5 |
| $LC$ | 0.0082 | 3 | 7 | 0.2174 | 5 | 5 |

that crisp simulation will not directly approximate the results of the two-step calculation, but can give insights into fuzziness in our results.

## 4.3. Spreadsheet Calculations and Simulations

The immediately preceding computations represent a systematic approach to solutions: we take known crisp solutions (or approximations in some cases) and enter them into an optimizing program, to determine constrained min and max values used to approximate end points of a fuzzy output's alpha-cuts. Along with other resources, perhaps in another package, we obtain graphical and tabular outputs. No mathematical analysis work is needed beyond that which went into developing the formulae that start this calculation (optimization) process.

Optimization methods are either not required at all in the basic cases or play a secondary role, e.g., in the probability case we will see immediately below. Consider again $\lambda \in [3, 5]$ and $\mu \in [5, 7]$. Absent further complexities such as balking or preemption, we obtain min and max values for $U$,$N$,$X$,$R$ and $LC$ by reasoning such as: a "low" ("high") arrival rate coupled with a "high" ("low") service rate produces "low" ("high') values of all performance values except $X$. We extrapolate from low (high) values to min (max). We need also to recognize that a min (max) value of $X$ occurs when both $\lambda$ and $\mu$ are min (max) values. The results of a (first) round of reasoning along these lines are given in Table 4, where the desired results are first posted on the rows followed by the $\lambda$ and $\mu$ choices that effect these results.

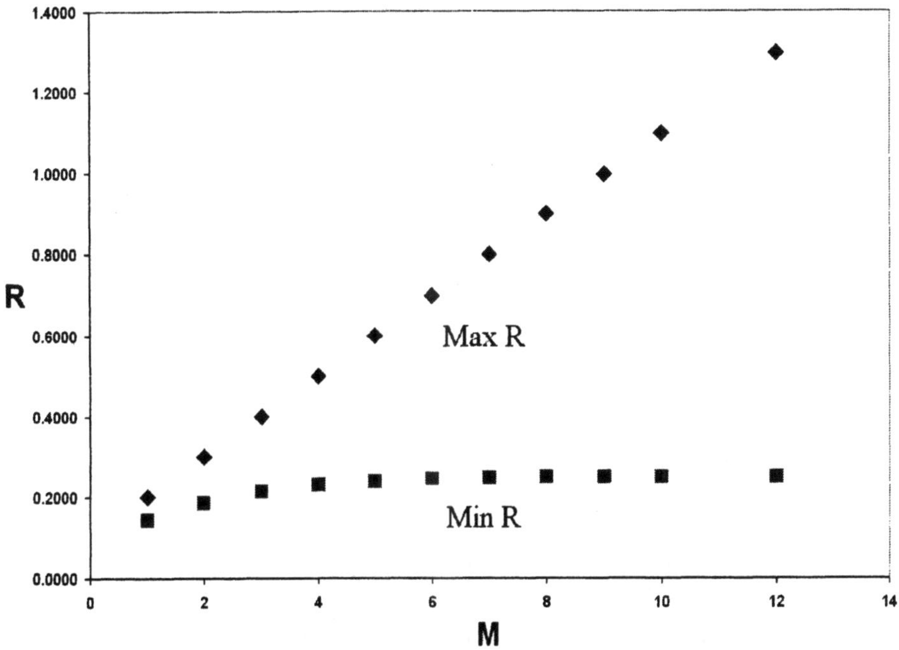


Figure 4: $\overline{R}[0]$ (Represented by R) is Plotted Against M (c=1).

The methodology extends to developing values for steady state probabilities (min and max values estimating fuzzy probabilities). The results are the same as those obtained by optimization and other methods so we do not display them here. Note that generating the performance values from these probabilities can be accomplished conveniently via optimization methods already discussed.

We can illustrate the "what if" simulation-style argumentation provided by the spreadsheet models. The concept of maximum system capacity (M) is a focus point in our studies. The alpha=0 cut of the (average) response time as a function of maximum capacity, M, is given in Figure 4 and the alpha=0 cut of the throughput rate vs. M is given in Figure 5. The characteristics of the performances are important, e.g., growth (and divergence of fuzzy values) in response time, adhering to mathematical results for the infinite M case; and the asymptotic behaviors in $X$, providing bounds to establish or confirm X

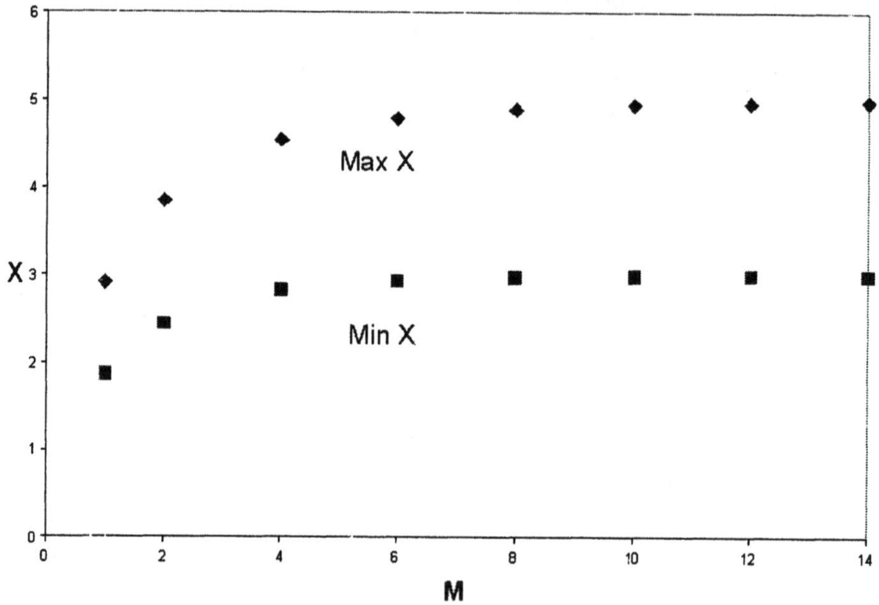


Figure 5: $\overline{X}[0]$ (Represented by X) is Plotted Against M (c=1).

constraints such as those imposed on max X in section 4.1.

Trends in four of the principal performance variables as a function of the service time (parameterized by arrival rate at its end points and middle value) are given in Figure 6. The functional dependency of performance variables on $\lambda$, $\mu$, and their ratio $\rho$ differs, e.g., several of the variables are a function of $\rho$ only whereas $\overline{X}$ and $\overline{R}$ are functions of both $\rho$ and $\mu$. $\overline{X}$ is the somewhat unusual response variable as these figures show and attention to it seems warranted; moreover, $\overline{X}$ appears to change the most from one- to two-step modeling. Note that these trends allow estimations of $\alpha$-cut values, e.g., the end points for $\alpha = 0$ values, with interpolated results for other cuts.

The mode of reasoning of this section carries over to general purpose simulation; we will devise simulations in a similar (min and max) fashion, along with modeling the $\alpha = 1$ case and projecting results onto other $\alpha$-cuts. Spreadsheet analysis can be viewed as a flexible support scheme to simulations.

# 5. GENERAL PURPOSE SIMULATION

In this section, we discuss some foundational issues relating to arrival/service rate simulation and utilization of standard/conventional (crisp) simulation pack-

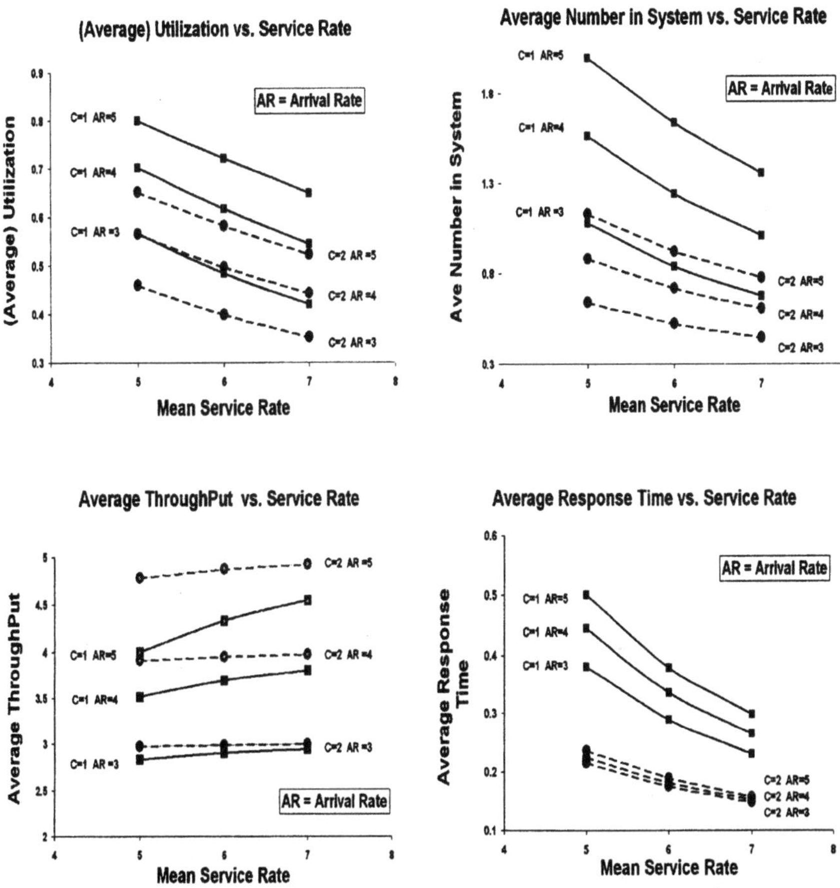


Figure 6: Trend Lines for Performance Variables Plotted Against Mean Service Rate and Parameterized by the Arrival Rate and Number of Servers (c=1 and c=2) ($M = 4$).

ages, specifically discrete event systems, to achieve results identical to or approximations of fuzzy calculations. Several purposes can be identified for using these simulation systems in the study of fuzzy probability computations. A first is to provide solutions in software that is widely available and efficient both in human model development time and in machine terms. Such software, together with approaches on how to use it, could then form the basis for a system development and programming methodology for fuzzy probability modeling.

For models based on arrival/service rates, the simulation systems provide a quite natural formulation. The model formulations (usually some kind of building block description) give insight into how a queuing system works internally. Simulation is sometimes called a "white/transparent box" approach for this reason (in contrast to black box approaches). The transparent box approach makes it relatively easy to provide flexibilities, e.g., a mix of fast and slow servers in a multiple server system, in contrast to some work here and elsewhere where multiple server systems are restricted to multiple identical servers. Similarly, accommodating dynamic effects in models is possible without excessive effort. A vast variety of stochastic processes is made available in a declarative mode within these packages and ready schemes exist for handling complexities such as balking and preemption.

Simulation languages provide much automatic output of model statistics and can provide additional results that fuzzy computation, as we have been describing it, does not. Full distributions of output variables and graphical displays of them are one such example. We will be exploiting these in obtaining some of our fuzzy results. The combination of automatic output and transparent box allows us to collect desired statistics according to model needs.

Both GPSS/H [13] and SLX [12] have been utilized in our studies; these are "sister" languages developed by Wolverine Software, Inc. [13]. In some cases we have used both systems to verify results, as each provides somewhat of a variant world view; potential directions for future work were also assessed. SLX is an object-based language that provides an "extensibility" element, signified by the "X" in its title (SL signifies "Simulation Language") - it includes several core GPSS/H elements in files that can be imported. GPSS/H (and its predecessors) provide perhaps the simpler approach to model our kind of systems. Our first case study will utilize it. After it, we turn to SLX, providing some programming descriptions, serving to dramatize our claims about how these packages, together with our basic (fuzzy modeling) approaches, are great simplifiers of needed computations. These SLX descriptions also serve to give insight into what the GPSS/H effort is like. In this chapter's context we employ only a few features of these packages and others we do not use may be part of future studies expanding on the present coverage.

The mathematical formulations seen in the sections on arrival/service rate models are based on Poisson arrival/service rates. The approximate equivalence of the Poisson counting process and the exponential interval process [9] means that we rendered these distributions in terms of exponential intervals, i.e., we utilize exponential inter-arrival and service time distributions.

## 5.1. Case Study: Long-Term Runs

Our targets for this simulation are estimates of $\overline{U}$, $\overline{N}$, $\overline{X}$, $\overline{R}$ and $\overline{LC}$. They will be determined by their $\alpha$-cuts. We will consider calculating $\overline{R}[\alpha]$, $0 \leq \alpha \leq 1$, since the others are computed in a similar manner.

Let **SIM** denote some simulation software that can be used to simulate our crisp queuing system with $c = 2$, $M = 4$. From this simulation we obtain a value for $R$. Actually, the simulation produces a distribution of values for time in the system and the $R$ we use is the expected value (mean value) of this distribution. Input to **SIM** will be $\lambda$ and $\mu$. We summarize all this as

$$R = \mathbf{SIM}(\lambda, \mu). \tag{36}$$

Now, going to the fuzzy system, we obtain fuzzy number estimators $\overline{\lambda}$, $\overline{\mu}$. Then

$$\overline{R}[\alpha] = \{R | R = \mathbf{SIM}(\lambda, \mu), \quad \mathcal{S} \ \}, \tag{37}$$

where

$$\mathcal{S} = \lambda \in \overline{\lambda}[\alpha] \ and \ \mu \in \overline{\mu}[\alpha], \tag{38}$$

for $0 \leq \alpha \leq 1$. Alpha-cuts of $\overline{R}$ will be intervals so let $\overline{R}[\alpha] = [r_1(\alpha), r_2(\alpha)]$. Then

$$r_1(\alpha) = min\{R | R = \mathbf{SIM}(\lambda, \mu), \quad \mathcal{S} \ \}, \tag{39}$$

and

$$r_2(\alpha) = max\{R | R = \mathbf{SIM}(\lambda, \mu), \quad \mathcal{S} \ \}. \tag{40}$$

So we solve these last two equations for selected values of $\alpha$ and we have estimated the fuzzy number for the expected time an item spends in the system. Also, $\overline{R}[0]$ is like a 99% confidence interval for this time. We compute the other fuzzy numbers ($\overline{X}$, $\overline{N}$,$\overline{U}$, $\overline{LC}$) the same way.

How are we to solve equations (39) and (40)? We suggest a simulation optimization method. The search space is given by $\mathcal{S}$. As we shall see, using the end points of the intervals for the $\alpha$-cuts of $\overline{\lambda}$, $\overline{\mu}$ will solve many of these problems. In this chapter there are only two places in a fuzzy queuing system that depend on a probability distribution: (1) arrivals; and (2) service stations. We now look at each of these and how we plan to solve the optimization problem.

Table 5: Optimal Values of $\lambda \in [\lambda_l, \lambda_u] = \overline{\lambda}[\alpha]$ for the Exponential Arrivals in Simulation

| Min | $\lambda$ | Max | $\lambda$ |
|---|---|---|---|
| $U$ | $\lambda_l$ | $U$ | $\lambda_u$ |
| $N$ | $\lambda_l$ | $N$ | $\lambda_u$ |
| $R$ | $\lambda_l$ | $R$ | $\lambda_u$ |
| $X$ | $\lambda_l$ | $X$ | $\lambda_u$ |
| $LC$ | $\lambda_l$ | $LC$ | $\lambda_u$ |

### 5.1.1. Arrivals

All arrivals will be governed by the exponential distribution. To describe arrivals we use $\lambda$ = the number of items (customers) arriving per unit time and $1/\lambda$ = the mean time between arrivals. The solution to the optimization problem is given in Table 5. We have assumed that all parameters in the fuzzy queuing system are held fixed except $\lambda$ for this arrival. If $\overline{\lambda}$ is our fuzzy estimator of $\lambda$, let $[\lambda_l, \lambda_u] = [\lambda_1(\alpha), \lambda_2(\alpha)] = \overline{\lambda}[\alpha]$ for some fixed $\alpha \in [0, 1)$.

We interpret Table 5 as follows: (1) if we want to approximate the left end point of the interval $\overline{X}[\alpha]$ use for $\lambda$ the left end point of the interval $\overline{\lambda}[\alpha]$; and (2) if you want to approximate the right end point of the interval $\overline{R}$ use for $\lambda$ the right end point of the interval $\overline{\lambda}[\alpha]$.

These results are rather intuitive. Let $\lambda$ increase from its minimum $\lambda_l$ to its maximum $\lambda_u$, holding other system parameters fixed. The system becomes more congested. We will call arrivals "customers" or "requests" in this section. When server utilization (mean value=$U$) increases, the number of customers in the system (mean value=$N$) increases and the time a customer spends in the system (mean value=$R$) also increases. At first it is not clear what happens to the number of customers leaving the system per unit time (mean=$X$) since $X = N/R$. But the number leaving per unit time will not decrease, so $X$ will also increase. If the queue has finite capacity, then the number of customers lost per unit time ($LC$) increases as $\lambda$ increases, illustrated by Table 5.

### 5.1.2. Service

The number of service completions per unit time is $\mu$ so the mean service time is $1/\mu$. The exponential depends only on $\mu$. The optimization problem solution is given in Table 6. If $\overline{\mu}$ is our fuzzy estimator of $\mu$, let $[\mu_l, \mu_u] = [\mu_1(\alpha), \mu_2(\alpha)] = \overline{\mu}[\alpha]$ for some fixed $\alpha \in [0, 1)$.

Table 6: Optimal Values of $\mu \in [\mu_l, \mu_u] = \overline{\mu}[\alpha]$ in the Exponential for Service in Simulation

| Min | $\mu$ | Max | $\mu$ |
|---|---|---|---|
| $U$ | $\mu_u$ | $U$ | $\mu_l$ |
| $N$ | $\mu_u$ | $N$ | $\mu_l$ |
| $R$ | $\mu_u$ | $R$ | $\mu_l$ |
| $X$ | $\mu_l$ | $X$ | $\mu_u$ |
| $LC$ | $\mu_u$ | $LC$ | $\mu_l$ |

We interpret Table 6 as follows: (1) if we want to approximate the left end point of the interval $\overline{X}[\alpha]$ use for $\mu$ the left end point of the interval $\overline{\mu}[\alpha]$; and (2) if you want to approximate the right end point of the interval $\overline{R}$ use for $\mu$ the left end point of the interval $\overline{\mu}[\alpha]$.

These results are also intuitive. Let $\mu$ increase from its minimum $\mu_l$ to its maximum $\mu_u$. All other parameters of the system are held fixed. This service station may have $c$ ($c \geq 1$) parallel and identical servers. The system consisting of this server and what comes before it becomes less congested. Fewer and fewer customers fill the queue in front of this server. This means that server utilization (mean value=$U$) decreases, the number of customers in the system (mean value=$N$) decreases and the time a customer spends in the system (mean value=$R$) also decreases. It is not clear what happens to the number of customers leaving the system per unit time (mean=$X$) since $X = N/R$. But the number leaving per unit time will not decrease, so $X$ will increase. If the queue has finite capacity, then the number of customers lost per unit time ($LC$) decreases as $\mu$ increases. These results are in Table 6.

### 5.1.3. Results

The results of our simulation, using GPSS/H ([12],[13]), are in Table 7. The simulation was run until $500,000$ items left the servers. Notice that they are almost identical to the one-step results in Table 1. The key is to choose the values of $\lambda$ and $\mu$ correctly. We could use Table 4 to pick $\lambda$ and $\mu$. However, for more complicated fuzzy systems we would not have the one-step results as in Table 4 and instead we would employ Tables 5 and 6.

Table 7: Simulation Values of $\overline{U}$,$\overline{N}$,$\overline{X}$,$\overline{R}$,$\overline{LC}$.($c = 2, M = 4$).

| | Min | $\lambda$ | $\mu$ | Max | $\lambda$ | $\mu$ |
|---|---|---|---|---|---|---|
| $U$ | 0.0745 | 3 | 7 | 0.3043 | 5 | 5 |
| $N$ | 0.444 | 3 | 7 | 1.133 | 5 | 5 |
| $X$ | 2.9702 | 3 | 5 | 4.9168 | 5 | 7 |
| $R$ | 0.149 | 3 | 7 | 0.237 | 5 | 5 |
| $LC$ | 0.0083 | 3 | 7 | 0.2181 | 5 | 5 |

## 5.2. Additional Case Studies

In this and subsequent pages we relate studies which provide some insight into possible future effort. Some of them involve distributions of the mean values we have been observing and some remarks are given on programming detail. From here on, the study primarily involves SLX. Since some details are shared between it and GPSS/H, our discussion serves to illustrate a portion of the programming style used throughout the chapter. An SLX programming formulation for a single-server model appears in Figure 7

```
arrivals: customer
iat = rv_expo (random-stream1,arrival_rate)
until_time = stop_time;
enqueue queue;
seize server;
        // enter replaces seize for multiple servers
        // (a capacity is defined in these cases)
depart queue;
advance rv_expo(random-stream2,service_rate);
release server;
        // leave replaces release for multiple servers
```

Figure 7: Core Portion of SLX Program for Single-Server Queue with FIFO Queue Discipline. Comments Indicate Changes for Multi-Server Cases.

The three lines of code starting at "arrivals: customer" are an SLX macro, which associates with an object type, here a "customer" object, and generates these objects according to the (random time) variable (rvexpo()), the exponential service time associated with the Poisson arrival rate. These objects represent the customers/requests for our system. Note that random stream control is in place here; the service rate utilizes another, different stream. Thus, we can e.g. change the arrival pattern (rate), for a new simulation run, while maintaining the service pattern (rate) from the first run; we thereby secure a stream of requests that experience the same service response pattern and thus eliminate spurious variation from also (haphazardly) varying the service. A stop time is a final feature of this macro, shutting off (unnecessary) generation of requests/customers which have no possible chance to be served.

Several statements in the remaining lines of code can be seen to conjugate each other, defining the starting and ending events of an activity: joining a queue, represented by the enqueue statement (with user-defined queue name) and leaving the same queue represented by a depart block (with the queue name). Similarly occupying the server (seize) and leaving the server (release) are conjugates. These conjugates are almost identical to ones used in GPSS/H.

Carrying out our simulations requires deciding two main issues: 1) length of each model run and 2) replications. In the previous case, we used only one very long duration run in seeking to approximate a steady state or long-duration result, which we saw was a successful endeavor. In such a case replications are not used. Replications accommodate randomness of individual runs, especially important in shorter runs, but more importantly provide distributions of our performance variables, which turn out to be useful in a number of ways. Sometimes, a few replications establish a solid mean with associated confidence intervals, making simulation effort minimal. A most important fact is that the knowledge we gain from these distributions is either not attainable or difficult to obtain in some, if not most, of the other methods presented in this paper.

The mean value approach of this section employs the min and max approach to the arrival/service rates that we used in the previous section and in the spreadsheet cases. In fact, these results were developed first and thus actually preceded the other ones.

Figure 8 relates a small part of a study with $\overline{\lambda} = (3/4/5)$, $\overline{\mu} = (5/6/7)$, $c = 1$ and a maximum number in the system fixed at M=4. The results aimed at the max throughput rate (X). Basic graphical output in Figure 8 (mean distributions) is provided via one-line declarations (code not presented in the figure). The model uses short run times and multiple replications, achieving results quickly and accurately to two decimal places. It also gives deviation about the mean and min and max values. Because the runs are short, values beyond those permitted in the long-term runs appear; these are due to randomness of the situation and provide some insights. In some cases, observations must be over a short period and so results like these help provide what is typical. It may be no coincidence that the max value (6.46) is similar to that from optimization models when the constraint on X's max is ignored.

```
Random Variable  #Obs  Mean  Std Dev  Minimum  Maximum
   Throughput     100  4.51   0.84     2.45     6.46

Lo   Hi Freq
2.0 2.5  1 | *
2.5 3.0  6 | **********
3.0 3.5  6 | **********
3.5 4.0 16 | ****************************
4.0 4.5 12 | *********************
4.5 5.0 29 | ****************************************************
5.0 5.5 20 | ***********************************
5.5 6.0  7 | ************
6.0 6.5  3 | *****
```

Figure 8: Graphical Depiction (Edited) of Results from a Model Designed to Provide Maximum Throughput Rate (100 Model Replications). Each Run Involves Approximately 100 Customers/Requests; $c = 1$, $M = 4$.

```
FREQUENCY

 30 +           +          100  REPLICATIONS
    +           +         "Short Term Runs"
    +           +         (Min/Max X Regions)
    +         + +
    +         + +
 20 +         + +                 +
    +         + +                 +   +
    +         + + +               + + +
    +         + + + +           + + + +
    +         + + + +           + + + + +
 10 +       + + + + +         + + + + + +
    +       + + + + + +       + + + + + +
    +       + + + + + +       + + + + + +
    +       + + + + + +     + + + + + + + +
    +   + + + + + + + +   + + + + + + + + + +   +
  0 +++++++++++++++++++++++++++++++++++++++++++++++++
    200       300       400       500        600
     AVERAGE THROUGHPUT (X) - Crisp Model Results
         Min X Region         Max X Region
```

Figure 9: Distributions of Min and Max Values of Modeled $\overline{X}$ (100 Model Replications; Approximately 100 Customers/Requests); $c = 1$, $M = 4$. X Values are Scaled by a Factor of 100.

Figure 9 shows the model result from a GPSS/H run and includes the mean distribution about the min value of X, as well as about the max, for $c = 1$ and $M = 4$. X values are scaled by a factor of 100; the peaks around 300 and 500 are actually at (a little less than) 3.0 and 5.0 and the reader can thus see the correspondence between this figure and its predecessor at the (higher value range of) points where they address the same situation. Short-term runs in this case involve approximately 100 customers/requests to complete service, complemented by 100 replications.

Table 8 provides more details and includes another study with M=10. If we combine Figure 8's results with an analogous computation designed to obtain the minimum of $\overline{X}$, again using the resulting mean of these computations, we get [2.82,4.52] for an $\alpha = 0$ output for $\overline{X}$. Similar arguments apply to the other variables. Table 8 contains $\alpha = 0$ cut values determined by all styles of simulation reported here, optimization based methods, spreadsheet computations and (general, package-based) simulation. All of these methods produce essentially the same results.

Table 8: Alpha=0 Cut for Key Variables $\overline{U}, \overline{N}, \overline{X}$ and $\overline{R}$ for $c = 1$ with $M = 4$ and $M = 10$.

| CASE: Single Server | Max Requests | Max Requests |
|---|---|---|
| | 4 | 10 |
| $\overline{U}$ | [0.4201,0.7998] | [0.4297,0.9089] |
| $\overline{N}$ | [0.6767,1.9965] | [0.7549,4.9928] |
| $\overline{X}$ | [2.8187,4.5186] | [2.9811,4.9283] |
| $\overline{R}$ | [0.2304,0.5000] | [0.2521,1.1003] |

### 5.2.1. Exploring Potential in $\alpha = 1$ Cut Models

In this section, we seek a method whereby we can exploit distributions from $\alpha = 1$ cut models to estimate all alpha cuts. Our first attack is to effect $\alpha = 0$ cut values for the performance values. We employ a heuristic procedure, using a 99% confidence interval mimicking section 2's assumption on raw data used in characterizing input. The duration of the model runs allowed approximately 100 customers/requests and the number of replications was 3000. This gives us the results in Figures 10 and 11. Figure 11 shows U in a close-up view where the reader can see the cut-off action. Notice that the 99% confidence interval is a good approximation to $\overline{U}[0]$ in Table 8. This heuristic undoubtedly needs a better base than a "common" ground. In our premier example, however, it works fairly well as Figures 10 and 11 indicate and helps to establish that simulations have the potential to accomplish the desired task.

### 5.2.2. Multiple $\alpha$ Cut (Spreadsheet-based) Models

In our final effort, we developed multiple (five) $\alpha$-cuts for the performance variables we have been stressing: $\overline{U}, \overline{N}, \overline{X}$ and $\overline{R}$, using the spreadsheet method and applying the same approach we used for the $\alpha = 0$ calculations in section 4.3, but now at the multiple cuts (equally spaced) on the fuzzy (triangular) inputs. This gives us the fuzzy triangular shaped results portrayted in Figure 12.

## 6. CONCLUSIONS AND FUTURE RESEARCH

We have considered several means for employing crisp analysis, modeling and simulation to estimate fuzzy outputs for fuzzy queuing systems models based on arrival/service rates (inter-event (time) intervals). The approach is complemented in a companion chapter, which builds models for similar ends but from a different starting point, i.e., a fuzzy (state) transition probability matrix. General purpose simulation models constitute the central theme in this chapter, but other modeling tools play important roles. Table 9 provides an overview. (In it we abbreviate "Properties" as "Props" and ues "values" "variables.")

Table 9: Overview of Simulations. See Text for Details and Abbreviations.

| Arrival/Service Rate Simulations | |
|---|---|
| One-Step Optimization | Generate Performance Values |
| Two-Step Optimization | Generate Steady State Probabilities |
| Queue Props Spreadsheet | Generate Performance Values |
| Queue Props Spreadsheet | Generate Steady State Probabilities |
| (Package) Simulation | Queue Systems Performance Values |
| (Package) Simulation | Queue Systems Steady State Probabilities |
| (Package) Simulation | Obtain $\alpha$=0 cut & others from $\alpha$=1 cut |

In our first round of modeling, we employed constrained optimization, using established optimization packages, to produce fuzzy probability and fuzzy performance values from arrival and service rates. One consequence of these models was that they clearly depicted where minimum and maximum values

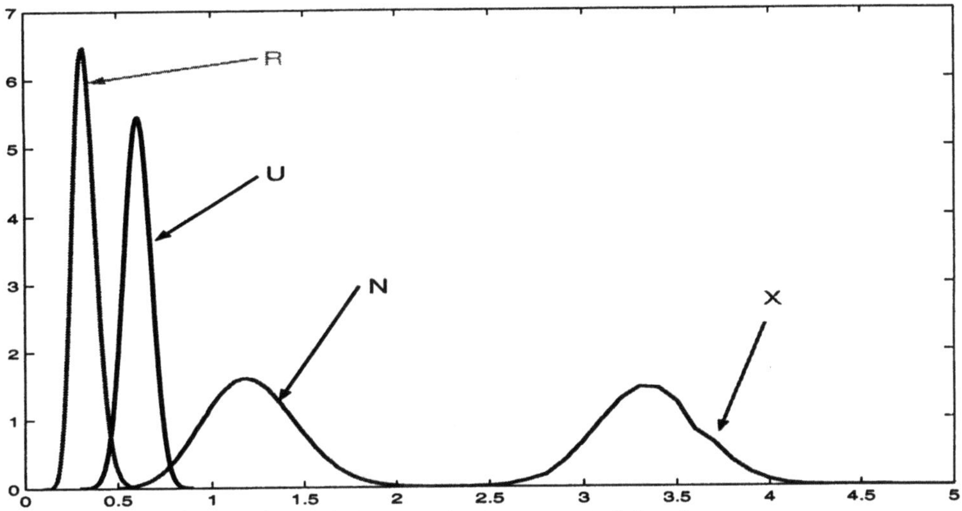


Figure 10: U,N,X,R Distributions ($c = 1, M = 4$).

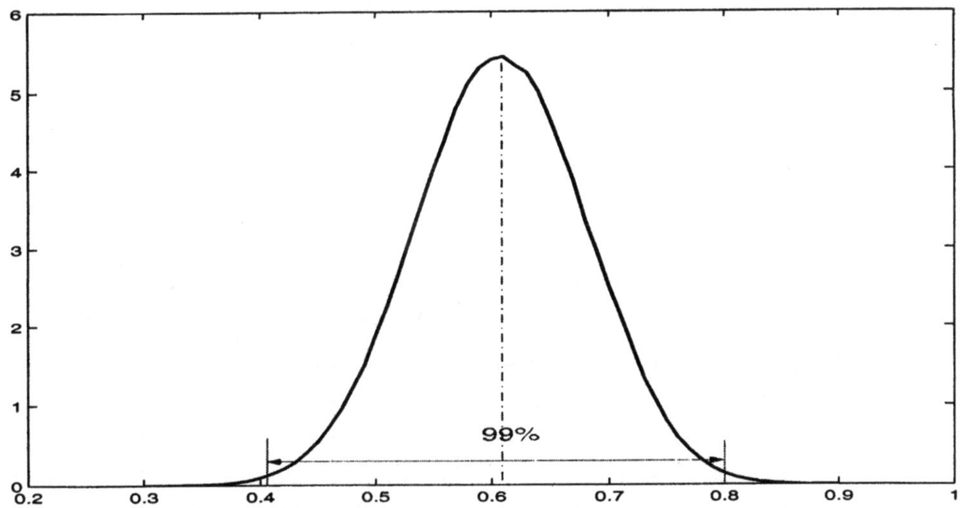


Figure 11: Distribution of Utilization ($c = 1, M = 4$).

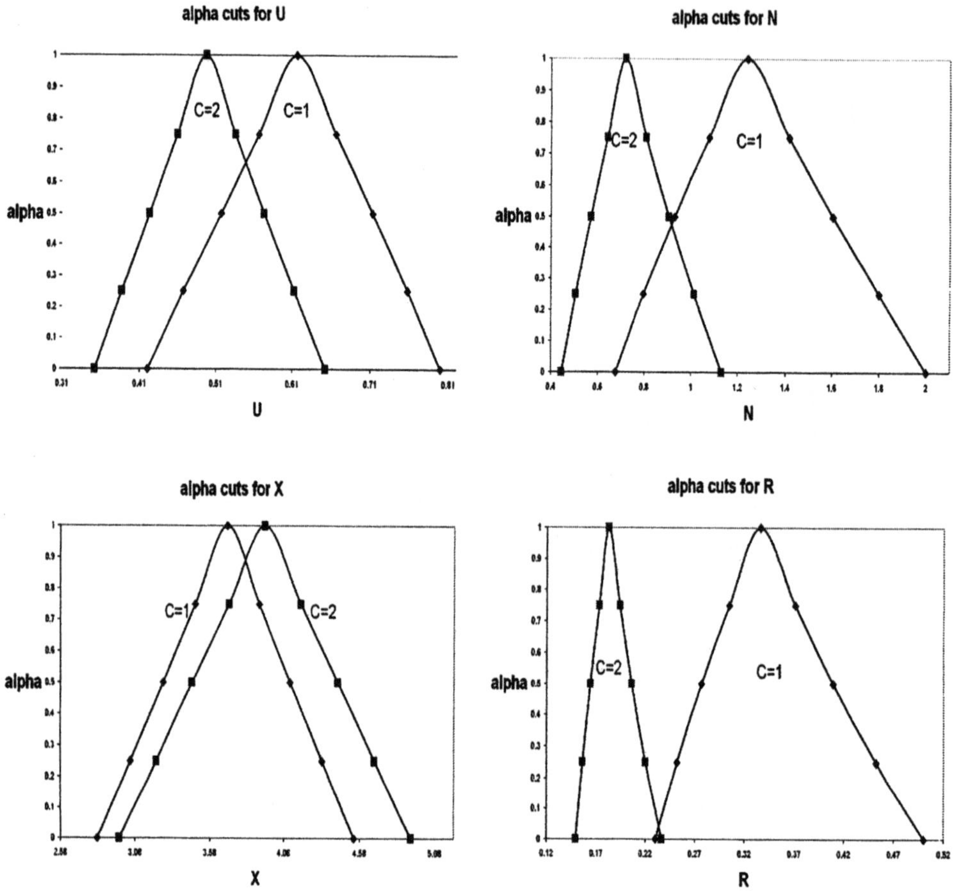


Figure 12: Alpha Cut Estimates for Performance Variables, $\overline{U}, \overline{N}, \overline{X}, \overline{R}$ for $c = 1$ and $c = 2$, with $M = 4$).

of fuzzy (range) estimates lie. Using this information we were able to effect simpler calculations in some cases, among them spreadsheet-based methods.

In this second round of modeling, we developed dependencies of (average) response time and (average) throughput rate on maximum number of requests allowed in the system. We also developed trend lines for four of the performance variables on service rate, parameterized by number of servers and selected values of arrival rate (determined the end and middle points of our triangular fuzzy inputs). Finally, we developed candidate alpha cuts for these same performance variables but placed them at the end of the paper since a portion of the values seen there also were established in some of the simulation runs. The methods in these spreadsheet cases as well as in original optimization depended heavily on the presence of pre-existing mathematical expressions which we could directly exploit. The simulation methods are not so restricted and can, with appropriate care, be used to develop results for cases in which there are no formulae in place. Some such work has already begun, with plans for more. A final note we make is that there are approximate mathematical results which seemingly would constitute a kind of intermediate state between closed form mathematics-based operations and simulation; such could well constitute a research topic in and of itself.

Another satisfying result in our spreadsheet calculations was that they work not only for direct computation of performance values but also for the probabilities that precede them. We alluded to this development, stating that the results we obtained matched those of the optimization cases, and thus we dropped further discussion of the topic, noting that the results could be carried forward to the performance variables using the same optimization methods employed earlier in the paper (where we acquired performance values directly from the arrival/service rates by optimization methods). A possible research item lies in simplifying this step.

In the first simulation model discussions, we developed a basis for modeling systems employing long-term runs (to simulate long-term or steady (equilibrium) state solutions). These results, we saw, were well-founded in that they agreed with those produced in the other approaches. Their greatest benefit lies in providing base case(s) for more complex models for which no mathematical results are available. Current research in this direction is underway.

In perhaps a more speculative, even heuristic mode, we sought to produce results with a minimum effort and certain added dimensions (for example, more transient behaviors). In some cases, very short runs with relatively few replications have proved to be sufficient for baseline estimates. We also demonstrated, e.g., that shorter-term simulations readily reproduced values equal to and sometimes beyond the range of fuzzy calculations (where in the latter case we may, e.g., have simplified an optimization by relaxing (a) constraint(s)). These kinds of results typically involved exploiting opportunities in simulation packages' (output) frequency generation capabilities. These packages are "distribution oriented" in their approach so that explorations beyond mean values, in many cases, are expedited. In a similar vein, we demonstrated a process of carrying out a single simulation, an $\alpha = 1$ cut model, and utilizing it to obtain not only the $\alpha = 0$ cut estimates but as many other cuts as we may desire; our demonstration was limited to five cuts. Simulations based on short-term model runs required associated replication schemes and we utilized relatively few replications, e.g., 100, in several cases, though we used many more (3000) in collateral runs to assess adequacy of the (former) choice.

Based on our study, crisp simulation has proved to be productive for fuzzy probability modeling. The simulation models are compact and relatively simple to implement. Computation time has generally been much shorter compared to fuzzy computations they compete with. We hope that the methodology details we provided support other researchers' attempts to reproduce and extend our results.

Other potentials of the simulation method are currently being explored, e.g., continuing the demonstrated value of simulation packages' output distribution generation capabilities. That shorter-term runs can be titrated for fuzziness means that we can match fuzzy method results and, if desired, we can produce results that are more (less) fuzzy. A method applied to the conventional/crisp data (used first in this chapter on real world data to get modeling started) employed confidence intervals; justifications and choice of appropriate intervals need more research. Operating on cuts made on top of frequency distributions emerging from stochastic modeling to project $\alpha$-cuts of fuzzy numbers may secure results for all cuts in a systematic procedure and is also a potential subject for additional research.

Similarly, a future set of activities lies in reconciling the transition probability approach (covered in the companion chapter in this volume) with this chapter's arrival/service rate approach. The transition probability matrices we have used in the companion chapter are general, i.e., they admit values *anywhere* in the matrix, whereas the matrices corresponding to the arrival/service rate models are banded. In the latter set of circumstances, exact equivalence in results can be expected from the two approaches. Developing approximations for the general (transition matrix) setting (perhaps in conjunction with similar accommodations in the arrival/service rate setting) may help us realize how the two styles of modeling can complement each other under a broader range of options.

Finally, challenges extend to other queuing situations, e.g., cases where servers operate at different rates and where dynamic effects occur, e.g., varying numbers of servers or service rate changes based on moment-to-moment demand. In contrast to work reported here where our calculation schemes were related primarily to mathematic results (with projections under strict control), future work will relate simulation models among themselves, rooting them back, of course, to the mathematical bases. So far, fuzzy models have preceded the crisp analogs and have set a standard for the crisp systems to match. In the future, the situation may well be that crisp simulation will set some standards with fuzzy modeling enjoying the task to emulate.

## REFERENCES

1. J.J.Buckley: Fuzzy Probabilities: New Approach and Applications, Physica-Verlag, Heidelberg, 2003.

2. J.J.Buckley: Fuzzy Probability and Fuzzy Sets for Web Planning, Physica Verlag, Heidelberg, 2004.

3. J.J.Buckley and Y.Hayashi: Applications of Fuzzy Chaos to Fuzzy Simulation, Fuzzy Sets and Systems, 99(1998), pp. 151-157.

4. J.J.Buckley and Y.Qu: On Using Alpha-Cuts to Evaluate Fuzzy Equations, Fuzzy Sets and Systems, 38(1990), pp. 309-312.

5. J.J.Buckley, K.D.Reilly and X.Zheng: Fuzzy Probabilities for Web Planning, Soft Computing. Published on-line, to appear in print.

6. J.J.Buckley, K.D.Reilly and X.Zheng: Crisp Simulation of Fuzzy Computations, Proc. 4th Int'l Symp. on Uncertainty Modeling and Analysis - ISUMA-2003, (2003), pp. 50-55.

7. S.Chanas and M.Nowakowski: Single Value Simulation of a Fuzzy Variable, Fuzzy Sets and Systems, 25(1988), pp. 43-57.

8. S.Chanas and S.Heilpern: Single Value Simulation of Fuzzy Variables - Some Further Results, Fuzzy Sets and Systems, 33(1989), pp. 29-36.

9. D.R.Cox and W. Smith: Queues, Methuen, 1961.

10. Frontline Systems (www.frontsys.com).

11. P.A.Fishwick: Fuzzy Simulation: Specifying and Identifying Qualitative Models, Int. J. General Systems, 19(1991), pp.295-316.

12. J.O.Henriksen: SLX-Pyramid Power, Proc. Winter Simulation Conference, Society for Modeling and Simulation Int'l, San Diego, CA, 1999.

13. J.O.Henriksen and R.C.Crain: GPSS/H Reference Manual, 4th ed., Alexandria, VA: Wolverine Software, Inc., 1996.

14. R.V.Hogg and E.A.Tanis: Probability and Statistical Inference, Sixth Edition, Prentice Hall, Upper Saddle River, N.J., 2001.

15. G.J.Klir : Fuzzy Arithmetic With Requisite Constraints, Fuzzy Sets and Systems, 91(1997), pp. 147-161.

16. G.J.Klir and J.A.Cooper: On Constrained Fuzzy Arithmetic, Proc. 5th Int. IEEE Conf. on Fuzzy Systems, New Orleans, 1996, pp. 1285-1290.

17. G.J.Klir and Y.Pan: Constrained Fuzzy Arithmetic: Basic Questions and Some Answers, Soft Computing, 21(1998), pp. 100-108.

18. Maple 9, Waterloo Maple Inc., Waterloo, Canada.

19. D.A.Menasce and V.A.F.Almeida: Capacity Planning for Web Performance, Prentice Hall, Upper Saddle River, N.J., 1998.

20. D.S.Negi and E.S.Lee, Analysis and Simulation of Fuzzy Queues, Fuzzy Sets and Systems, 46(1992), pp. 321-330.

21. Y.Pan and G.J.Klir: Bayesian Inference Based on Intelligent and Fuzzy Systems, 5(1997), pp. 193-203.

22. Y.Pan and B.Yuan : Bayesian Inference of Fuzzy Probabilities, Int. J. General Systems, 26(1997), pp. 73-90.

23. K.D.Reilly, J.J.Buckley, X.Zheng, Fuzzy Probability Modeling and Applications, Proc. 2003 Huntsville Simulation Conference, Society for Computer Simulation, Int'l, San Diego, CA, (2003), pp. 579-584.

24. H.A.Taha: Operations Research, Fifth Edition, Macmillan, N.Y., 1992.

25. P.Volkner and B.Werners: A Simulation-Based Decision Support System for Business Process Planning, Fuzzy Sets and Systems, 125(2002), pp. 275-287.

26. F.Wenstop: Fuzzy Sets Simulation Models in a Systems Dynamics Perspective, Kybernetics, 6(1976), pp.209-218.

27. X.Zheng, K.D.Reilly and J.J.Buckley, Comparing Genetic Algorithms and Exhaustive Methods Used In Optimization Problems For Fuzzy Probability Based Web Planning Models, Proc. Int'l Conf. on Artifi cial Intelligence (IC-AI'03) - Vol. 1, (2003), pp. 463-468.

28. X.Zheng, K.D.Reilly and J.J.Buckley, Fuzzy Optimization and Normal Simulation for Solving Web Queuing System Problems, Proc. SCI 2004 - World Multiconference on Systemics, Cybernetics, and Informatics, (2004), To appear.

Chapter 4

# SIMULATION OF FUZZY SYSTEMS II

James J. Buckley, Kevin D. Reilly and Xidong Zheng

## 1. INTRODUCTION

We begin in the next section with a discussion of our transition matrix based fuzzy probability queuing system model. This model is discussed in ([2],[4]) and so we need only present an overview. This is followed by our simulation methodology and results. We go through these fuzzy computations in detail so the reader can appreciate how the simulation method simplifies the process of going from initial data to the optimization models.

However, the simulations sometimes produce fuzzy results at variance with the fuzzy calculations, ranging from less to equal on to more fuzzy. The matter will be discussed at appropriate points in succeeding sections and the reader will see that we have scored some significant successes while yet leaving open some fundamental questions for future research. Accordingly, the chapter's final section has, along with a summary, our plans for future research on the topic. This paper expands on previous work reported in conference papers [5],[17],[22],[23]. The notation used is the same as that employed in the previous chapter. Also, for a brief review of the literature on (fuzzy) simulation of fuzzy systems see the previous chapter.

## 2. FUZZY TRANSITION MATRIX MODEL

In this section the mathematical background is presented, with pointers to some options for obtaining closed form and computational results. In the following section simulation approaches are outlined. As we will see, these interweave with mathematical arguments in several ways.

## 2.1. Fuzzy Transitions Overview

In the web modeling context we may speak almost interchangeably of customers and requests in so far as requests initiate from web "customers." More generally, web customer requests generate additional requests but from several system points of view, e.g., load on the system, these can be addressed in a first approximation as new customers; linking requests tightly could lead to correlated phenomena that in principle add a great deal of difficulty to the modeling effort (fodder for additional research). In so far as all this is understood we can use "customers" without serious ambiguity.

In the fuzzy transition probability model: (1) we first need to construct fuzzy numbers for $p(i)$ = the probability that $i$ customers arrive at the system during time interval $\delta$ and for $p$ = the probability that a customer leaves a server during time interval $\delta$ given that the customer was in the server at the start of the time interval; (2) using the $\overline{p}(i)$ and $\overline{p}$ calculate the fuzzy transition probabilities in a fuzzy transition matrix for a fuzzy, (usually) regular, Markov chain; (3) using the fuzzy transition matrix determine the fuzzy steady state probabilities; (4) using the fuzzy steady state probabilities compute the fuzzy numbers for system performance $\overline{U}$ = server utilization, $\overline{N}$ = number of customers in the system, $\overline{X}$ = throughput, $\overline{R}$ = response time and $\overline{LC}$ = lost customers due to finite system capacity (see [14]); and (5) input the fuzzy numbers for system performance into optimization models to find the optimal mix of the variables (number of servers, capacity,...) to $min\overline{R}$, $max\overline{U}$, max fuzzy profit, etc.

## 2.2. Fuzzy Numbers for Probabilities

We will measure changes in our system at time intervals $\delta$. This time interval may be one second, or 0.1 second, etc. We first need to gather data about the system, like the probability that $i$ customers arrive during time interval $\delta$. Suppose we observe the system during $N$ time periods and find that there have been $n_i$ times that $i$ customers have arrived for service, $i = 0, 1, 2, 3....$ We would expect, from practical considerations, that there is some positive integer $L$ so that $n_i = 0$ for $i > L$. Let $p(i)$ be the probability that $i$ customers arrive during time period $\delta$, $i = 0, 1, 2, 3, ..., L$. Then a point estimate of $p(i)$ is simply $n_i/N$. However, to show our uncertainty in this estimate we may also compute a confidence interval for $p(i)$.

We propose to find the $(1 - \beta)100\%$ confidence interval for $p(i)$, for all $0.01 \leq \beta < 1$. Starting at 0.01 is arbitrary and you could begin at 0.001, or 0.005, etc. Denote these confidence intervals as

$$[p(i)_1(\beta), p(i)_2(\beta)], \tag{1}$$

for $0.01 \leq \beta < 1$. Add to this the interval $[n_i/N, n_i/N]$ for the 0% confidence interval for $p(i)$. Then we have a $(1 - \beta)100\%$ confidence interval for $p(i)$ for $0.01 \leq \beta \leq 1$.

Now place these confidence intervals, one on top of the other, to produce a triangular shaped fuzzy number $\overline{p}(i)$ whose $\alpha$-cuts are the confidence intervals. We have

$$\overline{p}(i)[\alpha] = [p(i)_1(\alpha), p(i)_2(\alpha)], \tag{2}$$

for $0.01 \leq \alpha \leq 1$. All that is needed is to finish the "bottom" of $\overline{p}(i)$ to make it a complete fuzzy number. We will simply drop the graph of $\overline{p}(i)$ straight down to complete its $\alpha$-cuts so

$$\overline{p}(i)[\alpha] = [p(i)_1(0.01), p(i)_2(0.01)], \tag{3}$$

for $0 \leq \alpha < 0.01$. In this way we are using more information in $\overline{p}(i)$ than just a point estimate, or just a single interval estimate. Notice that $\overline{p}(i)[0]$ is the 99% confidence interval for $p(i)$. See also section 2.1 and 2.2 in the previous chapter.

We may do the same to obtain $\overline{p}$. For simplicity, throughout the rest of this paper we will always use triangular fuzzy numbers for the fuzzy values of uncertain probabilities.

## 2.3. Fuzzy Transition Probabilities

We first need to explain how the system can change at the end of each time interval $\delta$. System changes can occur only at the end of a time interval $\delta$. This time interval may be one second, one minute, one hour, etc. During a time interval $\delta$: (1) customers may arrive at the system but are only allowed into the system at the end of the time interval; (2) customers may leave the servers but are allowed to return to the calling source only at the end of the time interval; (3) at the end of the time interval all customers in the queue (in the system but not in the servers) are allowed to fill the empty servers; and (4) all customers who arrived are allowed into the system to fill empty servers or go into queue up to capacity $M$ with all others, it may be assumed, returning to the calling source. System changes can occur only at times $t = \delta, 2\delta, 3\delta, ...$

Let $p(i)$ be the crisp probability that $i$ customers arrive at the system during a time interval $\delta$, $i = 0, 1, 2, 3, ...$ Then $\sum_{i=0}^{\infty} p(i) = 1$. Next let $q(l|s)$ be the probability that, during a time interval $\delta$, $l$ customers in the servers complete service and are waiting to return to the calling source at the end of the time interval, given that $s$ servers are full of customers at the start of the time period, for $l = 0, 1, 2, ..., s$ and $s = 0, 1, 2, 3, ...c$. $c$ = the number of parallel and identical servers in the system. Then $\sum_{l=0}^{s} q(l|s) = 1$ for each $s$. Next we construct the transition matrix $P$. The rows of $P$ are labeled $0, 1, 2, 3, ..., M$ representing the system state at the start of the time period and the columns of $P$ are labeled $0, 1, 2, 3, ..., M$ representing the system state at the beginning of the next period. (See a $c = 1$ and $M = 4$ case in Table 1.)

Let us now assume that $c = 2$ and $M = 10$. So $P$ is $11 \times 11$. Let $P = (p_{ij})$ and we first need expressions for all the $p_{ij}$ which are fuzzified to $\overline{p}_{ij}$ and the fuzzy transition matrix is $\overline{P} = (\overline{p}_{ij})$. As an example let us look at

$$p_{6,10} = p(4)q(0|2) + p(5)q(1|2) + p(6)q(2|2). \tag{4}$$

Then

$$\overline{p}_{6,10} = \overline{p}(4)\overline{q}(0|2) + \overline{p}(5)\overline{q}(1|2) + \overline{p}(6)\overline{q}(2|2). \tag{5}$$

Fuzzy numbers for the $\overline{q}(i|2)$, $i = 0, 1, 2$, must be computed, assuming we have the fuzzy numbers for the $\overline{p}(i)$.

Let $p$ be the crisp probability that a customer leaves a server during time period $\delta$. Then we have a binomial probability distribution for the crisp $q(i|s)$

$$q(i|s) = \binom{s}{i} p^i (1-p)^{s-i}. \tag{6}$$

But now we have fuzzy probability $\overline{p}$ so we get the fuzzy binomial [1]. Hence

$$\overline{q}(i|s) = \binom{s}{i} \overline{p}^i (1-\overline{p})^{s-i}, \tag{7}$$

which is computed by $\alpha$-cuts

$$\overline{q}(i|s)[\alpha] = \{\binom{s}{i} p^i (1-p)^{s-i} | p \in \overline{p}[\alpha]\}, \tag{8}$$

for all $\alpha$ in $[0, 1]$.

Now we return to $\overline{p}_{6,10}$ which is evaluated by $\alpha$-cuts

$$\overline{p}_{6,10}[\alpha] = \{p(4)q(0|2) + p(5)q(1|2) + p(6)q(2|2) \,|\, \mathbf{S}\}, \tag{9}$$

for all alpha, where **S** is the statement "$p(i) \in \overline{p}(i)[\alpha]$, $0 \leq i \leq 10$, $p(0) + ... + p(10) = 1$, $q(i|2) \in \overline{q}(i|2)[\alpha]$, $i = 1, 2, 3$, $q(0|2) + q(1|2) + q(2|2) = 1$". This is restricted fuzzy arithmetic where we require the sums of the $p(i)$, and the $q(i|2)$, to equal one (see [1],[2],[10]-[12],[15],[16]). Let $\overline{p}_{6,10}[\alpha] = [\tau_1(\alpha), \tau_2(\alpha)]$ Then we have an optimization problem

$$\tau_1(\alpha) = min\{p(4)q(0|2) + p(5)q(1|2) + p(6)q(2|2)\}, \tag{10}$$

and

$$\tau_2(\alpha) = max\{p(4)q(0|2) + p(5)q(1|2) + p(6)q(2|2)\}, \tag{11}$$

all alpha, subject to the linear constraints in **S**. The solution to these optimization problems can be: (1) easy (common sense); (2) easy using differential calculus (if a partial on a variable is positive, then it is an increasing function of that variable); (3) solved by linear programming (using, e.g., Matlab or Maple [13]); (4) solved by non-linear programming using, e.g., the Premium Solver Platform V5.0 from Frontline Systems [7] or (5) solved by methods such as genetic algorithms (GA). In general, the Premium Solver Platform and GAs can handle many of these problems. The GAs have played a kind of "uniform" approach to a wide variety of problems; they often play a lead role in explorations because of their flexibility. Also, they can be modified to solve potentially more complex problems than we have addressed so far. Thus, having software available that has been used by a team of researchers is a benefit in and of itself. An agent "theme" [18]-[21] comes to mind (and is mentioned briefly at other points of the paper) with an agent "node" in a ready state to participate in overall computational schemes, e.g., to confirm other solutions and to play the lead role in establishing results that other methods can later aim at, with a goal of more efficient and compact solutions. We foresee a triumvirate of calculation schemes all bearing a simulation mode of operation as their prime or secondary focus, complementing mathematical analysis, as a means to attack the problems of interest.

We now assume we have the needed $\alpha$-cuts of all the $\overline{p}_{ij}$ in $\overline{P}$.

## 2.4 Fuzzy Steady State Probabilities

Let $P$ be the transition matrix for a regular Markov chain. Here we number the rows/columns $0, 1, 2, ..., M$ for $M$ system capacity. We say that the Markov chain is regular if $P^k > 0$ for some $k$, which is $p_{ij}^{(k)} > 0$ for all $i, j$. This means that it is possible to go from any state $S_i$ to any state $S_j$ in $k$ steps. A property of regular Markov chains is that powers of $P$ converge, or $\lim_{n\to\infty} P^n = \Pi$, where the rows of $\Pi$ are identical. Let $w$ be the unique left eigenvalue of $P$ corresponding to eigenvalue one, so that $w_i > 0$ all $i$ and $\sum_{i=0}^{M} w_i = 1$. That is $wP = w$ for $1 \times (M+1)$ vector $w$. Each row in $\Pi$ is equal to $w$ and $p^{(n)} \to p^{(0)}\Pi = w$. In this last expression $p^{(n)}$ is the vector of probabilities of being in state $S_i$ after $n$ steps and $p^{(0)}$ is the vector of initial probabilities. After a long time, regarding each step as a time interval, the probability of being in state $S_i$ is $w_i$, $0 \leq i \leq M$, independent of the initial conditions $p^{(0)}$. In a regular Markov chain the process goes on forever jumping from state to state, to state, ...

Now proceed to a fuzzy finite, regular, Markov chain by substituting the $\overline{p}_{ij}$, from the previous subsection, for the $p_{ij}$ producing a fuzzy transition matrix $\overline{P}$.

The uncertainty is in some of the $p_{ij}$ values but not in the fact that the rows in the transition matrix must be discrete probability distributions (row sums equal one). So we now put the following restriction on the $\overline{p}_{ij}$ values: there are $p_{ij} \in \overline{p}_{ij}[1]$ so that $P = (p_{ij})$ is the transition matrix for a finite Markov chain (row sums one). At this point $\overline{P}$ is a $(M+1) \times (M+1)$ matrix with rows/columns numbered $0, 1, 2, ..., M$. We will need the following definitions for our restricted fuzzy matrix multiplication. Pick and fix an $\alpha$ in $[0, 1]$. Define $Dom[\alpha]$ as the set of all $p_{ij} \in \overline{p}_{ij}[\alpha]$, $0 \leq i, j \leq M$, so that if we form a transition matrix $P = (p_{ij})$ with these $p_{ij}$ all the row sums equal one. Define $v = (p_{00}, p_{01}, ..., p_{MM})$. Row vector $v$ is just all the $p_{ij}$ in a transition matrix $P = (p_{ij})$. Then $Dom[\alpha]$ is all the vectors $v$, where the $p_{ij}$ are in the alpha-cut of $\overline{p}_{ij}$ all $i, j$, so that $P$ is the transition matrix for a finite Markov chain. In this chapter let us assume that $P$ is regular.

For each $v \in Dom[\alpha]$ set $P = (p_{ij})$ and we get $P^n \rightarrow \Pi$. Let $\Gamma(\alpha) = \{w | wP = w, 0 < w_i < 1, w_0 + ... + w_M = 1, v \in Dom[\alpha]\}$. $\Gamma(\alpha)$ consists of all vectors $w$, which are the rows in $\Pi$, for all $v \in Dom[\alpha]$. Now the rows in $\overline{\Pi}$ will be all the same so let $\overline{w} = (\overline{w}_0 ...., \overline{w}_M)$ be a row in $\overline{\Pi}$. Also, let $\overline{w}_j[\alpha] = [w_{j1}(\alpha), w_{j2}(\alpha)]$, for $0 \leq j \leq M$. Then [3]

$$w_{j1}(\alpha) = min\{w_j | w \in \Gamma(\alpha)\}, \tag{12}$$

and

$$w_{j2}(\alpha) = max\{w_j | w \in \Gamma(\alpha)\}, \tag{13}$$

where $w_j$ is the $j^{th}$ component in the vector $w$. The steady state fuzzy probabilities are : (1) $\overline{w}_0$ = the fuzzy probability of the system being empty; (2) $\overline{w}_1$ = the fuzzy probability of one customer in the system; etc.

In general, the solutions to equations (12) and (13) will be computationally difficult. We used both a genetic algorithm [2] and the Premium Solver Platform V5.0 [7] to get $\alpha$-cuts for the fuzzy steady state probabilities.

## 2.5 Fuzzy System Performance Variables

We first discuss the computing of $\overline{U}$ =server utilization, $\overline{N}$ =expected number of customers in the system and $\overline{X}$ =average server throughput because they all involve solving a linear programming problem. Then $\overline{R}$ = average response time is simply $\overline{N}/\overline{X}$ evaluated using alpha-cuts. Finally we see how to get $\overline{LC}$ = the expected number of lost customers per unit time due to finite system capacity. To motivate the fuzzy calculations we will first present the crisp definition. The crisp fuzzy steady state probabilities are $w_i$, $0 \leq i \leq M$ and crisp $p$ is the probability that a customer leaves a server during the time interval. Let us assume that there are now two servers ($c = 2$) in the system.

A crisp definition of $U$ could be expressed as

$$U = \sum_{i=2}^{M} w_i, \tag{14}$$

which is the probability that *both* servers are busy. An alternative is for either one of the servers *or* both to be busy and we calculate it as well at times. A further wrinkle occurs, i.e., a simulation package may centralize on average contents of its multiple server components. For the most part we need not get bogged down by these nuances and just go with one definition until another is called for, e.g., by some model user. Accordingly, we continue from our first definition: $U \times 100$ then gives the percentage of time we expect both servers to be busy. In the fuzzy case

$$\overline{U} = \sum_{i=2}^{M} \overline{w}_i, \tag{15}$$

and is evaluated by $\alpha$-cuts

$$\overline{U}[\alpha] = \{\sum_{i=2}^{M} w_i | \quad \mathbf{S} \quad \}, \tag{16}$$

for all $\alpha$, where **S** is "$w_i \in \overline{w}_i[\alpha]$, $0 \leq i \leq M$, $w_0 + ...w_M = 1$". This is restricted fuzzy arithmetic, first presented in ([10]-[12],[15],[16]). Notice that we cannot simply add up the $\overline{w}_i$ for $i = 2, 3, ..., M$ because it may result in a fuzzy number not in the interval $[0, 1]$. The restriction is that the sum of the $w_i$ must equal one so that it is a discrete probability distribution. Also see ([1],[2],[4]) for more details on restricted fuzzy arithmetic applied to this type of calculation. We compute the alpha-cuts of $\overline{U}$ by solving a linear programming problem. Let $\overline{w}_i[\alpha] = [w_{i1}(\alpha), w_{i2}(\alpha)]$, $0 \leq i \leq M$. Let $\overline{U}[\alpha] = [u_1(\alpha), u_2(\alpha)]$ The objective functions are

$$max/min[w_2 + ... + w_4], \tag{17}$$

subject to constraints

$$w_{i1}(\alpha) \leq w_i \leq w_{i2}(\alpha), i = 0, ..., M, w_0 + ... + w_M = 1. \tag{18}$$

The solution to the min (max) problem gives $u_1(\alpha)$ $(u_2(\alpha))$.

$N$ is just the expected number of customers in the system

$$N = \sum_{k=0}^{M} k w_k, \tag{19}$$

and $\overline{N}$ is determined by its $\alpha$-cuts

$$\overline{N}[\alpha] = \{\sum_{k=0}^{M} k w_k | \quad \mathbf{S} \quad \}. \tag{20}$$

The end points of the interval $\overline{N}[\alpha] = [n_1(\alpha), n_2(\alpha)]$ may also be found by solving linear programming problems

$$max/min[w_1 + 2w_2 + 3w_3 + ... + Mw_M], \tag{21}$$

subject to the same constraints given above. We get $n_1(\alpha)$ $(n_2(\alpha))$ from the min (max) problem.

$X$ is the expected number of customers leaving the system per time period $\delta$. We first derive a crisp expression for $X$ and then fuzzify it. Define $L(i)$, $i = 0, 1, 2$, to be the probability that $i$ customers leave a server at the end of the time period, with no conditions on how many servers were busy at the start of the time period. Then we see that

$$L(0) = w_0 + (1-p)w_1 + (1-p)^2 U_2, \tag{22}$$

where $U_2 = w_2 + ... + w_M$. Also

$$L(1) = pw_1 + 2p(1-p)U_2, \tag{23}$$

and

$$L(2) = p^2 U_2. \tag{24}$$

In the above equations the factors $(1-p)^2$, $2p(1-p)$ and $p^2$ come from the binomial probability distribution. In the binomial probability distribution $b(n,p)$, $n$ is the number of independent experiments and $p$ is the probability of a "success". Here $n = 2$ and a success is for a customer to leave a server. So, $p^2$ is the probability of two successes in two trials, $2p(1-p)$ is the probability of one success in two trials and $(1-p)^2$ is the probability of no successes in two trials. Then

$$X = 0L(0) + 1L(1) + 2L(2), \tag{25}$$

which simplifies to

$$X = pw_1 + 2pU_2. \tag{26}$$

Therefore, in the fuzzy case $\overline{X} = \overline{p}\overline{w}_1 + 2\overline{p}\overline{U}_2$ which is evaluated by $\alpha$-cuts

$$\overline{X}[\alpha] = \{pw_1 + 2pU_2 |\ \mathbf{S}\ \}, \tag{27}$$

where **S** is "$p \in \overline{p}[\alpha]$, $w_i \in \overline{w}_i[\alpha]$, $0 \leq i \leq M$, $w_0 + ... + w_M = 1$" with $U_2 = w_2 + ... + w_M$.

Let $\overline{X}[\alpha] = [x_1(\alpha), x_2(\alpha)]$ and let $\overline{p}[\alpha] = [p_1(\alpha), p_2(\alpha)]$. Then we may obtain the end points of the alpha-cut interval of $\overline{X}$ by solving the following non-linear programming problems

$$x_1(\alpha) = min\{p_1(\alpha)w_1 + 2p_1(\alpha)[w_2 + ...w_M]\}, \tag{28}$$

and

$$x_2(\alpha) = max\{p_2(\alpha)w_1 + 2p_2(\alpha)[w_2 + ... + w_M]\}, \tag{29}$$

for all alpha, subject to the linear constraints in **S**.

Finally we need to determine $\overline{LC}$. We first find the crisp expression for the expected number of requests (customers) rejected per unit time due to finite system capacity and then fuzzify it. Now $w_M \times 100$, $w_M$ being the probability of the system being full, is the percent of requests rejected per unit time. Let $\Lambda$ be the expected number of customers arriving at the system per unit time. Then the expected number of customers lost per unit time (LC) would be $w_M\Lambda$. It follows that

$$LC = w_M\Lambda = w_M \sum_{i=0}^{L} ip(i), \tag{30}$$

where $p(i)$ is the probability that $i$ customers arrive during the unit time interval and we have assumed that $p(i) = 0$ for $i > L$. Hence

$$\overline{LC} = [\overline{w}_M][\sum_{i=0}^{L} i\overline{p}(i)], \tag{31}$$

to be evaluated by $\alpha$-cuts and restricted fuzzy arithmetic. We get $\overline{w}_M$ from the fuzzy steady state probabilities and the $\overline{p}(i)$ are calculated from data. $M$ is system capacity and $\overline{w}_M$ is the fuzzy steady state probability that the system (servers and queue) is full.

Let $\overline{LG}[\alpha] = [lc_1(\alpha), lc_2(\alpha)]$ and $w_M[\alpha] = [w_{M1}(\alpha), w_{M2}(\alpha)]$. All we need to get is $\overline{\Lambda}[\alpha] = [\lambda_1(\alpha), \lambda_2(\alpha)]$ because then $lc_i(\alpha) = w_{Mi}(\alpha)\lambda_i(\alpha)$,

$i = 1, 2$, all alpha. We obtain values for $\lambda_i(\alpha)$ from linear programming computation.

$$\lambda_1(\alpha) = min\{1p(1) + 2p(2) + ... + Mp(M)\} \tag{32}$$

and

$$\lambda_2(\alpha) = max\{1p(1) + 2p(2) + ... + Mp(M)\}, \tag{33}$$

all alpha, subject to the constraints

$$p(i) \in \overline{p}(i)[\alpha], 0 \leq i \leq M, p(0) + ... + p(M) = 1. \tag{34}$$

Having fuzzy numbers for system performance we go on to final models for costs, benefits, and related quantities.

### 2.6 Final Fuzzy Optimizations

The final computations involve optimizations that, since ([2],[4]) cover them, we need only briefly overview them here. Input variables over which these optimizations occur include some already seen above and others: (1) $c$, the number of servers; (2) type of server (different values for $\overline{p}$); (3) $M$ system capacity; and (4) different arrival rates ($\overline{p}(i)$) due to advertising the web site. The system operates under different "times" : (1) normal time; (2) bursty time ([14], Chapter 8); and (3) long tailed distribution time ([14], Chapter 8). We want to, e.g.: (1) $min\overline{R}$ and $max\overline{U}$; (2) max fuzzy profit; and (3) $min\overline{LC}$. Different optimization methods are used: (1) analytical; (2) ranking the fuzzy numbers; and (3) using "ideal" points.

## 3. SIMULATION-OPTIMIZATION VIA GAs

In this section we describe several simulation approaches based on fuzzy transition matrices, with the principal targets from the previous discourse being: estimates of $\alpha$-cuts of fuzzy system performance values such as $\overline{U}$, $\overline{N}$, $\overline{X}$, $\overline{R}$, and $\overline{LC}$. The methods either compute the steady state probabilities on the way to the performance variables and can, thereby, be targeted at the probabilities themselves in some circumstances where these values can compare with extant results and/or pave the way to new ones.

The approaches include: (1) a genetic algorithm (GA) simulation followed by a linear optimization, developing first intermediate (fuzzy) steady state probabilities and then the performance values; and (2) an "ab initio" simulation using $\alpha$ =1 cuts of the fuzzy transition probability matrix values. The simulation, in this case, is a crisp one. It can therefore be developed with simpler software configurations. We have chosen the same software we used in the previous chapter. Most important, however, is how replicated crisp simulations, at an approximate level, can provide estimates, from output (frequency) distributions, for the fuzzy performance variables for any $\alpha$-cut. The approach is direct and its "cost," relative to the first approach, is low.

Section 4.1 addresses the first (GA) case, whereas section 5.1 addresses the crisp $\alpha = 1$ case. A topic of interest, we reiterate, concerns the relative fuzziness of simulation results and those from other (fuzzy) methods. Section 5.2 treats this matter.

## 3.1 A Genetic Algorithm Based Simulation Approach

Let us first describe the feasible set $\mathcal{F}$ for the genetic algorithm. We first set $v = (p_{00}, p_{12}, ..., p_{MM})$ a $1 \times (M+1)^2$ vector of all the probabilities in a $(M+1) \times (M+1)$ transition matrix $P = (p_{ij})$ for a regular Markov chain. Let $\overline{p}_{ij}[\alpha] = [p_{ij1}(\alpha), p_{ij2}(\alpha)]$. Now $\mathcal{F}$ consists of all $v$ so that

$$p_{ij1}(\alpha) \leq p_{ij} \leq p_{ij2}(\alpha), \tag{35}$$

for all $i, j$ and

$$p_{i0} + ... + p_{iM} = 1, \tag{36}$$

for $i = 0, ..., M$ (all the row sums are one). It is important that this $\mathcal{F}$ is convex. What this means is that if $v^a$ and $v^b$ are in $\mathcal{F}$, then so is $v^c$ where

$$v^c = \lambda v^a + (1-\lambda)v^b, \tag{37}$$

for all $0 \leq \lambda \leq 1$. This fact will be used in the crossover operation in the genetic algorithm.

What we need to describe is the initial population $\mathcal{P}_0$, the fitness function, crossover, mutation and the next generation $\mathcal{P}_1$. The initial population $\mathcal{P}_0$ is just a set of $K$ randomly generated $v_i \in \mathcal{F}$. To describe the construction of the next population let $v \in \mathcal{P}_0$ and using this $v$ construct the transition matrix $P = (p_{ij})$. Next compute the vector $w = (w_0, ..., w_M)$ so that

$$wP = w, \quad 0 \leq w_i \leq 1, \quad w_0 + ... + w_M = 1. \tag{38}$$

Let $w^{(i)} = (w_0^{(i)}, ..., w_M^{(i)})$, $1 \leq i \leq K$, be all the vectors $w$ obtained from equation (38) using all the $v \in \mathcal{P}_0$. Let $\overline{w}_i[\alpha] = [w_{i1}(\alpha), w_{i2}(\alpha)]$. Suppose in this run of the genetic algorithm we are looking for $w_{31}(\alpha)$ the left end point of the interval $\overline{w}_3[\alpha]$. We now sort the $w_3^{(i)}$ , $1 \leq i \leq K$, from smallest to largest.

The crossover operation generates possibly a new population member from two elements in $\mathcal{P}_0$. We will randomly choose two members of $\mathcal{P}_0$ for crossover. But first we randomly choose two $w$ from the set $w^{(i)}$, $1 \le i \le K$. In this random process it will be more likely that we pick a $w^{(i)}$ having a smaller value of $w_3^{(i)}$ than from those having the larger values of $w_3^{(i)}$ (this is easily accomplished from the sorting described above). Suppose we picked $w^a$ and $w^b$. These two then correspond to $v^a$ and $v^b$ in $\mathcal{P}_0$. From $v^a$ and $v^b$ we determine $v^c \in \mathcal{F}$ as follows:

$$v^c = \lambda v^a + (1 - \lambda) v^b. \tag{39}$$

This $v^c$ is in $\mathcal{F}$ for any $\lambda \in [0, 1]$. A value of $\lambda$ is randomly generated to get $v^c$. Generate around $I$ of the $v^c$ in this manner and put them all in $\mathcal{P}_0$. Calculate their corresponding vectors $w$. Notice that $\mathcal{P}_0$ has grown to be more than $K$ members.

Next we discuss mutation. To do this we need to consider the following equation:

$$1 - p_{iM2}(\alpha) \le \sum_{j=1}^{M-1} p_{ij} \le 1 - p_{iM1}(\alpha), \tag{40}$$

for $i = 0, ..., M$. We next randomly choose a few (maybe $J$) $v$ from $\mathcal{P}_0$ for mutation. Suppose $v = (p_{00}, ..., p_{MM})$ was chosen. Randomly choose an element in $v$ which is not $p_{iM}$, $0 \le i \le M$ (not the end of a row in $P$). Assume we picked $p_{46}$. Now

$$p_{461}(\alpha) \le p_{46} \le p_{462}(\alpha). \tag{41}$$

Randomly choose a value in this interval $[p_{461}(\alpha), p_{462}(\alpha)]$. Assume we got $p_{46}^*$. If equation (40) is satisfied for this $p_{46}^* = p_{46}$ we keep it, otherwise it is discarded and we randomly choose another value in the interval. So assume that we keep this $p_{46}^*$. Substitute $p_{46}^*$ for $p_{46}$ in $v$. Now adjust the value of $p_{4M}$ so that the fourth row sum is one (we may always do this since equation (40) was satisfied). We now have a new (mutated) $v$ in $\mathcal{P}_0$. After doing this maybe $J$ times we have introduced new mutated members into $\mathcal{P}_0$. For all the mutated $v$ compute their corresponding vectors $w$, discard the old values $w$.

Now sort all the $w_3^{(i)}$ from smallest to largest for all the $v$ in $\mathcal{P}_0$ and choose the $K$ smallest. Determine the $v$ in $\mathcal{P}_0$ corresponding to the $K$ smallest $w_3^{(i)}$, these $v$ are the next generation $\mathcal{P}_1$.

Continue through this process of getting $\mathcal{P}_i$, calculating the vectors $w$, sorting, crossover and mutation for $\Theta$ generations. We usually used around 200 generations. Then we will have a good estimate of $w_{31}(\alpha)$. This is repeated to estimate all the end points of the alpha-cuts of the fuzzy steady state probabilities.

In our initial applications of the genetic algorithm we experimented with different sizes for $\mathcal{P}_0$ and different values for $\Theta$. We usually used $K = 100$ for the size of the initial population and had values between 200 and 700 for $\Theta$.

In our previous experience with other genetic algorithms we noticed that too often crossover produces a result not in the feasible set. In fact the algorithm can spend too much time discarding the results of crossover because they are not in $\mathcal{F}$. The same result may occur in mutation. Since $\mathcal{F}$ is convex our crossover always gives a result in the feasible set.

Looking at the optimal solutions, especially the $11 \times 11$ case, we found that the optimal $v$ was quite often on the boundary of $\mathcal{F}$. The solution $v$ is on the boundary of $\mathcal{F}$ when one, or more, of the inequalities in equation (35) is an equality. So in the genetic algorithm we need population members $v \in \mathcal{P}_i$ on the boundary of $\mathcal{F}$. Notice that crossover, equation (39), always gives a $v^c$ "between" $v^a$ and $v^b$. So employing only crossover new populations will tend to migrate to the "center" of $\mathcal{F}$. Hence, the important operation of mutation is to make some $v$ exist on, or near, the boundary of $\mathcal{F}$. We can get $v$ from mutation on, or near, the boundary of $\mathcal{F}$ by having it more likely that the $p_{ij}$ we choose in $[p_{ij1}(\alpha), p_{ij2}(\alpha)]$ is at, or near, the end points of the interval.

# 4. GENETIC ALGORITHM EXAMPLES

We will now present two numerical examples. Both of them are single server (c=1) cases, but the maximum number of customers in the system differs. Both utilize fuzzy triangular transition matrices.

## 4.1 Case: c=1, M=4

We first consider a case where M=4 and c=1. The triangular fuzzy transition probabilities are given in Table 1. The table only shows the base of the triangle and the vertex is at, or near, the midpoint of the base, the former case when the triangular fuzzy transition probability is symmetric.

The computation first obtains the fuzzy steady state probabilities using the genetic algorithm (GA). We present these results in Table 2. For the problem at hand, we can see that transitions toward more occupied states occur and the resulting system is a very busy one, as we would expect.

After computing these values we progress to the results for the key performance entities $\overline{U}$, $\overline{N}$, $\overline{X}$, and $\overline{R}$. These results are displayed in Table 3.

Table 1: Fuzzy Probability Matrix (M=4, c=1).

| | | | | |
|---|---|---|---|---|
| [.070,.130] | [.260,.340] | [.170,.230] | [.070,.130] | [.200,.400] |
| [.021,.065] | [.127,.235] | [.215,.307] | [.120,.200] | [.250,.470] |
| [.000,.000] | [.021,.065] | [.127,.235] | [.215,.307] | [.415,.619] |
| [.000,.000] | [.000,.000] | [.021,.065] | [.127,.235] | [.700,.852] |
| [.000,.000] | [.000,.000] | [.000,.000] | [.021,.065] | [.935,.979] |

Table 2: GA Result: Alpha Zero Cut Fuzzy Steady State Probabilities (M=4, c=1).

$$\overline{w}_0[0] = [.0000, .0000]$$
$$\overline{w}_1[0] = [.0000, .0006]$$
$$\overline{w}_2[0] = [.0006, .0070]$$
$$\overline{w}_3[0] = [.0236, .0804]$$
$$\overline{w}_4[0] = [.9122, .9758]$$

The results that Table 3 portrays are (again) achieved via optimizations as outlined in section 2.5. According to remarks of the preceding paragraph we expect to and do indeed find a maximal steady state utilization. Correspondingly, $\overline{R}$ is near its maximum value too. And so on with other variables' final disposition.

## 4.2 Case: c=1, M=10

In another case, M=10 and c=1 (see Table 4 for the input), the computation takes slightly over 3 minutes. The fuzzy steady state probabilities and the $\overline{U}$, $\overline{N}$, $\overline{X}$, $\overline{R}$ results are given in Table 5 and Table 6, respectively. Several cases where the probabilities are near zero appear in the output as a result of the double precision arithmetic. Note that utilization, again, is very high. Recall that obtaining the fuzzy system performance result is a separate computation after securing the probabilities.

## 4.3 Computational Steps

The computations here utilize the Matlab optimization package. The computational details are important for more than one reason. A first is universal: to lay out matters so the calculations can be confirmed (and extended if desired) by other researchers. A second is that our current work is a predecessor of some agent style computing [18]-[21], a mode, we interpret, as calling for explicit solution patterns for the "domain" problem (to offset complexities in

Table 3: GA Result: Alpha Zero Cut Fuzzy System Performance Variables (M=4, c=1)

$$\overline{U}[0] = [1.0000, 1.0000] \qquad \overline{N}[0] = [3.9038, 3.9752]$$
$$\overline{X}[0] = [0.3000, 0.5000] \qquad \overline{R}[0] = [7.8076, 13.2513]$$

the networking and brokering features) and to facilitate a set of collaborators to engage in this kind of modeling effort. In addition, though we have not encountered time-consuming calculations so far, they are in the offing. Since some of our attempted calculations have already resisted facile solution, e.g., in one case, an inability to get a calculation initiated (might we add, with somewhat pricey software) we seek to promote dialog on better or alternative approaches to these computations.

The results we produce here are from successful use of the Matlab optimization toolkit, where another package failed on the larger of our two examples. We have yet to "push the envelope" on Matlab for these kinds of calculations so our previous paragraph comments must be nuanced to recognize this point. The calculation can be centered on a collection of (secondary storage) files, which are loaded into main memory at computation time:

1. a constraint file representing the law on probability sums;

2. an objective function file for computing directly from (the steady state) probabilities (e.g. equations (14), (19), (26) in section 2.5);

3. a file connecting Items 1 and 2, utilizing the fmincon() built-in function whose description follows.

Our Matlab call is:

[x, fval] = fmincon(@objfun, x0,[],[],[],[],lb,ub,@confun,options)

where • x is the array of steady state probabilities • fval is the max (or min) value sought, e.g., utilization, etc. • mincon is a noted in Item 3 • objfun is the objective function (Item 2) • x0 initializes x, (an array of all 0's suffices in our work) • [] ... [] are not used in these calculations • lb is the lower bound (array) for x • ub is the upper bound (array) for x • confun relates to Item 1 • options is not used (left at its default)

The GA (part of the) calculation in this example (computing the steady state probabilities from the fuzzy transition matrix) requires about 1/2 minute (elapsed time).

Table 4: Fuzzy Probability Matrix (M=10, c=1).

| | 0 | 1 | 2 | 3 | 4 | 5 |
|---|---|---|---|---|---|---|
| 0 | [.070,.130] | [.260,.340] | [.170,.230] | [.070,.130] | [.070,.130] | [.070,.130] |
| 1 | [.021,.065] | [.127,.235] | [.215,.307] | [.120,.200] | [.070,.130] | [.070,.130] |
| 2 | [.000,.000] | [.021,.065] | [.127,.235] | [.215,.307] | [.120,.200] | [.070,.130] |
| 3 | [.000,.000] | [.000,.000] | [.021,.065] | [.127,.235] | [.215,.307] | [.120,.200] |
| 4 | [.000,.000] | [.000,.000] | [.000,.000] | [.021,.065] | [.127,.235] | [.215,.307] |
| 5 | [.000,.000] | [.000,.000] | [.000,.000] | [.000,.000] | [.021,.065] | [.127,.235] |
| 6 | [.000,.000] | [.000,.000] | [.000,.000] | [.000,.000] | [.000,.000] | [.021,.065] |
| 7 | [.000,.000] | [.000,.000] | [.000,.000] | [.000,.000] | [.000,.000] | [.000,.000] |
| 8 | [.000,.000] | [.000,.000] | [.000,.000] | [.000,.000] | [.000,.000] | [.000,.000] |
| 9 | [.000,.000] | [.000,.000] | [.000,.000] | [.000,.000] | [.000,.000] | [.000,.000] |
| 10 | [.000,.000] | [.000,.000] | [.000,.000] | [.000,.000] | [.000,.000] | [.000,.000] |

| | 6 | 7 | 8 | 9 | 10 |
|---|---|---|---|---|---|
| 0 | [.030,.070] | [.030,.070] | [.000,.000] | [.000,.000] | [.000,.000] |
| 1 | [.050,.112] | [.030,.070] | [.015,.049] | [.000,.000] | [.000,.000] |
| 2 | [.070,.130] | [.050,.112] | [.030,.070] | [.015,.049] | [.000,.000] |
| 3 | [.070,.130] | [.070,.130] | [.050,.112] | [.030,.070] | [.015,.049] |
| 4 | [.120,.200] | [.070,.130] | [.070,.130] | [.050,.112] | [.045,.119] |
| 5 | [.215,.307] | [.120,.200] | [.070,.130] | [.070,.130] | [.095,.231] |
| 6 | [.127,.235] | [.215,.307] | [.120,.200] | [.070,.130] | [.165,.361] |
| 7 | [.021,.065] | [.127,.235] | [.215,.307] | [.120,.200] | [.250,.470] |
| 8 | [.000,.000] | [.021,.065] | [.127,.235] | [.215,.307] | [.415,.619] |
| 9 | [.000,.000] | [.000,.000] | [.021,.065] | [.127,.235] | [.700,.852] |
| 10 | [.000,.000] | [.000,.000] | [.000,.000] | [.021,.065] | [.935,.979] |

Table 5: GA Result: Alpha Zero Cut Fuzzy Steady State Probabilities (M=10, c=1).

$\overline{w}_0[0]$=[0.000000000000007105, 0.000000000003795853]
$\overline{w}_1[0]$=[0.000000000000091421, 0.000000000071027725]
$\overline{w}_2[0]$= [0.000000000001978444, 0.000000000971756618]
$\overline{w}_3[0]$= [0.000000000024610896, 0.000000013284027083]
$\overline{w}_4[0]$= [0.000000000684154316, 0.000000203856567872]
$\overline{w}_5[0]$= [0.000000015516770088, 0.000003296945473752]
$\overline{w}_6[0]$= [0.000000488650614631, 0.000043420172617435]
$\overline{w}_7[0]$= [0.000017131017363733, 0.000521541283974432]
$\overline{w}_8[0]$= [0.000603983775910370, 0.006225593230155810]
$\overline{w}_9[0]$= [0.023680832253123528, 0.079328046405422153]
$\overline{w}_{10}[0]$=[0.925054143853171710, 0.972922528990218140]

Table 6: GA Result: Alpha Zero Cut Fuzzy System Performance Variables (M=10, c=1).

$\overline{U}[0]$ = [1.000, 1.000] $\overline{N}[0]$ = [9.9139, 9.9723]
$\overline{X}[0]$ = [0.300, 0.500] $\overline{R}[0]$ = [19.828, 33.241]

# 5. CRISP SIMULATION (PACKAGE) APPROACHES

In this section we lay out an approach that is illustrated through a case study and has in it ingredients that can be used to solve a potentially large variety of problems. What we present is not the only attack we are making on the fuzzy matrix probability problem. Another approaches the fuzzy probability matrix through various sampling schemes; reference [22] shows some an earlier attempt that was successful in speed-up but not in accuracy (relative to the GAs). A new round of attack aims at improvements that are closer in spirit to the mainline attack of this section. With some still unresolved issues and scope limitations on the chapter we do not cover it here.

We also like to compare and contrast the present methods with those based on arrival and service rates in the companion chapter. We start with contrast. The latter, so far, are based primarily on (mathematical) foundations given by, e.g., [6] and [14], and remain strongly tied to them. Meanwhile, though, intense research is being pursued to move out into building a family of simulations that grow via relating new simulations to other, prior, simulations and ultimately back to models very close to the mathematical roots.

On the "compare" side, the same software is used here and in the companion paper, namely, SLX and GPSS/H [8],[9]. Though not (yet) a seamless integration, the potential exists for exploring models formulated in one purview to those in the other. SLX has been the prime software choice in the present study while the latter has been used repeatedly for supporting computation and confirmations. In the companion chapter, the two systems are in an almost equal balance. Much of calculation we report, then, has been done in replication mode and with variant styles — across platforms. Securing accuracy through confirmed results is not the whole story, support of the agent theme [18]-[21] being another.

The simulation system we will develop for transition probability based arguments is called $\mathbf{SIM}_{CS}$. It computes values of the performance variables mentioned above (i.e., utilization, number of customers in the system, throughput per unit time, and average response time). In the current discourse, we simply bypass the stage of developing the state probabilities.

The values that emerge from the simulations are stochastic and ("principled") means (still being sought) are needed to infer fuzzy values from them. Fortunately, the companion chapter provides a lead for a prime option we use here. It is not the first and simplest method to configure a calculation so that the end points of a desired fuzzy variable range are estimated from mean values emerging from the simulation distributions. While this method is most productive in models based on arrival and service rates, its utility for the transition matrix approach seems limited. The base problem is that there are far too many choices to be made from the values in the matrices we have worked with, i.e. not ones restricted to tridiagonal form most closely tied to arrival and service rate approach. (Review of the section on GAs in this chapter helps to substantiate this claim.)

The alternative of choice is to derive information from the *alpha*=1 situation, specifically, via the (frequency) distributions the solutions provide. Our proposed heuristic is that used in establishing alpha cuts from real-world, probabilistic data (see section 2.2) and used in a case study in the companion chapter. That is, we choose some confidence interval based on the output distributions. The lower the cut the broader the confidence interval, the aim being to utilize the *alpha*=1 cut to obtain any other cuts we may like. The simulated stochastic outputs result from replicated runs, the extent of which we must choose, along with the number of replications. These choices can be addressed by features such as stability, as our case study below demonstrates. The choices are being pursued in yet a few other ways, some of which we elude to in subsequent text. Let us now present some up-close action in this mode of modeling.

## 5.1 The $\alpha = 1$ Cut Model

We begin simulation modeling for the $\alpha = 1$ cut. In this case, the fuzzy input and output become crisp. We shall see soon how we employ these values for estimating other alpha cuts, particularly $\alpha = 0$. This simulation can be expressed as in (42), which expresses the expected response time $< R >$ in terms of our simulation system $\mathbf{SIM_{CS}}$:

$$< R >= \mathbf{SIM_{CS}}(inputs), \tag{42}$$

where inputs include c, $M, \overline{p}_{ij}$, where $c$ = the number of servers, $M$ = system capacity, and $\overline{p}_{ij}$ = the entries from the fuzzy transition matrix. More specifically, for the present effort dealing only with the $\alpha = 1$ cut, we can identify the center points of assumed triangular fuzzy numbers we are using as $\overline{p}_{ij}[1]$, or more simply, as $< p_{ij} >$, reflecting its crispness.

A transition diagram for a particular case may help. Figure 1 is for M=4; the transition probabilities in this figure, by their origin, are crisp values. The diagram is general with respect to choices for c, the number of servers.

A numerical example consistent with the figure is given in Table 7. The table is based on the earlier fuzzy transition matrix case in Table 1. Comparison of the tables reveals that the (crisp) entries in this (new) table are the midpoints, or near the midpoint, of the earlier (fuzzy) one. To develop the present table, we must set the value of $c$. Cursory examination of the table reveals we are actually depicting a $c = 1$ system. Recall, that under the earlier assumption there can be multiple arrivals but only a single departure over a $\delta$ interval. Thus, all transitions from higher number states to lower ones are zero except when the states differ by one customer (request).

These table values are employed in the (crisp) simulations. From these simulations we will establish frequency distributions of our quantities of interest. For the $\alpha = 1$ case, the means of these distributions will fall at the same points of $\alpha = 1$ case in the GA (and other optimization) approaches. The distributions themselves will be used heuristically for other alpha cuts (the zero cut being the emphasis here).

Note in passing, that we can get the $\alpha = 1$ cut's *mean* values more simply than from carrying out simulations. For example, the numbers of Table 7 were used directly in Matlab (matrix) calculations to obtain steady state $\alpha = 1$ cut results (e.g., using matrix powers or eigenvalue analysis; see [2] for example). These calculations provide solutions for the $\alpha = 1$ cut but not for other cuts. The solutions we present below, based on simulation output (frequency) distributions can estimate any of the cuts, say, our (primary) desired $\alpha = 0$ cut.

The transition diagram provides a guide to the core portion of the program describing the transitions. Figure 2 shows coding for our $M = 4$ case. A switch construct is used along with case enumeration for program control transfers; in larger examples other constructions can shorten the code.

Node_I can be taken as the current node and Node becomes node the next state node (determined by one of the random functions Trans_i() (see next paragraph). The tabulate expression is of interest in that the language of choice, SLX, provides an easy means to generate frequency distributions and plots associated with them: NodeTab, defined as random_variable (not shown) takes on a Node value which is then entered into a (pre-defined) frequency distribution.

The Trans1() ... Trans5() functions are similar and so we explain only one of them (refer to Figure 3). Trans1()'s "discrete_empirical" (SLX) function de-

Table 7: Crisp Transition Matrix for Alpha=One Simulation.

| | | | | |
|---|---|---|---|---|
| .10 | .30 | .20 | .10 | .30 |
| .04 | .18 | .26 | .16 | .36 |
| .0 | .04 | .18 | .26 | .52 |
| .0 | .0 | .04 | .18 | .78 |
| .0 | .0 | .0 | .04 | .96 |

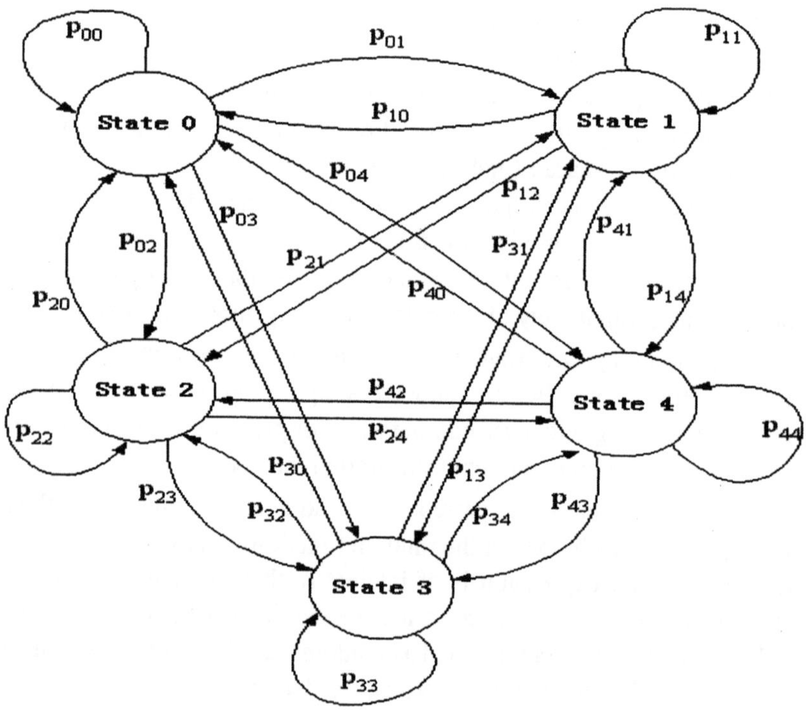


Figure 1: Transition Diagram for General Case, relative to c; M = 4.

```
switch (Node_I)
  {
   case 1:    Node = Trans1(streamT); break;
   case 2:    Node = Trans2(streamT); break;
   case 3:    Node = Trans3(streamT); break;
   case 4:    Node = Trans4(streamT); break;
   case 5:    Node = Trans5(streamT); break;
  }
tabulate  NodeTab=Node;
```

Figure 2: Core Part of (SLX) Simulation Model Handles Transitions (M=4) and Records Node Visitation Frequencies.

```
        discrete_empirical Trans1(

        // TRANSITION MATRIX (Row 1)
        // left  = input value (defined below)
        // right = state to transfer to
  T_M[1,1] 1, T_M[1,2] 2, T_M[1,3] 3, T_M[1,4] 4, T_M[1,5] 5);
                               . . .
                               . . .
        // Cumulative Probability (Row 1)
        // input point for discrete_empirical fct
         T_M[1,1]=0.1; T_M[1,2]=0.4; T_M[1,3]=0.6;
              T_M[1,4]=0.7; T_M[1,5]=1.0;
                               . . .
     rn_stream streamT          seed=100000;
```

Figure 3: Function Elements and Supporting Information Referencing Part of Apparatus to Implement the Transition Matrix of Table 7 in Conjunction with Figure 2's Switching Mechanism.

scribes input-output pairs, the first entry of the pair describing the probability axis (in cumulative form), described by (table) T_M(x,y) values just below the Trans1() function in the figure; the second entry of each pair defines the target node by its node number. The functions are driven by a random number stream on $[0, 1]$, whose seed can be chosen for random number control used in replications; the bottom line of the figure exhibits the key code line for "streamT," which is addressed in Figure 2.

It may be noted in passing that this code system opens doors for more complex kinds of simulations, e.g., in which the T_M(x,y) values change over time. Such a system is under investigation. These may result in quick but cruder approximations as we alluded to above in some earlier attempts [22]; however, at the time of this writing the idea seems to have a greater flexibility with potentially more options to explore.

Recall that $U$, $N$, $X$, $R$ and $LC$ are among the chief output variables of interest (performance values). In (purely) mathematical argument, these quantities are averages, typically, over a long period (not infrequently, infinite). In simulation, a model may be replicated several times, generating a distribution for each of these averages and in the work reported here this is what we have been doing. It might be worth noting, additionally, that some performance values produce useful distributed results with each run. Perhaps, the easiest case to envision is that of R: each individual customer (request) takes a certain amount of processing time and these times can be entered into a frequency distribution to get a distributed result. Other variables, such as U, are operative over an entire run; we would have to break the duration of the simulation into intervals and get results for each sub-interval (again tabulating them into a frequency distribution) to get a frequency distribution for it. In the present study, we did not exploit these finer features, simply determining averages for all the performance values (even for R); the distributions, then, arise from the replications. A seemingly lesser issue relates to smoothness of the output, though, as is generally the case, a larger number of replications typically leads to a smoother set of distributions for the results of interest. We provide information in this chapter, particularly, in Table 10 where a few choices for replication number are referenced. Worth mentioning in the same breath is that length of an individual run. In some systems there is a long build-up period, e.g., from a starting empty state; this can distort results (biasing them downward perhaps). However, in other cases, not atypical in queuing models, a system may cycle through alternating stages of empty and non-empty states. This means that in the general modeling picture, we should attend to runs of shorter duration as well as very long term ones. The problem is more a statistical one but variations in statistics and corresponding variations in fuzzy responses may be addressed in this manner. Still, in many cases, there are advantage in running a model a sufficiently long time so that some non-typical behavior does not have a serious impact on the results. The longer model's distributions, of course, become more compact.

## 5.2 Results

With the considerations outlined in previous paragraphs, we first chose 1000 replications, each embodying 1000 transitions. The elapsed time for this M=4, c=1 case was approximately twenty seconds. The time is shorter than for the GA; the latter's time given above also does not include the (final) optimization follow-up work. No systematic attempt has been made to extract a minimum

Table 8: Simulation Results for U, N, X, R (M=4, c=1). Reported are Deviation Measures and the Coeffi cient of Variation (cv).

| | mean | min | max | dev | var | cv |
|---|---|---|---|---|---|---|
| U: | 1.0000 | 0.9990 | 1.0000 | 0.0001 | 0.00000 | .00005 |
| N: | 3.9478 | 3.9180 | 3.9700 | 0.0100 | 0.00010 | .00254 |
| X: | 0.4022 | 0.2926 | 0.5108 | 0.0408 | 0.00166 | .10140 |
| R: | 9.9184 | 7.7750 | 13.4807 | 1.0400 | 1.08156 | .10485 |

Table 9: Simulation Results for U, N, X, R (M=10, c=1). The Same Variables are Reported Here as in the Previous Case.

| | mean | min | max | dev | var | cv |
|---|---|---|---|---|---|---|
| U: | 1.0000 | 1.0000 | 1.0000 | 0.0000 | 0.00000 | .00000 |
| N: | 9.9478 | 9.9170 | 9.9700 | 0.0101 | 0.00010 | .00102 |
| X: | 0.4022 | 0.2926 | 0.5108 | 0.0408 | 0.00166 | .10140 |
| R: | 24.9928 | 19.5613 | 33.9838 | 2.6222 | 6.87571 | .10492 |

for either of these values, though it could improve the elapsed time. We present two output displays (Figures 4 and 5 for this M=4, c=1 case study).

These figures were produced by Matlab on files produced by SLX programs. The first of these outputs represents the distributions for four of the key performance variables $\overline{U}$, $\overline{N}$, $\overline{X}$, and $\overline{R}$ under the conditions just mentioned. The second case shows $\overline{U}$ "up close," since in the first diagram its spread is difficult to appreciate. As a final remark, note that other alpha cuts can be extracted in this (simulation) approach in a manner similar to what was done in the companion chapter (i.e., as displayed in the final figure there).

Table 8 presents in tabular form our key (simulation) results for this case. The mean values of this table approximately match $\alpha$=1 cut values from the GA solution. Comparable results obtain for the M=10, c=1 case, again simulating 1000 transitions over 1000 replica. The results are reproduced in Table 9. These results have companion graphs (paralleling the situation for the previous M=4 case), Figures 6 and 7.

The results are quite close to the GA results and again the simulations give distributions that allow for a broader range of results (e.g., additional alpha cuts). The run time for the simulation is approximately 2 min, compared to GA's 3 min run (a figure that does not include the optimizations to get utilization, among details.

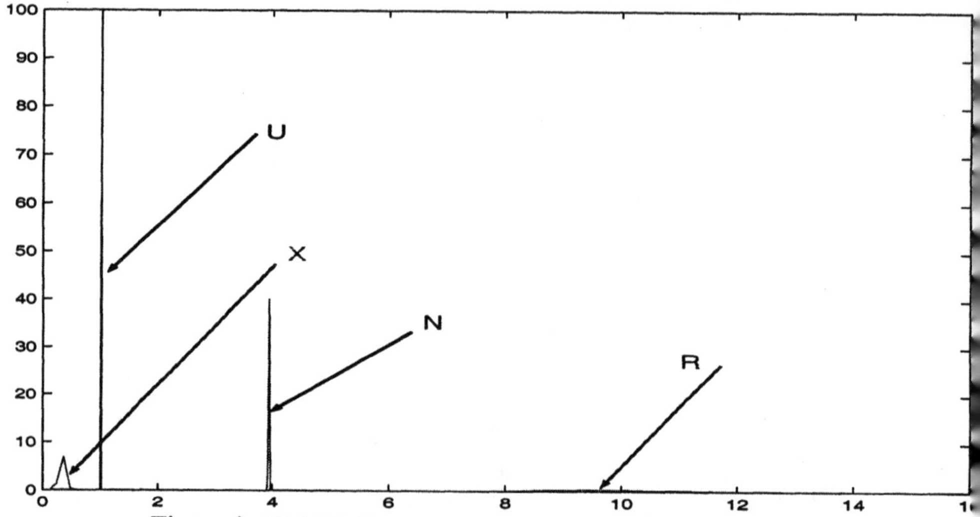


Figure 4: U,N,X,R Frequency Distributions (M=4, c=1).

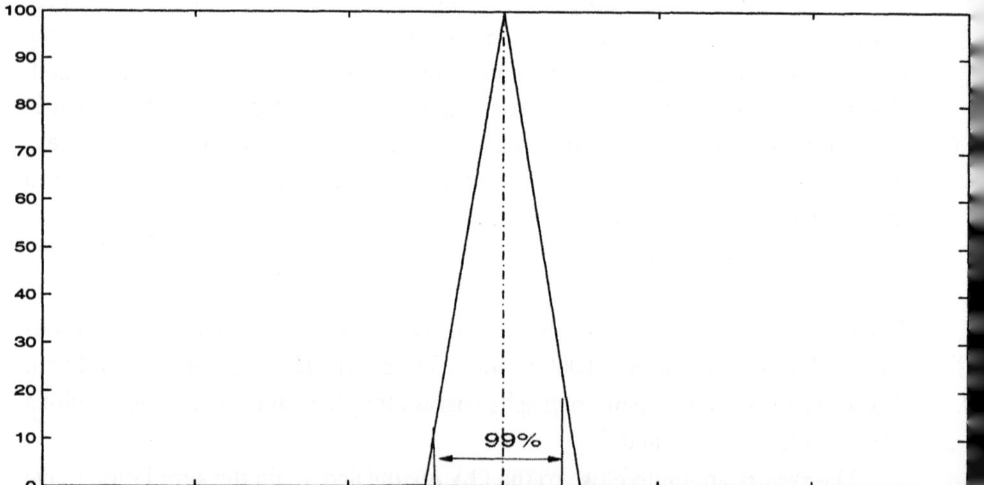


Figure 5: Frequency Distribution of Utilization (M=4, c=1).

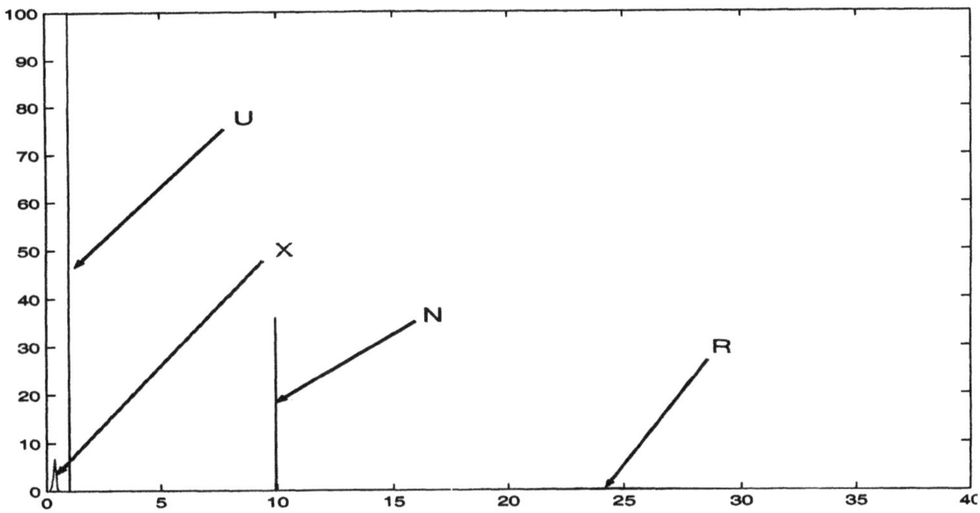


Figure 6: U,N,X,R Frequency Distributions (M=10, c=1).

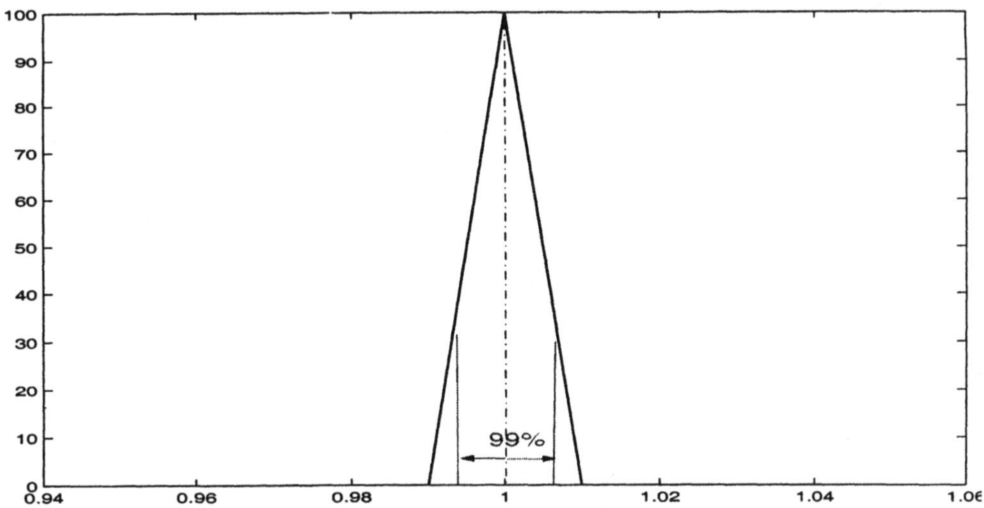


Figure 7: Frequency Distribution of Utilization (M=10, c=1).

Table 10: Runs Varying Transitions (Trans) and Replications (Reps) (M=4, c=1). GA Results are Printed at Top for Comparison Purposes.

| *Trans* | *Reps* | $\overline{U}$ | $\overline{N}$ | $\overline{X}$ | $\overline{R}$ |
|---|---|---|---|---|---|
| | | | Alpha-Zero Cut | | |
| GA Result-> | | [1.000,1.0] | [3.903,3.975] | [0.300,0.500] | [7.808,13.251] |
| 1000 | 1000 | [0.999,1.0] | [3.918,3.970] | [0.293,0.511] | [7.775,13.481] |
| 1000 | 2000 | [0.999,1.0] | [3.920,3.970] | [0.286,0.514] | [7.693,13.822] |
| 2000 | 1000 | [1.000,1.0] | [3.928,3.964] | [0.292,0.511] | [7.754,13.499] |
| 2000 | 2000 | [1.000,1.0] | [3.927,3.964] | [0.286,0.514] | [7.690,13.839] |
| | | | Alpha-One Cut | | |
| GA Result-> | | 1.0000 | 3.9478 | 0.4000 | 9.8697 |
| 1000 | 1000 | 1.0000 | 3.9478 | 0.4022 | 9.9184 |
| 1000 | 2000 | 1.0000 | 3.9477 | 0.4008 | 9.9613 |
| 2000 | 1000 | 1.0000 | 3.9477 | 0.4022 | 9.9182 |
| 2000 | 2000 | 1.0000 | 3.9477 | 0.4008 | 9.9613 |

Table 10 contains four runs (M=4 case, only) with varying numbers of transitions and replications (to make up the frequency distributions). The table also includes the GA results for comparison and gives the $\alpha$=1 cut solutions. The table's contents exhibit stability of the results over these variations. They also help to see, e.g., the response time result, calculated by dividing $\overline{N}$ by $\overline{X}$, introduces the least stable results, the fractional values for $\overline{X}$, and their variability, being contributing factors. Note that the simulation results are more fuzzy in all the cases ($\overline{U}$, $\overline{X}$, and $\overline{R}$), whereas in the case of $\overline{N}$ we get less fuzzy results. The two approaches mentioned in this chapter and its companion offer the reader a few discussion points on similar sets of phenomena and relative fuzziness in results, possibly provoking future research.

# 6. CONCLUSIONS AND FUTURE RESEARCH

In this chapter we considered how crisp and fuzzy methods can be employed, alone and in combination, to address problems involving fuzzy queuing systems (fuzzy probability models) defined in terms of (fuzzy) transition probability matrices.

The chapter began by outlining mathematical bases for the subject matter of interest. Attention then turned to key roles of optimization techniques and genetic algorithms in solving these kinds of systems and developing intermediate probability values and performance variables which depend upon them: utilization, number of customers/requests in the system, throughput rate, and response time. Some comments addressed lost customers (requests).

These solution techniques were found in the main to be effective, but the computational load was (also) found, on many occasions, to be high. Similarly, formulating models and getting them to work could be somewhat difficult, more so than, e.g., the models of our companion chapter, where the software is well developed and anticipates many users' needs (in crisp cases, of course). A simulation language (system, package) such as SLX, however, provides a bridge between the neatly formulated arrival and service rate models and more general kinds of modeling and simulation; this is due in part, to the language's object-directed world view.

Table 11 provides an overview for some of the main doings of the chapter. (In it we abbreviate "Properties" as "Props," and "Cost/Benefit" analysis values as "C/B;" we also use "values" where often we have used "variables.") In particular, it shows some key roles for different approaches, genetic algorithms, optimization packages and (package) simulations.

Table 11: Overview of Simulations. See Text for Details and Abbreviations.

| **Transition Matrix Based Simulations** | |
|---|---|
| Genetic Algorithm (GA) | GA computes $\overline{w_i}$ |
| Optimization Package | From Performance Values to C/B |
| Optimization Package | Alternative to GA for Performance Values |
| (Package) Simulation | Generate $\alpha$=0 cut from $\alpha$=1 cut |

There have been other lines of simulation research that parallel the work described here and several remarks throughout the paper have alluded to them. There are still others, not reported here, e.g., ones related to developing sampling procedures over the (fuzzy) ranges of the fuzzy probabilities in the transition matrix such that good approximations can be obtained at a comparably lower price to, say, the genetic algorithms.

This chapter focused mainly on one line of research that has shown deep potential in developing a performance variable's alpha cuts *at will* from a single simulation at the alpha one cut. A poor man's "existence proof," namely, that this *can* be done, is well established by the case studies we reported here. But, more work needs to be done to realize this potential fully and to develop a more rigorous procedure to direct the search for answers within model outputs.

It seems an obvious observation, based on several of our results and evaluations of them, that some fundamental research into correspondences between stochastic and fuzzy modeling approaches will be needed. Just as variance propagates in stochastic systems undergoing multiple stages of random operations, so too does fuzziness. The companion paper (the previous chapter) makes it very clear that this happens, through its explicitly designated "one-step" and "two-step" approaches.

The transition matrix approach seems a more difficult problem in that fuzziness is difficult to introduce at any stage other than transition matrix itself, since, unlike the arrival/service rate approach, we are absent explicit mathematical formulae with which to work. It may be possible to utilize regression techniques in some cases, a matter we leave on future research agenda.

There are broader picture connections worth noting in our final paragraphs. Putting our results which inherently involves either multiple solution techniques to precisely the same problem formulation or alternative problem formulations of the stated problem (or similarly stated) problems represents an agent perspective in the solutions domain. A "hot" topic today in computing science, agents, and their associated computational schemes, often going under names such as grid and collaborative computing, thus, may be opened for merger with our current research paths. An agent perspective would envision incorporating queuing systems both in terms of arrival and service rates as well as transition probability matrices along with a plethora of solution techniques.

Agent methods can be employed to explore options relating to fuzziness introduction, starting at early points such as the initial formulation (as we have seen in the transition matrices cases), at intermediate levels (e.g., at the state probability level and its successors, the performance variables, in the arrival/service rate formulations), or even at final stages (where costs and benefits are often assessed in (finite) optimization contexts).

In other studies referenced in the chapter, but not developed at any length, we explored combined systems such as crisp and fuzzy neural networks and logic programming-based and fuzzy rule systems. These studies execute a topic broadening of more or less conventional systems with strong fuzzy counterparts and extensions. The matters relating to the point of introduction of fuzziness into a train of mathematical developments, or into regression equations, evoke additional 'scientific computation' agent notions beckoning additional research. With these thoughts in mind this chapter can end on a note concordant with notes at the end of the companion chapter.

# REFERENCES


1. J.J.Buckley: Fuzzy Probabilities: New Approach and Applications, Physica-Verlag, Heidelberg, 2003.
2. J.J.Buckley: Fuzzy Probability and Fuzzy Sets for Web Planning, Physica-Verlag, Heidelberg, 2004.
3. J.J.Buckley and Y.Qu: On Using Alpha-Cuts to Evaluate Fuzzy Equations, Fuzzy Sets and Systems, 38 (1990), pp. 309-312.
4. J.J.Buckley,K.Reilly and X.Zheng: Fuzzy Probabilities for Web Planning, Soft Computing. Published on-line, to appear in print.
5. J.J.Buckley, K.Reilly and X.Zheng: Crisp Simulation of Fuzzy Computations, Proc. 4th Int'l Symp. on Uncertainty Modeling and Analysis (ISUMA 2003), (2003), pp. 50-55.
6. D. R. Cox and W. Smith: Queues. Methuen, 1961.
7. Frontline Systems (www.frontsys.com).
8. J.O.Henriksen: SLX-Pyramid Power, Proc. Winter Simulation Conference, Society for Modeling and Simulation Int'l, San Diego, CA, 1999.
9. J.O.Henriksen and R. C. Crain: GPSS-H Reference Manual, 4th ed., Alexandria, VA: Wolverine Software, Inc., 1996.
10. G.J.Klir : Fuzzy Arithmetic With Requisite Constraints, Fuzzy Sets and Systems, 91 (1997), pp. 147-161.
11. G.J.Klir and J.A.Cooper: On Constrained Fuzzy Arithmetic, Proc. 5th Int. IEEE Conf. on Fuzzy Systems, New Orleans, 1996, pp. 1285-1290.
12. G.J.Klir and Y.Pan: Constrained Fuzzy Arithmetic: Basic Questions and Some Answers, Soft Computing, 21 (1998), pp. 100-108.
13. Maple 9, Waterloo Maple Inc., Waterloo, Canada.
14. D.A.Menasce and V.A.F.Almeida: Capacity Planning for Web Performance, Prentice Hall, Upper Saddle River, N.J., 1998.
15. Y.Pan and G.J.Klir: Bayesian Inference Based on Intelligent and Fuzzy Systems, 5 (1997), pp. 193-203.
16. Y.Pan and B.Yuan : Bayesian Inference of Fuzzy Probabilities, Int. J. General Systems, 26 (1997), pp. 73-90.
17. K.D.Reilly, J. J. Buckley, X. Zheng, Fuzzy Probability Modeling and Applications, Proc. 2003 Huntsville Simulation Conference, Society for Computer Simulation, Int'l, San Diego, CA, (2003), pp. 579-584.
18. K.D.Reilly, Agent Computing Themes in Biologically Inspired Models of Learning and Development, Int. J. Dev. Neuroscience, 20 (2002), 269-285.

19. K.D.Reilly, Agents and Lightweight Use of Logic, Combined Simulation, Logic and Neural Agent Nodes, Proc. Huntsville Simulation Conference (2001), 132-137.

20. K.D.Reilly, A.Sprague and C.L.Plachco, Agents and Model Abstractions in Intelligent Systems Simulations, Proc. 2001 Summer Simulation Conference, Soc. for Computer Simulation, Int'l., San Diego, CA (2001) 153-158.

21. K.D.Reilly, A.Sprague and A.Fanning, Biologically Inspired Simulation: Agent Meta- phors and Immune Systems, Proc. 2001 Summer Simulation Conference, Soc. for Computer Simulation, Int'l., San Diego, CA (2001) 159-164.

22. X. Zheng, K. D. Reilly and J. J. Buckley, Comparing Genetic Algorithms and Exhaustive Methods Used In Optimization Problems For Fuzzy Probability-Based Web Planning Models, Proc. Int'l Conf. on Artifi cial Intelligence (IC-AI'03) - Vol. 1, (2003), pp. 463-468.

23. X. Zheng, K. D. Reilly and J. J. Buckley, Fuzzy Optimization and Normal Simulation for Solving Web Queuing System Problems, Proc. SCI 2004 - World Multiconference on Systemics, Cybernetics, and Informatics, (2004), To appear.

# Chapter 5

# EVENT-RELATED POTENTIAL NOISE REDUCTION USING THE HIDDEN MARKOV TREE MODEL

**Rafael E. Herrera, Mingui Sun, Ronald E. Dahl, Neal D. Ryan and Robert J. Sclabassi**

## 1. INTRODUCTION

Event-related potentials are neural responses embedded within EEG signals that are generated by presenting frequent and infrequent stimuli to a subject. These signals are usually small in amplitude and are embedded in spontaneous EEG activity. The latter is referred to in this context as background or noise. ERPs are widely used to study attention, memory and affective mechanisms of the nervous system [1]. To analyze them some signal processing is required. The most common method used in clinical settings is to perform a simple averaging of time aligned EEG segments.

This method requires several assumptions; first, that the ERP is deterministic and time invariant from trial to trial, and that the background EEG is a Gaussian random process drawn from an independent and identically distributed probability density function. Under these assumptions, theoretically, the noise variance is reduced by a factor equal to the number of averaged trials.

In reality, the ERP responses will change over time. If the observations are taken over a long period of time, the subject experiences *habituation* or *accommodation* and the response will wane and cease to appear. That is why these experiments have to be designed so that a large enough number of responses are acquired for a statistically meaningful average, but short enough so that the ERP response changes are minimal. Another weakness of the averaging model is that the EEG is correlated, non-stationary and not truly Gaussian. Through simple signal averaging we can measure only the ERP mean latency and its overall shape. It is obviously not the best estimator for this kind of signal.

Despite the usefulness of the ERP as a research and clinical tool, the mechanisms involved in their generation are still not well known. Observing the dynamic behavior of the ERP from trial to trial would allow us to gain a better understanding of these mechanisms in the brain. This paper presents a method to estimate the ERP signal from single trial segments.

In the literature, several methods based on the Wavelet Transform have been proposed. These methods are based on the wavelet coefficient thresholding [2, 3, 4, 5, 6]. For simplicity, however, the majority make similar assumptions as the averaging method mentioned above regarding the EEG.

These methods, however, may fall short in producing the best possible results when applied to EEG signals. These signals are fairly complex and contain substructures that the conventional methods do not account for. In particular, the statistical assumptions generally used result in concise mathematically terms, but do not approximate well real world signals.

In this work we propose a method to estimate an event-related potential signal from a single trial EEG segment. This method is based on the Hidden Markov Tree model [7] applied to wavelet coefficients and builds on the previous thresholding work. This paper is organized as follows: the next section will describe the model and the algorithm used for the estimation of its parameters. The method used for recording the ERP signals is presented next. Following that, we show its application using synthetic and actual ERP signals. We include a comparison to the result of using the wavelet soft-thresholding method. We end the paper with a discussion and conclusions section.

## 2. THE HIDDEN TREE MODEL

Let $w$ be a single-trial EEG segment with an embedded ERP signal, the following additive model is assumed

$$w = y + \epsilon \tag{1}$$

where $y$ is the noise-free ERP signal and $\epsilon$ is the background EEG considered as a random signal of unknown probability.

The Hidden Markov Tree (HMT) is a variant of the Hidden Markov Model (HMM). It is used to generate an estimate of the signal $y$ based on the noisy observation $w$. An attractive application of this model is that it describes the correlations among the coefficients of the wavelet transformation. To support this approach a qualitative description of the properties of the wavelet transform should be mentioned first. The wavelet transform has these main properties [7]:

- **Locality**. Wavelet transformations are localized in time and frequency.
- **Multiresolution**. Wavelet transform have the ability to zoom in and out to measure signal variations at different scales.
- **Compression**. Wavelet representations tend to be sparse, i.e. the majority of signal information can be represented by a small number of significant coefficients.

The above are well known properties and are amply discussed in wavelet transform textbooks.

A common interpretation of the wavelet transform of a random process states that it has the property of being a "decorrelator", resulting in wavelet coefficients that are statistically independent of all others. The extreme case of these behavior is the wavelet transform of a Gaussian white processes, the wavelet coefficients of the transformed signal are also Gaussian and white. However, real-world signals contain correlations and may not be Gaussian. By virtue of the properties mentioned above, these real-world signals will be decomposed in well localized components in time and frequency, affording a degree of "decorrelation". That is, the wavelet coefficients will exhibit some dependencies across time and scales. This behavior gives rise to two other properties [8, 7]:

- **Clustering**. If a given wavelet coefficient is small/large, then the adjacent coefficients are likely to be also small/large.
- **Persistence**. Small/large wavelet coefficients tend to propagate across scales.

In order to illustrate these two properties, we have simulated a waveform in which we have inserted a transient distortion in the form of a discontinuity. The top plot of Figure 1 shows this waveform, the discontinuity occurs in the middle of the trace. The bottom plots are three levels of wavelet decomposition using the biorthogonal wavelet transform of order $(3, 1)$.

The property of *clustering* is illustrated in levels **L1** and **L2**, by the large wavelet coefficients grouped together around the middle of the trace, where the discontinuity occurs. Note that they are larger relative to the other coefficients on the same trace. On the other hand, along the intervals on each side of the discontinuity the coefficients are small in amplitude.

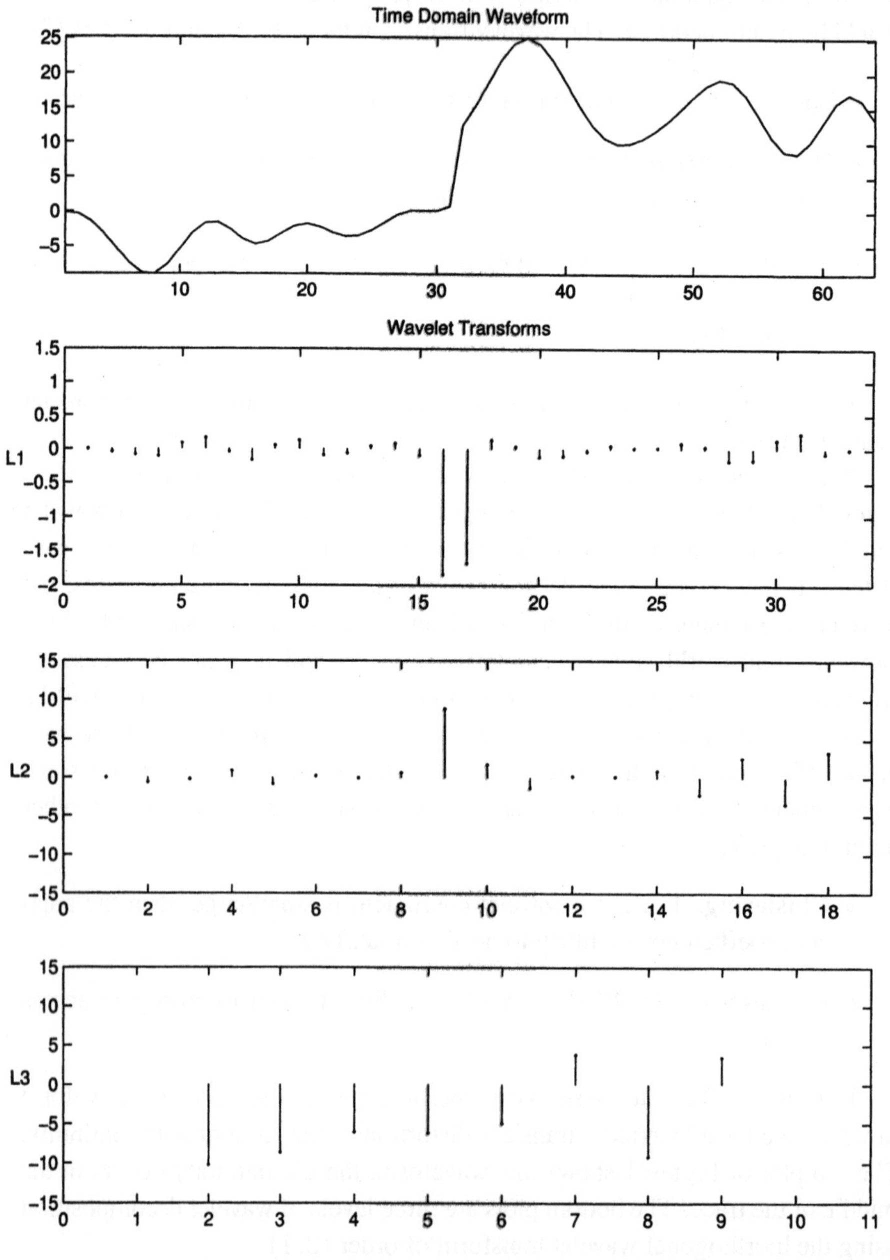


Figure 1: The properties of clustering and persistence are illustrated in this sample waveform. A transient discontinuity has been introduced in the middle of the waveform (top). Large coefficients in levels 1 and 2, related to the discontinuity, indicate persistence. Clustering is observed on all three levels.

The property of *persistence* is shown by the presence of large coefficients in levels **L1** and **L2**. In this example, large coefficients are present (relative to the surrounding coefficients) at the time of the discontinuity.

The Hidden Markov Tree attempts to capture the spatial interdependencies described by the property of persistence using a hidden Markov model. This HMT model is justified as follows: take a wavelet coefficient at, for example, level **L1**, if this coefficient is large, then we associate it with a class of coefficients with certain statistics (e.g., Gaussian, zero-mean and variance $V_1$.) If it is small, we associate it with to another class with different statistics (e.g., Gaussian, zero-mean and variance $V_2$.)

Furthermore, these classes are not directly observable, we can only observe them indirectly through the realization of the underlying stochastic process.

Due to the persistence property, at the next level it is likely that the corresponding wavelet coefficient will be also large (or small). If we focus on the wavelet coefficients associated with the discontinuity, we see that this is the case. Therefore, there exists a strong spatial correlation between coefficients of the same class across the scale.

The next step in this description is to formulate the Markovian relationships among the coefficients. First, the classes described above are defined by the values of discrete hidden state variables, one class for each state. Next, consider a wavelet coefficient at some scale level. It represents some information for a given time interval; the persistence property tells us that it may be correlated to a pair of coefficients at the next finer resolution level, in the same time interval.

However, each one of these two coefficients accounts for only one half of the time interval that the coarser coefficient represents. Therefore, the class associated with the coarser coefficient can be considered to affect the classes associated with the finer coefficients. This is the Markovian relationship between the hidden state variables that represent the classes.

To illustrate this model graphically, a dyadic wavelet decomposition is shown in Figure 2 on a *time-frequency* plane. Each tile represents the time-frequency support of each wavelet. The wavelet coefficient is represented by a filled node. Each wavelet coefficient node has a white node connected to it that represents the hidden state variable. Beginning at a coarse level each state node is linked to two nodes at the next finer resolution level, indicating the Markovian relationship. The resulting graph is a forest of binary trees with root nodes at the coarsest level. Finally, the tree is drawn so that the root node is placed in the upper side of the graph, as shown in Figure 2. The nodes at the top of the graph represent the wavelet approximation coefficients; they are not processed in this application.

Using the model of Eq. 1 and the linearity of the wavelet transform, let us

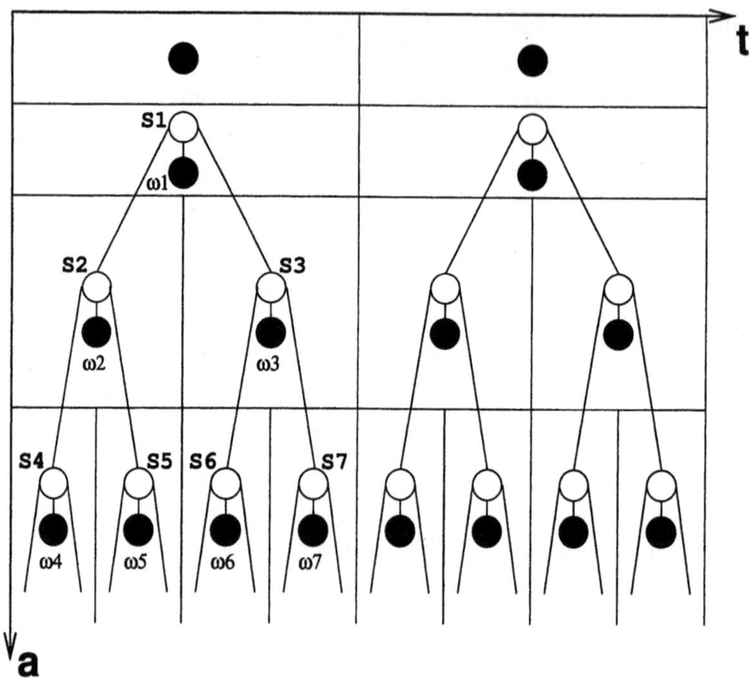


Figure 2: The coefficient dependencies are represented by connecting the state nodes vertically, producing a top-down (coarser-finer) binary tree graph.

represent the wavelet transform of the signal segment in 1 using the following equation

$$\omega_k^j = y_k^j + n_k^j \tag{2}$$

where $\omega_k^j$, $y_k^j$ and $n_k^j$ are the wavelet coefficients of the observed signal, the actual signal and the noise, respectively. The index $j$ indicates the scale level and $k$ indicates time. In order to simplify the notation in the rest of the paper we will change the indexing of the coefficients using the following mapping, $\omega_k^j \longrightarrow \omega_i$, and rewriting Eq. 2 into

$$\omega_i = y_i + n_i, \quad i = 1, \ldots, T. \tag{3}$$

where $T$ is the number of nodes in a sub-tree. The indexing used in Eq. 3 takes advantage of the binary tree structure shown in Figure 2. Using the root node as the starting point, the wavelet coefficients are indexed from left to right, top to bottom. Each sub-tree structure shown in Figure 2, then leads to two sequences. One sequence for the hidden state variables, denoted $\boldsymbol{S} = (S_i) = (S_1, S_2, \ldots, S_T)$. Another sequence for the corresponding wavelet coefficients, $\boldsymbol{\omega} = (\omega_i) = (\omega_1, \omega_2, \ldots, \omega_T)$.

Let us use the symbol $\rho(i)$ to denote the parent of node $i$ and $c(i)$ to denote the children of node $i$. For example, the parent of node 5, $\rho(5)$, is node 2; the children of node 3, $c(3)$, are the nodes 6 and 7. In addition, we will define $\mathcal{T}_i$ as the subtree of wavelet coefficients with root node $i$. $\mathcal{T}_1$ denotes the whole tree of observed coefficients. Also, if $\mathcal{T}_j$ is a subtree of $\mathcal{T}_i$, then $\mathcal{T}_{i\backslash j}$ is the set of wavelet coefficients obtained by removing the subtree $\mathcal{T}_j$ from $\mathcal{T}_i$.

The parameters of the hidden Markov tree model are specified as follows:

- The state probability of the root node

$$p_{S_1} = p[S_1 = m], \quad m = 1, \ldots, M. \tag{4}$$

  where $M$ is the total number of hidden states.

- The conditional probability that the state variable $S_i$ is in state $m$ given $S_{\rho(i)}$ is in state $n$

$$a^{mn}_{i,\rho(i)} = p[S_i = m | S_{\rho(i)} = n], \quad m, n = 1, \ldots, M. \tag{5}$$

- The probability density of the observed value given the state variable is in state $m$

$$b_m(\omega_i) = p[\omega_i | S_i = m], \quad m = 1, \ldots, M. \tag{6}$$

All these are grouped together into a parameters vector $\boldsymbol{\theta}$:

$$\boldsymbol{\theta} = \{a^{mn}_{i,\rho(i)}, b_m(\omega_i), p_{S_1}\} \quad \begin{aligned} i &= 1, \ldots, T; \\ m, n &= 1, \ldots, M. \end{aligned} \tag{7}$$

The densities $b_m(\omega_i)$ can be approximated using $M$ Gaussian mixtures of the form

$$b_m(\omega_i) \equiv N(\mu_{i,m}, \sigma^2_{i,m}), \quad m = 1, \ldots, M \tag{8}$$

which can be used to approximate an arbitrary probability density function, if a sufficient mixture number ($M$) is used. The parameter vector is as follows for the mixture case:

$$\boldsymbol{\theta} = \{a^{mn}_{i,\rho(i)}, \mu_{i,m}, \sigma^2_{i,m}, p_{S_1}\} \quad \begin{aligned} i &= 1, \ldots, T; \\ m, n &= 1, \ldots, M. \end{aligned} \tag{9}$$

Once we have established the model, the next step is training it to determine the optimal parameters that best describes the observed wavelet coefficients. To estimate the parameter vector using the maximum likelihood principle, we can use the algorithm called *expectation maximization* (EM) [9]. This work uses an implementation derived from [10] and [7].

## 2.1. The EM Algorithm

The EM algorithm was used in [10, 7] in order to estimate the model parameters in $\boldsymbol{\theta}$. Their implementation is referred as the *upward-downward procedure*, which is similar to the forward-backward procedure used to estimate the parameters of a hidden Markov model in [11].

For each subtree $\mathcal{T}_i$ the *downward variable* $\alpha_i(m)$ is defined as

$$\alpha_i(m) = p[S_i = m, \mathcal{T}_{1\backslash i}|\boldsymbol{\theta}] \tag{10}$$

the conditional probability of observing the nodes before node $i$ and state $S_i = m$, given the model.

Similarly, the *upward variable* is

$$\beta_i(m) = f(\mathcal{T}_i|S_i = m, \boldsymbol{\theta}) \tag{11}$$

the conditional likelihood of observing the subtree with root at node $i$, given the state $S_i$ at node $i$ and the model. Additionally,

$$\beta_{i,\rho(i)}(m) = f(\mathcal{T}_i|S_{\rho(i)} = m, \boldsymbol{\theta}) \tag{12}$$

$$\beta_{\rho(i)\backslash i}(m) = f(\mathcal{T}_{\rho(i)\backslash i}|S_{\rho(i)} = m, \boldsymbol{\theta}). \tag{13}$$

The purpose of these variables is to compute the state probabilities $p[S_i = m|\boldsymbol{\omega}, \boldsymbol{\theta}]$ and $p[S_i = m, S_{\rho(i)} = n|\boldsymbol{\omega}, \boldsymbol{\theta}]$. So,

$$p(S_i = m|\boldsymbol{\omega}, \boldsymbol{\theta}) = \frac{\alpha_i(m)\beta_i(m)}{\sum_{n=1}^{M} \alpha_i(m)\beta_i(m)} \tag{14}$$

$$p(S_i = m, S_{\rho(i)} = n|\boldsymbol{\omega}, \boldsymbol{\theta}) = \frac{\beta_i(m) a_{i,\rho(i)}^{mn} \alpha_{\rho(i)}(n) \beta_{\rho(i)\backslash i}(n)}{\sum_{n=1}^{M} \alpha_i(n)\beta_i(n)} \tag{15}$$

For the purpose of describing the EM algorithm, we will use Gaussian mixing components of the form:

$$g(\omega; \mu, \sigma^2) = \frac{1}{\sqrt{2\pi\sigma^2}} \exp\left[-\frac{(\omega - \mu)^2}{2\sigma^2}\right]. \tag{16}$$

In addition, each node $i$ in the tree is associated with a scale $J(i)$, where $J(i) \in \{1, \ldots, L\}$, with $J = 1$ being the finest scale and $L$ the coarsest level.

**The Expectation Step**

1. Select an initial parameter $\boldsymbol{\theta}^0$ and set $l = 0$.

2. *Initialize the upward variable.*

   For all $S_i$ at scale $J = 1$ and $m = 1, \ldots, M$:

$$\beta_i(m) = g(\omega_i; \mu_{i,m}, \sigma_{i,m}^2) \tag{17}$$

   For all $S_i$ at scale $J = 1, \ldots, L$ and $m = 1, \ldots, M$:

$$\beta_{i,\rho(i)}(m) = \sum_{r=1}^{M} a_{i,\rho(i)}^{rm} \beta_i(r) \tag{18}$$

$$\beta_{\rho(i)}(m) = g(\omega_{\rho(i)}; \mu_{\rho(i),m}, \sigma_{\rho(i),m}^2) \times \prod_{j \in c(\rho(i))} \beta_{j,\rho(j)}(m) \tag{19}$$

$$\beta_{\rho(i)\backslash i}(m) = \frac{\beta_{\rho(i)}(m)}{\beta_{i,\rho(i)}(m)} \tag{20}$$

3. *Initialize the downward variable.*

   For $S_1$ at scale $J = L$ and $m = 1, \ldots, M$:

$$\alpha_1(m) = p_{S_1}(m). \tag{21}$$

   For all states $S_i$ at scale $J = L - 1, \ldots, 1$ and $m = 1, \ldots, M$:

$$\alpha_i(m) = \sum_{n=1}^{M} a_{i,\rho(i)}^{mn} \alpha_{\rho(i)}(n) \beta_{\rho(i)\backslash i}(n) \tag{22}$$

**The Maximization Step**

1. Compute the probabilities

$$p(S_i = m|\boldsymbol{\omega}, \boldsymbol{\theta}) \quad = \quad \frac{\alpha_i(m)\beta_i(m)}{\sum_{n=1}^{M} \alpha_i(m)\beta_i(m)} \tag{23}$$

$$p(S_i = m, S_{\rho(i)} = n|\boldsymbol{\omega}, \boldsymbol{\theta}) \quad = \quad \frac{\beta_i(m) a_{i,\rho(i)}^{mn} \alpha_{\rho(i)}(n) \beta_{\rho(i)\backslash i}(n)}{\sum_{n=1}^{M} \alpha_i(n)\beta_i(n)} \tag{24}$$

2. Assuming there are $K$ trees apply the E step on each tree independently.

3. The next approximation of $\boldsymbol{\theta}^{l+1}$ is computed with

$$p_{S_i}(m) = \frac{1}{K}\sum_{k=1}^{K} p(S_i^k = m|\boldsymbol{\omega}^k, \boldsymbol{\theta}^l) \tag{25}$$

$$a_{i,\rho(i)}^{mn} = \frac{1}{K}\frac{\sum_{k=1}^{K} p(S_i^k = m, S_{\rho(i)}^k = n|\boldsymbol{\omega}^k, \boldsymbol{\theta}^l)}{p_{S_{\rho(i)}(n)}} \tag{26}$$

$$\mu_{i,m} = \frac{1}{K}\frac{\sum_{k=1}^{K} \omega_i^k p(S_i^k = m|\boldsymbol{\omega}^k, \boldsymbol{\theta}^l)}{p_{S_i}(m)} \tag{27}$$

$$\sigma_{i,m}^2 = \frac{1}{K}\frac{\sum_{k=1}^{K} (\omega_i^k - \mu_{i,m}^2 p(S_i^k = m|\boldsymbol{\omega}^k, \boldsymbol{\theta}^l)}{p_{S_i}(m)}. \tag{28}$$

4. Check the error and repeat if necessary.

## 2.2. Computational Underflows

The computation of the downward variable $\alpha_i(m)$ and upward variable, $\beta_i(m)$ is an iterative process that will lead rapidly to computational underflows. To avoid this problem, it is necessary to use of a scaling factor on the upward variable $\beta_i(m)$ so its computed value stays within the dynamic range of the computer.

We will denote the unscaled $\beta$ as $\beta_i(m)$; $\widehat{\beta}_i(m)$ the scaled $\beta$ and $\widehat{\widehat{\beta}}_i(m)$ the intermediate value before scaling. A similar notation will be used for $\alpha$.

The scaling factor $c_i$ is defined as

$$c_i = \frac{1}{\sum_{m=1}^{M} \widehat{\widehat{\beta}}_i(m)} \tag{29}$$

Then,

$$\widehat{\beta}_i(m) = c_i \widehat{\widehat{\beta}}_i(m) \tag{30}$$

$$\widehat{\alpha}_i(m) = c_i \widehat{\widehat{\alpha}}_i(m). \tag{31}$$

The effect of the scaling factor $c_i$ can be explored by following the upward-downward procedure specified in the previous section. Using the tree shown in Figure 3 we can express the terms of the procedure.

For node $i$ in level **L1** in Figure 3, we have

$$\beta_i(m) = \widehat{\widehat{\beta}}_i(m) = g(\omega_i; \mu_{i,m}, \sigma_{i,m}^2)$$

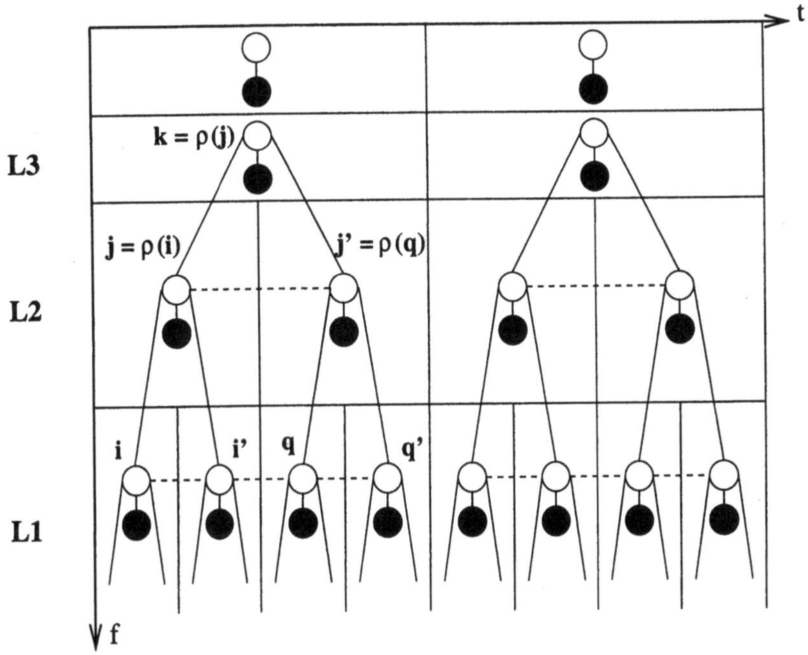


Figure 3: Hidden Markov tree model (HMT)

Then,

$$\widehat{\beta}_i(m) = c_i\widehat{\widehat{\beta}}_i(m) = c_i\beta_i(m) \tag{32}$$

Similarly, from Eqs. 18 and 19

$$\begin{aligned}\widehat{\beta}_{i,\rho(i)}(n) &= \sum_{r=1}^{M} a_{i,\rho(i)}^{rn}\widehat{\beta}_i(r) = \sum_{r=1}^{M} a_{i,\rho(i)}^{rn} c_i\beta_i(r) \\ &= c_i\beta_{i,\rho(i)}(n)\end{aligned} \tag{33}$$

$$\begin{aligned}\widehat{\widehat{\beta}}_{\rho(i)}(m) &= g(\omega_{\rho(i)}; \mu_{\rho(i),m}, \sigma^2_{\rho(i),m}) \times \prod_{r\in c(\rho(i))} \widehat{\beta}_{r,\rho(r)}(m) \\ &= g(\cdot) \times \prod_{r\in i,i'} \widehat{\beta}_{r,\rho(r)}(m) \\ &= g(\cdot)\widehat{\beta}_{i,\rho(i)}(m)\widehat{\beta}_{i',\rho(i')}(m) \\ &= g(\cdot)c_i\beta_{i,\rho(i)}(m)c_{i'}\beta_{i',\rho(i')}(m) \\ &= c_ic_{i'}\beta_{\rho(i)}(m)\end{aligned} \tag{34}$$

and

$$\widehat{\beta}_{\rho(i)}(m) = c_{\rho(i)} c_i c_{i'} \beta_{\rho(i)}(m). \tag{35}$$

Also, from Eq. 20

$$\begin{aligned} \widehat{\beta}_{\rho(i)\backslash i}(m) &= \frac{\widehat{\widehat{\beta}}_{\rho(i)}(m)}{\widehat{\beta}_{i,\rho(i)}(m)} = \frac{c_i c_{i'} \beta_{\rho(i)}(m)}{c_i \beta_{i,\rho(i)}(m)} = \frac{c_{i'} \beta_{\rho(i)}(m)}{\beta_{i,\rho(i)}(m)} \\ &= c_{i'} \beta_{\rho(i)\backslash i}(m) \end{aligned} \tag{36}$$

Now, on the node $j$ in level **L2**; noting that $\rho(i) = j$ and $\rho(q) = j'$,

$$\begin{aligned} \widehat{\beta}_{j,\rho(j)}(n) &= \sum_{r=1}^{M} a_{j,\rho(j)}^{rn} \widehat{\beta}_j(r) = \sum_{r=1}^{M} a_{j,\rho(i)}^{rn} c_j c_i c_{i'} \beta_j(r) \\ &= c_j c_i c_{i'} \beta_{j,\rho(j)}(n) \end{aligned} \tag{37}$$

Similarly,

$$\begin{aligned} \widehat{\widehat{\beta}}_{\rho(j)}(m) &= g(\cdot) \times \prod_{r \in c(\rho(j))} \widehat{\beta}_{r,\rho(r)}(m) \\ &= g(\cdot) \times \prod_{r \in j,j'} \widehat{\beta}_{r,\rho(r)}(m) \\ &= g(\cdot) \widehat{\beta}_{j,\rho(j)}(m) \widehat{\beta}_{j',\rho(j')}(m) \\ &= g(\cdot) c_j c_i c_{i'} \beta_{j,\rho(j)}(m) c_{j'} c_q c_{q'} \beta_{j',\rho(j')}(m) \\ &= c_j c_i c_{i'} c_{j'} c_q c_{q'} \beta_{\rho(j)}(m) \end{aligned} \tag{38}$$

and

$$\widehat{\beta}_{\rho(j)}(m) = c_{\rho(j)} c_j c_i c_{i'} c_{j'} c_q c_{q'} \beta_{\rho(i)}(m). \tag{39}$$

Finally,

$$\begin{aligned} \widehat{\beta}_{\rho(j)\backslash j}(m) &= \frac{\widehat{\widehat{\beta}}_{\rho(j)}(m)}{\widehat{\beta}_{j,\rho(j)}(m)} = \frac{c_j c_i c_{i'} c_{j'} c_q c_{q'} \beta_{\rho(j)}(m)}{c_j c_i c_i' \beta_{j,\rho(j)}(m)} = \frac{c_{j'} c_q c_{q'} \beta_{\rho(j)}(m)}{\beta_{j,\rho(j)}(m)} \\ &= c_{j'} c_q c_{q'} \beta_{\rho(j)\backslash j}(m) \end{aligned} \tag{40}$$

Now, lets derive the downward variables

$$\alpha_1(m) = p_{S_1}(m) = \widehat{\widehat{\alpha}}_1(m) \tag{41}$$

Noting that in this case, $1 = k = \rho(j)$,

$$\widehat{\alpha}_1(m) = c_k \widehat{\widehat{\alpha}}_1(m) = c_{\rho(j)} \alpha_1(m) \tag{42}$$

Then at the second level,

$$\begin{aligned} \widehat{\widehat{\alpha}}_j(m) &= \sum_{n=1}^{M} a_{j,\rho(j)}^{mn} \widehat{\alpha}_{\rho(j)}(m) \widehat{\beta}_{\rho(j)\backslash j}(m) \\ &= \sum_{n=1}^{M} a_{j,\rho(j)}^{mn} c_{\rho(j)} \alpha_{\rho(j)}(m) c_{j'} c_q c_{q'} \beta_{\rho(j)\backslash j}(m) \\ &= c_{\rho(j)} c_{j'} c_q c_{q'} \alpha_j(m) \end{aligned} \tag{43}$$

then

$$\begin{aligned} \widehat{\alpha}_j(m) &= c_{\rho(j)} c_j c_{j'} c_q c_{q'} \alpha_j(m) \\ \widehat{\alpha}_{j'}(m) &= c_{\rho(j)} c_j c_{j'} c_i c_{i'} \alpha_j(m) \end{aligned} \tag{44}$$

Using the scaled downward and upward variables, we can now compute the joint probability of Eq. 24

$$\begin{aligned} \widehat{p}(S_j = m, S_{\rho(j)} = n | \boldsymbol{\omega}, \boldsymbol{\theta}) &= \frac{\widehat{\beta}_j(m) a_{j,\rho(j)}^{mn} \widehat{\alpha}_{\rho(j)}(n) \widehat{\beta}_{\rho(j)\backslash j}(n)}{\sum_{r=1}^{M} \widehat{\alpha}_j(r) \widehat{\beta}_j(r)} \\ &= \frac{c_j c_i c_{i'} \beta_j(m) a_{j,\rho(j)}^{mn} c_{\rho(j)} \alpha_{\rho(j)}(n) c_{j'} c_q c_{q'} \beta_{\rho(j)\backslash j}(m)}{\sum_{r=1}^{M} c_{\rho(j)} c_j c_{j'} c_q c_{q'} \alpha_j(r) c_j c_i c_{i'} \beta_j(r)} \\ &= \frac{1}{c_j} \frac{\beta_j(m) a_{j,\rho(j)}^{mn} \alpha_{\rho(j)}(n) \beta_{\rho(j)\backslash j}(m)}{\sum_{r=1}^{M} \alpha_j(r) \beta_j(r)} \end{aligned} \tag{45}$$

Most of the factors cancel out from the numerator and denominator. The last remaining scaling coefficient can be easily removed. In general, the effect of the scaling factor in Eq. 24 is

$$p(S_j = m, S_{\rho(j)} = n | \boldsymbol{\omega}, \boldsymbol{\theta}) = c_j \cdot \widehat{p}(S_j = m, S_{\rho(j)} = n | \boldsymbol{\omega}, \boldsymbol{\theta}) \tag{46}$$

It is also easily shown that

$$\begin{aligned} \widehat{p}(S_i = m | \boldsymbol{\omega}, \boldsymbol{\theta}) &= \frac{\widehat{\alpha}_i(m) \widehat{\beta}_i(m)}{\sum_{n=1}^{M} \widehat{\alpha}_i(n) \widehat{\beta}_i(n)} \\ &= \frac{\alpha_i(m) \beta_i(m)}{\sum_{n=1}^{M} \alpha_i(n) \beta_i(n)} \\ &= p(S_i = m | \boldsymbol{\omega}, \boldsymbol{\theta}). \end{aligned} \tag{47}$$

With the inclusion of the scaling factor $c_i$, the upward-downward procedure is rewritten as follows:

**Scaled Expectation Step**

1. Select an initial parameter $\boldsymbol{\theta}^0$ and set $l = 0$.

2. *The upward step.*

   For all $S_i$ at scale $J = 1$ and $m = 1, \ldots, M$:

$$\beta_i(m) = g(\omega_i; \mu_{i,m}, \sigma_{i,m}^2) \tag{48}$$

   Apply the scaling factor. At $J = 1$, let $\widehat{\widehat{\beta}}_i(m) = \beta_i(m)$.

$$\widehat{\beta}_i(m) = c_i \widehat{\widehat{\beta}}_i(m) \tag{49}$$

   For all $S_i$ at scale $J = 1, \ldots, L$ and $m = 1, \ldots, M$:

$$\widehat{\beta}_{i,\rho(i)}(m) = \sum_{r=1}^{M} a_{i,\rho(i)}^{rm} \widehat{\beta}_i(r) \tag{50}$$

$$\widehat{\widehat{\beta}}_{\rho(i)}(m) = g(\omega_{\rho(i)}; \mu_{\rho(i),m}, \sigma_{\rho(i),m}^2) \times \prod_{j \in c(\rho(i))} \widehat{\beta}_{j,\rho(j)}(m) \tag{51}$$

$$\widehat{\beta}_{\rho(i)}(m) = c_{\rho(i)} \widehat{\widehat{\beta}}_{\rho(i)}(m) \tag{52}$$

$$\widehat{\beta}_{\rho(i)\backslash i}(m) = \frac{\widehat{\widehat{\beta}}_{\rho(i)}(m)}{\widehat{\beta}_{i,\rho(i)}(m)} \tag{53}$$

3. *The downward step.*

   For $S_1$ at scale $J = L$ and $m = 1, \ldots, M$:

$$\alpha_1(m) = p_{S_1}(m). \tag{54}$$

   Apply the scaling factor. At $J = L$, let $\widehat{\widehat{\alpha}}_1(m) = \alpha_1(m)$.

$$\widehat{\alpha}_1(m) = c_1 \alpha_1(m) \tag{55}$$

   For all states $S_i$ at scale $J = L - 1, \ldots, 1$ and $m = 1, \ldots, M$:

$$\widehat{\widehat{\alpha}}_i(m) = \sum_{n=1}^{M} a_{i,\rho(i)}^{mn} \widehat{\alpha}_{\rho(i)}(n) \widehat{\beta}_{\rho(i)\backslash i}(n) \tag{56}$$

$$\widehat{\alpha}_i(m) = c_i \widehat{\widehat{\alpha}}_i(m) \tag{57}$$

**Scaled Maximization Step**

1. First, compute the probabilities

$$p(S_i = m|\boldsymbol{\omega}, \boldsymbol{\theta}) = \frac{\widehat{\alpha}_i(m)\widehat{\beta}_i(m)}{\sum_{n=1}^{M} \widehat{\alpha}_i(n)\widehat{\beta}_i(n)} \tag{58}$$

$$p(S_i = m, S_{\rho(i)} = n|\boldsymbol{\omega}, \boldsymbol{\theta}) = c_j \cdot \frac{\widehat{\beta}_i(m) a^{mn}_{i,\rho(i)} \widehat{\alpha}_{\rho(i)}(n) \widehat{\beta}_{\rho(i)\backslash i}(n)}{\sum_{r=1}^{M} \widehat{\alpha}_i(r)\widehat{\beta}_i(r)} \tag{59}$$

2. Assuming there are $K$ trees apply the E step on each tree independently.

3. The next approximation of $\boldsymbol{\theta}^{l+1}$ is computed with

$$p_{S_i}(m) = \frac{1}{K}\sum_{k=1}^{K} p(S_i^k = m|\boldsymbol{\omega}^k, \boldsymbol{\theta}^l) \tag{60}$$

$$a^{mn}_{i,\rho(i)} = \frac{1}{K}\frac{\sum_{k=1}^{K} p(S_i^k = m, S^k_{\rho(i)} = n|\boldsymbol{\omega}^k, \boldsymbol{\theta}^l)}{p_{S_{\rho(i)}}(n)} \tag{61}$$

$$\mu_{i,m} = \frac{1}{K}\frac{\sum_{k=1}^{K} \omega_i^k p(S_i^k = m|\boldsymbol{\omega}^k, \boldsymbol{\theta}^l)}{p_{S_i}(m)} \tag{62}$$

$$\sigma^2_{i,m} = \frac{1}{K}\frac{\sum_{k=1}^{K} (\omega_i^k - \mu^2_{i,m} p(S_i^k = m|\boldsymbol{\omega}^k, \boldsymbol{\theta}^l)}{p_{S_i}(m)} \tag{63}$$

## 2.3. De-Noising Using the HMT

Using the model of Eq. 1 and an $M$-Gaussian mixture, the HMT model parameters can be estimated using the EM algorithm. This model is then utilized as a prior distribution for the signal $y_i$ and employed to calculate a conditional mean estimate of $y_i$ given the observations $\boldsymbol{\omega}$ and the model $\boldsymbol{\theta}$ [7].

The addition of the noise $\epsilon$ in Eq. 1 increases the mixture variance $\sigma^2_{i,m}$. If $\epsilon$ is Gaussian and white, of variance $\sigma_n^2$, then the observed signal variance would be $\gamma^2_{i,m} = \sigma^2_{i,m} + \sigma^2_n$. However, the other parameters of the model are unchanged. If after training, the estimated mixture variance of the noisy wavelet coefficient is $\gamma^2_{i,m}$, then

$$\sigma^2_{i,m} = \gamma^2_{i,m} - \sigma^2_n \tag{64}$$

The conditional mean estimate of $y_i$ can be obtained using the results of the EM algorithm.

$$E[y_i|\boldsymbol{\omega}, \boldsymbol{\theta}] = \sum_m p(S_i = m|\boldsymbol{\omega}, \boldsymbol{\theta}) \frac{\sigma^2_{i,m}}{\sigma^2_{i,m} + \sigma^2_n} \omega_i. \tag{65}$$

The last step in de-noising the signal is to apply the inverse wavelet transform of the estimated coefficients.

# 3. METHODS

Before discussing the experimental results of this method we will describe the methods used to acquire the EEG recordings. We developed a software tool to present visual and auditory stimuli to a subject in order to generate ERP signals. This program is used along with a data acquisition system to record both continuous EEG and the timing markers generated at each presentation of a stimulus. The timing markers were used in the post-processing to align the EEG segments. This program was developed using the X Window System and is currently implemented in the HP-UX and Linux operating systems.

The purpose of the program is to present a sequence of stimuli in a manner determined by the user. The stimuli is of two kinds, visual and auditory. Each stimulus is presented to the subject for an interval referred as *stimulus interval.* Stimuli are presented at regular intervals referred as *inter-stimulus interval.* These intervals can be controlled by the program. Visual stimuli consist of computer images rendered on the computer monitor. Auditory stimuli are generated using a Grass Click-Tone Control Module controlled remotely using the parallel interface. The tone frequencies of the Grass module can be remotely controlled using voltage levels available in the PC parallel interface.

We acquired continuous EEG recordings using electrodes attached to Fz, Cz and Pz. The signals were preconditioned before acquisition with a 0.1–100 Hz band pass filter. The sampling rate was 256 Hz. The subjects were asked to watch on a computer display several sequences of images. The content of these images were adult faces depicting neutral, happy and angry emotions. See Figure 4 for a sample picture of each emotion. A sequence of 360 images were used. In each one 68% were neutral faces and 16% were happy and 16% were angry faces. The images were arranged so their presentation appeared randomly. The stimulus interval for each image was displayed for 200 ms, followed by a inter-stimulus interval of 1200 ms. The analysis was performed offline with programs implemented in Matlab. Each single trial was extracted and stored on disk. In this study we used Daubechies wavelets and computed five levels of decomposition.

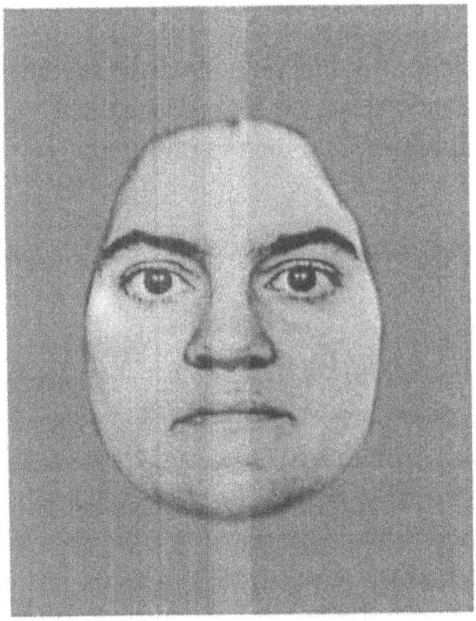

(a) Neutral

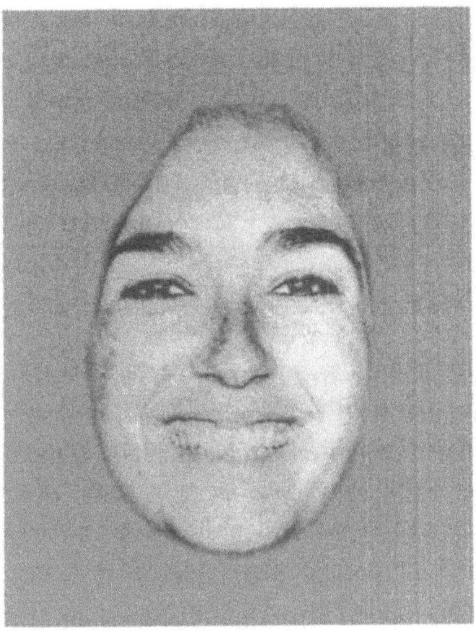

(b) Happy

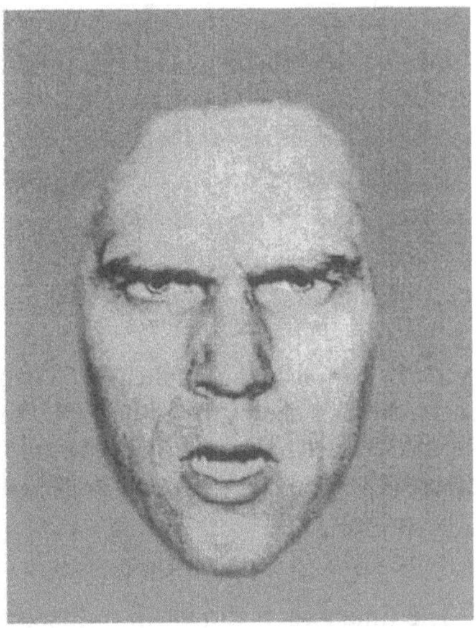

(c) Angry

Figure 4: Sample images of the faces used as stimuli. The top image shows a face with a neutral expression. The faces at the bottom express a happy and angry expression, respectively.

# 4. APPLICATION AND RESULTS

To demonstrate this method, a synthetic noisy ERP segment was generated by adding uniformly distributed white noise to a visual ERP signal considered to be noise-free, this signal is shown at the top of Figure 5.

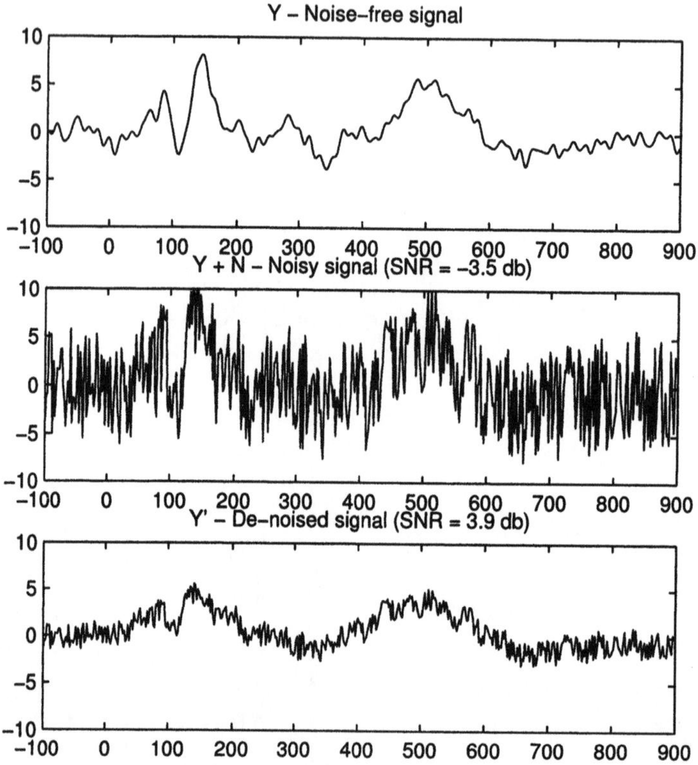


Figure 5: De-noising example using a hypothetical noise-free ERP signal (top plot). The middle plot shows the same ERP contaminated with uniform white noise at a SNR of -3.5 dB. The bottom plot shows the estimated de-noised signal, the SNR of the estimate is 3.9 db, an improvement of 7.4 dB.

The time axis is drawn in milliseconds, the visual stimulus was presented at time $t = 0$ ms. The features of interest in the plot are two, the first one arising at 100 ms indicating the activation of the visual cortex. This feature is an exogenous response and does not reflect cognitive process. The second feature is the positive deflection with an onset at around 450 ms terminating at 600 ms. This waveform is an endogenous response and reflects the cognitiveprocessing of the visual stimulus [1]. Endogenous responses have a typical onset at

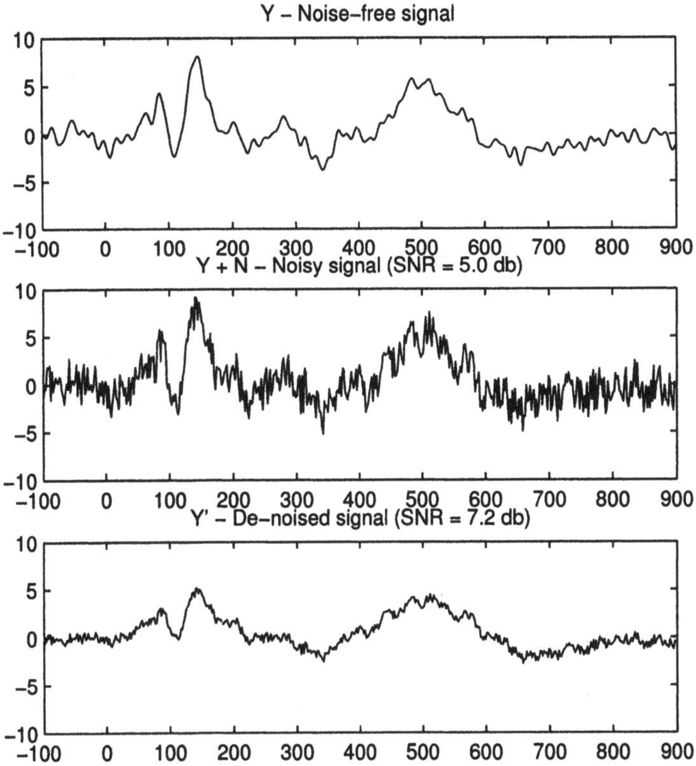


Figure 6: Synthetic noise-free ERP (top plot) contaminated with uniform white noise (middle plot.) The noisy segment has a SNR of 5 dB. The de-noised estimate (bottom plot) has a SNR of the estimate is 7.2 db, an improvement of 2.2 dB.

around 300 ms post-stimulus and are positive in amplitude, they are referred to in the literature as P300 waveforms.

In this demonstration, the test signals were created by adding noise scaled so that the SNR of each signal ranged from -3.5 dB to 5 dB. In Figure 5, the middle plot shows the noisy test signal with SNR of -3.5 dB. The de-noising result using the HMT model is shown at the bottom. The estimate measures a SNR of 3.9 dB, an improvement of 7.4 dB. As can be observed, only the principal features of the original ERP signal are preserved, namely the waveforms around 70 msec. and 450 msec. The exogenous and endogenous responses are still visible, but their amplitudes are reduced.

Figure 6 shows the estimate under a more favorable condition. In this case the SNR is 5dB, and the estimated signal measures a SNR of 7.2 dB. Even though in this example the SNR improvement is not as large as in the first example, the recovered signal preserves more subtle features of the original.

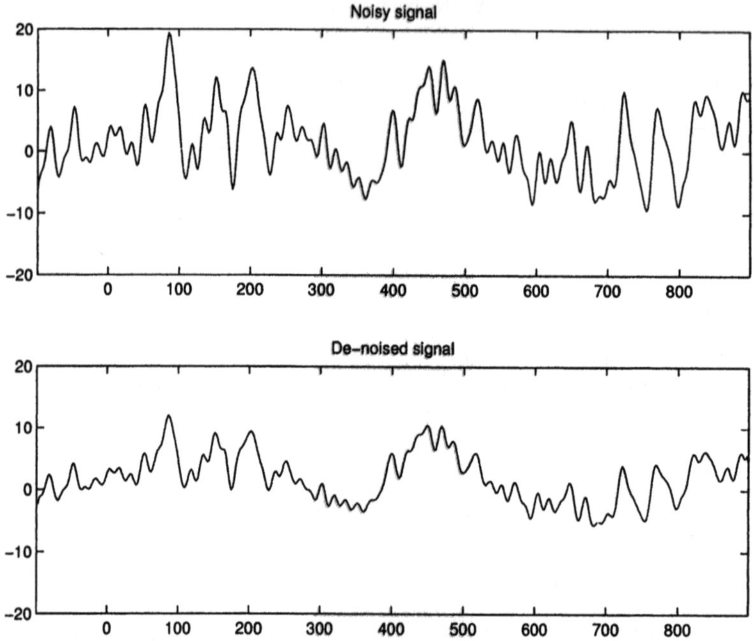


Figure 7: Sample de-noising results of an actual single-trial visual ERP signal using the HMT model. The stimulus occurs at time t=0 msec and the de-noised estimate shows the P300 waveform onset at t=400 msec.

This can be observed by comparing the endogenous responses on the top and bottom plots. The performance of the estimator at different levels of noise is summarized in Table 1. The table shows that wavelet-based HMT model consistently produces an estimate with an higher SNR.

The next example, shown in Figure 7, illustrates the results in de-noising a raw single-trial visual ERP signal. In this case the noise content is unknown. The exogenous response starts at around 50 msec. post-stimulus, and the P300 waveform begins at around 370 msec.

To illustrate the type of estimates that can be produced using a wavelet thresholding method to our sample single-trial ERP. Figure 8 shows the de-noising result obtained by applying the method of level dependent soft- thresholding [3]. The smoothness of the estimate indicates that most of the finer details were shrunk, although manual tuning of the thresholds may produce better results.

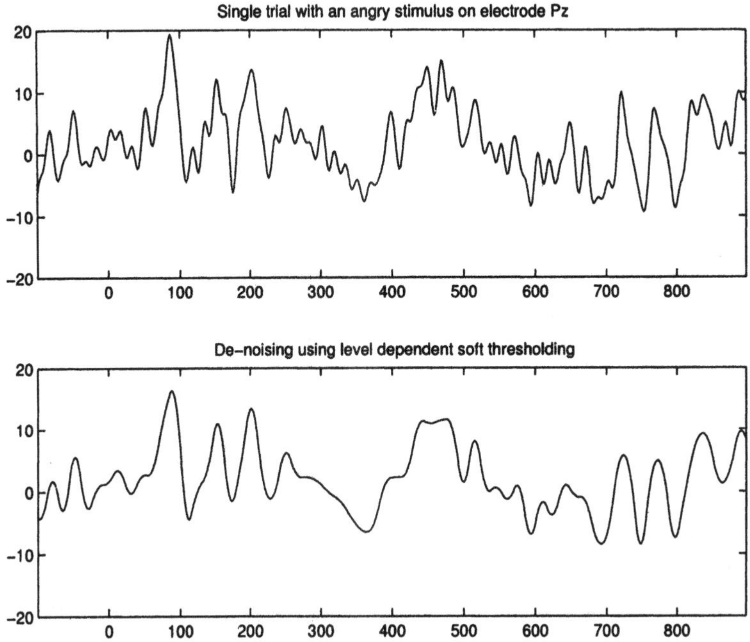


Figure 8: Sample de-noising results using a real single-trial visual ERP using level dependent soft-thresholding. The stimulus occurs at time t=0 msec and the de-noised estimate shows the P300 waveform onset at t=400 msec.

| **SNR (dB)** | **De-noised SNR (dB)** |
|---|---|
| 5.0 | 7.2 |
| 3.5 | 6.8 |
| 0.0 | 5.6 |
| -1.0 | 5.2 |
| -3.5 | 3.9 |

Table 1: Signal to noise ratios of the simulated ERP signals and the corresponding SNR of the de-noised estimates.

# 5. CONCLUSIONS

In this paper we have modeled a single trial ERP segment using the wavelet-based HMT model as an alternative to the averaging method for ERP analysis.

The objective is to capture the dynamic changes of the ERP that the averaging method does not capture by allowing the estimation of the ERP for each single-trial.

The HMT takes advantage of the persistence and clustering characteristics of the wavelet transform to model correlation structures within coefficients across scales, an approach that other wavelet-based methods do not use. These properties are suitable to model the statistical characteristics of an ERP signal. Due to the difficulty of knowing the signal distribution of an ERP segment, the use of a wavelet-based hidden Markov model is an advancement on the simple assumption of the Gaussian independent and identical distributions made by other noise reduction methods. This distribution is approximated using the Gaussian mixture included in the model.

Experimental results with synthetic signals show that the HMT tree model will achieve a measurable improvement in the SNR of the estimate. When used in real ERP signals, the HMT produces estimates which preserve many of the features that are apparent in the noisy signal. In contrast, another wavelet-based method, the soft-thresholding method, appears to be inferior in preserving these features.

This research was supported in part by the National Institutes of Health under grants no. MH41712, NS39047, MH38494, EB002099, EB002309 and Computational Diagnostics, Inc.

# REFERENCES

[1] D. Goodin and M. Aminoff, "Event-related potentials in the study of sensory discrimination and motor response in simple and choice reaction tasks," *J. Clinical Neurophysiology*, vol. 15, no. 1, pp. 34–43, 1998.

[2] D. L. Donoho, "De-Noising by Soft-Thresholding," *IEEE Trans. on Information Theory*, vol. 41, pp. 613–627, May 1995.

[3] I. Johnstone and B. Silverman, "Wavelet threshold estimators for data with correlated noise," *J.R. Statistic. Soc. B*, vol. 59, no. 2, pp. 319–351, 1997.

[4] M. Jansen, *Wavelet thresholding and noise reduction.* PhD thesis, Katholike Universiteit Leuven, 2000.

[5] G. Nason, "Wavelet shrinkage using cross-validation," *J. R. Statistic Soc. B*, vol. 58, no. 2, pp. 463–479, 1996.

[6] R. Herrera, M. Sun, R. Dahl, N. Ryan, and R. Sclabassi, "Removal of Non-White Noise From Single Trial Event-Related EEG Signals Using Soft-Thresholding," in *Proceedings of the World Congress on Medical Physics and Biomedical Engineering*, July 2000.

[7] M. Crouse, R. Nowak, and R. Baraniuk, "Wavelet-based statistical signal processing using hidden Markov models," *IEEE Trans. on Signal Processing*, vol. 46, no. 4, pp. 886–902, 1998.

[8] S. Mallat and W. L. Hwang, "Singularity Detection and Processing with Wavelets," *IEEE Trans. on Information Theory*, vol. 38, pp. 617–643, March 1992.

[9] R. Redner and H. Walker, "Mixture Densities, Maximum Likelihood and the EM Algorithm," *SIAM Review*, vol. 26, no. 2, pp. 195–239, 1984.

[10] O. Ronen, J. Rohlicek, and M. Ostendorf, "Parameter estimation of Dependence tree models using the EM Algorithm," *IEEE Signal Processing Letters*, vol. 2, no. 8, pp. 157–159, 1995.

[11] L. Rabiner and B. H. Juang, *Fundamentals of Speech Recognition.* Englewood Cliffs, New Jersey: Prentice Hall, Inc., 1993.

# Chapter 6

# CHANGE DETECTION IN IMAGE SEQUENCE BASED ON MARKOV RANDOM FIELD AND MEAN FIELD THEORY

**Qiang Liu, Mingui Sun, Ching-Chung Li and Robert J. Sclabassi**

## 1. INTRODUCTION

Change detection is carried out by comparing two or more images to distinguish their differences caused by changes of image contents from those by irrelevant disturbances. The applications of change detection are broad, including object-based video processing [1, 2, 3], remote sensing [17, 18], medical diagnosis [4, 5, 6, 7], driving and traffic assistance [8, 9, 10], etc.. In all these applications, the goal of change detection is to classify image pixels into two sets, "changed" and "unchanged". The former denotes "there are significant differences between the test image and the reference image(s) at the corresponding locations", and the latter denotes the opposite. The definition of "significant" is largely associated with human visual perception and may vary from application to application. In common cases, the image differences caused by relative motion between objects and camera, appearance/disappearance of objects, shape, color, and texture changes of objects, are considered to be "significant"; those caused by ambient and sensor noise, illumination variation, and registration error are "insignificant". The result of the classification is the so-called "change detection mask" (CDM), which is a binary image with "1" and "0" denoting "changed" and "unchanged" respectively.

In this paper, we consider intensity differences caused by device noise and illumination variation as irrelevant disturbance, otherwise as significant changes. Much research effort has been devoted to developing change detection algorithms in this scenario [11, 12, 13, 14, 15, 16, 17, 18]. One major category is single-threshold-based approaches that utilize certain test statistics adapted to noise and image models [11, 12, 13, 14, 16] to make decisions. There are usually two essential steps contained in these approaches, 1) defining a metric function of intensity variation, and 2) choosing a proper threshold to for the metric function. These methods are subjected to a quandary of either causing false alarms when the threshold is not large enough, or missing detection of significant changes when the threshold is overestimated. The reason is that the change detection is performed locally for each pixel on an image, but the single threshold to be applied is determined globally. In other words, this threshold is non-adaptive to the properties of a local region. Better results can be achieved if the threshold is increase/decreased when the contextual information of a local region suggests the test pixel be "unchanged"/"changed". This concept was investigated by Aach in [15]. He assumed that regions corresponding to moving objects were likely to have compact shape with smooth boundaries. Based on this assumption, a multiple-threshold approach was proposed, where the thresholds were functions of not only intensity difference but also "border pixel pairs" that represented the degree of smoothness of region boundary. However, as a direct extension from the single-threshold approach, this method does not generate results in an optimal sense.

Recently, change detection methods under maximum *a posteriori* (MAP) criterion have emerged to analyze remote-sensing images by satellites [17, 18]. These methods utilized Markov random field (MRF) theory to enforce spatial-contextual information. The prior knowledge of both "unchanged" and "changed" regions was modeled by energy functions, which represented the potential of a pixel being in the corresponding status ("unchanged"/"changed"). By minimizing the energy functions, the optimal results in the MAP sense could be obtained. We believe this is a powerful framework to model the uncertainty and minimize the error rate in detecting changes. In this paper, we present a new model under this framework aiming at detecting changes for image sequences. As an alternative to [17, 18], new energy functions are defined to reflect the prior belief on "unchanged"/"changed" pixels. In addition, an optimization process is carried out by utilizing Mean Field Theory (MFT). In [17], the simulated annealing [20] method was adopted with capability of localizing the global extremum, but requiring highly extensive computation. Another approach based on the iterative conditional mode algorithm [21], employed in [18], has a low computational cost, but may converge to a local extremum. Therefore, we propose the MFT based approach that renders a good trade-off between the two previous methods.

This chapter is organized as follows: Section 2. gives a brief review of the related theories; Section 3. describes the proposed methods and algorithms of change detection; Section 4. discusses the illumination invariant approach based on the proposed model; Section 5. presents the experimental results, and Section 6. concludes this chapter.

# 2. BACKGROUND THEORIES

Fundamentals of the MRF and the MFT are briefly introduced in this section.

## 2.1. Markov Random Field Theory in Change Detection

Let $\bar{F} = \{F_{1,2}, ..., F_{i,j}, ..., F_{m,n}\}$ be a 2-D random array, where $F_{i,j}$, $1 \leq i \leq m$, $1 \leq j \leq n$, is a random variable at site $(i,j)$. Let $\Omega = \{(i,j)|1 \leq i \leq m, 1 \leq j \leq n\}$ be the set of all sites. Let frame $\bar{f} = \{f_{i,j}, (i,j) \in \Omega\}$ be a realization of $\bar{F}$. Let $p(\bar{f})$ denote the joint pdf of $\bar{F} = \bar{f}$, where $p(\bar{f}) = p\{\bar{F} = \bar{f}\} = p\{F_{i,j} = f_{i,j}, (i,j) \in \Omega\}$. Then, with the same notation, $\bar{F}$ is an MRF if: (1) $p(\bar{f}) > 0, \forall \bar{f} \in \bar{F}$, and (2) $p(f_{i,j}|f_{\Omega'}) = p(f_{i,j}|f_{N_{i,j}})$, where $\Omega' = \Omega - (i,j)$, with symbol "$-$" denoting exclusion, and $N_{i,j} = \{(i',j')|(i-i')^2 + (j-j')^2 \leq k, (i',j') \in \Omega'\}$, with $k$ being a positive integer. $N_{i,j}$ defines the set of the $k$-th order neighboring sites of $(i,j)$. With the definition of $N_{i,j}$, a clique, denoted by $c$, is defined as a set containing single or multiple sites that are connected within $N_{i,j}$, $(i,j) \in \Omega$. Fig. 1 illustrates an example of cliques of a first-order neighborhood, where $c$ may be a collection of single-sites or double-sites. It was introduced in [22] that the joint pdf $p(\bar{f})$ may be approximated by the Gibbs distribution

$$p(\bar{f}) = \frac{e^{-\frac{1}{T}U(\bar{f})}}{\sum_{\bar{f}} e^{-\frac{1}{T}U(\bar{f})}}, \tag{1}$$

where $T$ is a constant and $U$ is an energy function of the MRF, given by

$$U(\bar{f}) = \sum_{c} V_c(\bar{f}) \tag{2}$$

with $V_c$ being the clique potential or clique function. The $V_c$ functions represent contributions to the total energy from single-site cliques, double-site cliques and so on. Note that (1) and (2) reflect the fact that the joint probability density function $p(\bar{f})$ is determined by the local activities, namely, the clique potentials.

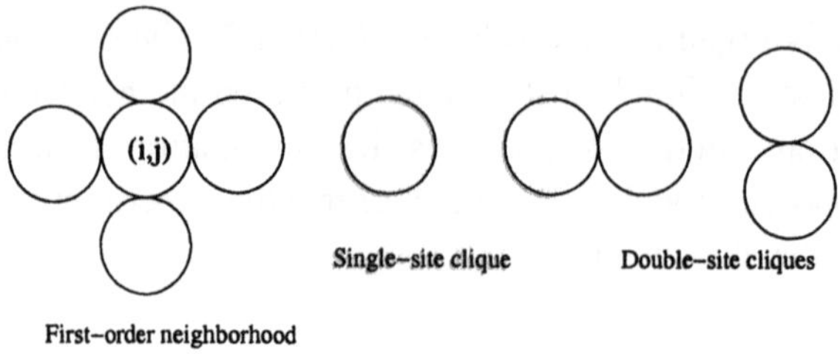


Figure 1: A first-order neighborhood system (first panel), single-site (second panel) and double-site cliques (third and fourth panels)

Considering the first-order neighborhood, we may rewrite (2) into the following form [24]

$$\begin{aligned} U(\bar{f}) = \sum_{(i,j)} \{ V_{(i,j)}(f_{i,j}) + V_{\{(i,j),(i+1,j)\}}(f_{i,j}, f_{i+1,j}) \\ + V_{\{(i,j),(i,j+1)\}}(f_{i,j}, f_{i,j+1}) \}, \end{aligned} \tag{3}$$

where the first, second, and third term are single-site, horizontal double-site and vertical double-site clique potentials, respectively. Notice that for a double-site clique $\{(i,j),(i',j')\}$, the associated clique potentials $V_{\{(i,j),(i',j')\}}(f_{i,j}, f_{i',j'})$ and $V_{\{(i',j'),(i,j)\}}(f_{i',j'}, f_{i,j})$ are equal. Therefore, (3) may be rearranged into

$$\begin{aligned} U(\bar{f}) &= \sum_{(i,j)} \{ V_{c_1}(f_{i,j}) + \frac{1}{2} \sum_{(i',j') \in N_{i,j}} V_{c_2}(f_{i,j}, f_{i',j'}) \} \\ &= \sum_{(i,j)} U_{i,j}(f_{i,j}), \end{aligned} \tag{4}$$

where $c_1$ and $c_2$ are single-site and double-site cliques in the defined neighborhood, and $U_{i,j}(f_{i,j})$ is the energy function associated with site $(i,j)$. As pointed out in [24], if $p(\bar{f})$ is a posterior distribution, minimizing the energy function $U(\bar{f})$ yields an *Maximum A Posteriori* (MAP) estimate of the joint pdf $p(\bar{f})$.

## 2.2. Mean Field Theory

To make the MRF theory more practical, we need to introduce the MFT. From the description of the MRF, we know that the value assigned to a random variable in the MRF is affected by the values at its neighboring sites, which are furtherly dependent on their neighbors. One way to calculate the interaction between one site and it neighbors is to apply the MFT [25],[26], which assumes that the impacts from the neighbors can be approximated by an average field. Let us denote the mean field for site $(i, j)$ by $f_{i,j}^{\text{mf}}$. As a result, if the first-order neighborhood is considered, one can write the energy function related to site $(i, j)$ in the following form [26]

$$U_{i,j}^{\text{mf}}(f_{i,j}) = V_{c_1}(f_{i,j}) + \sum_{(i',j')\in N_{i,j}} V_{c_2}(f_{i,j}, f_{i',j'}^{\text{mf}}), \tag{5}$$

where $V_{c_1}(\cdot)$ and $V_{c_2}(\cdot,\cdot)$ are potential functions of single-site and double-site cliques respectively; and, $f_{i',j'}^{\text{mf}}$ is the mean field for $f_{i',j'}$. Then, the marginal distribution of the MRF at site $(i, j)$ may be approximated by the following form [26],

$$p(f_{i,j}) = \frac{1}{\sum_{f_{i,j}} e^{-\frac{1}{T}U_{i,j}^{\text{mf}}(f_{i,j})}} e^{-\frac{1}{T}U_{i,j}^{\text{mf}}(f_{i,j})}. \tag{6}$$

As seen from (4) and (5), the energy function is decomposed into local clique functions, where each site is treated independently. Therefore, the joint pdf $p(\bar{f})$ can be approximated by

$$p(\bar{f}) \approx \prod_{i,j} p(f_{i,j}). \tag{7}$$

Then, maximizing $p(\bar{f})$ is equivalent to maximizing each $p(f_{i,j})$, or, to minimizing the corresponding $U_{i,j}^{\text{mf}}(f_{i,j})$.

In order to evaluate $U_{i,j}^{\text{mf}}(f_{i,j})$, the mean field values $f_{i',j'}^{\text{mf}}$ at the neighboring sites $(i', j')$ within $N_{i,j}$ must be computed. The general way to calculate a mean field value is by the following form

$$f_{i,j}^{\text{mf}} = \sum_{f_{i,j}} f_{i,j} \cdot p(f_{i,j}). \tag{8}$$

Note that (8) requires the evaluation of $p(f_{i,j})$, henceforth, $U_{i,j}^{\text{mf}}(f_{i,j})$. Therefore, the computation of the mean field value is usually carried out by iteration that stops when the change of the results from two consecutive iterations is sufficiently small.

# 3. MRF CHANGE DETECTION METHOD

## 3.1. MAP-MRF in Change Detection

Let us denote the CDM by $\bar{H} = \{H_{1,2}, ..., H_{i,j}, ..., H_{m,n}\}$, and a configuration of $\bar{H}$ by $\bar{h} = \{h_{1,2}, ..., h_{i,j}, ..., h_{m,n}\}$, where $h_{i,j} \in \{-1, 1\}, (i, j) \in \Omega$ with

"−1" denoting "unchanged" and "1" denoting "changed". Then, given two frames $\bar{f}^{(0)}$ and $\bar{f}^{(1)}$, our goal is to find the optimal $\bar{h}^*$ in the MAP sense, such that

$$\begin{aligned}\bar{h}^* &= \text{argmax}_{\bar{h}} p(\bar{h}|\bar{f}^{(0)}, \bar{f}^{(1)}) \\ &= \text{argmax}_{\bar{h}} \frac{p(\bar{f}^{(1)}|\bar{f}^{(0)}, \bar{h}) \cdot p(\bar{h}|\bar{f}^{(0)})}{p(\bar{f}^{(1)}|\bar{f}^{(0)})} \\ &= \text{argmax}_{\bar{h}} p(\bar{f}^{(1)}|\bar{f}^{(0)}, \bar{h}) \cdot p(\bar{h}|\bar{f}^{(0)}).\end{aligned} \tag{9}$$

Applying MRF assumption on both $\bar{F}$ and $\bar{H}$, maximizing $p(\bar{h}|\bar{f}^{(0)}, \bar{f}^{(1)})$ with respect to $\bar{h}$ is equivalent to minimizing the associated energy function $U(\bar{h}|\bar{f}^{(0)}, \bar{f}^{(1)})$. This, as suggested by (9), can be accomplished by minimizing $U(\bar{f}^{(1)}|\bar{h}, \bar{f}^{(0)})$ and $U(\bar{h}|\bar{f}^{(0)})$, which are associated with $p(\bar{f}^{(1)}|\bar{f}^{(0)})$ and $p(\bar{h}|\bar{f}^{(0)})$, respectively. $U(\bar{f}^{(1)}|\bar{h}, \bar{f}^{(0)})$ addresses the potential of the similarity between $\bar{f}^{(1)}$ and $\bar{f}^{(0)}$ with the knowledge of $\bar{h}$, i.e. whether the sites are changed or not. And, $U(\bar{h}|\bar{f}^{(0)})$ is always considered to represent the spatial domain constraints, e.g., the smoothness or similarity between neighboring sites. Therefore, a general form of the prior model of these energy functions is

$$U(\bar{h}|\bar{f}^{(0)}, \bar{f}^{(1)}) = \gamma_f U(\bar{f}^{(1)}|\bar{h}, \bar{f}^{(0)}) + \gamma_h U(\bar{h}|\bar{f}^{(0)}), \tag{10}$$

where $\gamma_f$ and $\gamma_h$ are regularization parameters. The larger the regularization parameter values, the more the corresponding constraint is emphasized.

Equivalently, we can write (10) by

$$U(\bar{h}|\bar{f}^{(0)}, \bar{f}^{(1)}) = \gamma_f [U(\bar{f}^{(1)}|\bar{h}, \bar{f}^{(0)}) + \gamma U(\bar{h}|\bar{f}^{(0)})], \tag{11}$$

where $\gamma = \frac{\gamma_h}{\gamma_f}$. It is noticed that minimizing $U(\bar{h}|\bar{f}^{(0)}, \bar{f}^{(1)})$ with respect to $\bar{h}$ is equivalent to minimizing $U(\bar{f}^{(1)}|\bar{h}, \bar{f}^{(0)}) + \gamma U(\bar{h}|\bar{f}^{(0)})$. Therefore, we define the energy function in the following form,

$$U(\bar{h}|\bar{f}^{(0)}, \bar{f}^{(1)}) = U(\bar{f}^{(1)}|\bar{h}, \bar{f}^{(0)}) + \gamma U(\bar{h}|\bar{f}^{(0)}). \tag{12}$$

In order to design the above energy functions, one needs to employ the prior knowledge. In our application, the prior knowledge includes the distributions of the frame difference in absence/presence of changes and the similarity between immediate sites (pixels). There are no specific routines in designing potential functions. In general, as indicated in [23], the formulation of a potential function should keep consistency with the prior knowledge: if the formulation of the regions in a clique tends to be consistent with the prior knowledge, the value of the energy function decreases; otherwise, the value increases.

In change detection, we interpret $U(\bar{f}^{(1)}|\bar{h}, \bar{f}^{(0)})$ as a sum of single-site clique potentials, which is

$$\begin{aligned} U(\bar{f}^{(1)}|\bar{h}, \bar{f}^{(0)}) &= \sum_{c_1} V_{c_1}(\bar{f}^{(1)}|\bar{h}, \bar{f}^{(0)}) \\ &= \sum_{i,j} V_{c_1}(f_{i,j}^{(1)}|h_{i,j}, f_{i,j}^{(0)}), \end{aligned} \tag{13}$$

where $V_{c_1}$ is selected to be

$$V_{c_1}(f_{i,j}^{(1)}|h_{i,j}, f_{i,j}^{(0)}) = -ln(p(d_{i,j} \mid h_{i,j})) \tag{14}$$

which is the negative of the natural logarithm of the pdf of the absolute frame difference $d_{i,j} = |f_{i,j}^{(1)} - f_{i,j}^{(0)}|$ at site $(i,j) \in \Omega$, given the knowledge of $h_{i,j}$. Therefore, if $d_{i,j}$ is consistent with the prior belief, the conditional probability will be high. As a result, its logarithm value will be low, and vice versa, as required by the design rules. Choosing the natural logarithm is instinctive. First, more penalty should be assigned when the probability value is near zero. In this case, the value of energy function should be extremely large. Second, considering $p(\bar{f}^{(1)}|\bar{f}^{(0)}, \bar{h} = -1)$, which is equivalent to the pdf of frame difference caused by noise, we may assume $p(\bar{f}^{(1)}|\bar{f}^{(0)}, \bar{h} = -1) = \prod_{i,j} p(d_{i,j} \mid h_{i,j} = -1)$, where $p(d_{i,j} \mid h_{i,j} = -1) = Z_{i,j} \cdot e^{-\frac{1}{T} V_{c_1}(f_{i,j}^{(1)}|h_{i,j}=-1, f_{i,j}^{(0)})}$, with $Z_{i,j}$ being normalization constants. We may take the natural logarithm on both sides of the above equation to obtain the potential function. For the case of $h_{i,j} = 1$, i.e., at the presence of change, the independence assumption may not hold in general. However, this assumption can be accepted as a reasonable simplification to trade off computational complexity [18]. Therefore, the above reasoning may also apply to the case $h_{i,j} = 1$. The collection of the *a prior* will be described in Section 3.2..

The other energy function $U(\bar{h}|\bar{f}^{(0)})$ in (10) addresses the contextual constraints on the neighboring sites. This can be explained as follows: with the knowledge of $\bar{f}^{(0)}$, we want to obtain $\bar{h}$ that complies with the properties of $\bar{f}^{(0)}$, for example, continuity of $\bar{h}$ if we assume smoothness of $\bar{f}^{(0)}$. Based upon this reasoning, we define

$$\begin{aligned} U(\bar{h}|\bar{f}^{(0)}) &= \sum_{i,j} \sum_{c_2 \subset N_{i,j}} V_{c_2}(\bar{h}|\bar{f}^{(0)}) \\ &= \sum_{i,j} \{\frac{1}{2} \sum_{(i',j') \in N_{i,j}} V_{c_2}(h_{i,j}, h_{i',j'})\} \end{aligned} \tag{15}$$

where $c_2$ is a double-site clique in a first-order neighborhood $N_{i,j}$ at site $(i,j) \in \Omega$. The scaling factor $\frac{1}{2}$ has been explained in (3) and (4). The clique potential

$V_{c_2}(\cdot,\cdot)$ is defined as

$$V_{c_2}(h_{i,j},h_{i',j'}) = -ln(1-0.5|h_{i,j}-\lambda\cdot h_{i',j'}|) \tag{16}$$

where $\lambda \in (0,1)$ is a constant representing the impact from site $(i',j')$ on site $(i,j)$ . The reasons behind this design are : (1) we encourage the state of site $(i,j)$ to agree with its neighboring sites; (2) the logarithm form is consistent with that in (14). The term $1-0.5|h_{i,j}-\lambda\cdot h_{i',j'}|$ acts as a probability of the random variable at site $(i,j)$ when its value agrees with those at its neighboring sites. Therefore, this definition also follows the design rules stated previously.

To minimize $U(\bar{h}|\bar{f}^{(0)},\bar{f}^{(1)})$, we must evaluate the clique potential functions. A question now is how to calculate $V_{c_2}(h_{i,j},h_{i',j'})$. As mentioned previously, we may apply MFT to simplify this calculation. If the first-order neighborhood system is assumed, we have the following approximation

$$U(\bar{h}|\bar{f}^{(0)}) \approx \sum_{i,j}\sum_{(i',j')\in N_{i,j}} V_{c_2}(h_{i,j},h^{\mathrm{mf}}_{i',j'}) \tag{17}$$

where

$$V_{c_2}(h_{i,j},h^{\mathrm{mf}}_{i',j'}) = -ln(1-0.5|h_{i,j}-\lambda\cdot h^{\mathrm{mf}}_{i',j'}|). \tag{18}$$

Combining (12) $\sim$ (18), we have

$$U(\bar{h}|\bar{f}^{(0)},\bar{f}^{(1)}) \approx \sum_{i,j} U^{\mathrm{mf}}_{i,j}(h_{i,j}|f^{(0)}_{i,j},f^{(1)}_{i,j}) \tag{19}$$

where

$$U^{\mathrm{mf}}_{i,j}(h_{i,j}|f^{(0)}_{i,j},f^{(1)}_{i,j}) = -ln(p(d_{i,j}\mid h_{i,j})) - \left[\gamma \sum_{(i',j')\in N_{i,j}} ln(1-0.5|h_{i,j}-\lambda\cdot h^{\mathrm{mf}}_{i',j'}|)\right]. \tag{20}$$

Essentially, to minimize $U(\bar{h}|\bar{f}^{(0)},\bar{f}^{(1)})$, we only need to evaluate $U^{\mathrm{mf}}_{i,j}(\cdot)$ at each site $(i,j)$, and choose $h_{i,j}$ between $-1$ and 1 to render a smaller value of $U^{\mathrm{mf}}_{i,j}(\cdot)$.

## 3.2. MRF-MFT Change Detection Algorithm

Eq. (20) requires evaluation of $p(d_{i,j}|h_{i,j}),(i,j)\in\Omega$. Instead of collecting the pdf for each site, we utilize the same pdf, denoted by $p(d|h)$, for all sites, where $d$ and $h$ have the same sample spaces as $d_{i,j}$ and $h_{i,j}$ respectively. This choice facilitates practical application since it would be extremely expensive to allocate memory for $p(d_{i,j}|h_{i,j})$ for each $(i,j)\in\Omega$. When $h(i,j)=-1$, this approximation can be justified because the states of unchanged sites are driven by noise which is usually considered to be independently and identically distributed (i.i.d.). For moving pixels, the above assumption is not true in general. However, if we assume that each pixel may experience the same or similar amounts of motion, the validity of using $p(d|1)$ for all the sites is also justifiable.

To train $p(d|-1)$, we utilize the video segments containing motionless scenes. This is relatively easy to accomplish in many applications, such as in surveillance and teleconference videos. In general, it is difficult to train $p(d|1)$; however, it is possible to train a prototype for specific applications. Practically, we adopt the following strategy to calculate $p(d|1)$: first, $p(d|1)$ is initialized to be a uniform distribution across the entire range of its sample space, i.e. $p(d|1) = \frac{1}{L+1}, d \in [0, L]$ for a discrete case; then, starting with the initial value, we adapt $p(d|1)$ during a detection process, using the following equation

$$p^{(r)}(d|1) = (1 - \epsilon \cdot \rho) \cdot p^{(r-1)}(d|1) + \epsilon \cdot \rho \cdot p^{(r)}_{d|1}, \tag{21}$$

where $p^{(r)}_{d|1}$ is the pdf of the "changed" pixels in frame $r$, $\rho$ is the ratio of the number of "changed" pixels to the total number of pixels in that frame, and $\epsilon \in (0, 1)$ is a control parameter. The term $\rho$ reflects the intuition that the more "changed" pixels there are, the more $p(d|1)$ should be adapted. Parameter $\epsilon$ is designed to control the rate of the adaptation.

An important question now is how the mean field value $h^{\mathrm{mf}}_{i,j}, (i, j) \in \Omega$ is evaluated. As mentioned previously, the mean field value is usually computed iteratively until it converges. As described in Section 2.2., $h^{\mathrm{mf}}_{i,j}$ can be evaluated based on the local energy function $U^{\mathrm{mf}}_{i,j}(h_{i,j}|f^{(0)}_{i,j}, f^{(1)}_{i,j})$

$$h^{\mathrm{mf}}_{i,j} = \sum_{h_{i,j}} h_{i,j} \cdot \frac{e^{-\frac{1}{T}U^{\mathrm{mf}}_{i,j}(h_{i,j}|f^{(0)}_{i,j}, f^{(1)}_{i,j})}}{\sum_{h_{i,j}} e^{-\frac{1}{T}U^{\mathrm{mf}}_{i,j}(h_{i,j}|f^{(0)}_{i,j}, f^{(1)}_{i,j})}}. \tag{22}$$

Applying (20), we have

$$e^{-\frac{1}{T}U^{\mathrm{mf}}_{i,j}(h_{i,j}|f^{(0)}_{i,j}, f^{(1)}_{i,j})} = [p(d|h) \cdot (\prod_{(i',j') \in N_{i,j}} [1 - 0.5(h_{i,j} - \lambda h^{\mathrm{mf}}_{i',j'})])^{\gamma}]^{\frac{1}{T}}. \tag{23}$$

Note that the computing time can be greatly reduced by using (23). With the evaluation of $h^{\mathrm{mf}}_{i,j}$, (23) can be re-evaluated with the new value of $h^{\mathrm{mf}}_{i,j}$. The iteration continues until the following condition is satisfied:

$$\frac{1}{m \cdot n} \sum_{i,j} |h^{\mathrm{mf}}_{i,j}(k+1) - h^{\mathrm{mf}}_{i,j}(k)| < \theta \tag{24}$$

where, $k$ is the index of iteration, $m \cdot n$ is the total number of pixels, and $\theta \in (0, 1)$ is a chosen threshold.

With these assumptions and simplifications, we present an algorithm to implement the proposed model as follows:

- Step 1 : Load $p(d|-1)$ and initialize $p(d|1) = 1/256$, for $d = 0 \sim 255$; Assign values to $\gamma$, $\lambda$, $\epsilon$ and $\theta$.

- Step 2 : Take two frames $\bar{f}^{(0)}$ and $\bar{f}^{(1)}$, and calculate $\bar{d} = |\bar{f}^{(0)} - \bar{f}^{(1)}|$; Initialize the mean field values $\bar{h}^{mf}$, where for each pixel $(i, j)$, $h_{i,j}^{\text{mf}} = 0$.

- Step 3 : For each pixel $(i, j)$, evaluate (20) with $h_{i,j} = -1$ or 1, and calculate the new mean field value by (22) and (23).

- Step 4 : Evaluate the difference between the new mean field value and the previous one as defined in (24); If the difference is less than $\theta$, then go to next step, otherwise go to step 3.

- Step 5 : For each pixel, if the local energy $U_{i,j}^{\text{mf}}(h_{i,j} = -1|f_{i,j}^{(0)}, f_{i,j}^{(1)}) > U_{i,j}^{\text{mf}}(h_{i,j} = 1|f_{i,j}^{(0)}, f_{i,j}^{(1)})$, then label pixel $(i, j)$ "unchanged", otherwise "changed".

- Step 6: Update $p(d|1)$ by (21); Finish if all the frames are done, otherwise go to step 2.

# 4. ILLUMINATION INVARIANT APPROACH

In the previous sections, we have presented an MRF-MFT model to identify changes exclusively due to noise. The disturbance caused by illumination changes have not been addressed. This type of disturbance usually appears in images as visually noticeable changes, but are most of the time uninteresting and should be discriminated or excluded by a change detection algorithm. Recently, research [14] has been conducted to develop approaches with "illumination-invariant" features. In the following, we describe a new construction of an illumination-invariant change detection algorithm by using the proposed MRF-MFT model.

### 4.1. Shading Model

The shading model [11, 19] formulates the gray level intensity of an image as the product of the illumination of a physical surface and its shading coefficients,

$$f_{i,j} = I_{i,j} S_{i,j}, \tag{25}$$

where $(i, j)$ is a particular pixel representing a point on the physical surface, $f_{i,j}$ is the obtained intensity, $I_{i,j}$ is the illumination, and $S_{i,j}$ is the shading coefficient at $(i, j)$. The shading coefficient is determined by a number of factors, such as the structure of physical surface, reflectance of the material, and angles of striking and reflected lights. A typical formulation of the shading coefficient was provided by Phong [29].

It is usually assumed that, for two given images containing the same objects, if there is no change in the physical structure of the object, the shading coefficient at the given location on two images are identical, i.e.,

$$S_{i,j}^{(0)} = S_{i,j}^{(1)}, \tag{26}$$

where the superscripts denote image indices. In addition, the illumination $I_{i,j}$ usually varies slowly in the spatial domain, which leads to the assumption that $I_{i,j}$ does not change within a small local region.

### 4.2. Illumination Invariant MRF-MFT Change Detection

Considering both the shading model and noise, we may formulate the intensity at pixel $(i, j)$ in image $k$ by

$$f_{i,j}^{(k)} = I_{i,j}^{(k)} S_{i,j}^{(k)} + \eta_{i,j}^{(k)}, \tag{27}$$

where $\eta_{i,j}^{(k)}$ are assumed to be $i.i.d.$ random variables due to noise. Therefore, the image difference can be modeled by

$$\hat{d}_{i,j} = (I_{i,j}^{(1)} S_{i,j}^{(1)} - I_{i,j}^{(0)} S_{i,j}^{(0)}) + (\eta_{i,j}^{(1)} - \eta_{i,j}^{(0)}). \tag{28}$$

Under the null hypothesis, namely, the object surface does not change, we have $S_{i,j}^{(0)} = S_{i,j}^{(1)}$, which leads to

$$\hat{d}_{i,j} = I_{i,j}^{(1)} S_{i,j}^{(1)} (1 - \mu_{i,j}) + (\eta_{i,j}^{(1)} - \eta_{i,j}^{(0)}), \tag{29}$$

where $\mu_{i,j} = I_{i,j}^{(0)} / I_{i,j}^{(1)}$ denotes the ratio of illumination on pixel $(i, j)$ in the two images. If there is no illumination change, then $\mu_{i,j} = 1$.

In order to extend the previously described model with consideration of illumination, let us define an adjusted image difference to reflect the illumination change

$$e_{i,j} = |f_{i,j}^{(1)} - \frac{1}{\mu_{i,j}} f_{i,j}^{(0)}|. \tag{30}$$

Under the null hypothesis, we have

$$e_{i,j} = |\eta_{i,j}^{(1)} - \frac{1}{\mu_{i,j}} \eta_{i,j}^{(0)}|. \tag{31}$$

Now, the single clique function defined in (14) is changed to

$$V_{c1}(f_{i,j}^{(1)}|h_{i,j}, f_{i,j}^{(0)}) = -ln(p(e_{i,j}|h_{i,j})). \tag{32}$$

If $\mu_{i,j}$ can be evaluated, so can the corresponding clique functions. A simple way is to use the image intensity values to estimate $\mu_{i,j}$. To do that, let us define

$$F_{i,j}^{(k)} = \frac{1}{M} \sum_{(p,q)\in W_{i,j}} f_{p,q}^{(k)}, \qquad k = 1, 2 \tag{33}$$

to compensate the noise effect, where $W_{i,j}$ is a window centered at pixel $(i, j)$, and $M$ is the number of pixels included in $\Omega_{i,j}$. If $M$ is sufficiently large, we have

$$F_{i,j}^{(k)} \approx \frac{1}{M} \sum_{(p,q)\in W_{i,j}} I_{p,q}^{(k)} S_{p,q}^{(k)}, \qquad k = 1, 2. \tag{34}$$

Considering that the illumination is usually a slow changing variable in the spatial domain, we may assume it a constant within $W_{i,j}$. Consequently, we have

$$F_{i,j}^{(k)} \approx \frac{1}{M} I_{i,j}^{(k)} \sum_{(p,q)\in W_{i,j}} S_{p,q}^{(k)}, \qquad k = 1, 2. \tag{35}$$

Then we can use $F_{i,j}^{(k)}$ to obtain an estimated $\mu_{i,j}$ by the following,

$$\hat{\mu}_{i,j} = \frac{F_{i,j}^{(0)}}{F_{i,j}^{(1)}} = \frac{I_{i,j}^{(0)}}{I_{i,j}^{(1)}} \cdot \frac{\sum_{(p,q)\in W_{i,j}} S_{p,q}^{(0)}}{\sum_{(p,q)\in W_{i,j}} S_{p,q}^{(1)}} = \mu_{i,j} \cdot \frac{\sum_{(p,q)\in W_{i,j}} S_{p,q}^{(0)}}{\sum_{(p,q)\in W_{i,j}} S_{p,q}^{(1)}}. \tag{36}$$

Under the null hypothesis, $\frac{\sum_{(p,q)\in W_{i,j}} S_{p,q}^{(0)}}{\sum_{(p,q)\in W_{i,j}} S_{p,q}^{(1)}} = 1$, henceforth, $\hat{\mu}_{i,j} = \mu_{i,j}$.

As a result, we have

$$p(e_{i,j}|h_{i,j} = -1) = p(|\eta_{i,j}^{(1)} - \frac{1}{\hat{\mu}_{i,j}} \eta_{i,j}^{(0)}|). \tag{37}$$

Therefore, if the distribution of $\eta_{i,j}^{(k)}$ is known, $p(e_{i,j}|h_{i,j} = -1)$ can be evaluated. Because $\eta_{i,j}^{(k)}$ represents a noise variable, for simplicity, let us assume it obeys a Gaussian distribution with a zero mean and a variance of $\delta_\eta^2$. Then, the function $\eta_{i,j}^{(1)} - \frac{1}{\hat{\mu}_{i,j}}\eta_{i,j}^{(0)}$ also has a Gaussian distribution with a zero mean and variance equal to $(1 + \frac{1}{\hat{\mu}_{i,j}^2})\delta_\eta^2$. Consequently, we have

$$p(e_{i,j}|h_{i,j} = -1) = \begin{cases} \frac{1}{\sqrt{2\pi(1+\frac{1}{\hat{\mu}_{i,j}^2})\delta_\eta^2}} & \text{if } e_{i,j} = 0, \\ \frac{2}{\sqrt{2\pi(1+\frac{1}{\hat{\mu}_{i,j}^2})\delta_\eta^2}} e^{-\frac{e_{i,j}^2}{2(1+\frac{1}{\hat{\mu}_{i,j}^2})\delta_\eta^2}} & \text{if } e_{i,j} > 0, \\ 0 & \text{otherwise} \end{cases} \tag{38}$$

Applying (38) to (32), we have the single clique function for "unchanged" pixels. For the "changed" case, $p(e_{i,j}|h_{i,j} = 1)$ can be calculated following the same procedure as described in 3.2., namely, being trained online. The adaptation of $p(e_{i,j}|h_{i,j} = 1)$ is still formulated by (21) except that that image difference $d$ is replaced by $e$.

# 5. IMPLEMENTATION AND EXPERIMENTS

In this section, the experimental results based on the proposed method are reported. We present results on three image sequences: 1) a simulated data generated by MATLAB (MathWorks Inc.), 2) *Hallway* sequence which is a popular test video clip containing multiple types of changes, and 3) a patient monitoring video sequence to demonstrate illumination-invariant function. All sequences are with QCIF format ( $144 \times 176$ pixels with Y component at 30 frames per second). Only the Y component is utilized to calculate frame differences.

As described in the previous sections, five controlling parameters are required, i.e., $T$, $\gamma$, $\lambda$, $\epsilon$, and $\theta$. A set of typical values of these parameters is listed in Table 1. These values are chosen experimentally.

- $T$: It is called "temperature" in MRF based methods, e.g. simulated annealing algorithm [20]. This parameter determines the spread of the Gibbs distribution. In the simulated annealing methods, $T$ is gradually decreased during an annealing process. However, as suggested by [27], a fixed $T$ is able to render a satisfactory result while reducing the computational cost. Therefore, a constant $T$ was utilized through out our experiments.

- $\gamma$: This is the regularizing parameter to balance the constraints introduced by different clique potentials. In our application, a larger $\gamma$ emphasizes the smoothness constraint.
- $\lambda$: This parameter models the impact between neighboring sites. In (16), $h_{i,j} - \lambda h_{i',j'}$ is utilized to represent the difference between neighboring sites $(i, j)$ and $(i', j')$. As is seen that the larger the $\lambda$, the more impact from $(i', j')$ is introduced.
- $\epsilon$: Its role is to control the adaptation of the pdf of $d$ in the presence of change. The larger $\epsilon$, the more the pdf adapts to each CDM, and the faster it adapts to the testing data. However, considering the risk of false detection, we assign a moderate $\epsilon$ value.
- $\theta$: The threshold to stop the iteration in calculating the mean field values.

Table 1: Typical control parameters

| parameter | $T$ | $\gamma$ | $\lambda$ | $\epsilon$ | $\theta$ |
|---|---|---|---|---|---|
| value | 2 | 1 | 0.99 | 0.5 | 0.01 |

## 5.1. Simulated Data

To evaluate our method quantitatively, we generated a synthetic image sequence by using MATLAB in the following way: a circle (with a radius of 20, line width of 3, both in pixels, gray level intensity of 5, and the coordinates of the origins randomly generated) is plotted in a frame; then, the white Gaussian noise with mean of 127 and standard deviation of 1 is added to the frame. It should be noted that the signal-to-noise (SNR) ratio of the simulating data is worse than the SNR of common videos. Two pairs of sample frames are shown in Fig. 2. Let us denote the known CDM by $\bar{h}^{(r)}$ and the detected CDM by $\bar{h}^{(t)}$. Then, the $\Omega_e = \{(i, j)|h_{i,j}^{(r)} \neq h_{i,j}^{(t)}, (i, j) \in \Omega\}$ denotes the set of sites with false labels. The error rate is then defined by

$$E_r = ||\Omega_e||/||\Omega|| \tag{39}$$

where $||\Omega_e||$ and $||\Omega||$ denote the number of sites in $\Omega_e$ and $\Omega$ respectively.

Fig. 3 demonstrates the results of the simulated data, with the parameters listed in Table 1. Fig. 3(a) shows the results obtained from frame 1 and 2. The

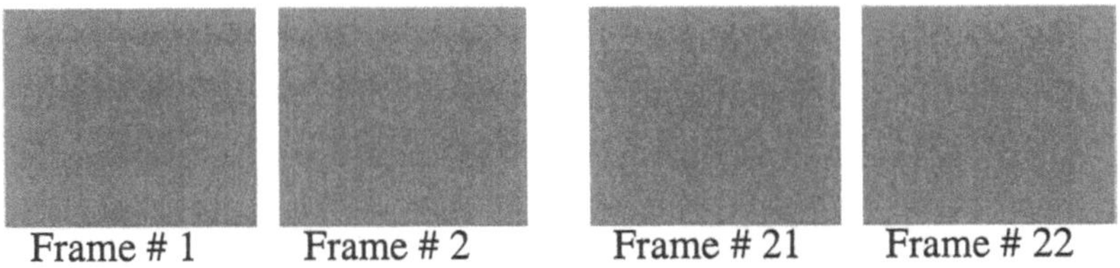


Figure 2: Two pairs of sample frames in the synthetic sequence: from left to right, frame 1 and 2 forming a pair; frame 21 and 22 forming another pair.

left, middle and right panels in Fig. 3(a) show the known CDM, the detected CDM, and $p(d|h = -1)$ and the initial $p(d|h = 1)$, respectively. Compared with the known CDM, the detected CDM has visible false detections. However, with the adaption of $p(d|h = 1)$, the false detections are reduced. As seen in Fig. 3(b), where the results are obtained from frames 21 and 22, the detected CDM (the middle panel) contains much less false detections. The error rate of each CDM is plotted in Fig.3(c). It is seen that the error rate decreases as frames $1 \sim 20$ being processed, then becomes stable after that. The reason is that $p(d|h = 1)$ adapts gradually to the testing data at the initial 20 frames, and then stabilizes. The adaptation speed is quite satisfactory for many common applications (e.g. video surveillance, video editing), as indicated by our results using natural videos.

## 5.2. Real-World Data

In the following, we report experimental results of two real-world video sequences. The first test sequence is called *Hallway* that can be found in the public domain. Sample frames are illustrated in Fig. 4, where the top panel shows frame 1 of *Hallway* sequence, and frame $25, 50, 100, 250, 275$ are shown on the bottom panel. It is seen that frame 1 contains only background scene, while the subsequent frames have appearances of new objects, including two walking persons and a suitcase placed at the left side of the hallway. The obtained CDMs by using the MRF-MFT algorithm described in Section 3.2. are illustrated in Fig. 5. One can see that the foreground was well separated from the background scene. The conditional probabilities required by the potential functions are shown in Fig. 6 (for the clearance of display, only part of the pdf's are displayed). The right panel shows a close-look of the pdf's on the left panel. The pdf of noise, i.e. $p(d|h = -1)$, was estimated from the intensity differences in manually selected regions, which contained no apparent changes. The pdf's of intensity differences caused by relevant changes were first initialized to be uniformly distributed within value range of $0 \sim 255$, then

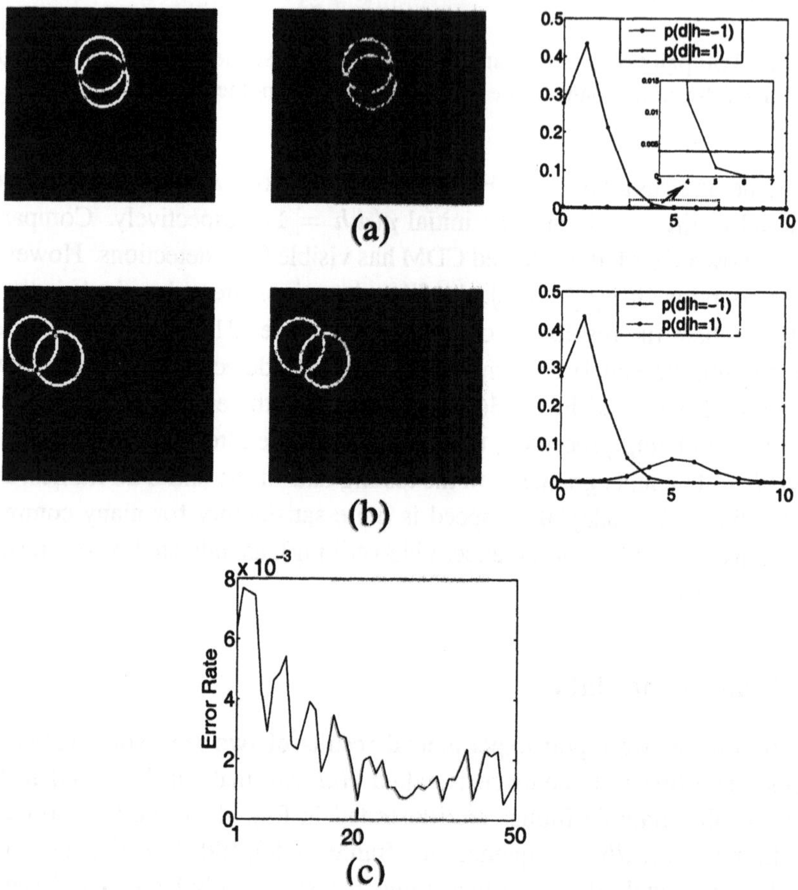


Figure 3: (a) The change detection results from frame 1 and 2. From left to right: the known CDM, the detected CDM and $p(d|h=-1)$ and initial $p(d|h=1)$, respectively. (b) The change detection results from frame 21 and 22. From left to right: the known CDM, the detected CDM and $p(d|h=-1)$ and $p(d|h=1)$ (adapted from frame 1 $\sim$ 20), respectively. (c) The error rate from frame 1 $\sim$ 50.

Figure 4: Frames 1, 25, 50, 100, 250 and 275 of *Hallway* sequence

Figure 5: The detected CDM's from the sample frames of *Hallway* sequence, with the parameter values listed in 1. The white ("1-pixel") regions denote "there are significant changes between the test image (containing moving objects) and the reference image (containing merely background scene)". It is seen that the significant changes caused by the moving subjects and the suitcase being placed in the hallway were well identified.

adapted in the process of change detection for the subsequent video frames.

The convergence of the mean field value took 5.09 number of iterations in average. The algorithm was implemented in C++ and compiled with Microsoft Visual C++6.0. Experiments were carried on an AMD Athlon 1900 (1.66 GHz) PC with 512M DDR2100 RAM. Our program performed an average of 9.4 milliseconds per iteration. For the presented data, this leads to a time consumption of 47.85 milliseconds per frame pair. It is an acceptable amount of cost when the frame pair is formed at 10 frames per second or less.

Next, we present experimental results on a video segment recorded at the Epilepsy Monitoring Unit at the University of Pittsburgh Medical Center. Sample video frames were shown in Fig. 7. This type of video is often recorded for patient care and diagnostic purposes. Automatic detection of changes can facilitate this application by 1) tracking the subject and controlling the camera to present an optimal view of the subject, and 2) segmenting the region of interest to advance transmission and archiving of the monitoring video. We'll demonstrate the effectiveness of the illumination invariant approach described in Section 4.. Firstly, we carried out experiment on the testing video without

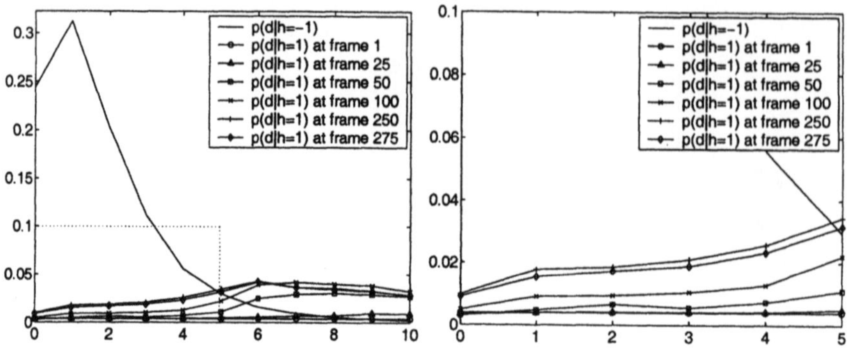


Figure 6: The pdf's obtained from *Hallway* sequence: the left panel shows $p(d|h = -1)$, and $p(d|h = 1)$ at the frames of 1, 25, 50, 100, 250 and 275; the right panel plots the pdf's in the marked range on the left panel, showing a close-look of the adaptation of $p(d|h = 1)$.

illumination invariance function. A typical result is illustrated in Fig. 7. The left panel shows a snapshot of the environment before the occupancy of the subject. The middle panel shows a video frame with the subject sitting in bed. Comparing these two images, we found that there were large intensity differences (in amplitude) contained in the background area. For example, the pixels in the marked regions on the right panel, which shows the detected CDM without concerning illumination variation, had a maximum intensity difference of 35. These intensity differences may be caused by shadow, light source change, and automated camera gain adjustment, which can all be considered as illumination variation. Thereafter, by using the algorithm described in Section 4., these irrelevant disturbance can all be greatly reduced. This is demonstrated by our experimental results shown in Fig. 8, where the image containing the background scene, sample video frames with presence of the subject, and the corresponding CDM are shown on the top, middle and bottom panels respectively. It is seen that the irrelevant changes in the background area were successfully eliminated, while the subtle changes of the bed caused by the movements of the subject were retained.

## 6. CONCLUSION

In this chapter, we have presented a new approach to the change detection problem in image sequences. This approach employed two well-established theories: MRF and MFT. Based upon the MRF theory, change detection is modeled as an optimization problem, namely, the CDM is calculated in the

Figure 7: Experimental result on patient monitoring video WITHOUT illumination invariance function. The left panel shows a snapshot of the monitoring unit before the patient's occupancy. The middle panel shows a sample video frame at the presence of patient. The right panel shows the detected CDM by the proposed method without concerning illumination variation. It is seen that the CDM was affected by the illumination change, for instance, the marked polygonal regions in the background area.

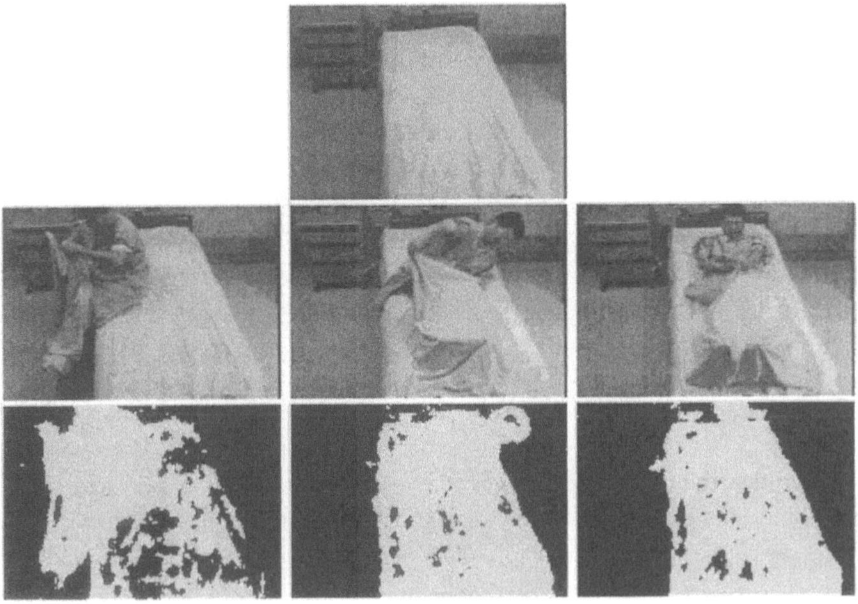

Figure 8: Experimental results based on algorithms described in Section 4. featured with illumination invariance. Top panel: the image containing the background scene; middle panels: sample video frames with presence of the subject; bottom panels: the corresponding CDM's. It is seen that the irrelevant changes in the background area were successfully eliminated, while the subtle changes of the bed caused by the movement of the subject were retained.

sense of MAP. In order to carry out an efficient computation, we utilized MFT, which simplifies the procedure of searching for the optimal CDM. Experimental results are reported based on this optimization approach. Both the synthetic and real-world data indicate that this approach accurately detects the changes between frame pairs. One remaining problem, however, is to determine the values of control parameters in the associated functions. Currently, the parameters are chosen in an experimental manner. In the future, a meaningful cost function of these parameters may be designed to provide the values in a certain optimal sense.

This work was supported in part by National Institutes of Health Grants MH38494 and EB002309 and Computational Diagnostics, Inc.

# REFERENCES

[1] Thomas Sikora, "The MPEG-7 Visual Standard for Content Description An Overview," *IEEE Transactions on Circuits and Systems for Video Technology*, Vol.11, No.6, pp. 696-702, June. 2001.

[2] T. Meier and K. N. Ngan,"Segmentation and tracking of moving objects for content-based video coding," *IEEE Transactions on Circuits and Systems for Video Technology*, Vol.9, No.8, pp. 1190-1203, Dec 1999.

[3] M. Kim, J. G. Choi, D. Kim, etc., "A vop generation tool: automatic segmentation of moving objects in image sequences based on spatio-temporal information," *IEEE Transactions on Circuits and Systems for Video Technology*, Vol.9, No.8, pp. 1216-1226, Dec 1999.

[4] M. Bosc, F. Heitz, J. P. Armspach, I. Namer, D. Gounot, and L. Rumbach, "Automatic change detection in multimodal serial MRI: application to multiple sclerosis lesion evolution," *Neuroimage*, vol. 20, pp. 643-656, 2003.

[5] M. J. Dumskyj, S. J. Aldington, C. J. Dore, and E. M. Kohner, "The accurate assessment of changes in retinal vessel diameter using multiple frame electrocardiograph synchronized fundus photography," *Current Eye Research*, vol. 15, no. 6, pp. 632-652, June 1996.

[6] J.-P. Thirion and G. Calmon, "Deformation analysis to detect and quantify active lesions in three-dimensional medical image sequences," *IEEE Transactions on Medical Image Analysis*, vol. 18, no. 5, pp. 429-441, 1999.

[7] L. Lemieux, U. Wieshmann, N. Moran, D. Fish, and S. Shorvon, "The detection and significance of subtle changes in mixed-signal brain lesions by serial MRI scan matching and spatial normalization," *Medical Image Analysis*, vol. 2, no. 3, pp. 227-242, 1998.

[8] C.-Y. Fang, S.-W Chen, and C.-S. Fuh, "Automatic change detection of driving environments in a vision-based driver assistance system," *IEEE Trans. Neural Networks,* vol. 14, no. 3, pp. 646-657, May 2003.

[9] W. Y. Kan, J. V. Krogmeier, and P. C. Doerschuk, "Model-based vehicle tracking from image sequences with an application to road surveillance," *Opt. Eng.*, vol. 35, no. 6, pp. 1723-1729, 1996.

[10] M. J. Black, D. J. Fleet, and Y. Yacoob, "Robustly estimating changes in image appearance," *Computer Vision and Image Understanding*, vol. 78, no. 1, pp. 8-31, 2000.

[11] K. Skifstad and R. Jain, "Illumination independent change detection for real world image sequences," *Comput. Vis. Graph. Image Process.*, Vol. 46, no. 3, pp. 387-399, 1989.

[12] Y. Z. Hsu et al.,"New likelihood test methods for change detection in image sequences," *Comput. Vis. Graph. Image Process.*, vol. 26, pp. 73-106, 1984.

[13] T. Aach, A. Kaup, and R. Mester,"Statistical model-based change detection in moving video," *Signal Processing*, vol. 31, pp. 165-180, Mar. 1993.

[14] Emrullah Durucan and Touradj Ebrahimi "Change Detection and Background Extraction by Linear Algebra," *Proceedings of the IEEE* vol.89, No.10, pp. 1368-1381, Oct. 2001.

[15] T. Aach, A. Kaup, and R. Mester,"Bayesian algorithms for adaptive change detection in image sequences using Markov random fields," *Signal Processing: Image Communication 7*, pp. 147-160, 1995.

[16] S. Liu, C. Fu, and S. Chang, "Statistical change detection with moments under time-varying illumination," *IEEE Trans. Image Processing*, vol. 7, no. 9, pp. 1258-1268, September 1998.

[17] T. Kasetkasem and P. K. Varshney, "An Image Change Detection Algorithm Based on Markov Random Field Models", *IEEE Trans. Geoscience and Remote Sensing*, vol. 40, No. 8, pp. 1815-1823, Aug. 2002.

[18] L. Bruzzone, and D. F. Prieto "An Adaptive Semiparametric and Context-Based Approach to Unsupervised Change Detection in Multitemporal Remote-Sensing Images," *IEEE Trans. on Image Processing*, Vol. 11, No. 4, Apr. 2002.

[19] A. V. Oppenheim, R. W. Schafer, and T. G. Stockham Jr, "Nonlinear filtering of multiplied and convolved signals," *Proc. IEEE*, vol. 56, pp. 1264-1291, Aug. 1968.

[20] S. Kirkpatrick, C. D. Gellatt, and M. P. Vecchi, "Optimization by simulated annealing," Science, vol. 220, pp. 671-680, 1983.

[21] J. Besag, "On the statistical analysis of dirty pictures (with discussions)," *J. Roy. Statist. Soc.,* ser. B, vol. 48, pp. 259-302, 1986.

[22] M. Hassner and J. Sklansky, "The use of Markov random fields as models of texture," *Comput. Graphics Image Processing*, vol. 12, pp.357-370, 1980.

[23] R. Chellappa and A. Jain, *Markov Random Fields Theory and Applications*, Boston : Academic Press, 1993.

[24] S. Geman and D. Geman, "Stochastic relaxation, Gibbs distributions, and the Bayesian restoration of images," *IEEE Trans. Pattern Anal. Machine Intell.*, vol. PAMI-6, pp. 721-741, Nov. 1984.

[25] David Chandler, *Introduction to modern statistical mechanics*,New York : Oxford University Press, 1987.

[26] Jun Zhang, "The mean field theory in EM procedures for blind markov random field image restoration," *IEEE Transactions on Image Processing*, Vol.2, No.1, pp. 27-40, Jan. 1993.

[27] Jun Zhang and Gerald G. Hanauer, "The Application of Mean Field Theory to Image Motion Estimation," *IEEE Transactions on Image Processing*, Vol.4, No.1, pp. 19-33, Jan. 1995.

[28] Jie Wei and Ze-nian Li "An efficient two-pass MAP-MRF algorithm for motion estimation based on mean field theory," *IEEE Transactions on Image Processing*, Vol.9, No.6, pp. 960-972, Sep. 1999.

[29] B. T. Phong, "Illumination for computer generated pictures," *Commun. ACM,* vol. 18, pp. 311-317, 1975.

# Chapter 7

# ANALYSIS OF MULTI-CHANNEL SUBDURAL EEG BY RECURRENT NEURAL NETWORKS

**Robyn R. Bates, Mingui Sun, Mark L. Scheuer, Robert J. Sclabassi**

## 1. INTRODUCTION

Epilepsy, a neurophysiological condition in which there is a disruption of the brain's normal electrical activity, is associated with symptoms which can vary from a brief lapse of attention to episodes of seizures [1]. It is one of the most common neurological disorders and up to about 1% of the population in the United States is afflicted. While the existence of seizures has been noted since ancient times, the progress of epilepsy characterization and treatment was minimal until the advent of electrophysiological monitoring techniques [2].

An epileptic seizure is caused by repetitive, abnormal, and synchronized, neuronal activity within the brain, with symptoms that depend both on the location of the seizure onset within the brain, and the spread of the activity. A diagnosis of epilepsy is primarily based on an analysis of the electrical activity in the patient's brain, recorded as an electroencephalogram (EEG) [3]. The EEG represents the summed potentials of a large number of cortical neuronal cells which are mostly oriented in parallel columns orthogonal to the cortical surface [4].

An EEG record from a patient with epilepsy can contain information about seizure genesis in both the inter-ictal activity and the immediate pre-seizure stage. Detection of epileptoform activity in the pre-seizure component of a multi-electrode EEG record is highly useful in the localization of epileptogenic foci.

For long-term monitoring of a patient with epilepsy, traditional methods whereby the entire EEG is reviewed by a trained technician, and significant portions by a specially trained neurologist, are inefficient and time consuming. The development of computer-based EEG analysis utilities has led to the automated examination of long-term EEG records; however those methods have not yet been widely accepted.

The use of automated methods in medicine, although not yet at the level of automated diagnosis, can be employed in an advisory capacity to isolate points of clinical interest for further analysis or detect clinically significant data. These results can then help to provide the clinician useful information with which to initiate appropriate interventions. With respect to the former application, analysis of the EEG record of a patient with intractable seizures can allow localization of epileptogenic foci – an important step in planning surgical treatment [5-8].

## 1.1 Neural Network Models

Artificial neural networks (ANN's), or simply neural networks (NN's), are powerful analytical tools designed from interconnected computing elements called neurons (also variously called neurodes, nodes, neural units, or merely units). Loosely inspired by the makeup of the nervous system, these interconnected elements examine patterns of data and learn to classify them. As the neuron is a basic component of the brain, a neural unit is the building block of a neural network. Although the two are far from being the same, or from performing the same functions, they still possess similarities that are notable. Neural networks in the brain consist of a large number of interconnected units that give them the ability to process information in a highly parallel way. An artificial neuron sums all inputs to it and creates an output that is carrying information to other neurons. The strength by which two neurons are influencing each other is called a synaptic weight. In most neural networks, all neurons are connected to all other neurons by synaptic weights that can have seemingly arbitrary values, but in reality, these weights show the effect of a stimulus on the neural network and the ability or lack of it to recognize that stimulus [9].

Neural networks have been used in a wide variety of signal processing and pattern recognition applications and have been successfully applied in such diverse fields as speech processing, handwritten character recognition, time series prediction, data compression, feature extraction and pattern recognition in general. Their benefit lies in the relative simplicity with which the networks can be designed for a specific problem along with their ability to perform nonlinear data processing.

## 1.2 Recurrent Neural Network for Seizure Detection

Recurrent neural networks have been proposed for time-series modeling and prediction, especially for those systems in which discrete time measurements are obtained. They are also appropriate for modeling real-world systems in which empirical measurements are obtained at discrete time points [10, 11].

This type of neural network can be trained on multiple temporal patterns, which may evolve on different time-scales and be sampled at non-uniform time intervals. Another advantage of the recurrent model is that, despite sparseness of the training data, the network is able not only to make good predictions at the final time step for temporal processes unseen in training, but also to reproduce an interval of the signal of interest at an earlier time. Furthermore, it may be possible for the network to predict the existence and role of a signal sample for which no target information is provided. The ability of the model to cope with outlier data points is likely to be useful in an application, such as EEG data, where there is a significant amount of noise, irrelevant components, spontaneous events, ambiguous artifacts, and distortion [12].

The research described in this chapter investigates the application of recurrent neural networks to the analysis of epileptic EEG records. There are three primary goals of these investigations:

1.To employ the recurrent neural network to detect the possible onset of seizure activity in multi-channel subdural EEG (SEEG) data;

2. To classify the EEG recording-electrodes in order of earliest detection of seizure-like activity;

3. From steps 1 and 2, make certain qualitative conclusions regarding the location of the epileptogenic foci.

# 2. METHODS

## 2.1 Data

EEG data were available from a five-year old white male patient who presented with intractable epilepsy. Epileptogenic foci were not well localized and ought to be several in number. A callosotomy, which resulted in only limited short-term improvement, was performed when the child was five. There is no history of other chronic illness.

One collection of the EEG data was recorded from implanted subdural electrodes (SEEG), while the second collection was obtained from scalp recordings. Both sets were obtained after the callosotomy had been performed. For this study, only forty-eight sets of the subdural (SEEG) data were used.

The SEEG data was obtained from those sets, which were determined, by expert analysis, to contain definitive seizure activity. Each SEEG data set was one minute of data (15000 data points per channel, sampled at 250 samples/sec) . Since one of the objectives of the this study is to examine the neural network's ability to detect pre-seizure epileptoform activity, a number of sets of SEEG data preceding those which contained definitive seizure activity were also selected for analysis.

The data had previously been low-pass filtered and demeaned. Since the data was obtained from multi-channel recordings, it was necessary to normalize the dataset over all channels before dividing up each channel by training or testing sets.

By making the assumption of slow time variation, one may divide a time-series sample into segments of appropriately small time-intervals (generally < 4 seconds) and assume each segment to be stationary in the interval. However, there are instances in which the variation of the EEG signal over even a small time interval cannot be estimated by a sequence of stationary epochs without making some assumption about the smoothness of the time variation – *viz.* the second-order differences.

Recurrent neural networks are particularly susceptible to large first and second-order differences because of the inherent short-term memory which is a consequence of the network architecture. By making the assumption of the interval stationarity, it is assumed that these differences are small within each interval and that the transition between intervals is smooth [13].

Pre-processing the data by filtering or smoothing is typically performed to minimize the effects of non-stationarity. For example, some methods that are employed include the use of linear or other types of filters. However, methods used for pre-processing might introduce their own set of difficulties into the analysis. For example, the optimal cut-off points for these filters are however determined on a trial and error basis, which may require considerable time and expertise.

## 2.2 The Recurrent Neural Network Architectures

The architecture selected for this study, the recurrent architecture, is able to process sequential temporal data, since it employs one or more feedback loops that relate the output of the network to subsequent inputs to the network. Thus, this architecture is thus particularly well-suited for EEG data analyses. There are a variety of recurrent architectures available for study which differ in the layers involved in the feedback connections as well as the number of feedback connections between the layers [14, 15].

Although there are number of recurrent architectures in which the hidden layer is only partially connected to the input layer, it was decided to employ a fully connected recurrent network for this study. Since each input node receives equal weight by the network, the use of a fully-connected network is reasonable to optimize the trade-off between the representation of non-linear dynamics and training speed and stability [16].

Since the seizure events are not well-defined when observing the data at a discrete point in time, it is necessary to analyze the EEG records in the context of the surrounding data. This requirement is met by designing the neural network to possess a short-term memory, so that each data point is analyzed with respect to each preceding point – and by designing the input layer to consist of nodes representing contiguous EEG channels. For the architectures used in this study, the following relationship describes the recurrent process:

$$x_{i+1} = 0.5\sum_{1}^{n}\left(x_i + y_{i-1}\right) \tag{3.1}$$

where: $x_i$ – input at time t;
$y_{i-1}$ – neural network output at time t-1

Since both the inputs and outputs of the network are scaled over [0,1], a weight of 0.5 is used to scale the feedback sum of the input and output to Thus, both values receive equal weight in the recurrent summation. In this relation, the node inputs are considered individually and not in some linear combination.

Topological considerations regarding the spatial and temporal characteristics of the seizure are of significant importance in the development of a methodology for studying epileptoform activity. The use of different recording montages can either enhance or disguise the presence of abnormal cortical activity. Also, the underlying anatomy of the cortex can affect the resolution of the signal recorded at an electrode by such factors as attenuation by tissue or bone, coupling of neighboring neuronal networks in localized segments of tissue, or by electrical conduction along specific neuronal pathways at a distance from the actual signal source. Artifacts and noise generated from the use of the recording equipment or process are also responsible for concealing both pre-seizure and seizure activity in the EEG record [17].

These factors have implications in the design of seizure detection methods, particularly, with respect to the choice and design of the appropriate neural network architecture. Localization and estimating the distribution of signal sources in the cortex from the EEG record requires the solution of an inverse problem in which the cortical potential fields are calculated from a subset of electrostatic potentials measured on the scalp or subdurally. The characterization and solution of this problem is difficult because of the possibility that different internal sources can produce similar-looking electrostatic fields when measured by a finite number of electrodes [18, 19].

A number of approximations are available for the solution of this source-localization problem, such as source imaging, but they fail to address the cases in which there might be more than one independent source. The general problem is the determination of the location and magnitudes of the sources of epileptoform activity within the cortex from the set of electrode potentials derived from specific recording sites on the scalp or from subdural measurements [20].

The subdural montage used to record the EEG data from the patient considered in this study consisted of 94 electrodes. Since this study did not employ the use of any pre-processing techniques to reduce the complexity of the data before processing by the neural network, a basic assumption was made concerning the design of the input layer for the neural network. It was assumed that the relatively large number (94) of recording electrodes that were placed subdurally would provide sufficient spatial resolution to pinpoint the location of seizure sources. It was further assumed that this resolution would be sufficient to allow the detection of pre-seizure activity in the EEG record by ensuring a recording of seizure activity sufficiently close to the epileptogenic foci so as prevent the attenuation effects mentioned above.

Also, because of the nature of the recurrent architecture, it was assumed that values of the EEG data depended on the immediately preceding values. Therefore, using too few inputs can result in inadequate modeling, whereas too many inputs can excessively complicate the network, in terms of the training times and dimensionality of the network. If, as in this study, each input node represented each EEG electrode, using too many input layers will involve the input from too many electrodes and therefore provide low spatial resolution with regard to the actual localization of the seizure-like activity, if any. In the extreme case, for example, if an input layer of 94 nodes was used, corresponding to the complete montage for the subdural EEG, the detection of seizure-like activity would indicate that a seizure might have occurred in time, but with no information as to where on the montage the seizure-like activity happened.

With these considerations in mind, it was decided to employ an input layer of five electrodes, which consisted of one central electrode and four closest neighbor electrodes, which would allow for the localization of seizure activity within a region covered by five electrodes. Larger network input layers and overall architectures were considered, but it was felt that a resolution of five electrodes would constitute a good tradeoff between localization and computer-processing time during the training of the network. In particular, the neural network should be able to provide a sufficiently low error during training, but not require a overly large amount of time to train.

With respect to the number of nodes in the hidden layer of the network, designing an optimal structure can employ empirical rule is that the number of weights should be less than a tenth of the number of training patterns. One method of determining the number of hidden-layer nodes is the Baum-Haussler rule:

$$\text{Number of Hidden Nodes} \leq (N_{train})(E_{tol})/(N_{data} + N_{output})$$

where $N_{train}$ is the number of the training data points, $E_{tol}$ is the relative error tolerance (the value is typically in the region of 0.01), $N_{data}$ is the number of data points per training example, and $N_{output}$ is the number of output neurons [20]. Thus, for a network in which there are 100,000 training sequences of data available, 250 data points per training sequence and 1 output neuron, there should be 4 nodes in the hidden layer. Although this heuristic does not take into account the effect of the short-term memory provided by the recurrent connection, the result of four nodes was considered an approximation and recurrent architectures possessing five and ten nodes per hidden layer were tested in this study.

The single output node for both the 5-5-1 and 5-10-1 architectures is intended to represent the activity "center-of-gravity" for the five input nodes. Thus, it may not represent any one of the five electrode signals being examined, but rather a spatial point in between the set of five. Schematics for both the 5-5-1 and 5-10-1 architectures appears below (Figures 1, 2).

## 2.3. Training

For this particular study, two collections of training data, not from the testing data, were obtained -- one confirmed set of seizure data; one confirmed set of non-seizure data. The presence of seizure events was confirmed by expert epilptologists. From each of these two data collections, a total of 1000 data sets, 500 sets consisting of 94 channels, each channel with 250 points (1 second) of confirmed seizure activity, and 500 sets of 94 channels with 250 points of non-seizure data per channel, were extracted.

The targets for these training sets are determined as follows: if the set of raw data corresponds to seizure data, its target is "1". If it corresponds to non-seizure data, its target is "0".

Each sample of 250 sample points of raw seizure and non-seizure data is used only once. The sample sets were presented until the training goal had been satisfied.

The resilient backpropagation training algorithm was used during training. This algorithm is a local adaptive learning scheme, in which only the sign of the derivative is considered to indicate the direction of the weight update (see figure 3.)

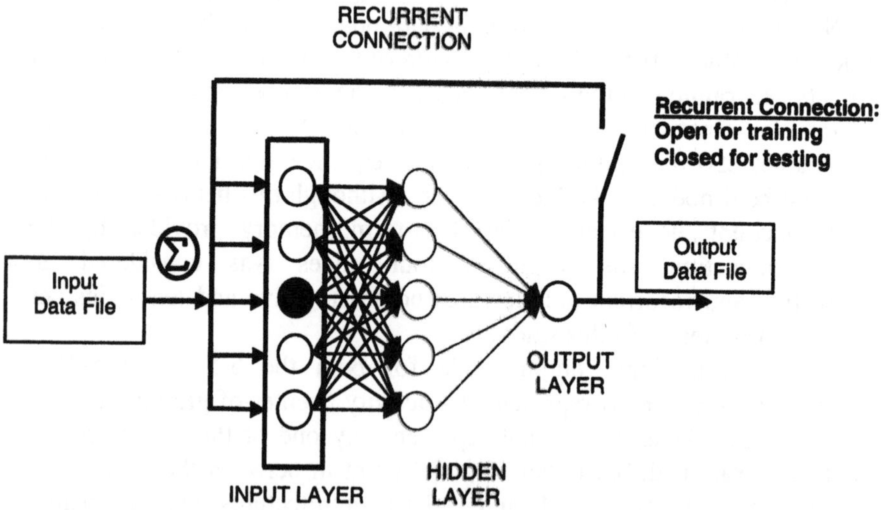


*Figure 1*: 5-5-1 Recurrent Neural Network Architecture: five input nodes (one central input signal and four closest-neighbor signals); five hidden-layer nodes, and one output node. The recurrent connection provides feedback, with a one-sample time delay, from the output node to the input layer. The output layer of a single node represents the degree of seizure-like activity for the five input nodes. Thus, it does not indicate the degree of seizure-like activity for any one of the input nodes – but rather the "center-of-gravity" for the five input nodes. This architecture is fully-connected and the recurrent connection feeds into all the nodes in the input layer. The 5-10-1 architecture is similar, but with 10 hidden nodes.

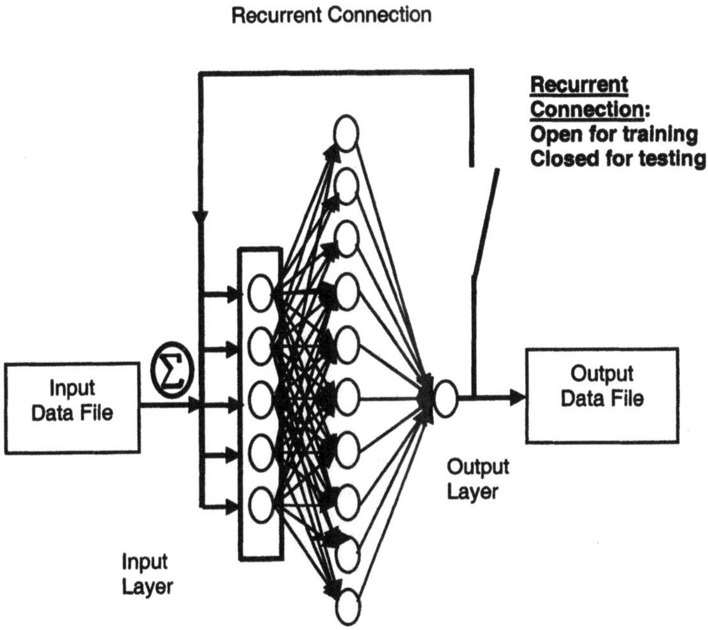


*Figure 2*: 5-10-1 Recurrent Neural Network Architecture: five input nodes (one central input signal and four closest-neighbor signals); ten hidden-layer nodes, and one output node. The recurrent connection provides feedback, with a one-sample time delay, from the output node to the input layer. The output layer of a single node represents the degree of seizure-like activity for the five input nodes. Thus, it does not indicate the degree of seizure-like activity for any one of the input nodes – but rather the "center-of-gravity" for the five input nodes. Like the 5-5-1 network, this architecture is fully-connected and the recurrent connection feeds into all the nodes in the input layer.

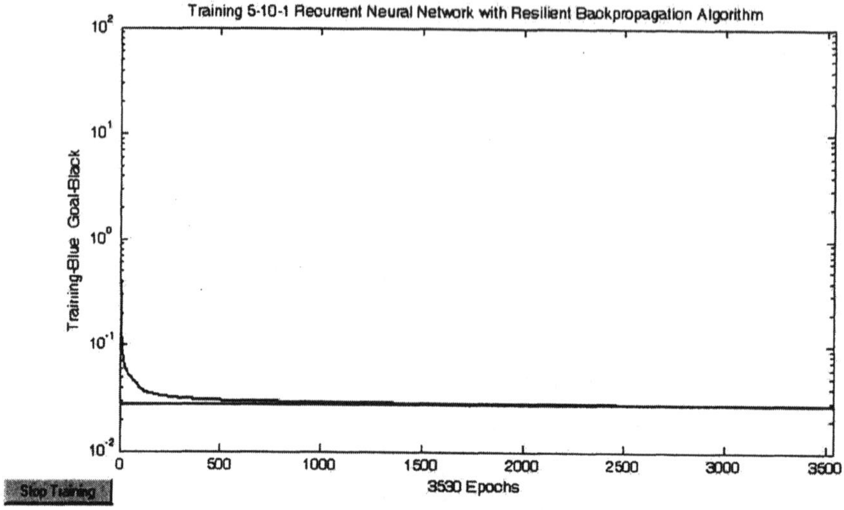


*Figure 3*: MATLAB™ plot of resilient backpropagation training for the 5-10-1 recurrent neural network. This network was trained on random sets of seizure and non-seizure data. Each set of data consisted of 250 samples of data obtained from confirmed intervals of seizure and non-seizure subdural EEG data. The seizure data was given a target value of 1; the non-seizure data was given a target value of 0. It can be noted that the training reached the training goal of 0.0275 within 3530 epochs without any encountering any local minima.

A training set for the neural network was constructed by selecting ten-thousand points from the SEEG data. The data points from the known seizure components of the data were given a corresponding target value of "1"; those data points from the known non-seizure component were given a target value of "0." A collection of training samples was then constructed using a uniformly distributed random-number generator in MATLAB™. During this process the non-seizure and seizure samples were mixed. This procedure was followed to ensure that the neural network did not receive any particular bias of pattern during the training.

Since the purpose of the neural network was to detect either "seizure activity" or "no seizure activity," the neural networks were trained with the recurrent connection open. Thus, each sample of 250 points, representing either seizure activity or non-seizure activity was presented randomly to the network and the architectures were allowed to learn specific patterns as seizure or non-seizure, but prevented from learning a sequence of patterns. In this way, the network was trained to be flexible in its recognition of sequences of samples.

## 2.4 Validation

During the validation stage of the design, these recurrent architectures were tested with EEG data containing confirmed seizure activity that were not part of the pool of EEG data to be tested for seizure activity.

After preliminary tests to determine which size network would still provide appropriate seizure discrimination, as well acceptable computational speed, the 5-5-1 and 5-10-1 architectures were determined to produce acceptable results.

## 2.5 Testing

After the network has been satisfactorily trained, the recurrent connection was connected and the network run with the recurrent connection closed .

The data to be analyzed was given to the network in the following manner: for five channels at a time, sequential blocks of 250 sample points (corresponding to 1 second of data) in each channel were presented to the network. After training had been completed, the network was used to analyze the EEG data sets. The data sets -- 94 channels, each with 15000 data points, for the SEEG data, were each run in their entirety before going on to the next data set. Since the input layer of the neural network had five nodes, five channels of data from each set were presented to the network at a time. Thus, channels 1 through 5 were run until all the data points had been analyzed by the neural network, then the next five channels, 6 through 10, were run, and so on, until all the channels had been analyzed by the network. Thus, for each data set, the total output of the neural network for the subdural data (e.g.), $Y$ , is represented by:

$$\bar{Y} = \sum_{1}^{94} \sum_{1}^{15000} y_{ij} \tag{3.2}$$

Thus, an array of 94 x 15000 was obtained for the subdural and the scalp data, respectively (in practice, the SEEG record contained 96 channels, the last two channels were EKG data and were not used in the data analyses. 95 channels were used for programming convenience). Each SEEG data set took roughly forty minutes to analyze.

## 2.6 Analysis and Display of Neural Network Output

The neural networks generated outputs over the interval [0,1], where output values close to 0 were considered to represent points with low seizure-like activity. Output values close to +1 were considered to represent regions of high seizure-like activity; values between 0 and +1 indicate the degree of seizure-like activity.

The [0,1] scaled values for each data set were then displayed as MATLAB™ contour plots, using an appropriate colormap function, to provide an optimal contrast for visualization. Detection of seizure-like activity was accomplished by visual inspection of these plots, which presented the neural network output as data points per channel (see figure 6).

Since the results of the 5-10-1 architecture proved the most accurate on test-data with expert confirmed seizure activity, further analysis of these results was performed. The standard-deviation of the neural network activity was then calculated for all channels at each time-point. A decrease in the standard-deviation of the neural network output at each seizure-like event was noted. Also, by spectrogram analysis, a large increase in the low-frequency components of the variation at the time of the seizure event in each sample was seen (figure 7). This suggests that, at the onset of seizure-like activity: 1) the magnitude of the signal standard-deviation across the subdural space is small; 2) the time-dependent change of the standard-deviation is small; 3) the signal power is concentrated in a small range of low-frequency components. These results suggest the entrainment of epileptogenic neural components during seizure onset.

## 3. RESULTS

In the analysis of the forty-nine sets of SEEG data, it was found that certain channels in a number of the data sets consistently showed relatively strong seizure-like activity -- above 0.5 on a scale of 0 to +1 despite which of the three architectures was used. The output values obtained from all three neural network architectures do not represent the level of seizure-like activity found at each channel – rather, each output value represents the level of seizure-like activity obtained for the group of five channels making up the nodes of the input layer. Thus, the value obtained at the output can be considered as the activity at a location which is the "center-of-gravity" of the five electrode locations corresponding to each group of the five channels.

It was found that the neural network identified strong seizure-like activity on two of the recording channels for almost all the samples tested. In the same samples, three other recording channels also showed relatively high seizure-like activity in many of the samples (see figure 7) These five recording channels, outlined below, are depicted on the subsequent montage in the shaded areas. A sample output from the 5-10-1 recurrent neural network, shown with its corresponding 94-channel sample of SEEG raw data is shown in figure 5.

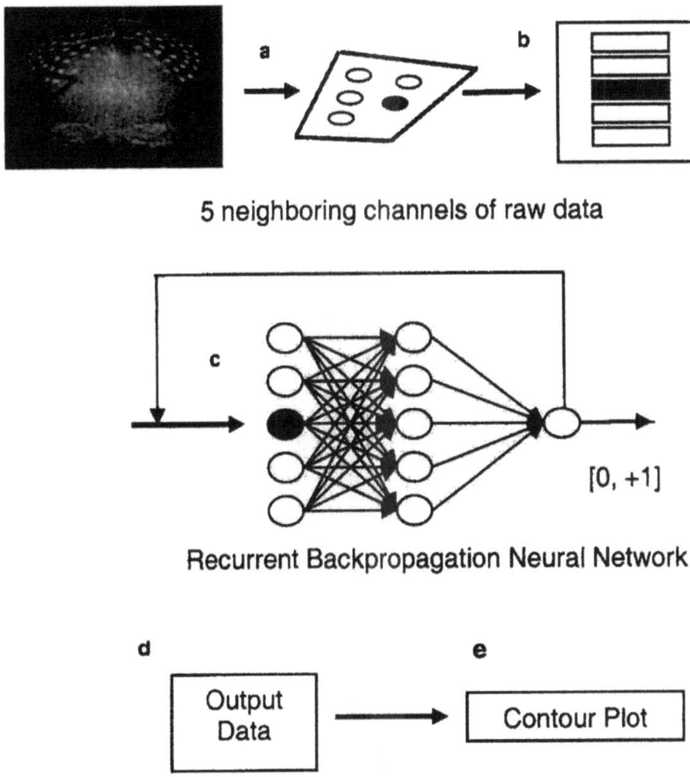


*Figure 4:* 5-5-1 Recurrent "Neural Network" Architecture in Testing Configuration: five neighboring channels of raw data are selected in sequence from the total data set (a). One point is selected as the central point; four neighboring points are included (b). The neural network is run in recurrent fashion during the testing phase (c). The output values range from 0 (no seizure-like activity) to +1 (high seizure-like activity) (d). These values are scaled from 0 (no seizure-like activity) to 1 (high seizure-like activity) for the convenience of contour plotting (f).

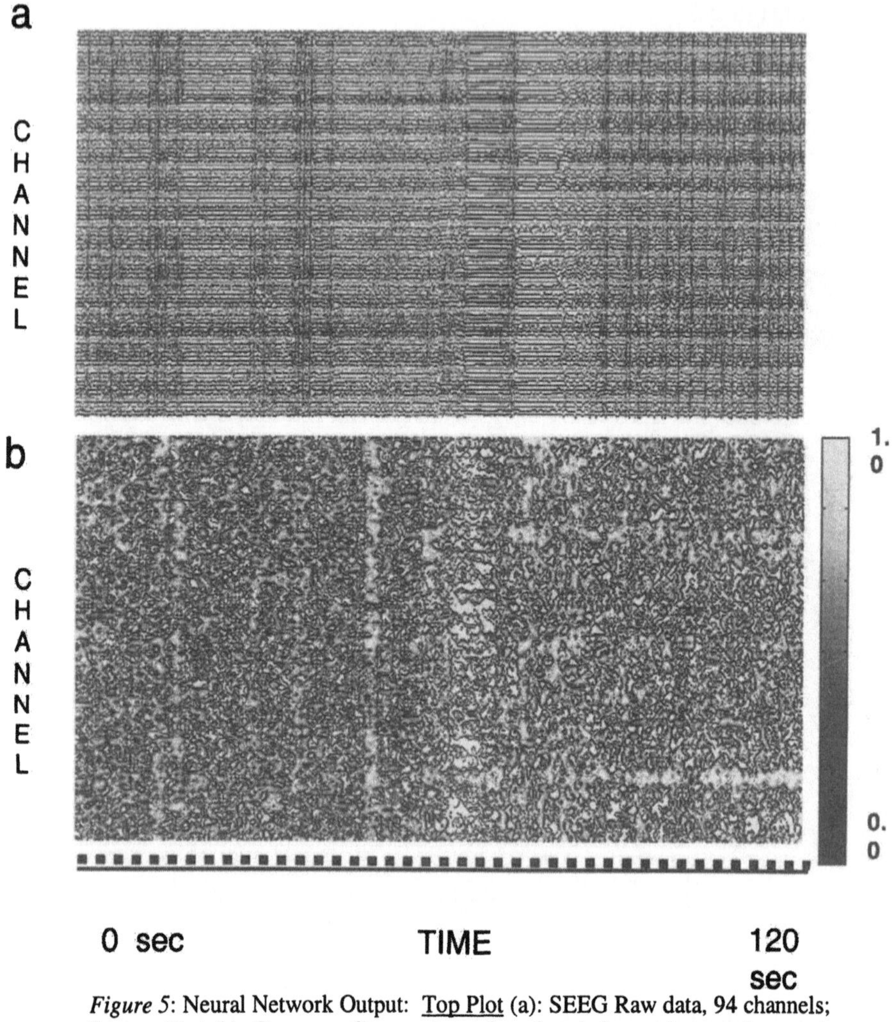


*Figure 5*: Neural Network Output: Top Plot (a): SEEG Raw data, 94 channels; Bottom Plot (b): Corresponding contour plot of scaled neural network output: 120 seconds total output from 5-10-1 neural network. High degree of seizure-like activity is seen 70-80 seconds into the output contour plot across all 94 channels.

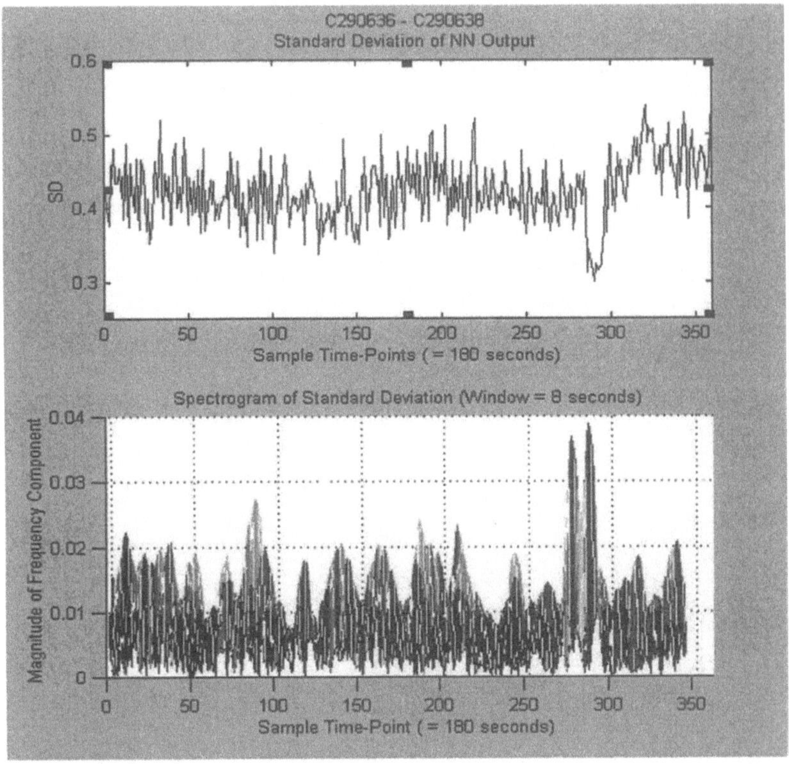


*Figure 6:* Statistical Analysis of Neural Network Output

<u>Top Plot</u>
Decrease in magnitude of the standard deviation of neural network activity around the onset seizure-like activity

<u>Bottom Plot</u>

Spectrogram of the standard deviation data seen in the top figure. It is seen that there was a large increase in the low-frequency components of the standard-deviation at the time of the seizure event in the original SEEG data and the decrease of the magnitude of the corresponding standard deviation.

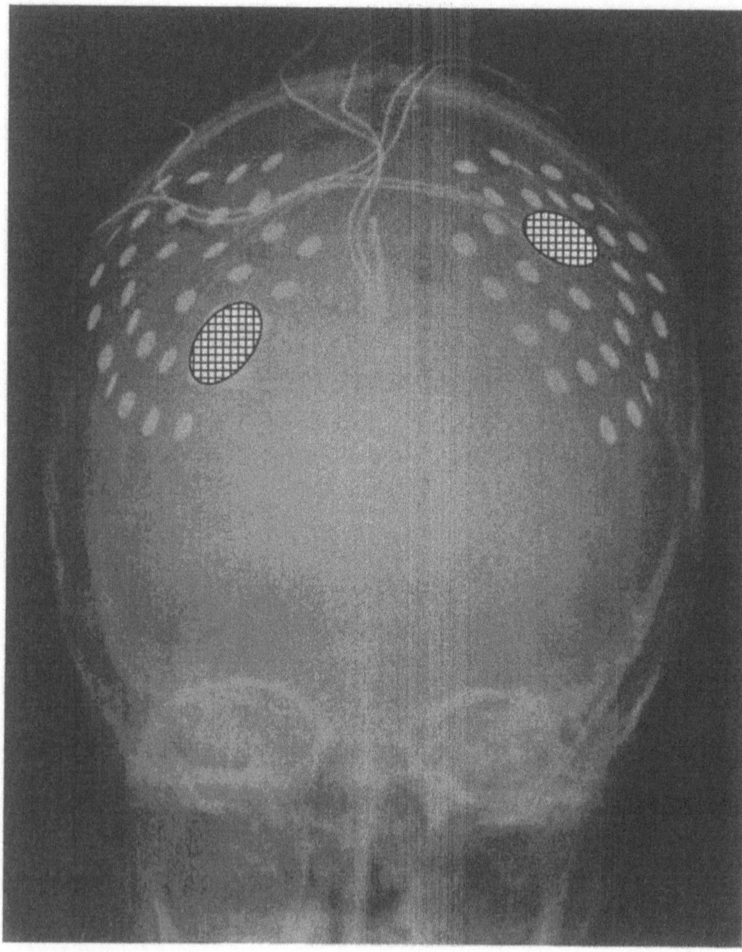

*Figure 7*: Subdural electrode montage showing the sites of the highest seizure-like activity detected by the neural network.

# 4. DISCUSSION

This work showed that a recurrent neural network is useful in detecting seizure-like activity in epileptic EEG records. The consistency of results obtained from a number of subdural EEG data sets verifies that the neural network accomplished what it was designed to do within the bounds of a number of limitations. These limitations include the nature of the neural network training and test data, the particular complexity of the data used, the restrictions associated with the type of neural network used, and the approximations inherent in the recording techniques.

The primary limitation to this study was the lack of a suitable number of patient samples. Data were obtained from a single individual with idiopathic epilepsy. The EEG data obtained from this patient were particularly complex, often showing seizure activity on all 94 recording channels. In addition, the focus (or foci) of epileptogenic activity was not known, nor could be it discovered from the data records.

Seizure onset patterns are highly variable throughout the patient population and the onset of seizure activity in one patient can be indistinguishable from non-seizure activity in another patient. Furthermore, the onset of seizure activity can involve small changes, undetectable by human expert or automatic means. Two or more types of seizures can occur repeatedly in the same patient.

Obviously, for a utility of the type proposed in this study to have clinical usefulness, it would need to be tested on a patient population of significant size. Furthermore, in order to be an effective clinical tool, the neural network would have to be able to recognize a wide variety of epileptic EEG's – not merely the complex types of the patient in this study. Also, the data obtained would have to be from an appropriate sample of the population, reflecting specific characteristics with respect to age, gender, diagnosis, etc., and who are accessible for study. Ideally, all inferences applied to the target population should apply to the broader population (*viz.* external validity).

A significant limitation of any computer-based analysis technique is how it performs with respect to problem size and complexity. There are considerable trade-offs between the complexity of the data under consideration and the characteristics of the neural network used to analyze them. In general, the quality of the results obtained from the neural network depends on: (1) the complexity of the data being analyzed, (2) the extent of noise in the training data resulting from recording techniques, and (3) the size and architecture of the network in relation to the size required for an optimal or clinically acceptable solution.

With respect to (1), EEG signals are difficult to analyze and the characterization of a certain series of epochs given by one expert may differ considerably than that obtained by another expert, especially in the designation of seizure activity.

Other general assumptions made regarding the unprocessed EEG data are reasonable approximations. In the EEG frequency range, the combination of signals received at the electrodes can be assumed to be linear. Also, during a seizure event, correlated activity in two components occurring in the same time-interval period is not attenuated by other independent activity, including baseline noise. Therefore, periods of correlated activity reflects the emergence of temporarily coupled sources that integrate synchronously active network.

Addressing the second limitation of neural network analysis of EEG, there are a number of approximations introduced into the results by the techniques used to obtain the EEG signals during clinical recordings.

However, certain trends can be taken into account when undertaking an iterative design of the network. In particular, adjustment of the network training rate is one parameter that is readily adjusted within the MATLAB code used to design the neural network. Since the recurrent architecture involves short-term memory, the lag-time in the response of the neural network to abruptly changing patterns (viz. from non-seizure-like to seizure-like) must be considered. For the architecture used in this study, one iteration of the network represented $1/250^{\text{th}}$ of a second (the sampling rate), so the total lag-time of this particular network was considered to be insignificant.

The training algorithm itself, the resilient back-propagation algorithm, is not typically used to train recurrent neural networks. The most widely-used gradient-based algorithms for training these type of networks are the backpropagation-through-time, recurrent backpropagation, and real-time recurrent learning algorithms. However, these algorithms tend to be affected by slow convergence. Most backpropagation algorithms are restricted by the potential problem resulting from the weight changes being a function of the gradient magnitude. Since the resilient backpropagation algorithm employs only the sign of the derivative to specify the direction of weight update, the potential problem arising from the increasing magnitude of the weight step partial derivative is avoided. This results in a typically quicker training time. Since the ultimate goal of this project is to provide a clinically useful utility, a relatively rapid training time is preferred.

## 5. CONCLUSIONS

It was shown that recurrent neural networks with relatively simple architectures are capable of detecting seizure-like activity in multi-channel EEG records. Furthermore, the results obtained from the three different architectures studied were fairly consistent in locating the generators of seizure-like activity based on the patient's subdural EEG montage. If these results are verified by other studies, then these neural networks may serve as effective functional engines in more sophisticated seizure detection utilities. Additional work to increase the detection ability of these networks needs to be undertaken, especially with regard to the separation of the EEG data recorded at each electrode. Also, a means by which a utility using these neural networks can be made more adaptive to a variety of seizures from a wide patient population needs to be examined.

## 6. ACKNOWLEDGEMENTS

This work was supported in part by National Institutes of Health Grants No. NS39047 and No. NS38494, and by Computational Diagnostics, Inc..

# REFERENCES

Niedermeyer, Ernst, and Da Silva, Fernando, Electroencephalography: Basic Principles, Clinical Applications, and Related Fields, Baltimore: Williams & Wilkins, 1993, p. 8-9, 228-235.

Ebersole, John, S., "Defining Epileptogenic Foci: Past, Present, and Future," J. Clin. Neurophys., 14(6) (1997), pp. 470-483.

Gutnick, M. J., Connors, R. W., and Prince, D.A., "Mechanisms of Neocortical Epileptogenesis in Vitro," J. Neurophys. 48:6 (December 1982), pp. 1321 – 1335.

Nuñez, P.L., Neocortical Dynamics and Human EEG Rhythms, New York, Oxford University Press, (1994).

Blum, D.E., "Computer-based Electroencephalography: Technical Basics, Basis for New Applications, and Potential Pitfalls," Electroenceph. and Clin. Neurophys. 106 (1998), pp. 118 - 126.

Gotman, J. "Automatic Recognition of Epileptic Seizures in the EEG," Electroenceph. and Clin. Neurophys. 54 (1982), pp. 530-540.

Gotman, J., "Automatic Seizure Detection: Improvements and Evaluation," Electroenceph. and Clin. Neurophys. 76 (1990), pp. 317-324

Gotman, J., and Gloor, P., "Automatic Recognition and Quantification of Interictal Epileptic Activity in the Human Scalp EEG," Electroenceph. and Clin. Neurophys., 41 (1976), pp. 513 – 529.

Haykin, S., Neural Networks, A Comprehensive Foundation, New York: Macmillan College Publishing Company Books (1994).

Elman, J.L., "Finding Structure in Time," Cognitive Science 14, (1990), pp. 179-211.

Casey, M., "How Discrete-Time Recurrent Neural Networks Work," Department of Mathematics 0112, University of California at San Diego, La Jolla, CA., Technical Report INC-9503, (April 11, 1995), pp. 1 – 41.

Petrosian, A., Prokhorov, D., Homan, R., Dashieff, and Wunsch, D., "Recurrent Neural Network Based Prediction of Epileptic Seizures in Intra- and Extracranial EEG," Neurocomputing 30 (2000), pp. 201 – 218.

Palus M., "Nonlinearity in Normal Human EEG: Cycles, Temporal Asymmetry, Nonstationarity and Randomness, Not Chaos," Biological Cybernetics. 75:5 (November 1996), pp. 389 – 396.

Cleermans, A., McClelland, J.L., and Servan-Schreiber, D., "Encoding Sequential Structure in Simple Recurrent Networks," School of Computer Science, Carnegie-Mellon University, Pittsburgh, PA, CMU-CS-88-183, (November, 1988).

Williams, R.J. and Zipser, D., "Experimental Analysis of the Real-time Recurrent Learning Algorithm", Connection Science, 1989 1:1, pp. 87-111.

Haykin, ibid., pp. 21-22.

Niedermeyer, Ernst, and Da Silva. ibid., pp. 108 – 111.

Haykin, ibid., pp. 265 – 266.

Niedermeyer, Ernst, and Da Silva, ibid., pp. 1066 – 1068.

Zhukov, Leonid, Weinstein, David, and Johnson, Chris, "Independent Component Analysis for EEG Source Localization," IEEE Eng. Med. Bio., 19:3, pp. 87 – 96.

Baum, E.B., Haussler, D., "What size net gives valid generalization?" Neural Computation, vol. 1, no. 1, pp. 151-160, 1989.

# Chapter 8

# MULTICRITERIA OPTIMIZATION UNDER PARAMETRIC UNCERTAINTY

**Luke E. K. Achenie, G.M. Ostrovsky**

## 1. INTRODUCTION

Often the performance of chemical processes cannot be estimated only by one objective function and it is necessary to take into account several conflicting criteria (Sophos et al., 1980), for example (a) process economics and environmental requirements, and (b) integration of process design and control. Therefore, multicriteria optimization (MCO) has evolved as an important problem in chemical process analysis and many other engineering disciplines (see for example, Keeney and Raiffa, 1976 and Caballero et al., 1997). MCO methods have been used for solving the process optimization. Luyben and Floudas (1994) used the multicriteria optimization (MCO) approach to combine economic and control objectives incorporating open-loop controllability measure. Sophos et al. (1980) considered multicriteria optimization within the petrochemical industry. Clark and Westerberg (1983) showed that the MCO problem can be reduced to a bi-level optimization problem. Palazoglu and Arkun (1987) considered MCO in the design of a robust chemical plant under uncertainty; they employed an economic objective function and dynamic operability measure as criteria. They also formulated a two-stage optimization problem in which the inner problem is a one-criterion optimization problem, using the $\varepsilon$-constraint method (Haimes et al., 1975). Using multicriteria optimization, Chakraborty and Linninger (2003) investigated the trade-off between expected cost and flexibility of plant-wide waste management.

MCO can be formulated as

$$\min_{x \in D_x}(f_1(x),\ldots,f_p(x)) \tag{1}$$

where $D_x = \{x : g(x) \leq 0\}$ be the feasible region of (1) and g(x) is of dimension m. Note that in general, the separate NLP's

$$
\begin{aligned}
f_i^* &= \min_x f_i(x) \\
g(x) &\le 0
\end{aligned}
\tag{2}
$$

have different minimizers $x_i^* (x_i^* \neq x_j^*, i \neq j)$. Therefore, the MCO problem as stated in (1) is not well defined unless we define what constitutes a solution.

## 2. PARETO SET

We will consider a space of dimension p, in which each coordinate axis corresponds to a separate criterion. For illustration, we will consider the case $p = 2$ (Figure 1). There is the concept of a utopia point $U(f_1^*,...f_p^*)$ such that each coordinate $f_i^*$ is a solution to a one criterion optimization (OCO) problem (9.2). There is also the concept of a Pareto Set (PS, or non-inferior set of points). Any point $\overline{f} = f(\overline{x})$ $((g(\overline{x}) \le 0)$ belongs to PS if in the small vicinity of $\overline{x}$ one cannot find a point $\overline{\overline{x}}$ $(g(\overline{\overline{x}}) \le 0)$, in which there is at least a j such that

$$
\begin{aligned}
\overline{\overline{f}}_j &\equiv f_j(\overline{\overline{x}}) < f_j(\overline{x}) \\
\overline{\overline{f}}_i &\equiv f_i(\overline{\overline{x}}) \le f_i(\overline{x}), i \neq j
\end{aligned}
\tag{3}
$$

This means that at any point in a PS, one cannot improve a criterion $f_i(x)$ without making another criterion $f_j(x)$ $(j \neq i)$ worse.

Consider geometric interpretation of PS. For this we consider the mapping of x-space to f-space through the relations

$$
\begin{aligned}
& f_i = f_i(x), \quad i = 1,...p \\
& \forall x \in D_x.
\end{aligned}
\tag{4}
$$

Let $D_f$ be the region in the f-space to which $D_x$ is mapped; see Figures 1 and 2 for the case p=2. For each criterion $f_i(x)$ there is the concept of a lower boundary surface determined as

$$\min_x f_l(x)$$
$$f_i(x) = c_i, i \in \{1,..., p\}\ i \neq l$$
$$g_j(x) \leq 0, \qquad j = 1,...,m$$
$$c_i \in D_f$$

for different sets of $\{c_i\}$. For $D_f$ represented as Figure. 2, the lower

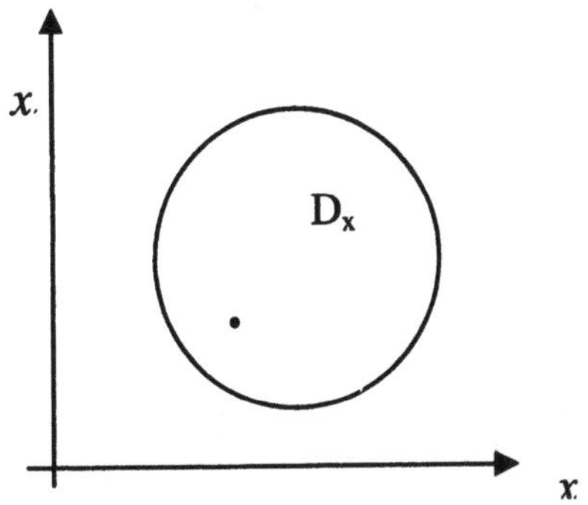


*Figure 1* *x*-space

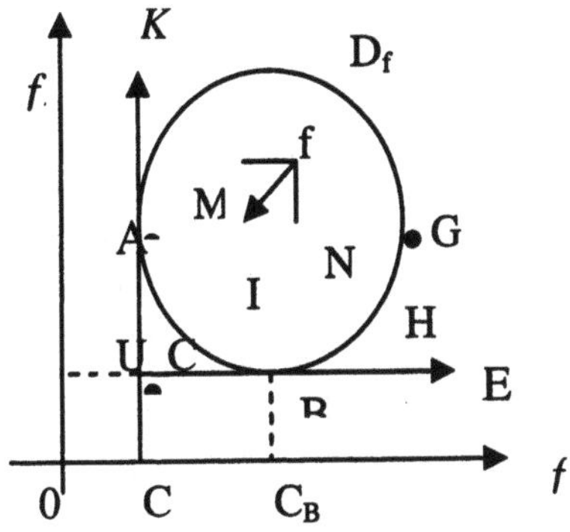


*Figure 2* Criteria-space

boundary surface of $f_2(x)$ is the curve ABG. One can show that if a point f belongs to PS then it belongs to the lower boundary curve. Indeed if a point f is an interior point of $D_f$ (see Figure 2) then in a vicinity of f we can always find a point, at which all criteria take better (i.e. lower) values. In order to find such a point we must move the point f in any direction inside the angle MfN (for example along the direction I). Consider the PS for the region $D_f$ represented on Figure 2. Let U be a utopia point. Draw the straight line UE parallel to the abscissa. Similarly draw UK parallel to the ordinate.

Let the points B and A be the tangent points of the straight lines UE and UK with $D_f$, respectively. It is easy to show that the following inequalities hold

$$f_2(B) \leq f_2(f) \quad \forall f \in D_f$$
$$f_1(A) \leq f_1(f) \quad \forall f \in D_f.$$

The point B ( $f_1^B, f_2^B$ ) is obtained by solving

$$\min_x f_2(x)$$
$$g(x) \leq 0.$$

Then $f_1^B = f_1(x_B), f_2^B = f_2(x_B)$ where $x_B = \arg\min\{f_2(x)/g(x) \leq 0\}$
On the other hand point A is obtained by solving

$$\min_x f_1(x)$$
$$g(x) \leq 0.$$

Here $f_1^A = f_1(x_A), f_2^A = f_2(x_A)$ where $x_A = \arg\min\{f_1(x)/g(x) \leq 0\}$

The lower boundary curve in Figure 9.2 is the curve ABG. It is clear that the arc BG does not belong to the PS since at each point of the arc we can improve simultaneously both criteria $f_1$ and $f_2$. In addition, PS coincides with the arc ACB of the lower boundary curve. Indeed, take a point $f_1^C, f_2^C$ belonging to the arc ACB of the lower boundary curve. At this point, (3) cannot be satisfied. This means that if we decrease $f_1$ then $f_2$ increases and vice versa. Therefore, all the points of the arc ACB represent the PS.

## 3. SOLUTION STRATEGIES

We consider some approaches, which reduce the multi-criteria optimization problem to one criterion optimization problem. One of such approach consists of construction of some convolution of original criteria $f_1(x),..., f_p(x)$ (i=1,...,p).

### 3.1. Convolution of Criteria

We consider here the following two methods: an average criterion and the worst-case strategy method.

## Minimization of Average Criterion

Very often, some weighted average of the multiple criteria is employed. Specifically each criterion $f_i(x)$ is assigned a weight coefficient $a_i$, reflecting its importance relative to the other criteria. The resulting problem is given by

$$f^{1*} = \min_x f^1(x,a)$$
$$g(x) \leq 0 \tag{5}$$

where

$$f^1(x,a) = \sum_{i=1}^{p} a_i f_i(x)$$
$$\sum_{i=1}^{p} a_i = 1$$
$$a_i \geq 0. \tag{6}$$

We need to show that a solution $[\bar{x}, \bar{f}_1 ..., \bar{f}_p]$ of (5) belongs to the PS. Suppose instead that $\bar{f} \notin PS$. This implies that in a vicinity of $\bar{x}$ there is a point $\bar{\bar{x}}$ for which condition (3) is met. However, for such a point we have

$$\bar{\bar{f}} = \sum_{i=1}^{p} a_i \bar{\bar{f}}_i < \sum_{i=1}^{p} a_i \bar{f}_i$$

Thus, we have obtained the point, for which the value of the objective function is less than $\bar{f}$. However, this contradicts the fact that $\bar{f}$ is the minimum of (5). Changing the values of the weight coefficient $a_i$ and solving problem (5) will generate different points of the PS. Thus a fixed value of $[a_i$, d] corresponds to each point in the PS. One can show that the method can obtain all the points of the PS if $D_f$ is a convex region. On the other hand if the region is not convex then there are points in the PS, which cannot be obtained by the technique. Consider for example the Pareto set AREQB in Figure. 3. Let the straight line KL be tangent to the PS at the

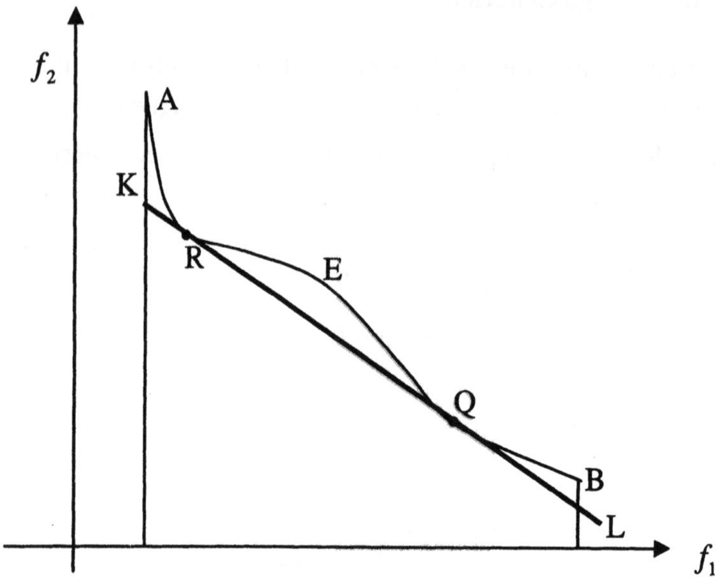


*Figure.3* Pareto set for a non-convex $D_f$

points R and Q. One can show that the method cannot determine the points, belonging to the section REQ. This is similar to the duality gap when using the Lagrange multiplier method.

**Worst Case Strategy**

In the worst case strategy (Clark and Westerberg, 1983), as in Section 9.3.1, each criterion $f_i(z)$ is assigned a weight coefficient $a_i$, reflecting its importance relative to the other criteria. The key difference is that the worst weighted criterion is minimized. The resulting problem is given by

$$f^{2*}(a) = \min_x f^2(x,a) \tag{7}$$

$$g_j(x) \le 0, j = 1,...,m$$

where

$$f^2(x,a) = \max_{j \in J}(a_j f_j(x)) \;, \quad \sum_{j=1}^{p} a_j = 1, \; a_j \ge 0, \; J = (1,..., p)$$

Define y to be an auxiliary variable; then according to Theorem A.4, problem (7) can be reformulated as

$$\min_{x,y} y$$
$$\max_{i \in I} a_i f_i(x) \le y$$
$$g_j(x) \le 0,\ j = 1,\dots,m$$

where I=(1,…,p). Furthermore, using equivalent relations (A7), we can reduce the problem to

$$\min_{x,y} y$$
$$a_i f_i(x) \le y \quad i = 1,\dots,p$$
$$g(x) \le 0 \tag{8}$$

Let $[\bar{y}, \bar{x}, \bar{f}_j]$ $(\bar{f}_j = f_j(\bar{x}))$ be the single global solution to (9.8); that is the following conditions hold

$$\text{for } x \in D_x\ x \ne \bar{x}\ \ y > \bar{y}\quad,\ a_i f_i(x) \le y,\ i = 1,\dots,p \tag{9}$$

One of the constraints in (8) will be active and

$$\bar{y} = \max_{i \in J} a_i \bar{f}_i$$

We next show that the point $\bar{f} = (\bar{f}_1,\dots,\bar{f}_p)$ belongs to PS. Suppose instead that $\bar{f} \notin PS$. Then there exists a point $\bar{\bar{f}} = (\bar{\bar{f}}_1,\dots,\bar{\bar{f}}_p)$ in the vicinity of $\bar{f}$ for which at least one $\bar{\bar{f}}_i$ is strictly less than $\bar{f}_i$; let it be $\bar{\bar{f}}_1$, thus

$$\bar{\bar{f}}_1 < \bar{f}_1,\ \bar{\bar{f}}_i = \bar{f}_i, i \ne 1 \tag{10}$$

Let $\bar{\bar{y}} = \max_i a_i \bar{\bar{f}}_i$. It follows from (10) that

$$\bar{\bar{y}} = \max_i a_i \bar{\bar{f}}_i \le \max_i a_i \bar{f} = \bar{y}$$

Consequently, we have found the point, at which all constraints in (8) are met and the objective function $\bar{\bar{y}}$ is less than or equal to $\bar{y}$. However, this contradicts (9). Consequently, the point $\bar{f}$ belongs to PS. Solving (8) for different values of $a_i$ one can obtain different points in the PS. One can show that the method permits to obtain all the points of the PS. Note however that for $p > 3$ the operation will be computationally intensive.

## 3.2. ε-Constraint Method

The idea is to minimize one of the performance criteria, while requiring the remaining criteria to satisfy some target values. For at least two criteria, we solve the following problem (Haimes, 1975)

$$\min_x f_p(x) \tag{11}$$

$$\begin{aligned} &f_i(x) \le \varepsilon_i \quad i = 1,2,\ldots,(p-1) \\ &g_j(x) \le 0, \qquad j = 1,\ldots,m \end{aligned} \tag{12}$$

Here the values $\varepsilon_j$ are arbitrary values. A solution of the problem belongs to PS if all constraints (12) are active (Sophos et al., 1980). In general, solving problem (11) for different values of the parameter $\varepsilon_i$ determines different points in the PS. The method permits to obtain all the points of the PS (Clark and Westerberg, 1983).

## 3.3. Method of Consecutive Conciliations

In this approach, the performance criteria are arranged in order of importance, with $f_1(x)$ being the most important. Then the problem to be solved is

$$\begin{aligned} &\min_x f_1(x) \\ &g(x) \le 0 \end{aligned} \tag{13}$$

Assuming $[x^{(1)}, f_1^{(1)}]$ $(f_1^{(1)} = f_1(x^{(1)})$ is the global solution of the problem. For a given set of scalars $\varepsilon_i > 0 \; \{i = 1, ..., (p-1)\}$, and for $k = 2, ..., (p-1)$ we solve

$$\min_x f_k(x)$$

$$g(x) \leq 0 \qquad (14)$$

$$f_i(x) \leq f_i^{(i)} + \varepsilon_i, \quad i = 1, ..., (k-1) \qquad (15)$$

Let $[x^{(k)}, f_k^{(k)}]$ $(f_k^{(k)} = f_k(x^{(k)}))$ be the solution of the problem. The value $\varepsilon_i$ is the allowable deterioration of the optimal value $f_i^{(i)}$ of the i-th criterion $f_i(x)$. Consider problem (14) for k=2.

$$f_2^{(2)} = \min_x f_2(x)$$

$$g_j(x) \leq 0, \qquad j = 1, ..., m \qquad (16)$$

$$f_1(x) \leq f_1^{(1)} + \varepsilon_1 \qquad (17)$$

The inequality (17) allows a violation $\varepsilon_1$ of $f_1^{(1)}$.

Now we will give a geometric interpretation of the method for the case. The point A (Figure 4) corresponds to the solution of problem (13). After that we must solve (16). Condition (17) determines the region between $f_1 = f_1^*$ and $f_1 = f_1^{(1)} + \varepsilon_1$ in which the solution of (16) must be found (Figure 4). It is easy to see that the point C at the intersection of $f_1 = f_1^{(1)} + \varepsilon_1$ and the Pareto set will be the solution to the problem. One can show that the method permits to obtain all the points of PS by solving (16) for different values of $\varepsilon_1$. In this case the method of consecutive conciliations is identical to the ε-Constrained Method.

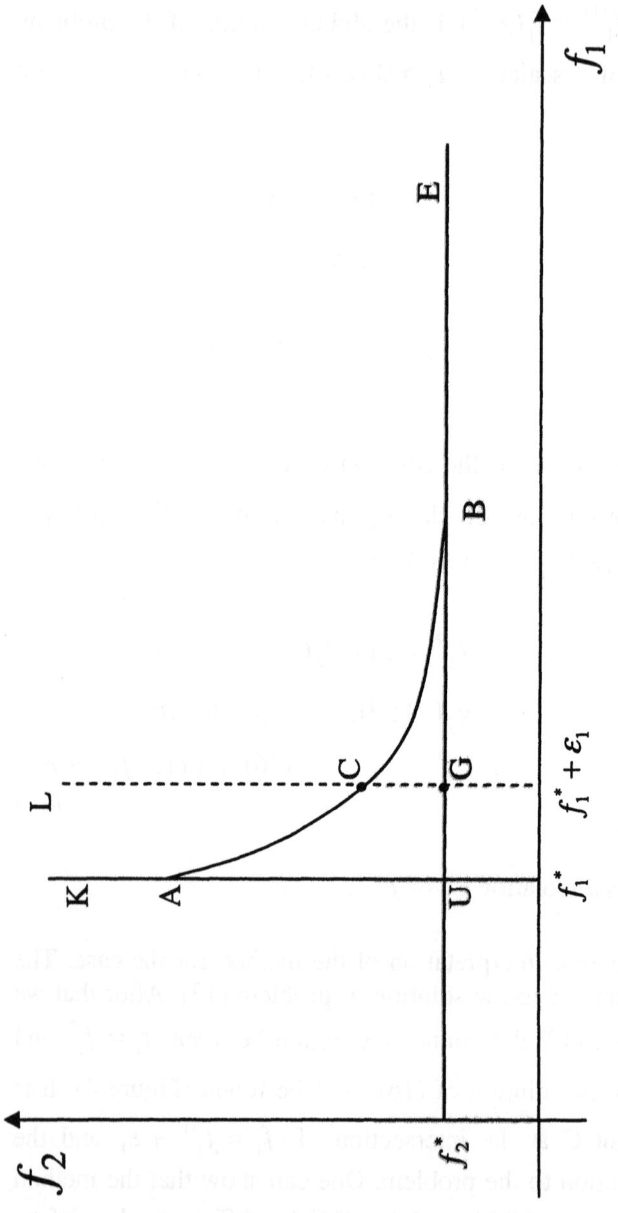


*Figure 4* Method of consecutive conciliations: geometric interpretation for $p=2$

# 4. USING THE PARETO SET FOR DECISION MAKING

It is clear from description of the methods of MCO that we can obtain only point-wise representation of the PS (as a multidimensional table). Suppose a PS has been found. The decision maker can use this information in one of two ways: (a) Using engineering consideration, select one of the points in PS as a final solution of the MCO problem; (b) Formulate a new criteria $F(f_1,\ldots,f_p)$, which account for the relative significance of the individual criteria. In case (b) we look for the best point in the PS using the criterion $F(f_1,\ldots,f_p)$. Therefore, solving MCO problem is reduced to solving the problem

$$\min_x F(f_1(x),\ldots,f_p(x)) \tag{18}$$

$$G(f_1(x),\ldots,f_p(x)) = 0 \tag{19}$$

where (19) is the surface of the PS in criteria space. Problem (18) must determine the best value of weight coefficient $a_i$ in (5) or (7). Since we construct the PS point-wise (as a multidimensional table) we do not have an explicit expression for $G(f_1,\ldots,f_p)$.

Consider the particular case

$$F(f_1,\ldots,f_p) = \sum_{i=1}^{p}(f_i^* - f_i(x))^2$$

where $f_i^*$ is the solution of (2)

By solving (18) we will find the point in PS nearest to the "utopia" point. For p=2 it will be the point at which the level curve

$$\sum_{i=1}^{p}(f_i^* - f_i(x))^2 = a$$

is tangent to PS (Figure 5).

It is clear that for p>3 both approaches are very computationally intensive. In connection with this we consider solving the MCO problem using a bi-level optimization method (Clark and Westerberg, 1983). For the vectors $x_1$ and $x_2$, let us consider the bi-level optimization problem.

$$\min_{x_1} \Phi_1(x_1, x_2)$$
$$g_1(x_1, x_2) \le 0$$
$$h_1(x_1, x_2) = 0$$
(20)

$$x_2 = \arg \min_{w} \Phi_2(x_1, w)$$
$$g_2(x_1, w) \le 0$$
$$h_2(x_1, w) = 0$$
(21)

Problem (20) is an outer optimization problem, while (21) is an inner optimization problem. Let us stress that the search variables of the inner optimization problem differs from those of the outer optimization.

If the points of the surface G are determined by solving (8) then (20) can be represented by the following bi-level optimization problem

$$\min_{a} F[f_1(x), \dots, f_p(x)]$$
(22)

$$a_i \geq 0, \quad \sum_{i=1}^{p} a_i = 1$$

$$x = \arg\min_{y, w_i} y$$

$$a_j f_j(w) \leq y \quad j = 1, \ldots, p$$

where a is a vector $a = (a_1, \ldots, a_p)$. In fact if we use the designation

$$\Phi_1 \equiv F[f_1(z), \ldots, f_p(z)], \ a \equiv x_1, \ \Phi_2 \equiv y, \ x \equiv x_2$$

then we will obtain the bi-level problem (20). This is the full mathematical formulation of MCO problem. However, one must remember that the decision

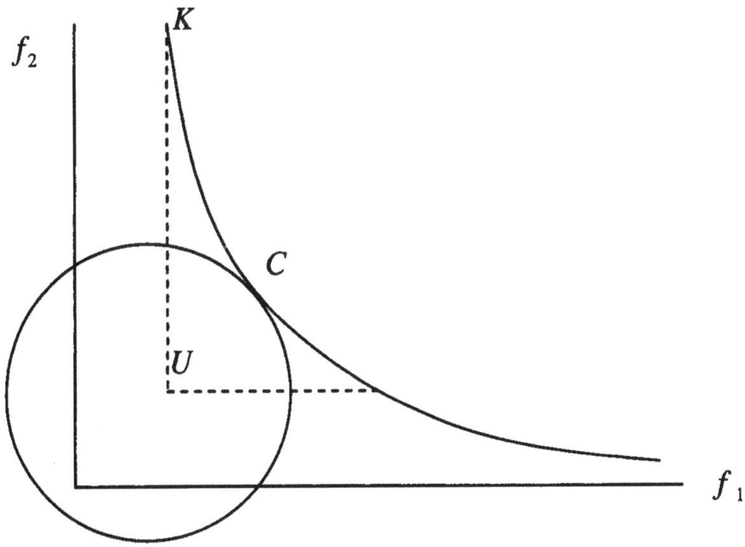


*Figure 5* Point in PS nearest to the "utopia" point for $p$=2

maker must provide the form of the function F, which determines the importance of each separate criterion. Similarly, one can determine the equivalent bi-level optimization problems for the "minimization of an average criterion" and "ε-constraint" methods for determination of the PS.

Let us consider the approaches to solving bi-level optimization problems. A direct strategy for solving the problem consists in solving inner optimization problem for each search point of the outer optimization algorithm. This simple minded approach causes several difficulties. To see this, consider the following example (Clark and Westerberg, 1983)

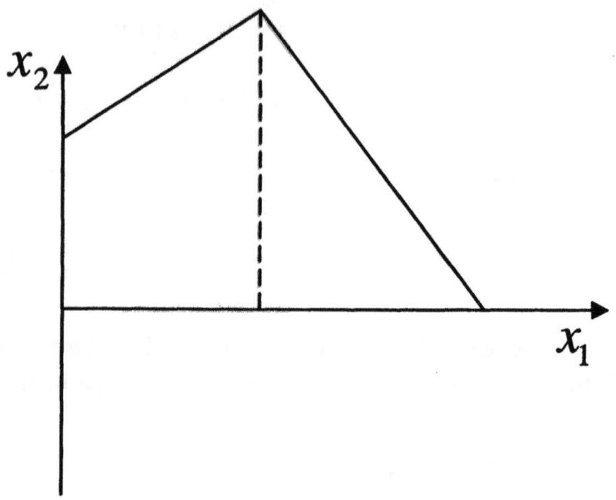


*Figure 6* Dependence of $x_2$ on $x_1$

$$\min_{x_1} (x_1 - 2) + (x_2 - 5)^2$$
$$0 \leq x_1 \leq 5$$
$$x_2 = \arg \min_{y} (x_1 + 1)^2 + (y - 5)^2$$
$$-3x_1 + 2y \leq 2$$
$$x_1 + 3y \leq 14.$$

We can obtain an explicit expression for $x_2$ with respect to $x_1$ (Figure 6) as

$$x_2(x_1) = \begin{cases} 1.5x_1 + 1, \text{ if } x_1 \leq 2 \\ -0.33x_1 + 4.47, \text{ if } x_1 \geq 2. \end{cases}$$

Subsequently we obtain the bi-level optimization problem as

$$\begin{aligned}&\min_{x_1}(x_1+x_2)^2+(x_2-5)^2\\&0\le x_1\le 5\\&x_2-x_2(x_1)=0.\end{aligned} \tag{23}$$

At $x_1=2$ the function $x_2(x_1)$ has no derivative resulting in a non-differentiable problem. Since in problem (23) there is a nonlinear equality, in general the problem can be multi-extremal.

A better approach to solving the bi-level optimization problem is as follows. Use the Karush-Kuhn-Tucker necessary conditions for the inner optimization problem and solve the problem

$$\begin{aligned}&\min_{x_1}\Phi_1(x_1,x_2)\\&g_1(x_1,x_2)\le 0\\&h_1(x_1,x_2)=0\\&g_2(x_1,x_2)\le 0\\&h_2(x_1,x_2)=0\\&\nabla_{x_2}L(x_1,x_2)=0\\&\mu_j\ge 0\quad j=1,\dots,n_{g_2}\\&\mu_j g_{2,j}=0\quad j=1,\dots,n_{g_2}\end{aligned}$$

Here $L=\Phi_2(x_1,x_2)+\lambda^T h_2+\mu^T g_2$ , $g_{2,j}$ is the $j$ - th component of the vector $g_2$ and $n_{g_2}$ is dimensionality of the vector-function $g_2(x)$ .In this case we obtain a differentiable optimization problem. However, we need second order information (i.e. Hessian involving both the objective and constraints) if we decide to employ SQP for the solution. In addition, multi-extremality of the problem remains.

# 5. MULTI-CRITERIA OPTIMIZATION UNDER UNCERTAINTY

## 5.1. Formulation of the Problem

The multi-criteria optimization problem under parametric uncertainty can be posed as

$$\min_{d,z}(f_1(d,z,\theta),\ldots,f_p(d,z,\theta))$$
$$g(d,z,\theta)\le 0$$

Here we give extensions of some methods for solving MCO problems considered in Section 9.3. We will consider one-step and two-step optimization formulations of MCO problems.

## 5.2. One-Step MCO Problem

The one-step MCO problem can easily be reduced to the previous case. Consider the case when we choose to minimize the mean value of the criteria while satisfying some mean values of the constraints. Suppose the mean values are given by the expected values

$$\begin{aligned}\bar{f}_i(d,z) &= E_\theta\{f_i(d,z,\theta)\} = \int_T f_i(d,z,\theta)p(\theta)d\theta \\ \bar{g}_j(d,z) &= E_\theta\{g_j(d,z,\theta)\} = \int_T g_j(d,z,\theta)\rho(\theta)d\theta.\end{aligned} \tag{24}$$

In this case we must solve the problem

$$\min_{d,z}(\bar{f}_1(d,z),\ldots,\bar{f}_p(d,z))$$
$$\bar{g}(d,z)\le 0.$$

This problem has the form (1) and therefore for solving the MCO problem we can use the concept of a Pareto set and the MCO method considered in Section 3. Note however that the MCO becomes considerably complicated because of the need to calculate the multi-dimensional integral (24) for each d and z.

## 5.3 Two-Stage MCO Problem

In the two-stage MCO problem, the complexity consists in taking into account the different characteristics of the design and control variables. Here we will consider extensions of the average criterion (AC) method, the worst-case strategy (WCS) method and the method of consecutive conciliations for solving MCO problem. All the methods will take into account the ability to tune the control variables.

We will use the following general approach for the extension of the AC method and the WCS method. First, we will transform each of the criteria $f_i(d,z,\theta)$ to a new function $\bar{f}_i(d)$, which will depend only on design variables. With $\bar{f}_i(d)$ (i=1,...,p) we will be able to use the AC method or the WCS method for solving MCO problems under uncertainty. First consider the possibility of using the transformation used in the construction of the objective function in TSOP1. In this case, using (5.47) we will transform each criterion $f_i(d,z,\theta)$, $(i=1,...,p)$ to a new function

$$\bar{f}_i(d) = \int_T f_i^*(d,\theta)\rho(\theta)d\theta \tag{25}$$

where

$$f_i^*(d,\theta) = \min_z f_i(d,z,\theta)$$
$$g(d,z,\theta) \leq 0. \tag{26}$$

Each $\bar{f}_i(d)$ depends only on design variables. Let $z^{i*}(d,\theta)$ be the solution of (9.26). Note that each $\bar{f}_i(d)$ has its own internal optimization problem (26). Therefore, each $z^{i*}(d,\theta)$, $(i=1,...,p)$ is different. Using $\bar{f}_i(d)$ we can construct the PS with the help of the AC method or the WCS method. Suppose we construct the PS and the decision maker selects a point in the PS. Let $\overline{f_i}(\overline{d})$, $(i=1,...,p)$ correspond to the point. However, each $\bar{f}_i(d)$ depends on its own control variables, which cannot be realized simultaneously. Consequently, we cannot implement the results and the approach cannot be used for solving the MCO problem under uncertainty. As such to solve the MCO, we will consider another approach in which a single internal optimization problem will be used for construction of all $\bar{f}_i(\bar{d}),(i=1,...,p)$.

Consider the following optimization problem

$$\min_z F(f_1,...,f_p,a)$$
$$g(d,z,\theta)\leq 0. \tag{27}$$

Here $F(f_1,...,f_p,a)$ is a convolution of p criteria $f_1,...,f_p$, which is constructed using the average criterion method or the worst-case strategy method. Here a is a vector of parameters (see (5) and (7)). We will suppose that the convexity condition is met; this permits to find all points of PS if the average criterion method is used. Let us construct $\bar{f}_i(d,a)$, which will employ the optimal solution $z^*(d,\theta,a)$ from (27) as follows

$$\bar{f}_i(d,a)=\int_T f_i(d,z^*(d,\theta,a),\theta)\rho(\theta)d\theta. \tag{28}$$

The function $\bar{f}_i(d,a)$ (which is independent of $z$ ) is a mean value of the original criterion $f_i(d,z,\theta)$ at the operation stage since for each $\theta$ problem (27) is solved as an internal optimization problem. Again we can use the same AC or the WCS method for the construction of a convolution of $\bar{f}_i(d,a)$, $(i=1,...,p)$. Here we will use the same parameters a, used in the construction of the convolution $F(f_1,...,f_p,a)$. Designate the solution of the AC or the WCS problems (when using the functions $\bar{f}_i(d,a)$) as $[d^*,\bar{f}_i^*]$ $(\bar{f}_i^*=\bar{f}_i(d^*,a))$. If we solve the AC or the WCS problems with the functions $\bar{f}_i(d,a)$ $(i=1,...,p)$ for all values of parameters $a$ satisfying (6), we will construct some curve (surface) in the space of the $\bar{f}_i$, $(i=1,...,p)$. This curve is an analog of the usual PS in the sense that the decision maker (DM) must make final decision using the curve. From engineering consideration he must select a point $[\bar{d},\bar{a}]$ from this curve as the solution of the MCO problem. We will refer to the curve as the DM curve.

Let us analyze the obtained result. For each $\theta^l$ the control variables are obtained by solving (9.27) where $a = \bar{a}$ and $d = \bar{d}$ using the average criterion method or the worst-case strategy (i.e. we solve a conventional MCO problem). Thus, the found values of the variables z correspond to one of the points on the Pareto set for the functions $f_i(\bar{d}, z, \theta^l)$. Now consider the values $\bar{f}_i(\bar{d}, \bar{a})$, $(i = 1,\dots, p)$. These are obtained by solving (5) or (7) where the functions $\bar{f}_i(d, \bar{a})$ are used. Again we obtain a solution, which corresponds to one of the points of the conventional PS for $\bar{f}_i(d, \bar{a})$. Thus for the functions $\bar{f}_i(d, \bar{a})$ one cannot find a better MCO solution than $\bar{f}_i(\bar{d}, \bar{a})$. It is clear that the solution can be realized, if at each time instance the internal optimization problem (27) is solved since the same $z^*(d, \theta, \bar{a})$ is used for construction of $z^*(d, \theta, a)$.

## Minimization of Average Criteria

Formulate internal optimization problem (9.27) using as the objective function the weighted sum (9.5)

$$\begin{aligned} &f^{1*}(d, \theta, a) = \min_z f^1(d, z, \theta, a) \\ &g(d, z, \theta) \le 0 \end{aligned} \tag{29}$$

where

$$\begin{aligned} &f^1(d, z, \theta, a) = \sum_{k=1}^{p} a_k f_k(d, z, \theta) \\ &\sum_{k=1}^{p} a_k = 1 \\ &a_k \ge 0. \end{aligned}$$

Let $z^*(d, \theta, a)$ be the solution to the problem. Then $\bar{f}_i(d, a)$ has the form (28).

The new criteria $\bar{f}_i(d, a)$ $(i = 1,\dots, p)$ do not depend on the control variables z. Now we can directly use the method of minimization of the weighted average criterion

$$\min_d \bar{f}(d,a) \tag{30}$$

where

$$\bar{f}(d,a) = \sum_{k=1}^{p} a_k \bar{f}_k(d,a). \tag{31}$$

This is a bi-level optimization problem since for calculation of $\bar{f}_i(d,a)$ we must use $z^*(d,\theta,a)$, which is the solution of (28). We saw in a previous section that it is very computationally intensive, requiring the use of global, nondifferentiable optimization methods. To make matters worse, during the calculation of the objective function of (30), we must calculate p multidimensional integrals at each value of d. In connection with this we reduce the problem to a simpler problem as follows. Substitute in $\bar{f}(d,a)$ expressions for $\bar{f}_i(d,a)$ from (28) to obtain

$$\bar{f}(d,a) = \sum_{k=1}^{p} a_k E\{f_k(d,z^*(d,\theta,a),\theta)\} = \sum_{k=1}^{p} a_k \int_T f_k(d,z^*(d,\theta,a),\theta)\rho(\theta)d\theta$$

This is equivalent to

$$\bar{f}(d,a) = \int_T [\sum_{k=1}^{p} a_k f_k(d,z^*(d,\theta,a),\theta)]\rho(\theta)d\theta. \tag{32}$$

The term in the square brackets is the optimal value of the objective function of the internal optimization problem (29). Therefore, we can rewrite (32) as

$$\bar{f}(d,a) = \int_T \min_z (\sum_{k=1}^{p} a_k f_k(d,z,\theta) / g(d,z,\theta) \le 0)\rho(\theta)d\theta.$$

Since for a given θ the optimal value at z does not depend on the values of z for other $\theta$ , we can rewrite the above as

$$\bar{f}(d,a) = \min_{z(\theta)} \int_T (\sum_{k=1}^{p} \alpha_k f_k(d,z,\theta) / g(d,z,\theta) \le 0)) \rho(\theta) d\theta \, .$$

Here z(θ) is a multivariable function with respect to the uncertain parameters θ. Substitute the expression for $\bar{f}(d,a)$ in problem (30) to obtain

$$\min_{d,z(\theta)} \int_T (\sum_{k=1}^{p} a_k f_k(d,z,\theta)) \rho(\theta) d\theta \tag{33}$$

$$g(d,z(\theta),\theta) \le 0 \;\; \forall \theta \in T \, . \tag{34}$$

From condition (34) it follows that

$$\chi_1(d) \le 0 \, . \tag{35}$$

The system (33) to (35) has the form (5.51). Therefore, we can use the SB method for solving the problem. It is interesting to note that during the search for each d we must calculate only one multidimensional integral. Using $\bar{f}_i(d,a)$ one can construct DM curve.

Suppose the decision maker selects the point $[\bar{d},\bar{a}]$ from the DM curve as the solution of MCO problem. This means that if we solve the internal optimization problem (29) at each time instance during the operation stage, the mean of $f_i(\bar{d},z,\theta)$ (j=1,...,p) will be equal to $\bar{f}_i(\bar{d},\bar{a})$.

## The Worst Case Strategy

Here we formulate the internal optimization problem using the worst-case strategy as

$$f^{2*}(d,\theta,a) = \min_z f^2(d,z,\theta,a)$$

$$g(d,z,\theta) \le 0 \tag{36}$$

where

$$f^2(d,z,\theta,a) = \max_k (a_k f_k(d,z,\theta))$$
$$\sum_{k=1}^{p} a_k = 1, a_k \geq 0. \tag{37}$$

Let $z^*(d,\theta,a)$ be the solution of the problem. As in the previous case $\bar{f}_i(d,a)$ has the form (28). Again the new criteria $\bar{f}_i(d,a)$ $(i = 1,\dots, p)$ do not depend on the control variables z and we can directly use the method of the worst-case strategy for construction of the PS. In this case we must solve the problem

$$\bar{f}^{2*}(a) = \min_d \bar{f}^2(d,a) \tag{38}$$

where

$$\bar{f}^2(d,a) = \max_k a_k \bar{f}_k(d,a).$$

Again we have obtained a very computationally intensive bi-level optimization problem, which requires calculation of p multidimensional integrals for calculation of the objective function. We cannot simplify the problem the same way as we did in the case of the average criterion strategy. Now consider the problem

$$\bar{\bar{f}}^{2*}(a) = \min_d \bar{\bar{f}}^2(d,a) \tag{39}$$

where

$$\bar{\bar{f}}^2(d,a) = \int_T [\max_k a_k f_k(d,z^*(d,\theta,a),\theta)] \rho(\theta) d\theta. \tag{40}$$

There exists the following known inequality

$$\max_k \sum_i f_k(x,i) \leq \sum_i \max_k f_k(x,i) . \tag{41}$$

Since an integral can be approximated with Gaussian quadrature, the inequality leads to

$$\max_k a_k \int_T f_k(d, z^*(d,\theta,a),\theta)]\rho(\theta)d\theta \le \int_T [\max_k a_k f_k(d, z^*(d,\theta,a),\theta)]\rho(\theta)d\theta$$

c

$$\bar{f}^2(d,a) \le \bar{\bar{f}}^2(d,a), \forall d, \forall a.$$

Here $\bar{\bar{f}}^2(d,a)$ is an upper bound of $\bar{f}^2(d,a)$ for any d and a. From here we have

$$\bar{f}^{2*}(a) \le \bar{\bar{f}}^{2*}(a).$$

Consequently, $\bar{\bar{f}}^{2*}(a)$ is an upper bound of the optimal value of the objective function of (38). The term in the square brackets in (40) is the optimal value of the objective function of the internal optimization problem (36). Therefore, we can rewrite (40) as

$$\bar{\bar{f}}^2(d,a) = \int_T \min_z [\max_k \alpha_k f_k(d,z,\theta) / g(d,z,\theta) \le 0]\rho(\theta)d\theta.$$

This is equivalent to

$$\bar{\bar{f}}^2(d,a) = \min_{z(\theta)} \int_T [\max_k \alpha_k f_k(d, z(\theta),\theta)]\rho(\theta)d\theta$$
$$g(d, z(\theta),\theta) \le 0, \forall \theta \in T.$$

Substitute the expression of $\bar{\bar{f}}^2(d,a)$ in (39) to obtain

$$\min_{d,z(\theta)} \int_T [\max_k a_k f_k(d, z(\theta),\theta)]\rho(\theta)d\theta \qquad (42)$$
$$g(d, z(\theta),\theta) \le 0, \quad \forall \theta \in T.$$

Constraint (35) follows from (42). Using Gaussian quadrature we obtain the discrete variant of the problem

$$\min_{d,z^i} \sum_{i \in I_1} w^i \max_{k \in J} a_k f_k(d, z^i, \theta^i)$$
$$g_j(d, z^i, \theta^i) \le 0, i \in I_1, j = 1, ..., m$$
$$\chi_1(d) \le 0$$

where $J = (1,..., p)$.With the help of Theorem A.4 we can transform the problem to

$$\min_{d,z^i,y^i} \sum_{i \in I_1} w_i y_i$$
$$g_j(d, z^i, \theta^i) \le 0 \quad i \in I_1 \quad j = 1, ..., m$$
$$\max_{k \in J} a_k f_k(d, z^i, \theta^i) \le y^i \quad J = (1, ..., p)$$
$$\chi_1(d) \le 0.$$

Note that $y^i$ $(i = 1,..., p)$ are new auxiliary variables. This is the same as

$$\min_{d,z^i,y^i} \sum_{i \in I_1} w_i y_i$$
$$g_j(d, z^i, \theta^i) \le 0, i \in I_1, j = 1, ..., m \tag{43}$$
$$a_k f_k(d, z^i, \theta^i) \le y^i, \ k = 1, ..., p$$
$$\chi_1(d) \le 0.$$

Consider the implications of the results. Suppose the decision maker assigns values to the parameters $a_i$, which reflect the relative importance of the corresponding criterion. Then (9.43) provides an upper bound of the objective function of (38). Average values of each criterion will have the form (28) in which $z^*(d, \theta, a)$ is the solution of (36).

## Method of Consecutive Conciliations

We showed already that if we construct Pareto set with the method of consecutive conciliations then it coincides with the $\varepsilon$-constraint method (in the absence of uncertainty). As discussed earlier, Palazoglu and Arkun (1987) used the $\varepsilon$-constraint method for solution of the MCO problem under uncertainty. For fixed $\theta$ they reduced the MCO problem to one criterion optimization problem in which one criterion (for example $f_p(x)$ is employed as the objective function and the other criteria become additional constraints (see (11) and (12)). Using (11) as the internal optimization problem they formulated a TSOP1. However, with this approach the equal status of the criteria $f_1(x),..., f_p(x)$ are lost. To avoid this drawback, we will discuss another approach also based on the method of consecutive conciliations (the $\varepsilon$-constraint method).

For development of the approach we need to formulate two-stage analogs of problems (13) and (14). The analog of problem (13) is of the usual one-criterion optimization problem

$$f_1^{(1)} = \min_d E\{f_1^*(d,\theta)\}$$
$$\chi_1(d,\theta) \le 0 \tag{43}$$

where $f_1^*(d,\theta)$ is the solution of

$$f_1^*(d,\theta) = \min_z f_1(d,z,\theta)$$
$$g(d,z,\theta) \le 0$$

Let $z_1^*(\theta)$ be the solution of the problem. It is clear that the optimal value of the objective function in (43) can be written as

$$E\{f_1\ (d, z_1^*(\theta), \theta)\}. \tag{44}$$

Now consider a two-stage analog of (16), for which the internal optimization problem is

$$f_2^*(d,\theta) = \min_z f_2(d,z,\theta)$$
$$g(d,z,\theta) \le 0. \tag{45}$$

Let $z_2^*(d,\theta)$ be the solution of the problem. Let us formulate an analog of the constraint (17). Consider

$$E\{f_1(d,z_2^*(d,\theta),\theta)\}. \tag{46}$$

This gives the mean value of the first criterion when $z(\theta)$ is the control variable vector obtained by solving problem (45). From here it is naturally required that the value (46) would not exceed $f_1^{(1)} + \varepsilon_1$ ; in other words the following inequality must be met

$$E\{f_1\ (d,z_2^*(d,\theta),\theta)\} \le f_1^{(1)} + \varepsilon_1.$$

Finally, the two-stage analog of (16) will be

$$\begin{aligned} &\min_d E\{f_2^*(d,\theta)\} \\ &\chi_1(d) \le 0 \\ &E\{f_1(d,z_2^*(d,\theta),\theta)\} \le f_1^{(1)} + \varepsilon_1 \end{aligned}$$

Using Gauss quadrature we can obtain discrete variant of the problem as

$$\begin{aligned} &\min_{d,z^i} \sum_{i\in I_1} w_i f_2(d,z^i,\theta^i) \\ &g_j(d,z^i,\theta^i) \le 0, j=1,\dots,m,\ i\in I_1 \\ &\sum_{i\in I_1} w_i f_1(d,z^i,\theta^i) \le f_1^{(1)} + \varepsilon_1 \\ &\chi_1(d) \le 0 \end{aligned} \tag{47}$$

Similarly we can formulate two-stage analog of (14) for j=3. It has the form

$$\begin{aligned} &\min_d E\{f_3^*(d,\theta)\} \\ &\chi_1(d) \le 0 \\ &E\{f_1(d,z_3^*(d,\theta),\theta)\} \le f_1^{(1)} + \varepsilon_1 \\ &E\{f_2(d,z_3^*(d,\theta),\theta)\} \le f_2^{(2)} + \varepsilon_2 \end{aligned} \tag{48}$$

where [ $z_3^*(d,\theta)$ $f_3^*(d,\theta)$ ] is the solution of the internal optimization problem

$$\min_{z} f_3(d,z,\theta)$$
$$g(d,z,\theta) \le 0.$$

Using $E\{f_1(d,z_2^*(d,\theta),\theta)\}$ and $E\{f_2^*(d,\theta)\}$ obtained by solving (47) for different values of $\varepsilon_1$ one can construct the DM curve for p=2.

Comparison of the methods

The extension of the average criterion method permits to obtain some points of a DM curve by solving the system (33) to (35). However, for obtaining all points of the DM curve it is required convexity of the region $D_f$. The extension of the worst-case strategy requires solving a very computationally intensive problem. To avoid this we must solve problem (39) which give only an upper bound of $\bar{f}^{2*}(a)$ (see (38). The method of consecutive conciliations does not have the drawbacks of the first two methods.

## Computational Experiment

Example 9.1 Consider the MCO problem for a three-stage flow sheet (Example 7.12). We will suppose that products C and D are hazardous to the environment. Therefore, it is desirable to decrease the exit flowrate these products. Thus, here we will have two criteria, which characterize performance of the chemical process. One criterion $(f_1)$ will represent the economics of the CP. It will have the form (7.138). The other criterion $(f_2)$ is

$$f_2 = 24F(C_C^3 + C_D^3).$$

Using the average criterion strategy, the worst-case strategy and consecutive conciliation (see Section 3), we construct the PS for the case when uncertain parameters take nominal values. In agreement with the theory (see Section (9.3) the points obtained by all methods lie on one curve ABC (Figure 7). For the case when uncertainty is taken into account, we construct the DM curve using the extensions of the AC method and consecutive conciliation method. With the worst constraint strategy we construct only the curve which is obtained by solving the upper bound problem (39) for the set of parameters a, satisfying condition (6). All the methods gave the same curve $A^*B^*C^*$.

Example 9.2 We consider the reactor-separator problem (see example 6.3). Here we suppose that the products X and Y are hazardous to the environment. Therefore it is desirable to minimize the exit flowrate of these products. We consider the two-criterion optimization problem, in which one criterion is $f_1$ (see Example 6.3) which takes into account capital and operating costs. The other criterion is

$$f_2 = 10(1-\beta)F(x_X + x_Y).$$

Using the average criterion strategy, the worst-case strategy and consecutive conciliation (see Section 9.3), we construct the PS for the case when uncertain parameters take nominal values. They gave the same curve $A^1B^1C^1$ (see Figure 8). For the case when uncertainty is taken into account, we construct DM curve using extensions of the AC method and consecutive conciliation method. Using the worst constraint strategy we construct only the curve which is obtained by solving the upper bound problem (9.39) for the set of parameters a, satisfying (6). The first two methods gave the same curve $A^2B^2C^2$ (see Figure 8). The last method gave the curve $A^3B^3C^3$, which is above $A^2B^2C^2$.s

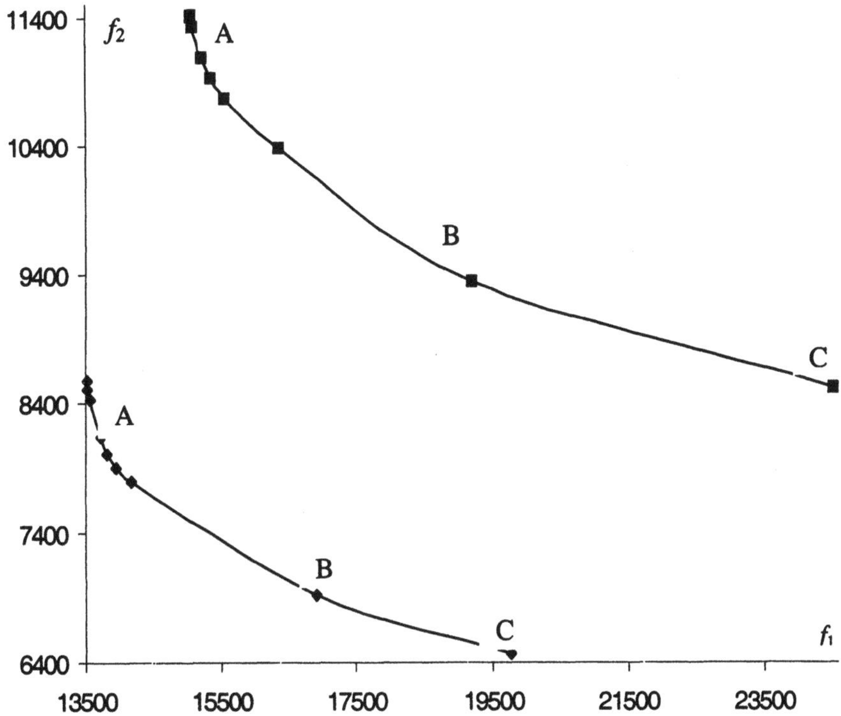


*Figure 7* Pareto set (for nominal values of uncertain parameters) and DM curve when accounting for uncertainty in Example 9.1.

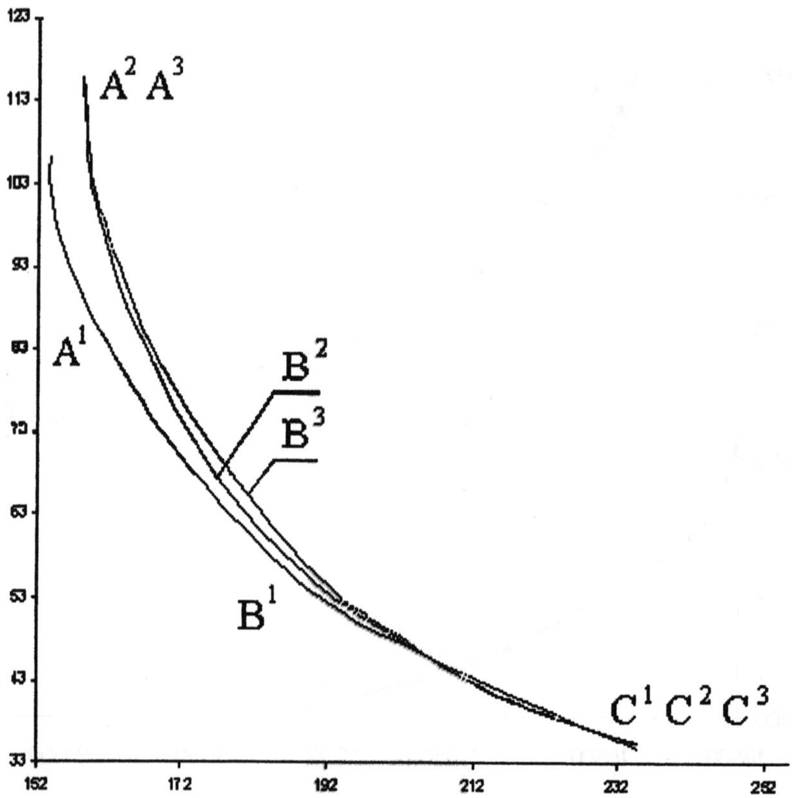


*Figure 8* Pareto sets for nominal values of uncertain parameters and DM curve in the case of taking into account uncertainty for Example 9.2.

# REFERENCES

Caballero R., Ruiz, F, Steuer R.E. *Advances in Multiple Objective and Goal programming*. Berlin: Springer-Verlag, 1997.

Chakraborty A., Linninger A.A. Plant-wide waste management. 2. Decision Making under uncertainty. Ind.Eng.Chem.Res. 2003; 42:357-369

Clark P.A., Westerberg A.W. Optimization for design problems having more than one Objective. Comp. Chem. Eng. 1983; 7:259-278.

Haimes Y., Hall W.A., Friedman N.J. *Multi-objective Optimization in Water Resource Systems: The Surrogate Worth Trade-Off Method*. Amsterdam: Elsevier, 1975.

Keeney R.L. Raiffa H. *Decisions with Multiple Objectives*. New York: John Wiley, 1976.

Luyben M.L., Floudas C.A. Analysing the interaction of design and control-1. A multiobjective framework and application to binary distillation synthesis. Comp. Chem.Eng. 1994; 18:933.

Palazoglu A., Arkun Y. Design of chemical plants with multiregime capabilities and robast dynamic operability characteristics. Comp. Chem. Eng. 1987; 11:205-216

Sophos A., Rodstein F., Stephanopoulos G. Multi-objective analysis in modeling the Petrochemical Industry. Chem. Eng. Sci. 1980; 35: 2415-2426

# Chapter 9

# DESIGN OF NEURAL NETWORKS FOR PAVEMENT RUTTING

**Rafiqul Alam Tarefder and Musharraf Zaman**

## 1. INTRODUCTION

Rutting is one of the major distresses of asphalt pavements. Currently, an Asphalt Pavement Analyzer (APA) can be used to evaluate the rut potential of asphalt pavements in the laboratory. Although it is preferable to conduct APA tests to predict the rutting potential of an asphalt pavement, such tests are not always feasible for a project due to economic reasons. A rut prediction model can be a useful tool in such situations. Prediction of rutting using a model is a rather challenging task. Traditional statistical models have often exhibited weaknesses in predicting reliable rut values (Tarefder et al. 2002). This study proposes Neural Networks (NNs) to predict rutting of asphalt pavements.

For a given set of data and a family of neural networks, design of neural networks (stage three) for pavement rutting involves selecting a NN from this family that best approximates the data with high probability. Therefore, NNs for rutting can be considered as a parameterized nonlinear function like polynomials, Fourier series, splines, etc. What distinguishes neural network for pavement rutting, as opposed to many other standard techniques, is that the neural network has a generalization capability. In other words, once the NN has been trained on a number of input-output pairs, it can then accurately predict outputs from inputs that the network has not seen previously.

The generalization performance of a NN is influenced by three factors: the physical complexity of problem, the size of the training set, and the architecture of the neural network (Fine 1998). Rutting is a complex problem. The physics, and in some cases the mechanics, of this problem is not fully understood and therefore, it is difficult to express in a differential, integral or variational formulation (Ramsamooj et al. 1998). Therefore, the issue of generalization can be viewed from two different perspectives. According to the second perspective, the architecture of the network is fixed, and the issue to be resolved is that of determining the size of the training set needed for a good generalization to occur. According to the third perspective, for a given family of NNs, the issue of interest is that of determining the best architecture of network for achieving good generalization. This study focuses on the third viewpoint. The task is to design a suitable architecture based on the limited data samples.

The difficulty of this task may be associated with the fact that a finite database is used for estimating the architecture or model parameters. It is often possible to find a model, which fits the available data perfectly by taking a large number of model parameters (Tarefder et al. 2004). It is shown later in this paper that an over parameterized model gives very poor results on new data (data which has not been used for estimating the parameters). On the other hand, a model with too few parameters gives poor results both on the data used for estimating the parameters (i.e. training data) and on fresh data (i.e. test data). Therefore, a good design of a NN is a tradeoff between the performance on the training data and performance on new data (test data). The objective is to find a suitable architecture (i.e. the model with the smallest number of parameters) that exhibits good performance on training data (i.e. data used for estimating the parameters) and on test data (i.e. data used for estimating the prediction performance).

A success in designing a NN not only depends on finding a suitable architecture but also on relevant input factors to the network, efficient training algorithm, appropriate performance index, and prediction approach. Steps required to a successful NN design, training, and estimation of its performance are the topics of discussion in this paper. In subsequent sections the NN preliminaries, input data, output data, data processing, NN architecture, and prediction results, and comparison of outputs from NNs to actual rut depths are discussed.

## NN Basics

The basic structural constituent of a NN is known as "neurons". A neuron is an information-processing unit that is fundamental to the operation of a NN. There are three elements of a neuron. These elements are: (i) synaptic weight or connection link that is characterized by a weight or strength of its own, (ii) adder that sums the input signals, weighted by the respective synapses of the neuron, and (iii) activation function that is used for limiting the amplitude of the output of a neuron. Neuron can also include an externally applied threshold, $b_j$ (also referred to as bias). Mathematically, a neuron j can be described by Eq. 1 and Eq. 2 as follows:

$$v_j = \sum_{k=1}^{k} w_{jk} x_k + b_j \tag{1}$$

$$y_j = \varphi(v_j) \tag{2}$$

where $x_1$, $x_2$..... $x_k$ are the input signals; $w_{j1}$, $w_{j2}$.... $w_{jk}$ are the synaptic weights converging to neuron j; $v_j$ is the cumulative effect of all the neurons connected to neuron j and the internal threshold of neuron j; $\varphi(.)$ is the activation function; and $y_j$ is the output signal of the neuron. Activation functions used in this study are the sigmoid function and the linear transfer function. Usually, neurons are organized in the form of layers. Depending on the number of layers, a NN can be classified as a single layered network or a multiple layered network. Based on its role, a layer is classified as either an input, hidden or output layer. The input layer receives inputs from an external source and the output layer passes its computed values to an external source. The remaining layers are called hidden layers (Hornik et al. 1994).

## NN Inputs

The NN input factors are the aggregate factors (shape, size, type), binder factors (grade, specific gravity), environmental factors (temperature, wet/dry condition), mix factors (asphalt content, gradation, voids in the mineral aggregate, air voids) and load factors (wheel load, hose pressure) (Tarefder et al. 2002). The factors considered are:

1. Percentage of materials passing through 25.0m sieve
2. Percentage of materials passing through 19.0 mm sieve
3. Percentage of materials passing through 12.5 mm sieve
4. Percentage of materials passing through 9.5 mm sieve
5. Percentage of materials passing through no. 4 sieve
6. Percentage of materials passing through no. 8 sieve
7. Percentage of materials passing through no. 16 sieve
8. Percentage of materials passing through no. 30 sieve
9. Percentage of materials passing through no. 50 sieve
10. Percentage of materials passing through no. 100 sieve
11. Percentage of materials passing through no. 200 sieve
12. Binder's Performance Grade (PG)
13. Aggregate's Fractured Face (FF)
14. Asphalt Content (% AC)
15. Percentage air voids (% Air)
16. Voids in Mineral Aggregates (VMA)
17. Fine Aggregate Angularity (FAA)
18. Temperature
19. Wheel load
20. Tire pressure
21. Wet/dry conditioning

All the mixes considered in this study are designed by the Superpave method (Table 1).

**Table 1.** Mix Information Used in Neural Network Design

| Properties | Mix Type | S2 | S3 | S3-rec | S4 | S6 |
|---|---|---|---|---|---|---|
| Sieve Size, mm (in) | 37.5 (1 ½ in.) | 100 | - | - | - | - |
| | 25.0 (1 in.) | 90-100 | 100 | 100 | - | - |
| | 19.0 (3/4 in.) | - | 90-100 | 90-100 | 100 | - |
| | 12.5 (1/2in.) | - | 90max | 90max | 90-100 | - |
| | 9.5 (3/8 in.) | - | - | - | 90 max | 100 |
| | 4.75 (No.4) | 40-40 | - | - | - | 80-100 |
| | 2.36 (No.8) | 19-45 | 23-49 | 23-49 | 28-58 | 54-90 |
| | 2.00 (No.10) | - | - | - | - | - |
| | 1.18 (N0.16) | 18-24 | 22-28 | 22-28 | 26-32 | 39-39 |
| | .60 (No.30) | 14-18 | 17-21 | 17-21 | 19-23 | 26-32 |
| | 0.425 (No.40) | - | - | - | - | - |
| | .30 (No.50) | 11-11 | 14-14 | 14-14 | 16-16 | 19-23 |
| | 0.15 (N.100) | - | - | - | - | 16-16 |
| | 0.075 (No.200) | 0.6 –1.2 $P_{eff}$ | 0.6 –1.2 $P_{eff}$ | 0.6 –1.2 $P_{eff}$ | 0.6 –1.2 $P_{eff}$ | 5-15 |
| Design Method | | Superpave | Superpave | Superpave | Superpave | Superpave |
| Nominal Maximum Size (NMS), mm | | 25 | 19 | 19 | 12.5 | 4.75 |
| Lift Thickness, mm | | 56-112 | 56-112 | 56-112 | 37.5-75 | 12.5-25 |
| Compaction Method | | SGC | SGC | SGC | SGC | SGC |
| Asphalt to Dust Ratio | | 1.2 | 0.9 | 1.1 | 1.1 | 0.9 |

Note: $P_{eff}$ = Effective percentage Binder, SGC = Superpave Gyratory Compactor, '-' = No value

The input factors 1 to 11 represent the aggregate gradation that is determined by sieve analysis and expressed as percent of aggregate passing through different sieve sizes. The sieves shown by factors 1 to 11 are used to define the aggregate gradations of Superpave mixes. Factor 12 represents the binder's stiffness; a stiffer binder produces a mix with low rut potential. The input factor 13 represents coarse aggregate angularity (or fractured face). Fine aggregate angularity is represented by factor 17. The mix factors are shown by the input factors 14 to 16. The VMA varies from 6.6-22.2% with an average of 16.2%, whereas the air voids vary from 2.5-11.1% with an average of 7.2%. The APA testing parameters used to simulate field pavement conditions are shown by input factors 18 to 21. In a typical APA rut test, asphalt samples are preconditioned at a test temperature of 64°C (for Oklahoma mix), the vertical wheel load is kept at 445 N (100 lbs), and the hose pressure is held at 700 kPa (100 psi) (OHD 2001). The numerical spreads of the above factors are shown in Figure 1.

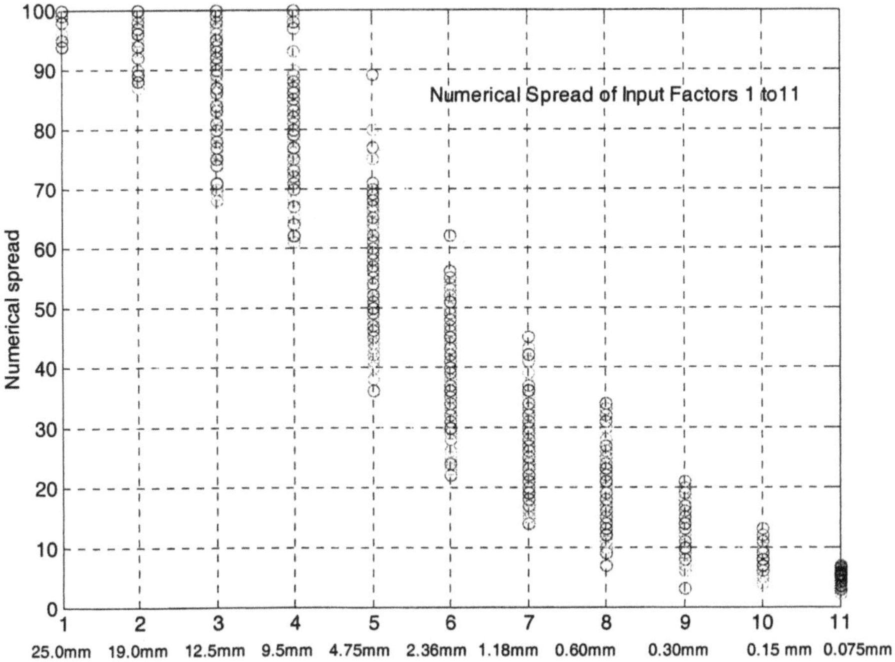

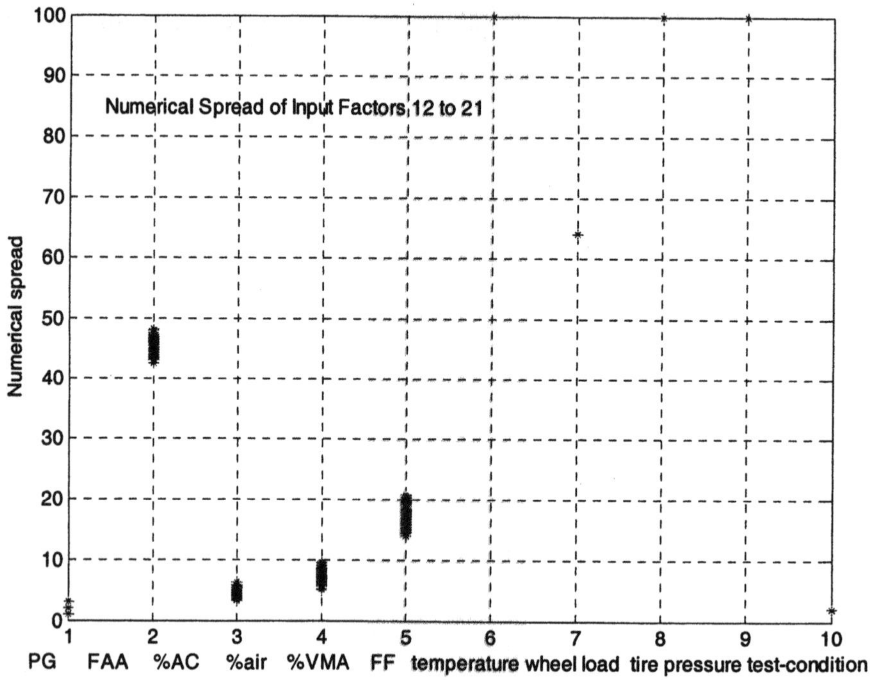


Figure 1. Numerical spread of the input factors.

## NN Outputs

The output data is obtained by means of cyclic rut tests using APA. In this equipment, rutting susceptibility is evaluated by subjecting HMA samples to moving wheel loads and measuring rutting (permanent deformation) at selected points along the wheel path as a function of the number of loading cycle. The deformations of samples or rutting are recorded over 8000 cycles. For the purpose of this study, it suffices to describe this time series of deformations by an interpolation with piecewise linear elements using only a few deformation values. Consequently, the domain of the neural network to be constructed and trained is a vector space of input factors whose range space consists of vectors obtained from a few values of deformation. Observations of deformations are made at eight selected cycles: 1, 500, 1000, 1500, 2000, 4000, 6000, and 8000. Since the deformation at cycle number 1 for all data is essentially the same (zero deformation), the target vector consists of 7 components. The range of 8000 cycle rut depth is 0.6 mm – 7.4 mm. Finally, a dataset consists of 21 inputs and 7 outputs (500, 1000, 1500, 2000, 4000, 6000, and 8000-cycle rut depths).

## Data Processing

As preprocessing steps, missing data and outlier data are removed, non-numeric data are transformed to numeric data, and data are scaled into the active range of the activation functions used. Initially, 573 data sets (a total of 1146 samples, each data set represents two samples) are available. After removing the missing data, 537 data sets are available. Data that deviates more than two times the standard deviation from the mean value of the corresponding data vectors is considered as outlier. 18 data sets are removed based on the outlier criterion and finally, 519 data sets are retained for normalization. All input values to a NN must be numeric. There are two non-numeric input parameters: one is performance grade (PG) and the other is testing conditioning. The PG has three different values, which are coded as three different numeric input parameters. The PG that corresponds to a grade of PG 64-22 is assigned a code of 1. Likewise, the codes of PG 70-28 and PG 76-28 binders are assigned as 2 and 3, respectively. Similarly, sample testing in dry conditions are given a value of 1, while samples cured under wet conditions are given a value of zero. All of the 16 input vectors are then normalized, so that each input factor averaged over the entire data sets has zero mean and unit standard deviation.

## Principal Component Analysis

The purpose of Principal Component Analysis (PCA) is to derive new variables (in decreasing order of importance) that are linear combinations of the original variables and are uncorrelated. Geometrically, principal components analysis can be thought of as a rotation of the axes of the original coordinate system to a new set of orthogonal axes that are ordered in terms of the amount of variation of the original data they account for (Engelbrecht 2002). Mathematically, a PCA orthogonalizes the components of the input vectors (so that they are uncorrelated to each other) and orders the resulting orthogonal components (principal components) so that the largest variation comes first, and it eliminates those components that contribute the least to the variation in the data set (Haykin 1994; Hertz et al. 1991). A total of three principal component analyses are conducted in which factors accounting for 0.1%, 1% and 2% of the variation of the input vectors are used. Using a variance of 0.1%, the number of input factors remained the same. Whereas, using a variance of 1%, the number of input factors reduces from 21 to 10; that is, input factors accounting for 99.0% of variation in the total data set leads to a reduction in input dimension. Using 98% of variation in the total data set, the number of input factors got reduced to 9 and is used to construct and train our neural network. Finally, a data set (training set) is designed that consists of data in the form of pairs of vectors and is composed of 9 input factors and 7 target vectors.

## NN Architecture

The manner in which the neurons are structured in a NN is called architecture. Usually, neurons are organized in the form of layers. NN architecture can be defined as: each of the 9-inputs is connected to each of the q-hidden neurons (either in one or two layers), and the outputs of the hidden neurons are fed into the 7-output neurons. The task is to determine the number of hidden neurons, q in the final NN.

The NN architecture is designed by a sequential algorithm in which at each step a new neural network is designed by adding a neuron to a specific hidden layer, trained by the Levenberg-Marquardt minimization algorithm, validated, and tested for generalization performance (Demuth and Beale 1998).

As the first step, NNs having one hidden layer (h1) are studied. A family of NN of architectures in which the number of hidden neurons varies from 1 to 40 {Aq; q = h1= 1:40} are evaluated for generalization performance. The number of neurons in the input and output layers are simply the number of reduced inputs from principal component analysis (i.e., 9) and outputs (i.e., 7), respectively. The family of feedforward NN is denoted by 9-h1-7 FNN.

In the second step, networks having two hidden layers (h1 & h2) are studied. The number of neurons in hidden layer one varied from h1=1:26 and in hidden layer two varied from h2 = 1:18. For each hidden neuron in layer two, neurons are added from 1 to 20 in the hidden layer one. Consequently, a total of 468 architectures {$A_q$; q = h1 x h2= 468} are investigated. All the architectures studied in this family have nine inputs (found in the PCA) and seven output neurons; therefore the NNs are denoted by 9-h1-h2-7 NN.

The architecture or topology of NN must be established before the training. In the subsequent discussion in this paper, it will be shown that a NN with 11 neurons in the first hidden layer and 11 neurons in the second hidden layer (9-h1-7, 9-h1-h2-7) has the best performance. To explain the training procedure in the next section, subsequent reference is made to a 9-h1-h2-7 NN shown in the Figure 2. This is a fully connected, "three layer", and feed-forward neural network. The input layer consists of 9 inputs, which is the reduced number of inputs after principal component analysis. No processing is done in the reduced input nodes; they only distribute the network inputs to 11 neurons in the h1-hidden layer. Consequently, this network has three processing layers (also called neuron layers) of which topology can be denoted by 9-11-11-7 NN.

## Training

In the training step, the hidden layer one takes a preconditioned input column of $n_i$=9 vectors and maps it to a column of $n_{h1}$=11 vectors by a tan-sigmoid transfer function. The tan-sigmoid transfer function, φ (v) is given by,

$$\varphi(v) = \frac{1}{1+e^{-2v}} \tag{3}$$

The resulting vectors are then taken as an input by the hidden layer two as inputs and mapped by a tan-sigmoid function to a column of $n_{h2}$=11. These vectors are then taken as input by the output layer neurons and mapped though a linear operation to an output consisting of a column vector with $n_o$=7 components. The network weights are randomly generated from a uniform distribution for the linear transfer function, whereas for the tangent sigmoid transfer function the random weights are processed in accordance with the algorithm developed by Nguyen and Widrow (Hagan et al. 1996). The weights are continuously updated based on error (difference between the NN outputs and target vector) determined by the Levenberg-Marquardt algorithm. The trained network is tested for its performance.

## Performance Index

The final network is selected based on the NN performance measure on the test data. The most common measure of performance of NN is the Mean Squared Error (MSE), expressed as:

$$MSE = \frac{\sum_{j=1}^{p}\sum_{i=1}^{n}(o_{i,j} - t_{i,j})^2}{n.p} \tag{4}$$

where,
n = total number of data set.
o = network output.
p = number of outputs.
t = target output.

Instead of mean square error, an Average Relative Error (ARE) can be used to measure the performance of a NN. The average relative error is calculated using the $L_2$-norm of error vector normalized by the $L_2$-norm of output vector as shown below,

$$ARE = \frac{1}{n}\sum_{i=1}^{n}\left(\frac{1}{p}\sum_{j=1}^{p}\frac{(o_{j,i} - t_{j,i})^2}{t_{j,i}^2}\right) \tag{5}$$

Although the above two indexes are most common for measuring performance of a NN, an additional measure of NN performance, the correlation (R-value) between the output and target values for all data sets, is also used for architecture selection (Hornik et al. 1989).

**Performance Estimation**

In this study, NN architecture is selected based on the performance measured by Eq. 4. However, the MSE performance of a NN of fixed architecture varies with initial choice of weight vectors to start the minimization algorithm (Fine 1998). Quantifying a performance representative of a family of weights is one of the most central problems in designing an optimal architecture. This study settles for architectures that work satisfactorily "most of the time" (likelihood estimation). The likelihood method is simply what is used to generate simulated data after the unknown parameters (weights) are guessed. Ideally, the performance of a NN for a given architecture and m number of weight initializations $\{w = [w_1, w_2 \ldots\ldots w_m]\}$, the expected value is approximated by a mean or a maximum likelihood estimator.

**Mean Estimation.** The performance (MSE, ARE, R-value) of a fixed architecture is approximated by taking the mean of the performances of randomly initialized networks.

**Maximum Likelihood Estimation.** The principle behind the maximum likelihood method involves multisampling of weight, w. If $NN(w_1)$, $NN(w_2)$, ......$NN(w_m)$ are the m observed performances of the network, then the estimated performance of the NN is the most likely to produce or represent these observed values. The probability density function of NN(w) is determined. Then the one with maximum probability density is considered the final performance.

**Data Set Division**

One of the problems that occur during the NN training phase is called overfitting (Kearn 1997). The error on the training set is driven to a very small value, but when new data is presented to the network the error can become large. In such case, the network memorizes the training examples, but it does not learn to generalize to new situations. Overfitting occurs when the NN architecture is too large or NN is trained for too long. Evaluation of generalization error (validation error) during training can be used to eliminate NN overfitting. In order to eliminate overfitting, this study divides the original data set (input-output) into three sets: the training set, the validation set, and the test set. The training data set is used for computing the gradient and updating the network weights and biases. The error on the validation set is monitored during the training process. The test set is not used during training, but is used to compare the performance of different models (architectures).

There is no rule to divide the total data sets. This study divides 519 data sets into three parts based on the training and validation performance of 9-11-11-7 NN. The MSE, ARE, and R-values for different data divisions are summarized in Table 2. The difference between Set A and Set B is that the number of data in the validation set in Set B is higher than that in Set A. Obviously, the MSE performance of Set B is better (lower MSE value) than that of Set A for the validation set, whereas the MSE performance of Set B and Set A are almost equal for the training set. Due to increased number of unknown data in the validation data set, the R-square of Set B is less than that of Set A. There is little difference between the MSE performances of these two sets (Set A and Set B) on the test data. There is a R-square value improvement in the performance of Set C compared to Set A and Set B. Also, the Set D has highest R-square value. This is because most of available data are used to train the network and the calculation of R-square value involves all data. However, the Set D is rejected because the MSE and AVE errors are high. For similar reasons, the data division of Set E is rejected. From Table 2, it can be seen that the MSE, and ARV errors of Set C are smaller than those of the other data sets. Therefore, the data division of data Set C is chosen for designing the NN in this study.

**Table 2.** Data Set Division and Network Performance

| Data Set | | Mean Square Error (MSE) | | | Average Relative Error (ARE) | | | R-value |
|---|---|---|---|---|---|---|---|---|
| | | Training | Validation | Test | Training | Validation | Test | All Data |
| A | $D_T = 260$<br>$D_V = 130$<br>$D_G = 129$ | 0.1308 | 0.3568 | 0.4008 | 0.2020 | 0.2729 | 0.2878 | 0.7995 |
| B | $D_T = 260$<br>$D_V = 207$<br>$D_G = 52$ | 0.1271 | 0.3312 | 0.4162 | 0.2035 | 0.2783 | 0.2516 | 0.8123 |
| C | $D_T = 312$<br>$D_V = 155$<br>$D_G = 52$ | 0.1333 | 0.3409 | 0.3942 | 0.1996 | 0.2888 | 0.2454 | 0.8246 |
| D | $D_T = 415$<br>$D_V = 52$<br>$D_G = 52$ | 0.1432 | 0.3569 | 0.4508 | 0.1934 | 0.2914 | 0.2503 | 0.8417 |
| E | $D_T = 52$<br>$D_V = 415$<br>$D_G = 52$ | 0.3891 | 0.7480 | 0.7556 | 0.3506 | 0.4971 | 0.3812 | 0.3189 |

Note: $D_T$ = Training Data Sets, $D_V$ = Validation Data Sets, $D_G$ = Test Data Sets.

The adequacy of data set division can be examined by plotting the test data set performance during the training process. If the error in the test set reaches a minimum at a significantly different iteration number than the validation set error, this indicates a poor division of data set. Figure 3 shows the MSE performance of the training, validation and test data sets as a function of epochs (an epoch is defined as the complete representation of all the data sets to the NN) for the division of set C. The result is reasonable. Since the test set error and the validation set error have similar characteristics, it does not appear that any significant overfitting has occurred. This also confirms that the selected Set C eliminates the dependence of NN performance on the training set and thereby ensuring that the division in the data sets is not affecting the selection of network architecture.

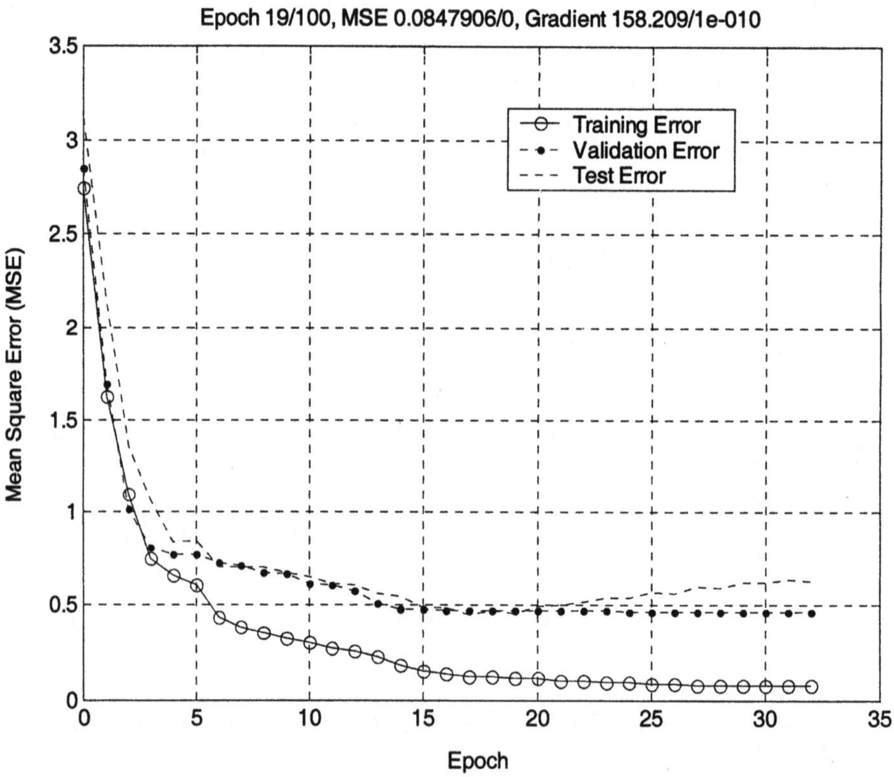


Figure 3. Performance of a neural network (NN) on different data sets.

## Architecture of One hidden layer NNs

First, a two layer feedforward network with 1 hidden neuron (9-1-7) is initialized and trained using a total of 312 training data sets and a total of 155 validation data sets. Before the training, principle components that contributed less than 2% to the total variation in the data set are eliminated. As a result of this step, the dimension of the input space reduced from 16 to 9. A total of 50 trials are performed with different random initializations of network weight and bias values. In each trial, each of the subsets (training, validation, test data sets) is randomly chosen so that the sequence of data in an epoch differed from one trial to another. The average of MSE performances from 50 trials is then computed. Next, a second NN with two neurons in the hidden layer (9-2-7) is chosen, trained and used to determine the MSE performances. The procedure of designing and training up to 40 more NNs continued before the average MSE performances on the test data sets are determined, as shown in Figure 4. The standard deviation for each NN is also plotted.

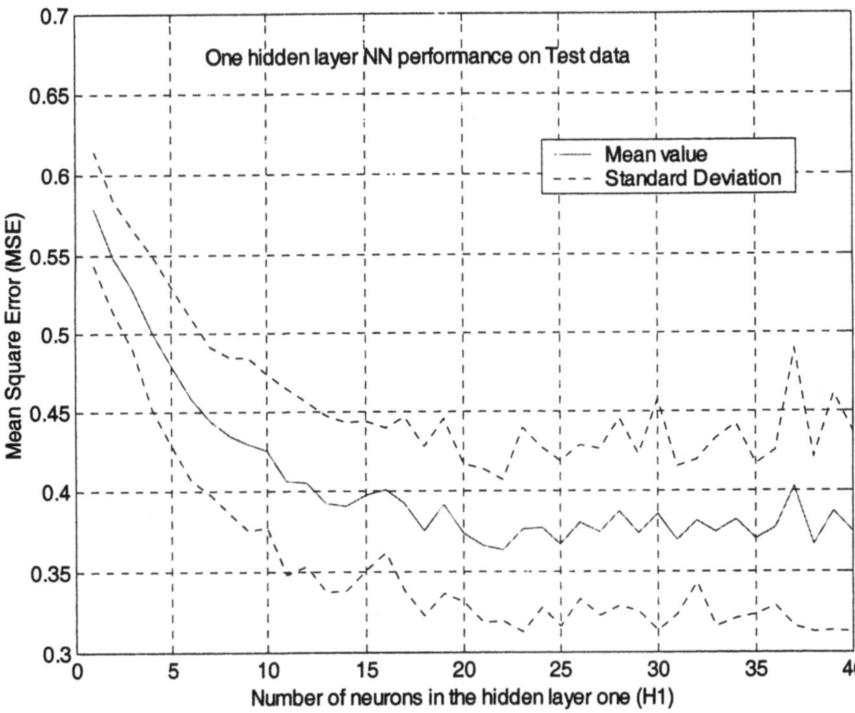


Figure 4. Test Sets MSE performances of NNs of one hidden layer.

It is evident that as the number of neurons in hidden layer one increases the average MSE error decreases until it reaches 22 neurons. After 22 neurons, an increase in the hidden neurons of NNs does not improve the NN performance, but rather the standard deviation of MSE increases. Therefore, the NN with 22 hidden neurons are selected (i.e., 9-22-7 NN) from this category of NNs.

To further investigate our above selection of 9-22-7 NN, the correlation between the NN rut output and the actual rut depths is examined. The R-square between the NN predicted rut and the actual rut along with the MSE and AVR are shown in Figure 5.

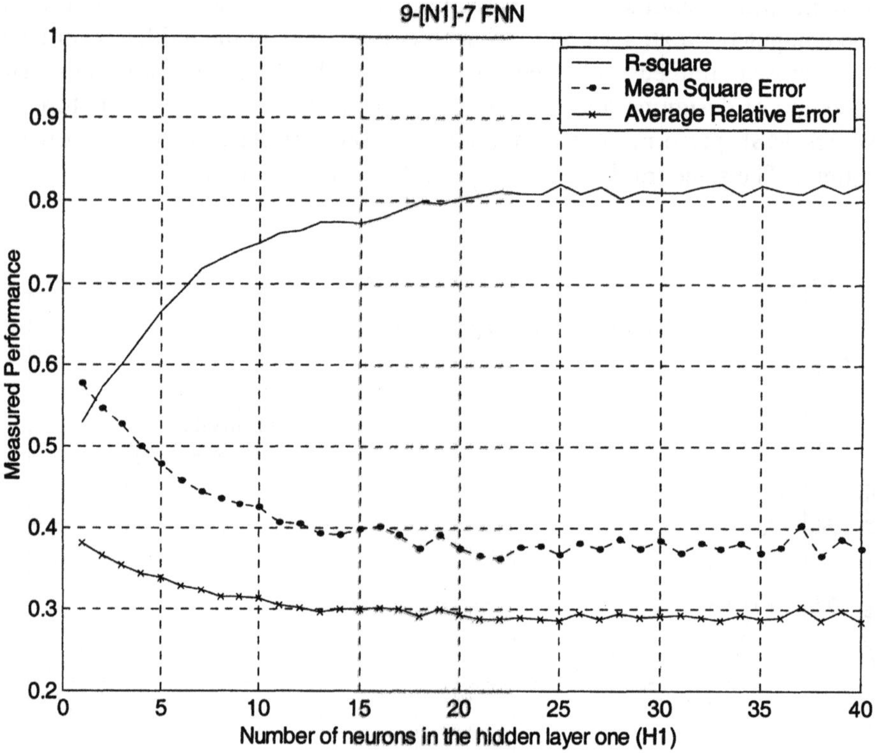


Figure 5. R-square, Mean Square Error (MSE) and Average Relative Error (ARE) performances of NNs with one hidden layer

The R-square shown is measured using all the data sets available, whereas the MSE and ARR test shown are measured on test data sets. The R-square value increases as the number of hidden neurons is increased in the trial feedforward neural network. The nearly maximum R-square of 0.8111 with a standard deviation 0.0287 can be seen when the number of hidden neurons is 22 in the trial network (9-22-7). Although the trial network 9-25-7 has shown a higher

R-square value of 0.8211, it was not selected as our final one layer NN due to its higher standard deviation of 0.0331. In addition, 9-25-7 FNN has a higher MSE value of 0.3671 compared to the MSE value of 0.3632 in the 9-22-7 FNN. Therefore, the final selection of 9-22-7 NN is reasonable.

## Architecture of Two hidden layer NNs

The performance of a NN having one layer of hidden neurons can be improved to a certain extent by using two layers of hidden neurons. A NN with two hidden layers may have performance better than that of a NN with one hidden layer. A trial and error approach similar to that in the previous section is adopted, except, in this case, the number of hidden neurons in one layer is increased while the number of neurons in the other layer of the NN remained constant. The input and output layers are kept same as the previous. That is, the input layer takes 9 inputs and the output layer has 9 neurons. The number of nodes in the first layer is arbitrarily chosen to vary from 1 to 20, whereas the number of nodes in the second layer is kept between 1 and 20. A total of 400 NNs are trained to find a NN that shows better performance over the others. For a selected configuration, a network is trained several times (selected arbitrarily) and then a simulation is performed on the trained NN using the training data set, validation data set, and test data set as well as the total data set. Results are reported by average and standard deviation of MSE, as shown in Table 3.

**Table 3.** Training Performance of Trial Neural Networks

| Neurons in h1 layer | Neurons in h2 layer | Training Data Set | | Validation Data Set | | Test Data Set | | Total Data Set | |
|---|---|---|---|---|---|---|---|---|---|
| | | Mean MSE | Std dev. MSE | Mean ARE | Std dev. ARE | Mean R-value | Std dev. R-value | Mean MSE | Std dev. MSE |
| 12 | 11 | 0.1308 | 0.0779 | 0.3629 | 0.0451 | 0.3784 | 0.0740 | 0.2505 | 0.0604 |
| 15 | 12 | 0.1232 | 0.0557 | 0.3476 | 0.0384 | 0.3810 | 0.0607 | 0.2435 | 0.0402 |
| 14 | 12 | 0.1100 | 0.0534 | 0.3520 | 0.0409 | 0.3850 | 0.0676 | 0.2390 | 0.0409 |
| 11 | 11 | 0.1269 | 0.0442 | 0.3631 | 0.0402 | 0.3874 | 0.0588 | 0.2508 | 0.0333 |
| 10 | 12 | 0.1327 | 0.0512 | 0.3634 | 0.0468 | 0.3882 | 0.0568 | 0.2540 | 0.0368 |
| 8 | 12 | 0.1612 | 0.0557 | 0.3715 | 0.0479 | 0.3891 | 0.0504 | 0.2705 | 0.0432 |
| 10 | 11 | 0.1356 | 0.0576 | 0.3604 | 0.0474 | 0.3897 | 0.0632 | 0.2551 | 0.0447 |
| 15 | 10 | 0.1314 | 0.0925 | 0.3622 | 0.0633 | 0.3909 | 0.0872 | 0.2537 | 0.0765 |
| 14 | 9 | 0.1125 | 0.0576 | 0.3552 | 0.0470 | 0.3922 | 0.0694 | 0.2428 | 0.0460 |
| 13 | 10 | 0.1260 | 0.0501 | 0.3658 | 0.0342 | 0.3928 | 0.0595 | 0.2524 | 0.0342 |

Note: MSE = Mean Square Error, ARE = Average Relative Error, R-value = Correlation of Determination, Std dev.= Standard Deviation, h1 = hidden layer one, h2 = hidden layer two

Column 2 and column 3 show the number of hidden neurons in hidden layer one and two, respectively. The results are presented in ascending order of test sets MSE. The average performance of the first four NNs (first 4 rows) over all simulations are close to each other. The mean MSE error (0.3784) on test data is lowest in 9-12-11-7 NN, whereas the validation MSE value (0.3476) is minimum in 9-15-12-7 NN. However, it can be seen that 9-11-11-7 NN has lower variance in performance compared to that of any NN in the first four rows. The variance or standard deviation is very important in selection of NNs. If we compare the test set performance of 9-12-11-7 NN to that of 9-11-11-7 NN, where the MSE for 9-12-11-7 NN is 0.3784±0.0740, and that of 9-11-11-7 is 0.3874±0.0588, then the latter NN is preferred even though the former has a smaller MSE. The NN 9-11-11-7 has a smaller variance, having MSE values in the range [0.3286, 0.4462], while the NN 9-12-11-7 has MSE values in a larger range [0.3044, 0.4524]. Using the results shown for NN 9-11-11-7, the interval associated with confidence level of $\alpha$=0.01 is estimated to be in the range [0.2522, 0.5226]. It means that 99% of the observation reported for 9-11-11-7 NN in Table 3 (based on the MSE of test data sets) is in this interval. Further, analyses of R-value performances can show that 9-13-10-7 NN, 9-11-11-7 NN, and 9-14-12-7 NN can reach a R-value of 0.8141 with 347 (i.e., weights = 13x9+10x13+7x10 plus bias = 13+10+7), 326, and 411 parameters, respectively. The final selection of network in two hidden layer family is 9-11-11-7 NN, as it has the lowest number of adjustable parameters.

## NN Prediction

At this stage, the trained and tested (validated) network (11-11-7 NN) is used to map or simulate new set of inputs. The difference between the testing and prediction is that the target output is known during testing, whereas in prediction steps we use the tested NN to find the unknown (target) rutting.

**Mean and Maximum Likelihood Prediction.** A final simulation output is obtained through the development of ensemble networks, where the aim is to optimize NN outputs through a combination of a number of individual network outputs, trained on same data sets, using the same architecture and learning algorithm. As stated above, the training and simulation procedure is carried out for several times and the resulting output vector is compiled. For example, Figure 6 shows a histogram plot of RD (7) of an unknown (target) rutting. Similarly, histograms of the deformations or rut depths, RD (1)-RD (7) for the each test sets are compiled. Estimators of the deformations are calculated from the histograms. In particular, deformations are predicted based on estimators of the mean and maximum likelihood estimator.

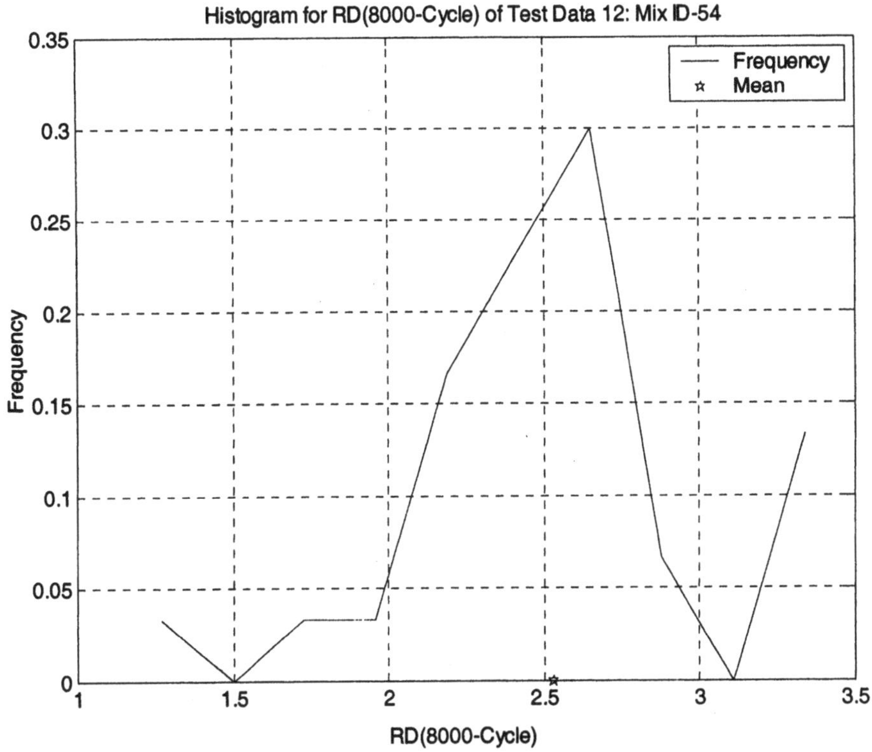


Figure 6. Histogram for rut depth, RD (7) of test data.

**Best Net Prediction.** Among 100 trials, the NN (trained on same data sets, using the same architecture and learning algorithm) that provides the lowest error (MSE or ARE) on the validation data sets is used in simulation or prediction. In this paper, the estimation from a NN with the lowest MSE is termed 'best MSE net' estimation, whereas the estimation by NN with lowest ARE is termed as 'best ARE net' estimation.

**Analysis of Prediction Results.** The deformations based on mean, maximum likelihood, best net ARE, and best net ARE estimations using 9-11-11-7 NN are depicted in Figure 7. An excellent agreement is observed between the predicted and observer test data. Also, a regression analysis of the network and predicted deformations is performed. The entire data set is applied through the network and a linear regression between the network outputs and the corresponding targets is performed. For 7 outputs, seven regressions can be performed. The results for 8000-cycle rut depth for a test data set 12 using 9-11-11-7 NN are shown in Figure 8. The best linear fit is indicated by the dashed line. The perfect fit (output equal to target) is indicated by the solid line. As the best linear fit line comes close to the perfect fit line, the NN simulation is evaluated as better. Similarly, deformation responses obtained from a single best net, based on the minimum MSE and ARE, are also shown in Figure 8. It can be seen that the maximum likelihood prediction is close to the mean prediction where as the best net simulations do not have generalization capability. That is the use of families of networks trained on different initial conditions can improve NN performance through either linear combinations of the trained networks compared to by simply choosing the single best network. A possible explanation of this can be that the linear combination of network results in a new, more complex network that can explain the improved fit to the training data. The total error from simulation over the test data sets is determined as follows:

Total relative error of mean estimator = 0.2666

Total relative error of the maximum likelihood mean estimators = 0.2503

Total relative error of best net based on the minimum validation error (MSE) =0.3329

Total relative error of best net based on the minimum validation error (ARE) =0.3429

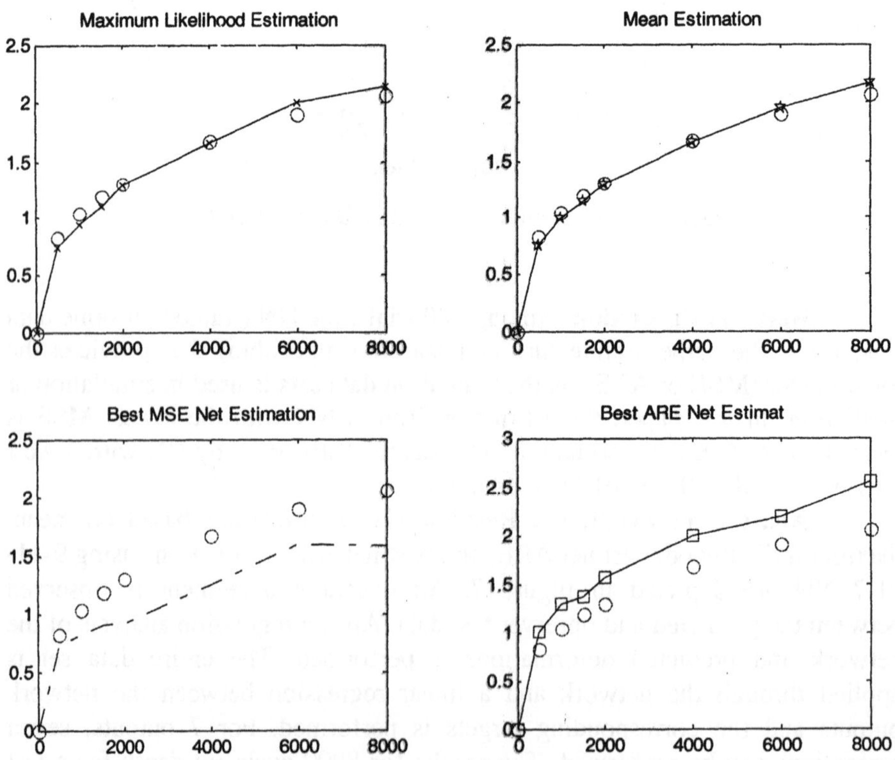


Figure 7. Observed rutting versus neural network predicted rutting.

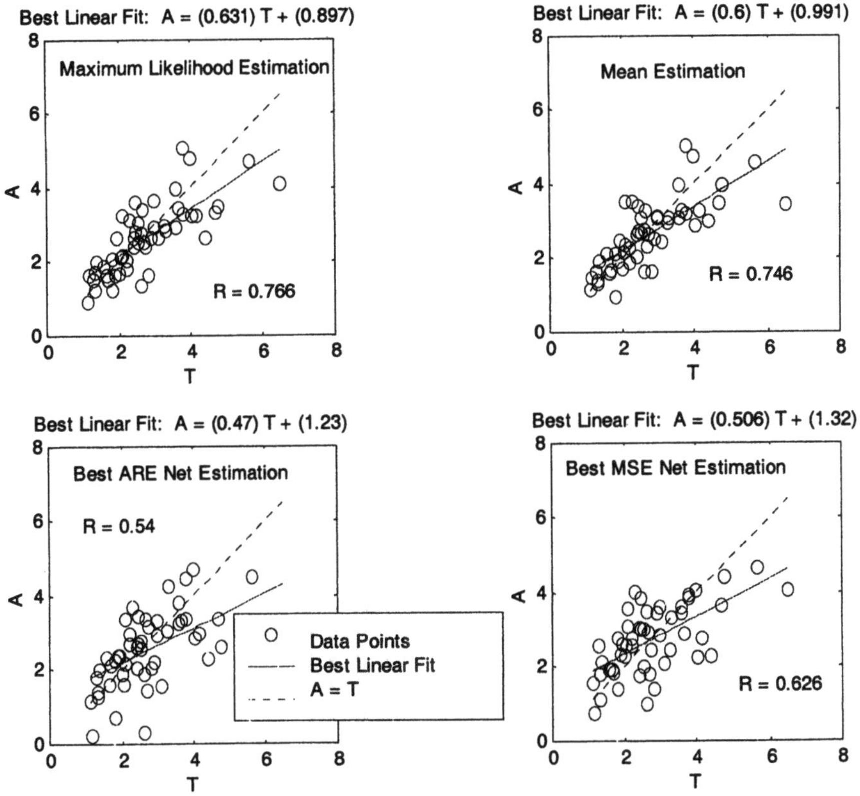


Figure 8. Regression plot for 8000-cycle rut estimation using 9-11-11-7 NN (A=actual rut depth, T= target rut depth)

# CONCLUSIONS

In this study, 3-layer and 4-layer neural networks are designed to determine the rutting performance of asphalt concrete. A total of 519 sets of processed data obtained from mix design information and laboratory tests are used for developing this NN model. Finally, the NN selected has 11 neurons in each hidden layer, whereas the output layer uses a total of 7 neurons. Using a total of 21 inputs, the developed model produces outputs (rut depths) at 7 different cycles. The time series of deformation recorded over 8000 cycles are determined by an interpolation with piecewise linear elements, using these few outputs. Preprocessing and principal component analyses are applied, and the network trained using the Levenberg-Marquardt algorithm. Using randomly generated weight factors to initialize the training algorithm, histograms are compiled and outputs estimated using statistical estimators. An excellent agreement is observed between test data and simulations. It is believed that the developed NN design procedure will be a useful tool in the study of pavement design and wear.

## ACKNOWLEDGEMENT

The financial support for this study was provided in part by the Oklahoma Department of Transportation, in cooperation with the Federal Highway Administration. The authors are thankful to Dr. Luther White, Professor of Mathematics, University of Oklahoma, for his advice and technical assistance.

# REFERENCES

Demuth, H., and Beale, M. (1998). "Matlab Neural Network Toolbox." *Math Works, Inc.*, Novi, Mi.

Engelbrecht, A. P., (2002) *Computational Intelligent*, John Wiley & Sons, Ltd., 2002.

Fine, T., (1998). *Feedforward Neural Network Methodology.* Springer-Verlag, NY.

Hagan, M., Demuth, H., and Beale, M. (1996). "Neural Network Design." *PWS Publishing Co.*, Boston, Ma.

Haykin S., (1994). *Neural Networks: A Comprehensive Foundation*, MacMillan, 1994.

Hertz, J., Krogh, A., and Palmer, R. G., *Introduction to the Theory of Neural Computation*, Addison-Wesley Publishing Company, New York, pp. 130–141, 1991.

Hornik, K., Stinchcombe, M., White, H., and Auer, P., (1994). "Degree of approximation Results for Feedforward Networks Approximating Unknown Mappings and Their Derivatives," *Neural Computing*, Vol 6, pp. 1262-1275.

Kearns, M., (1997). "A Bound on The Error of Cross Validation Using the Approximation and Estimation Rates, with Consequences for the Training-Test Split," *Neural Computation*, Vol. 9, 1143-1161.

OHD, Oklahoma Highway Department (2001). "Method of Test for Determining Rutting Susceptibility Using the Asphalt Pavement Analyzer," *Oklahoma Highway Department Laboratory Test No. 43 (OHD L 43)*, Oklahoma City, Oklahoma.

Ramsamooj, D. V., Ramadan, J., and Lin, G. S. (1998). "Model Prediction of Rutting in Asphalt Concrete." *Journal of Transportation Engineering, Vol. 124, No. 5*, September/October, American Society of Civil Engineers, pp. 448-456.

Tarefder R. A. and Zaman, M., (2002). "Evaluation of Rutting Potential of HMA Using the Asphalt Pavement Analyzer," *Final Report*, ORA: 125-6660, ODOT.

Tarefder, R. A., Zaman, M. M., and Hobson, K. (2002). "Laboratory Assessment of Binders' Contribution to Rutting Susceptibility," *International Journal of Pavement (IJP)*, Vol. 1, No. 2, pp. 36-47.

Tarefder, R. A, White, L., and Zaman, M. M. (2004). 'Neural Network Modeling of Asphalt Concrete Permeability,' Transportation Research Board (TRB) 83rd Annual Meeting, Washington D.C., CD-ROM paper, 04-2878, pp. 221-243.

# Chapter 10

# NEURAL NETWORKS FOR RESIDENTIAL INFRASTRUCTURE MANAGEMENT

**Deidre E. Paris**

## 1. INTRODUCTION

Over the last decade, there has been a rapid acceptance of new technologies like neural networks for solving a wide range of business problems. While neural networks have developed from the field of artificial intelligence and brain modeling, neural networks are nothing more than function approximation tools that learn the relationship between independent variables and dependent variables, much like regression or other more traditional approaches. The principal difference between neural networks and statistical approaches is that neural networks make no assumptions about the statistical distribution or properties of the data, and therefore tend to be more useful in practical situations. Neural networks are also an inherently nonlinear approach giving them much accuracy when modeling complex data patterns. An Artificial Neural Network (ANN) is an information processing paradigm that is inspired by the way biological nervous systems, such as the brain, process information. The key element of this paradigm is the novel structure of the information processing system. It is composed of a large number of highly interconnected processing elements (neurons) working in unison to solve specific problems. ANNs, like people, learn by example. An ANN is configured for a specific application, such as pattern recognition or data classification, through a learning process. ANNs are based on the neural structure of the brain in that the brain basically learns from experience.

## 2. NEURAL NETWORK HISTORY

The study of the human brain is thousands of years old. In 1943, Warren McCulloch, a neurophysiologist, and a young mathematician, Walter Pitts, wrote a paper on how neurons might work. They modeled a simple neural network with electrical circuits. Reinforcing this concept of neurons and how they work was a book entitled *Organization of Behavior* by Donald

Hebb written in 1949. It pointed out that neural pathways are strengthened each time that they are used.

As computers advanced into their infancy of the 1950s, it became possible to begin to model the theories concerning human thought. Nathanial Rochester from the IBM research laboratories led the first effort to simulate a neural network. That first attempt failed. But later attempts were successful. It was during this time that traditional computing began to flower and, as it did, the emphasis in computing left the neural research in the background. Yet, throughout this time, advocates of "thinking machines" continued to argue their cases. In 1956 the Dartmouth Summer Research Project on Artificial Intelligence provided a boost to both artificial intelligence and neural networks. One of the outcomes of this process was to stimulate research in both the intelligent side, AI (Artificial Intelligence), as it is known throughout the industry, and in the much lower level neural processing part of the brain.

In the years following the Dartmouth Project, John von Neumann suggested imitating simple neuron functions by using telegraph relays or vacuum tubes. Also, Frank Rosenblatt, a neuro-biologist of Cornell, began work on the Perceptron. He was intrigued with the operation of the eye of a fly. Much of the processing which tells a fly to flee is done in its eye. The Perceptron, which resulted from this research, was built in hardware and is the oldest neural network still in use today. A single-layer perceptron was found to be useful in classifying a continuous-valued set of inputs into one of two classes. The perceptron computes a weighted sum of the inputs, subtracts a threshold, and passes one of two possible values out as the result. Unfortunately, the perceptron is limited and was proven as such during the "disillusioned years" in Marvin Minsky and Seymour Papert's 1969 book *Perceptrons*.

In 1959, Bernard Widrow and Marcian Hoff of Stanford developed models they called ADALINE and MADALINE. These models were named for their use of Multiple ADAptive LINear Elements. MADALINE was the first neural network to be applied to a real world problem. It is an adaptive filter which eliminates echoes on phone lines. This neural network is still in commercial use.

Unfortunately, these earlier successes caused people to exaggerate the potential of neural networks, particularly in light of the limitation in the electronics then available. This excessive hype, which flowed out of the academic and technical worlds, infected the general literature of the time. Disappointment set in as promises were unfilled. Also, a fear set in as writers began to ponder what effect "thinking machines" would have on man. Asimov's series on robots revealed the effects on man's morals and values when machines where capable of doing all of mankind's work. These fears, combined with unfulfilled, outrageous claims, caused respected voices to critique the neural network research. The result was to halt much of the funding. This period of stunted growth lasted through 1981.

In 1982 several events caused a renewed interest. John Hopfield of Caltech presented a paper to the national Academy of Sciences. Hopfield's approach was not to simply model brains but to create useful devices. With clarity and mathematical analysis, he showed how such networks could work and what they could do. Yet, Hopfield's biggest asset was his charisma. He was articulate, likeable, and a champion of a dormant technology.

At the same time, another event occurred. A conference was held in Kyoto, Japan. This conference was the US-Japan Joint Conference on Cooperative/Competitive Neural Networks. Japan subsequently announced their Fifth Generation effort. US periodicals picked up that story, generating a worry that the US could be left behind. Soon funding was flowing once again.

By 1985 the American Institute of Physics began what has become an annual meeting - Neural Networks for Computing. By 1987, the Institute of Electrical and Electronic Engineer's (IEEE) first International Conference on Neural Networks drew ore than 1,800 attendees.

By 1989 at the Neural Networks for Defense meeting Bernard Widrow told his audience that they were engaged in World War IV, "World War III never happened," where the battlefields are world trade and manufacturing. The 1990 US Department of Defense Small Business Innovation Research Program named 16 topics which specifically targeted neural networks with an additional 13 mentioning the possible use of neural networks. Today, neural networks discussions are occurring everywhere. Their promise seems very bright as nature itself is the proof that this kind of thing works. Yet, its future, indeed the very key to the whole technology, lies in hardware development. Currently most neural network development is simply proving that the principal works.

## 3. THE HUMAN BIOLOGICAL NEURON AND ARTIFICIAL NEURON

The fundamental processing element of a neural network is a neuron. Basically, a biological neuron receives inputs from other sources, combines them in some way, performs a generally nonlinear operation on the result, and then outputs the final result. As shown in Figure 1, in the human brain, a typical neuron collects signals from others through a host of fine structures called *dendrites*. The neuron sends out spikes of electrical activity through a long, thin stand known as an *axon*, which splits into thousands of branches. At the end of each branch, a structure called a *synapse* converts the activity from the axon into electrical effects that inhibit or excite activity from the axon into electrical effects that inhibit or excite activity in the connected neurons. When a neuron receives excitatory input that is sufficiently large compared with its inhibitory input, it sends a spike of electrical activity down its axon. Learning occurs by changing the effectiveness of the synapses so that the influence of one neuron on another changes. Basically, a biological neuron receives inputs from other sources, combines them in some way, performs a generally nonlinear operation on the result, and then output the final result. The figure below shows a simplified biological neuron and the relationship of its four components.

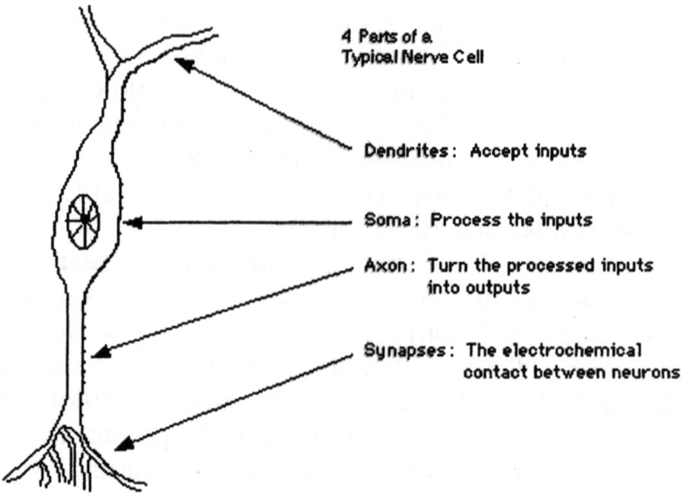


Figure 1. Parts of a typical human nerve cell

## 3.1 The Artificial Neuron

The basic unit of neural networks, the artificial neurons, simulates the four basic functions of natural neurons. Artificial neurons are much simpler than the biological neuron; Figure 2 below shows the basics of an artificial neuron.

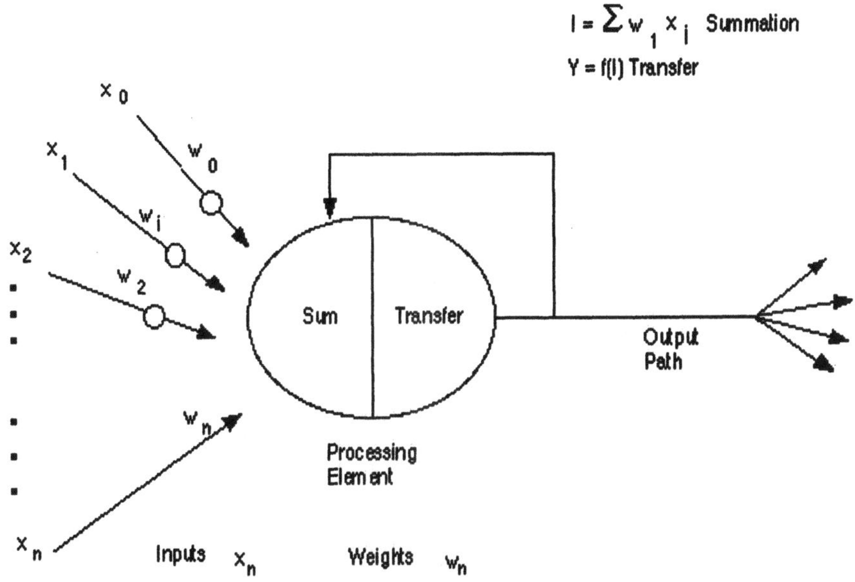


Figure 2. Model of an artificial neuron

Note that various inputs to the network are represented by the mathematical symbol, x(n). Each of these inputs are multiplied by a connection weight, these weights are represented by w(n). In the simplest case, these products are simply summed, fed through a transfer function to generate a result, and then output. Even though all artificial neural networks are constructed from this basic building block the fundamentals may vary in these building blocks and there are differences.

# 4. NEURAL NETWORK MODELS

In this section we provide details of three popular neural network models. The equations presented in this section and their derivation is taken from research on neural networks in business (Smith and Gupta, 2000). Each model is presented in terms of its purpose, architecture, and algorithm. Each of these models has some similarity to more traditional statistical and operations research techniques, and the relationships to the analogous traditional techniques are
discussed.

## 4.1 Neural Network Design

Designing a neural network consists of:

- Arranging neurons in various layers.
- Deciding the type of connections among neurons for different layers, as well as among the neurons within a layer.
- Deciding the way a neuron receives input and produces output.
- Determining the strength of connection within the network by allowing the network learn the appropriate values of connection weights by using a training data set.

There are several types of neural networks, each with a different purpose, architecture and learning algorithm, and these are outlined in the following sections.

### 4.1.1 Multilayered Feedforward Neural Networks

According to a recent study (Wong, Bodnovich and Selvi, 1997), approximately 95% of reported neural network business application studies utilize multilayered feedforward neural networks (MFNNs) with the back propagation learning rule. This type of neural network is popular because of its broad applicability to many problem domains of relevance to business: principally prediction, classification, and modeling. MFNNs are appropriate for solving problems that involve learning the relationships between a set of inputs and known outputs. They are a supervised learning technique in the sense that they require a set of training data in order to learn the relationships.

The MFNN architecture is shown in Figure 3 and consists of two or more layers of neurons connected by weights. The flow of information is from left to right, with inputs **x** being passed through the network via the hidden layer of neurons to the output layer. The weights connecting input element *i* to hidden neuron *j* are denoted by $W_{ji}$ , while the weights connecting hidden neuron *j* to output neuron *k* are denoted by $V_{kj}$ .

Each neuron calculates its output based on the amount of stimulation it receives from the given input vector **x**. More specifically, a neuron's net input is calculated as the weighted sum of its inputs, and the output of the neuron is based on a sigmoidal function indicating the magnitude of this net input. That is, for the jth hidden neuron

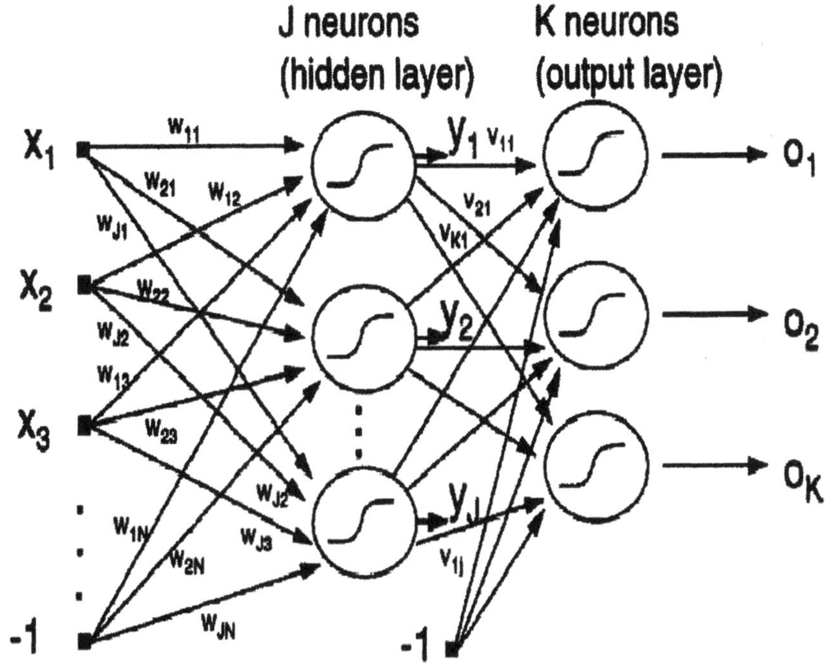


Figure 3. Architecture of Multilayered Feedforward Neural Networks (MFNNs)

$$net_j^h = \sum_{i=1}^{N} W_{ji} x_i \quad \text{and} \quad y_j = f(net_j^h), \tag{Eq. 4.1}$$

while for the $k$th output neuron

$$net_k^o = \sum_{j=1}^{J+1} V_{kj} y_j \quad \text{and} \quad o_k = f(net_k^o) \tag{Eq. 4.2}$$

Typically, the sigmoidal function f (net) is the well-known logistic function

$$f(net) = \frac{1}{1 + e^{-\lambda net}} \tag{Eq. 4.3}$$

where $\lambda$ is a parameter used to control the gradient of the function, although the only requirement is it to be bounded between 0 and 1, monotonically increasing, and differentiable.

For a given input pattern, the network produces an output (or set of outputs) $o_k$, and this response is compared to the known desired response of each neuron $d_k$. The weights of the network are then modified to correct or reduce the error, and the next pattern is presented. The weights are continually modified in this manner until the total error across all training patterns is reduced below some pre-defined tolerance level (or the network has started to "overtrain" as measured by deteriorating performance on the test set (Zurada, 1992).

The weight update rule for the output layer weights V is given by

$$V_{kj}(t+1) = v_{kj}(t) + c\lambda(d_k - o_k)o_k(1 - o_k)y_j(t)$$

(Eq. 4.4)

and for the hidden layer weights W by

$$W_{ji}(t+1) = w_{ji}(t) + c\lambda^2 y_j(1 - y_j)x_i(t)\left(\sum_{k=1}^{K}(d_k - o_k)o_k(1 - o_k)v_{kj}\right)$$

(Eq. 4.5)

Proof that the elect of these weight updates minimizes the total average-squared error

$$E = \frac{1}{2P}\sum_{p=1}^{P}\sum_{k=1}^{K}(d_{pk} - o_{pk})^2$$

(Eq. 4.6)

where $d_{pk}$ is the desired output of neuron $k$ for input pattern $p$, and $o_{pk}$ is the actual network output of neuron $k$ for input pattern $p$, relies on the fact that the algorithm (known as the backpropagation learning algorithm) performs steepest descent on this error function (Zurada, 1992).

There are many training issues involved in applying MFNNs successfully, including ensuring that the learned relationships generalize well to new data. To ensure this, data are typically divided into a training and a test set, where the performance on the test set is used to indicate the generalization of the neural network results. Other issues involve optimal selection of the many training parameters including the number of hidden neurons, the learning rate $c$, the initial weights, and the slope of the sigmoidal function $\lambda$. Convergence to local minima of the error function (Eq. 4.6) is also a concern, since this means that the final combination of weights will always produce an error. Researchers have recently started using heuristics approaches like genetic algorithms instead of the backpropagation learning rule to determine the optimal weights for the MFNN to minimize the total average-squared error (Gupta and Sexton, 1999; Montana, 1995; Sexton, Gupta, Smith and Montagno, 1998).

The MFNN, with an algorithm for determining the optimal weights for a given training set of data (backpropagation or heuristic algorithm), can be seen as similar to any function approximation technique like regression, where the weights are analogous to regression coefficients estimated by least squares. The difference of course is the improved potential of the function approximation when learning highly complex and nonlinear data due to the increased number of free parameters.

### 4.1.2 Hopfield Neural Networks

While MFNNs learn the relationships between inputs and outputs in a supervised manner, Hopfield neural networks are completely different, in function, architecture and approach. With MFNNs, the neurons are connected in layers, and the weights are modified throughout the algorithm to reflect the learning process. With Hopfield networks however, there is no layer structure to the architecture, and the weights do not change. Hopfield networks (Hopfield, 1982) are a fully interconnected system of N neurons as shown in Figure 4 for N=4. The weights of the network $W_{ij}$ are fixed and symmetric ($W_{ij}=W_{ji}$), and store information about the memories or stable states of the network. Each neuron has a state $x_i$ which is bounded between 0 and 1. Neurons are updated according to a differential equation, and over time an energy function is minimized. The local minima of this energy function correspond to the stable states of the network.

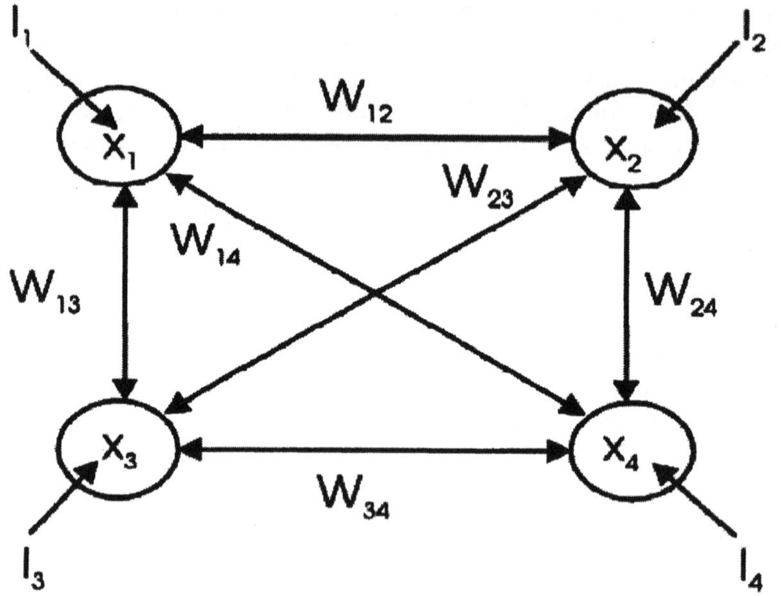


Figure 4. Architecture of Hopfield neural network

Hopfield networks are principally used to solve optimization problems of the kind familiar to the operations researcher. Hopfield and Tank (Hopfield and Tank, 1985) showed that the weights of a Hopfield network can be chosen so that the process of neurons updating simultaneously minimizes the Hopfield energy function and the optimization problem. Each neuron $i$ updates itself according to the following differential equation

$$\frac{dnet_i}{dt} = -\frac{net_i}{\tau} + \sum_{j=1}^{N} W_{ij}x_j + I_i,$$

$$x_i = f(net_i), \qquad \text{(Eq. 4.7)}$$

where f (.) is a sigmoidal output function bounded by 0 and 1 like (Eq. 4.3) and $\tau$ is a constant. These equations are similar to the calculation of a neuron output in the MFNN except that a constant term $I$ has been added to the net input of each neuron, and the time dynamics are now continuous (although the process is usually simulated with a discrete Euler approximation). Each time a neuron is updated in this manner, the energy function

$$E = -\frac{1}{2}\sum_{i=1}^{N}\sum_{j=1}^{N} W_{ij}x_i x_j - \sum_{i=1}^{N} I_i x_i$$

(Eq. 4.8)

is reduced. In fact, this energy function is a Liapunov function for the system and is guaranteed not to increase (Hopfield, 1982). This proof relies on the fact that the neuron update rules (Eq. 4.7) result in steepest descent of the energy function (Eq. 4.8), just like the weight update rules (Eq. 4.4) and (Eq. 4.5) of the MFNN with backpropagation result in steepest descent of the error function (Eq. 4.6).

The approach to solving optimization problems using Hopfield networks is to choose the weights $W_{ij}$ and constant terms $I_i$ to force the energy function and the optimization objective function to be equivalent. The optimization problem is expressed as a single function to be minimized, which incorporates all costs and constraints of the problem using a penalty function approach. Notice that the weights $W_{ij}$ are simply the coefficients of the quadratic terms $x_i x_j$ in the energy function, while the constant terms $I_i$ are the coefficients of the linear terms $x_i$ . Once the network weights and constants have been chosen, the neuron states $x_i$ are randomly initialized, and the neurons begin updating in a random sequence according to differential (Eq. 4.7). Over time, the energy function minimizes until the neuron states have stabilized, and the final neuron states correspond to a local minimum solution of the optimization problem. This solution may not necessarily be a feasible one or a good one since the penalty function treatment of the cost and constraints means that a balance needs to be found between which components of the energy function are minimized. Penalty function parameters need to be selected to reflect the relative degree of difficulty in minimizing each component of the energy function. Numerous researchers have tried to alleviate this problem by modifying the energy function form (Brandt, Wang, Laub and Mitra, 1988), or by analytically choosing values for the penalty parameters (Hegde, Sweet and Levy, 1988; Lai and Coghill, 1992).

Clearly, Hopfield networks are a steepest descent technique for solving an optimization problem using a penalty function approach. The performance of Hopfield networks has been improved by incorporating hill-climbing strategies into the neuron update equations (Eq. 4.7), like simulated annealing (Smith, Palaniswami and Krishnamoorthy, 1996). Variations of the Hopfield network include Boltzmann machines (Ackley, Hinton and Sejnowski, 1985) and mean-field annealing (Van Den Bout and Miller III,

1989). Enhancements to these approaches such as neuron normalization have enabled certain hard constraints to be enforced by the neuron updating, rather than relying on a penalty function approach (Van Den Bout and Miller III, 1989).

### 4.1.3 Self-Organizing Neural Networks

For many decades, statisticians have used discriminant analysis and regression to model the patterns within data when there are labeled training data (with inputs and known outputs) available, and clustering techniques when no such data are available. These techniques find analogies in neural networks, where MFNNs are used with back propagation when training data are available, and self-organizing neural networks are used as a clustering technique when no training data are available. Clustering has always been used to group the data based upon the natural structure of the data. The objective of an appropriate clustering algorithm is that the degree of similarity of patterns within a cluster is maximized, while the similarity these patterns have with patterns belonging to different clusters is minimized.

Often patterns in a high-dimensional input space have a very complicated structure, but this structure is made more transparent and simple when they are clustered in a one, two or three dimensional feature space. Kohonen developed self-organizing feature maps (SOFMs) as a way of automatically detecting strong features in large data sets. SOFMs find a mapping from the high-dimensional input space to low-dimensional feature space, so the clusters that form become visible in this reduced dimensionality (Kohonen, 1982; 1988). In comparison with the two previous neural network models discussed, the SOFM involves adapting the weights to reflect learning (like the MFNN with back propagation) but the learning is unsupervised since the desired network outputs are unknown. Another significant difference between the SOFM and the previous models is the architecture and the role of neuron locations in the learning process. In the SOFM, input vectors are connected to an array of neurons, usually one-dimensional (a row) or two-dimensional (a lattice). Figure 5 shows this architecture for $n$ inputs and a square array of nine neurons.

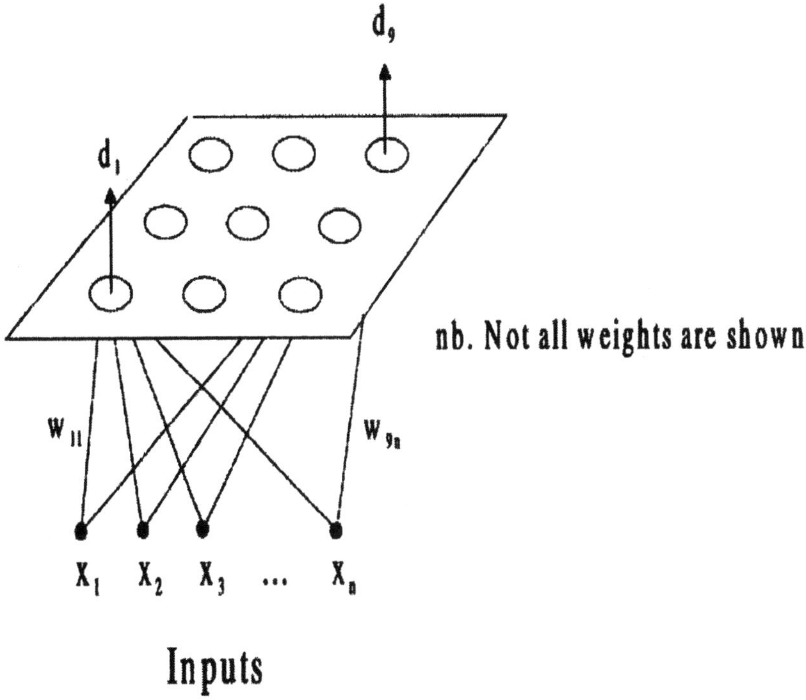


Figure 5. Architecture of a SOFM with nine neurons

When an input pattern is presented to the SOFM, certain regions of the array will become active, and the weights connecting the inputs to those regions will be strengthened. Once learning is complete, similar inputs will result in the same region of the array becoming active or "firing". Central to this idea is the notion of the ordering and physical arrangement of the neurons. With SOFMs the ordering of the neurons is important since we are referring to regions of neurons firing. If a neuron fires, it is likely that its neighbors will also fire, and thus for the first time we are concerned with the physical location of the neurons. This idea has more biological justification than the other neural models, since the human brain involves large regions of neurons operating in a centralized and localized manner to achieve tasks. In the human brain, as in the SOFM, there is usually a clear "winning neuron" which fires the most upon receiving an input signal, but the surrounding neurons also get affected by this, firing a little, and the entire region becomes active.

In order to replicate the response of the human brain in the SOFM, the learning process is modified so that the winning neuron (defined as the neuron whose weights are most similar to the input pattern) receives the most learning, but the weights of neurons in the neighborhood of the winning neuron are also strengthened, although not as much. It is appropriate at this point to define the concept of a neighborhood in relation to the architecture of the SOFM. For a linear array of neurons, the neighbors are simply the neurons to the left and right of the winner. This is called a neighborhood size of one. To achieve the effect of an active region of neurons, we need to consider larger neighborhood sizes, as shown in Figure 6 for rectangular array of neurons, with a hexagonal neighborhood structure.

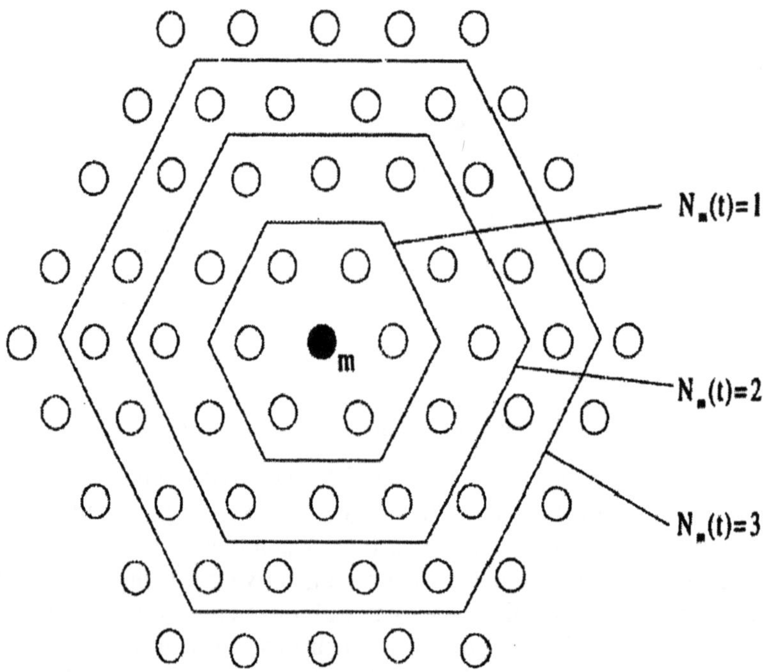


Figure 6. Concept of neighborhood size for a rectangular array of neurons

Initially the neighborhood size around a winning neuron is allowed to be quite large to encourage the regional response to inputs, but as the learning proceeds, the neighborhood size is slowly decreased so that the response of the network becomes more localized. The localized response, which is needed to help clearly differentiate distinct input patterns, is also encouraged by varying the amount of learning received by each neuron within the winning neighborhood. The winning neuron receives the most learning at any stage; with neighbors receiving less the further away they are from the winning neuron.

The size of the neighborhood around winning neuron *m* at time *t* is denoted by $N_m(t)$. The amount of learning that every neuron *I within* the neighborhood of *m* receives is determined by

$$c = \alpha(t)\exp(-\|r_i - r_m\|/\sigma^2(t))$$

(Eq. 4.9)

where $\|r_i - r_m\|$ is the physical distance (number of neurons) between neuron *i* and the winning neuron *m*. The two functions $\alpha(t)$ and $\sigma^2(t)$ are used to control the amount of learning each neuron receives in relation to the winning neuron. These functions can be slowly decreased over time. The amount of learning is greatest at the winning neuron (where $i=m$ and $r_i = r_m$) and decreases the further away a neuron is from the winning neuron, as a result of the exponential function. Neurons outside the neighborhood of the winning neuron receive no learning.

Like the other neural network models considered thus far, the learning algorithm for the SOFM follows the basic steps of presenting input patterns, calculating neuron outputs, and updating weights. The differences lie in the method used to calculate the neuron output (this time based on the similarity between the weights and the input), and the concept of a neighborhood of weight updates. The steps of the algorithm are as follows:

***Step 1*:** Initialize

- weights to small random values
- neighborhood size $N_m(0)$ to be large (but less than the number of neurons in one dimension of the array)
- parameter functions $\alpha(t)$ and $\sigma^2(t)$ to be between 0 and 1

***Step 2*:** Present an input pattern x through the input layer and calculate the closeness
(distance) of this input to the weights of each neuron *j*:

$$d_j = \|\mathbf{x} - \mathbf{w}_j\| = \sqrt{\sum_{i=1}^{n} (x_i - w_{ij})^2}$$

(Eq. 4.10)

***Step 3*:** Select the neuron with minimum distance as the winner *m*

***Step 4*:** Update the weights connecting the input layer to the winning neuron and its neighboring neurons according to the learning rule

$$w_{ji}(t+1) = w_{ji}(t) + c[x_i - w_{ji}(t)]$$

where

$$c = \alpha(t)\exp(-\|r_i - r_m\|/\sigma^2(t))$$

for all neurons j in $N_m(t)$
***Step 5:*** Continue from STEP 2 for $\Omega$ epochs; then decrease neighborhood size, $\alpha(t)$ and $\sigma^2(t)$: Repeat until weights have stabilized.

SOFMs have been predominantly used for clustering and feature extraction, finding application as a data mining technique. As such, they are comparable to traditional clustering techniques like the k-means algorithm (Hartigan, 1975). There has also been quite a significant amount of research undertaken in using SOFMs for solving optimization problems as an alternative to the Hopfield neural networks discussed in the previous section. This involves combining the ideas of the SOFM with the elastic net algorithm (Durbin and Willshaw, 1987) to solve Euclidean problems like the traveling salesman problem (Favata and Walker, 1991; Goldstein, 1990). In recent work, a modified SOFM has been used to solve broad classes of optimization problems by freeing the technique from the Euclidean plane.

### 4.1.3 Other Neural Network Models

There are many other different types of neural network models, each with their own purpose and application areas. Most of these are extensions of the three main models we have discussed in this section. Their potential application to problems of concern to the business world and the operations researcher is unclear, but they are referenced here for completeness. These other neural network models include adaptive resonance networks (Carpenter and Grossberg, 1988), radial basis networks (Broomhead and Lowe, 1988), modular networks (Jacobs and Jordon, 1991), neocognitron (Fukushima, 1980), brain-state-in-a-box (Anderson, Silverstein, Ritz and Jones, 1977), to name just a few.

## 5. RESIDENTIAL SATISFACTION DECISION SUPPORT SYSTEM MODEL

The residential satisfaction decision support system is a multilayered feedforward neural network. The neural network is trained using Defoors train dataset. The data is divided into two groups: input variables and an output variable. The inputs are the independent research variables specified in the model; the output variable SATIS is the dependent variable. The train dataset is made up of data rows, which makes up a set of corresponding independent variables and a dependent variable. These data rows are also referred to as cases.

The decision support system is developed by first training the neural network. Training a neural network refers to the process of the model "learning" the patterns in the training dataset in order to make classifications. The training dataset includes many sets of input variables and a corresponding output variable. When the value of an input variable is fed into an input neuron, the network begins by finding linear relationships between the input variables and the output variable. Weight values are assigned to the links between the input and output neurons; every link has a weight that indicates the strength of the connection. The weights of the network are set randomly when it is first being trained. After all the rows of Defoors' dataset are passed through the network, the answer the network is producing is repeatedly compared with correct answers, and each time the connecting weights are adjusted slightly in the direction of the correct answer. If the total of the errors of all cases in the dataset is too large, then a hidden neuron is added between the inputs and outputs. The training process is repeated until the average error is within an acceptable range. The errors between the network and the actual result are reduced as more hidden neurons are added. The network has learned the data sufficiently when it has reached an acceptable error and is ready to produce the desired results, which are called classifications, for all of the data rows. The effectiveness of neural networks is demonstrated when the trained network is able to produce good results for data that the network has never seen before. This is examined using the trained network on Moores Mill test dataset.

The neural network output variable is SATIS which indicates residents overall living satisfaction. This variable had four categories that respondents could select from to describe their satisfaction level: 1=very dissatisfied, 2=somewhat dissatisfied, 3=somewhat satisfied and 4=very satisfied. These categories were collapsed into two categories to simplify the neural network model; 1 & 2=NOT SATISFIED and 3 & 4=SATISIFIED. Thus, the residential satisfaction train dataset is clustered into 2 categories: NOT SATISFIED and SATISFIED. Table 1 provides definitions of the input variables that were used to train the neural network.

# 6. NEURAL NETWORK ANALYSIS RESULTS

There were 18 input variables used to train the neural network and SATIS was the output variable. The neural network generated 79 hidden neurons during training; in which 56 neurons was the optimal number of hidden neurons that best solves the classification problem. The training time, or time it took for the network to learn before it was able to make accurate classifications was 49 seconds.

Figure 7 shows the number of hidden neurons graphed against the percentage of correct classifications. The vertical line between the curve and the x-axis shows that the network needed 56 hidden neurons during training before it can make correct classifications on the dataset.

*Table 1. Input data for neural network.*

| Variable Name | Definition | Variable Name | Definition |
|---|---|---|---|
| SATPROMAN | How satisfied residents are with the property management staff. | SATUNIT | How satisfied residents are with their apartment units. |
| TENANTPOLICIES | How satisfied residents are with property management's tenant selection policies. | QUICKCOMPLY | How satisfied residents are with the expediency of staff responding to complaints. |
| RFAIRLY | How satisfied residents are with property management enforcing rules fairly. | REPAIRSQUALITY | How satisfied residents are with the quality of maintenance repairs. |
| TALK | How satisfied residents are with availability of property management staff to address residents' concerns. | CLEANNESS | How satisfied residents are with the overall cleanliness of the property. |
| COOPERATIVE | How satisfied residents are with the ability of property management staff to cooperate with residents. | COMMUNCLEAN | How satisfied residents are with the cleanliness of the community that surrounds the apartment complex. |
| FRIENDLY | How satisfied residents are with property management level of friendliness towards residents. | SATCOM | How satisfied residents are with the community that surrounds the apartment complex. |
| RECOMMEND[1] | If residents will recommend their apartment complex to a friend as a place to live. | SAFENIGHTHOOD[3] | How safe residents feel during the night in their neighborhood. |
| QUALLIFE[2] | Residents' quality of life after renovations. | SATMAINTEN | How satisfied residents are with the property's maintenance staff. |
| BLDQUALITY | How satisfied residents are with the quality of the apartment buildings on the property. | SAFENIGHT[3] | How safe residents feel during the night at their apartment complex. |

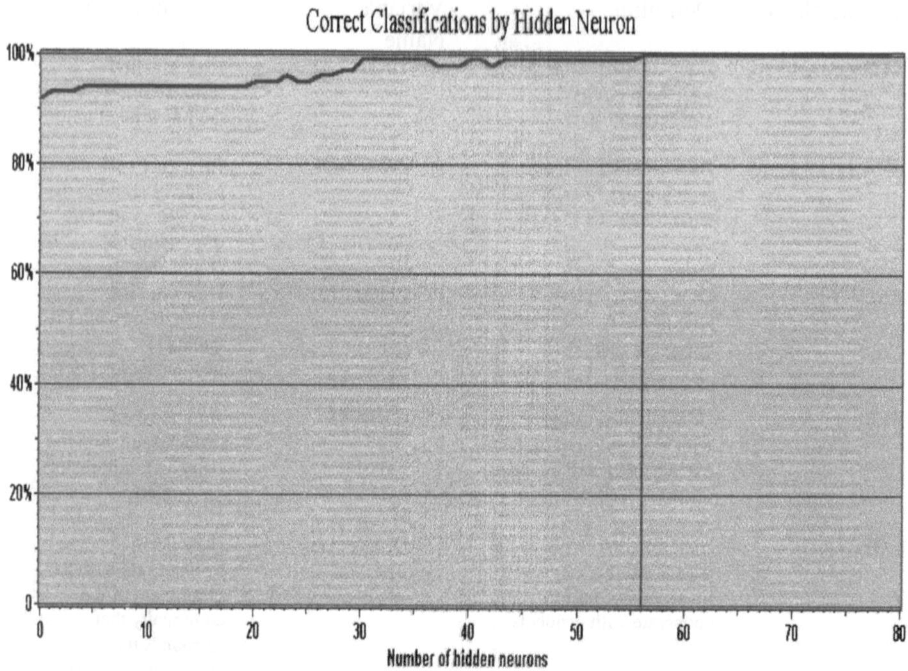


Figure 7. Graphical display of correct classifications by number of hidden neurons

## 6.1 Actual and Predicted Outputs

Table 2 displays the actual and classified outputs for all the data rows in the trained dataset. This table displays results for every row in the data file to which the net was applied. The Row Number column is the number of the row in the data file for each example. An asterisk is displayed beside the row number that the model makes an incorrect classification. The Actual column displays the category classification as it appears in the data file. The Classified column displays the category classification predicted by the network; the classification is either satisfied or not satisfied. The Not Satisf. and Satisf. columns are output classification categories and display the network's classification strength for each category. This value is the neuron activation strength for each category based on that set of input values. This value can loosely be thought of as a probability; the values for all categories add up to 1. When the value is close to 1 in a category, the network is more confident that the example set of inputs belongs to that particular category. As shown in the Table 2 below, there were only 2 data rows (rows #15 & #64) that the network classified incorrectly. These two rows were classified as satisfied with a weight value of .998 (row #15) and .749 (row #64).

*Table 2.* Actual and classified outputs for all rows of trained data.

| Row Number | Actual | Classified | Not Satisf. | Satisf. | Row Number | Actual | Classified | Not Satisf. | Satisf. |
|---|---|---|---|---|---|---|---|---|---|
| 1 | satisf. | satisf. | 0.000 | 1.000 | 2 | satisf. | satisf. | 0.003 | 0.997 |
| 3 | not sa. | not sa. | 0.999 | 0.001 | 4 | not sa. | not sa. | 0.995 | 0.005 |
| 5 | satisf. | satisf. | 0.000 | 1.000 | 6 | satisf. | satisf. | 0.000 | 1.000 |
| 7 | satisf. | satisf. | 0.000 | 1.000 | 8 | not sa. | not sa. | 0.992 | 0.008 |
| 9 | not sa. | not sa. | 0.997 | 0.003 | 10 | satisf. | satisf. | 0.000 | 1.000 |
| 11 | satisf. | satisf. | 0.000 | 1.000 | 12 | satisf. | satisf. | 0.000 | 1.000 |
| 13 | not sa. | not sa. | 0.999 | 0.001 | 14 | satisf. | satisf. | 0.000 | 1.000 |
| 15 * | not sa. | satisf. | 0.002 | 0.998 | 16 | satisf. | satisf. | 0.002 | 0.998 |
| 17 | satisf. | satisf. | 0.000 | 1.000 | 18 | satisf. | satisf. | 0.000 | 1.000 |
| 19 | satisf. | satisf. | 0.000 | 1.000 | 20 | satisf. | satisf. | 0.000 | 1.000 |
| 21 | not sa. | not sa. | 0.999 | 0.001 | 22 | satisf. | satisf. | 0.000 | 1.000 |
| 23 | satisf. | satisf. | 0.000 | 1.000 | 24 | satisf. | satisf. | 0.004 | 0.996 |
| 25 | not sa. | not sa. | 0.999 | 0.001 | 26 | satisf. | satisf. | 0.007 | 0.993 |
| 27 | not sa. | not sa. | 0.984 | 0.016 | 28 | satisf. | satisf. | 0.014 | 0.986 |
| 29 | not sa. | not sa. | 0.999 | 0.001 | 30 | satisf. | satisf. | 0.001 | 0.999 |
| 31 | satisf. | satisf. | 0.000 | 1.000 | 32 | satisf. | satisf. | 0.000 | 1.000 |
| 33 | satisf. | satisf. | 0.021 | 0..979 | 34 | not sa. | not sa. | 1.000 | 0.000 |
| 35 | satisf. | satisf. | 0.000 | 1.000 | 36 | satisf. | satisf. | 0.000 | 1.000 |
| 37 | satisf. | satisf. | 0.000 | 1.000 | 38 | satisf. | satisf. | 0.000 | 1.000 |
| 39 | not sa. | not sa. | 0.999 | 0.001 | 40 | satisf. | satisf. | 0.001 | 0.999 |
| 41 | satisf. | satisf. | 0.000 | 1.000 | 42 | satisf. | satisf. | 0.000 | 1.000 |
| 43 | satisf. | satisf. | 0.000 | 1.000 | 44 | satisf. | satisf. | 0.003 | 0.997 |
| 45 | not sa. | not sa. | 1.000 | 0.000 | 46 | satisf. | satisf. | 0.000 | 1.000 |
| 47 | satisf. | satisf. | 0.000 | 1.000 | 48 | satisf. | satisf. | 0.000 | 1.000 |
| 49 | not sa. | not sa. | 0.829 | 0.171 | 50 | satisf. | satisf. | 0.018 | 0.982 |
| 51 | satisf. | satisf. | 0.001 | 0.999 | 52 | satisf. | satisf. | 0.021 | 0.979 |
| 53 | satisf. | satisf. | 0.000 | 1.000 | 54 | satisf. | satisf. | 0.045 | 0.955 |
| 55 | satisf. | satisf. | 0.000 | 1.000 | 56 | satisf. | satisf. | 0.000 | 1.000 |
| 57 | satisf. | satisf. | 0.000 | 1.000 | 58 | satisf. | satisf. | 0.000 | 1.000 |
| 59 | satisf. | satisf. | 0.008 | 0.992 | 60 | satisf. | satisf. | 0.009 | 0.991 |
| 61 | satisf. | satisf. | 0.000 | 1.000 | 62 | satisf. | satisf. | 0.000 | 1.000 |
| 63 | satisf. | satisf. | 0.002 | 0.998 | 64 * | not sa. | satisf. | 0.251 | 0.749 |
| 65 | not sa. | not sa. | 0.947 | 0.053 | 66 | satisf. | satisf. | 0.095 | 0.905 |
| 67 | not sa. | not sa. | 0.790 | 0.210 | 68 | satisf. | satisf. | 0.000 | 1.000 |
| 69 | satisf. | satisf. | 0.001 | 0.999 | 70 | satisf. | satisf. | 0.014 | 0.986 |
| 71 | satisf. | satisf. | 0.000 | 1.000 | 72 | satisf. | satisf. | 0.000 | 1.000 |
| 73 | not sa. | not sa. | 0.742 | 0.258 | 74 | satisf. | satisf. | 0.003 | 0.997 |
| 75 | satisf. | satisf. | 0.000 | 1.000 | 76 | satisf. | satisf. | 0.003 | 0.997 |
| 77 | satisf. | satisf. | 0.066 | 0.934 | 78 | satisf. | satisf. | 0.000 | 1.000 |
| 79 | satisf. | satisf. | 0.000 | 1.000 | 80 | satisf. | satisf. | 0.011 | 0.989 |
| 81 | not sa. | not sa. | 1.000 | 0.000 | 82 | not sa. | not sa. | 1.000 | 0.000 |
| 83 | not sa. | not sa. | 0.996 | 0.004 | 84 | not sa. | not sa. | 0.944 | 0.056 |
| 85 | satisf. | satisf. | 0.000 | 1.000 | 86 | satisf. | satisf. | 0.001 | 0.999 |
| 87 | satisf. | satisf. | 0.000 | 1.000 | 88 | satisf. | satisf. | 0.000 | 1.000 |
| 89 | satisf. | satisf. | 0.001 | 0.999 | 90 | satisf. | satisf. | 0.000 | 1.000 |
| 91 | satisf. | satisf. | 0.000 | 1.000 | 92 | satisf. | satisf. | 0.001 | 0.999 |
| 93 | satisf. | satisf. | 0.000 | 1.000 | 94 | satisf. | satisf. | 0.000 | 1.000 |
| 95 | satisf. | satisf. | 0.000 | 1.000 | 96 | satisf. | satisf. | 0.000 | 1.000 |
| 97 | satisf. | satisf. | 0.016 | 0.984 | 98 | satisf. | satisf. | 0.000 | 1.000 |
| 99 | satisf. | satisf. | 0.087 | 0.913 | | | | | |

*denotes a data row that was classified incorrectly.

## 6.2 Agreement Matrix for Training Network

The agreement matrix shows how the network's classifications compare to the actual classification in the Defoors data file in which the network was applied. Table 3 is the agreement matrix for the trained networking using Defoors data file. Column labels Actual "NOT SATISFIED" and Actual "SATISFIED" refer to the category classification in the data file. The row labels Classified as "NOT SATISFIED" and Classified as "SATISFIED" refer to the network's predictions.

When the network was applied to 99 rows of training data, there were 22 actual examples of residents being "NOT SATISFIED", but the network classified 2 of those cases as "SATISFIED" and 20 as "NOT SATISFIED". There were 77 actual cases of residents being SATISFIED, which the network confirmed.

***Table 3.** Agreement matrix for trained network using Defoors data file*

| | *ACTUAL "NOT SATISFIED"* | *ACTUAL "SATISFIED"* | *TOTAL* |
|---|---|---|---|
| Classified as "NOT SATISFIED" | 20 | 0 | 20 |
| Classified as "SATISFIED" | 2 | 77 | 79 |
| TOTAL | 22 | 77 | 99 |
| True-Positive Ratio | 0.91 | 1.00 | n/a |
| False-Positive Ratio | 0.00 | 0.09 | n/a |
| True-Negative Ratio | 1.00 | 0.90 | n/a |
| False-Negative Ratio | 0.09 | 0.00 | n/a |
| Sensitivity | 90.91% | 100.00% | n/a |
| Specificity | 100.00% | 90.91% | n/a |

### 6.2.1 Explanation of Classifier Statistical Parameters

There are statistical parameters that are specific to the classifier. They reflect the neural network performance compared to the actual classification. These parameters apply to each output classification (SATISFIED and NOT SATISFIED) separately. The following classification parameters are calculated from the comparison of the actual and neural network classification. The neural network classification can be considered as the predicted classification from the network. The actual classification can be considered as the true classification, which comes from the Defoors train database. Below is an explanation for the classifier parameters for *ACTUAL SATISFIED* cases. When the category is *ACTUAL NOT SATISFIED*, the terms are reversed.

*True-Positive Ratio* (also known as Sensitivity): is equal to the number of residents *classified as SATISFIED* by the network that were *actually confirmed to be SATISFIED* (77) through the Defoors train dataset, divided by the *total number of SATISFIED* (77) residents as confirmed by the Defoors train dataset. It is also equal to one minus the *False-Negative ratio.* 77/77=1.00

*False-Positive Ratio*: is equal to the number of residents *classified as SATISFIED* by the network that were *actually confirmed to be NOT SATISFIED* (2) by the Defoors train dataset, divided by the *total number of NOT SATISFIED* (22) residents as confirmed by the Defoors train dataset. It is also equal to one minus the *True-Negative ratio.* 2/22=0.09

*True-Negative Ratio* (also known as Specificity): is equal to the number of residents *classified as "NOT SATISFIED"* by the network that were *actually confirmed to be "NOT SATISFIED"* (20) by the Defoors train dataset, divided by *the total number of "NOT SATISFIED"* (22) residents as confirmed by the Defoors train dataset. It is also equal to one minus the *False-Positive ratio.* 20/22=0.91

*False-Negative Ratio*: is equal to the number of residents *classified as "NOT SATISFIED"* by the network that were *actually confirmed to be "SATISFIED"* (0) by the Defoors train dataset, divided by *the total number of "SATISFIED"* (77) residents as confirmed by the Defoors train dataset. It is also equal to one minus the *True-Positive ratio.* 0/77=0.00

*Sensitivity and Specificity:* The terms sensitivity and specificity come from medical literature, but are now being used for neural network classification problems. Sensitivity and specificity are calculated by comparing the network's results with the 99 rows of training data for all possible output categories (SATISFIED and NOT SATISFIED).

Sensitivity is a concept that can be thought of as the probability that the mode will detect the condition when it is present. Sensitivity (true positives) equals 1 minus the number of false negatives. Examining the column labeled *Actual SATISFIED*:

*Sensitivity (true positives)*: is equal to the number of residents the network *classifies as SATISFIED* that are also confirmed as *SATISFIED* by the Defoors train dataset (77) divided by the total number of residents confirmed as *SATISFIED* by the Defoors train dataset (77). 77/77=1.00 or 100%. This number implies that the sensitivity of the model for satisfaction is 100.00%.

Specificity is a concept that can be thought of as the probability that the network model will detect the absence of a condition. Specificity (true negatives) equals 1 minus the number of false-positives. Examining the column labeled "Actual Satisfied":

*Specificity (true negatives):* equals the number of residents the network *classifies as NOT SATISFIED* that are also confirmed by the Defoors train dataset as *NOT SATISFIED* (20) divided by the total number of residents confirmed as *NOT SATISFIED* by the Defoors train dataset (22). 20/22=.9091 or 90.91%. This number implies that the specificity for the model is 90.91%.

The calculations above for sensitivity and specificity were for the category *Actual SATISFIED*. When the category is *Actual NOT SATISFIED*, the terms are reversed.

## 6.3. ROC (Receiver Operating Characteristic or Relative Operating Characteristic) Curve Graphs For Trained Network

The ROC graphs the false-positive ratio on the x-axis and the true-positive ratio on the y-axis for each classification category. The circle plotted on the curve shows the intersection of the true-positive and the false-positive ratio on the y-axis for each classification category, and converts continuous probabilities to binary classifications for the trained network.

The area under the curve represents how well the network is performing. A value close to 1 means that the network is discriminating very well between the different output categories. The area under ROC curves shown in Figure 8 and Figure 9 below for both, NOT SATISFIED and SATISIED categories, is .9740 which implies that there is a 97.40% chance that the network will make correct classifications.

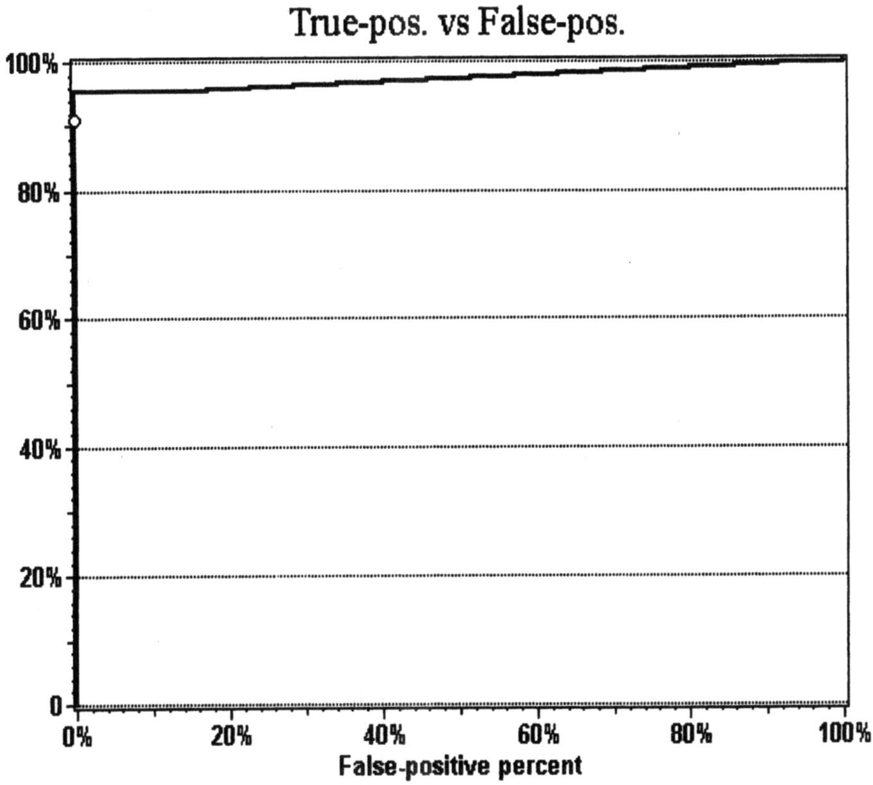


Figure 8. ROC for NOT Satisfied classification for trained data

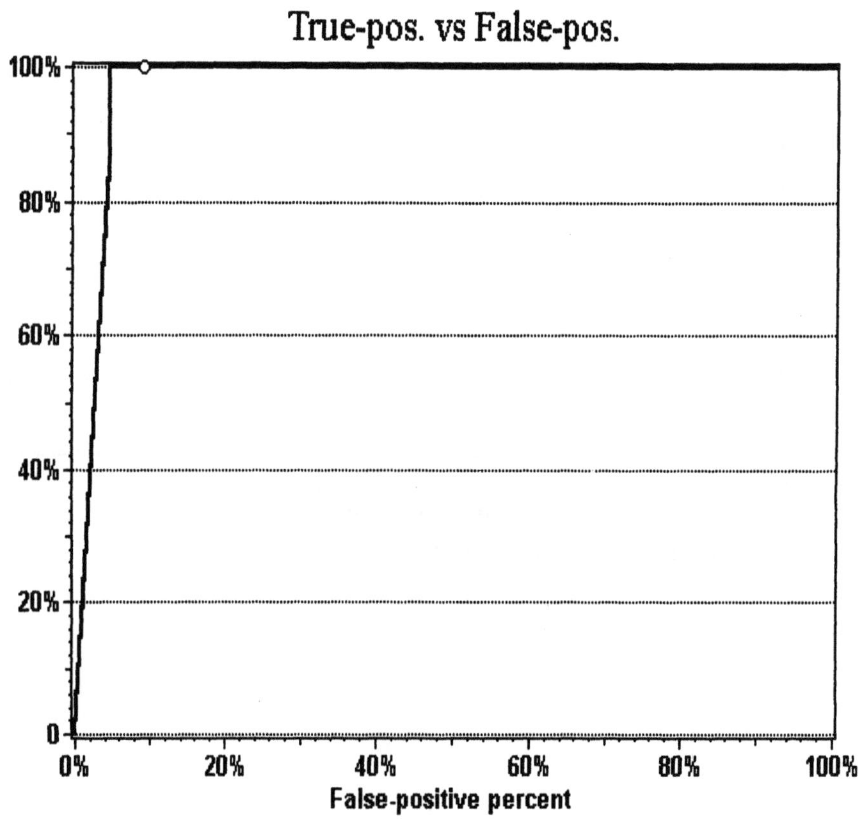


Figure 9. ROC for SATISFIED classification for trained data

# 7. VALIDATION OF NEURAL NETWORK

After the residential satisfaction decision support system was trained using data from Defoors train dataset, the model was validated by running the model on Moores Mill test data and observing how efficient the model was in discriminating between different output categories (NOT SATISFIED and SATISFIED). The Moores Mill test dataset has the same input variables and output variable as the train dataset. There are 80 data rows in the Moores Mill train dataset. Out of the 80 data rows, 70 residents were SATISFIED; 10 were NOT SATISFIED. This section will present similar model validation statistical information that was presented on training the network model.

## 7.1 Actual and Predicted Outputs

Table 4 displays the actual and classified outputs for all the data rows in the test dataset. As shown in this table, there were 4 rows that were classified incorrectly: row numbers 25, 30, 46, and 63. All of these data rows were actually NOT SATISFIED, but the network classified them as SATISFIED. The weights that were assigned to these rows for the SATISFIED classification were respectively, 1.000, 0.814, 0.989, and 0.921.

## 7.2 Network Agreement Matrix for Validating Network

Table 5 is the agreement matrix for validating the network model using Moores Mill data file. When the network was applied to 80 rows of data, there were 10 actual cases of residents being "NOT SATISFIED", but the network classified 4 of those cases as "SATISFIED" and 6 as "NOT SATISFIED". There were 70 actual cases of residents being "SATISFIED", which the network confirmed. A true-positive ratio of 1.00 and a false-positive ratio of .40 were given for the actual SATISFIED classification. The sensitivity which is also refer to as true positive is 100% which implies that there is a 100% chance that the network will detect when a resident is satisfied. On the other hand, the actual NOT SATISFIED classification has a true-positive ratio of .6 and a false- negative ratio of 0.0. The sensitivity for the

*Table 4. Actual and classified output for all of test data.*

| Row Number | Actual | Classified | Not Satisf. | Satisf. | Row Number | Actual | Classified | Not Satisf. | Satisf. |
|---|---|---|---|---|---|---|---|---|---|
| 1 | satisf. | satisf | 0.004 | 0.996 | 2 | not sa. | not sa. | 0.905 | 0.095 |
| 3 | satisf. | satisf. | 0.005 | 0.995 | 4 | satisf. | satisf. | 0.000 | 1.000 |
| 5 | satisf. | satisf. | 0.000 | 1.000 | 6 | satisf. | satisf. | 0.003 | 0.997 |
| 7 | satisf. | satisf. | 0.000 | 1.000 | 8 | satisf. | satisf. | 0.004 | 0.996 |
| 9 | satisf. | satisf. | 0.000 | 1.000 | 10 | satisf. | satisf. | 0.001 | 0.999 |
| 11 | satisf. | satisf. | 0.000 | 1.000 | 12 | satisf. | satisf. | 0.256 | 0.744 |
| 13 | satisf. | satisf. | 0.000 | 1.000 | 14 | satisf. | satisf. | 0.000 | 1.000 |
| 15 | satisf. | satisf. | 0.079 | 0.921 | 16 | satisf. | satisf. | 0.000 | 1.000 |
| 17 | satisf. | satisf. | 0.002 | 0.998 | 18 | satisf. | satisf. | 0.000 | 1.000 |
| 19 | satisf. | satisf. | 0.000 | 1.000 | 20 | not sa. | not sa. | 0.590 | 0.410 |
| 21 | not sa. | not sa. | 0.990 | 0.010 | 22 | satisf. | satisf. | 0.000 | 1.000 |
| 23 | not sa. | not sa. | 0.997 | 0.003 | 24 | satisf. | satisf. | 0.001 | 0.999 |
| 25 * | not sa. | satisf. | 0.000 | 1.000 | 26 | satisf. | satisf. | 0.000 | 1.000 |
| 27 | satisf. | satisf. | 0.000 | 1.000 | 28 | satisf. | satisf. | 0.000 | 1.000 |
| 29 | satisf. | satisf. | 0.000 | 1.000 | 30 * | not sa. | satisf. | 0.186 | 0.814 |
| 31 | satisf. | satisf. | 0.000 | 1.000 | 32 | satisf. | satisf. | 0.000 | 1.000 |
| 33 | satisf. | satisf. | 0.000 | 1.000 | 34 | satisf. | satisf. | 0.000 | 1.000 |
| 35 | satisf. | satisf. | 0.002 | 0.998 | 36 | satisf. | satisf. | 0.000 | 1.000 |
| 37 | satisf. | satisf. | 0.001 | 0.999 | 38 | satisf. | satisf. | 0.000 | 1.000 |
| 39 | satisf. | satisf. | 0.000 | 1.000 | 40 | satisf. | satisf | 0.079 | 0.921 |
| 41 | satisf. | satisf. | 0.001 | 0.999 | 42 | satisf. | satisf. | 0.003 | 0.997 |
| 43 | satisf. | satisf. | 0.107 | 0.893 | 44 | satisf. | satisf. | 0.001 | 0.999 |
| 45 | satisf. | satisf. | 0.000 | 1.000 | 46 * | not sa. | satisf. | 0.011 | 0.989 |
| 47 | satisf. | satisf. | 0.009 | 0.991 | 48 | satisf. | satisf. | 0.000 | 1.000 |
| 49 | satisf. | satisf. | 0.000 | 1.000 | 50 | satisf. | satisf. | 0.000 | 1.000 |
| 51 | satisf. | satisf. | 0.000 | 1.000 | 52 | satisf. | satisf. | 0.001 | 0.999 |
| 53 | satisf. | satisf. | 0.000 | 1.000 | 54 | satisf. | satisf. | 0.000 | 1.000 |
| 55 | satisf. | satisf. | 0.000 | 1.000 | 56 | satisf. | satisf. | 0.007 | 0.993 |
| 57 | satisf. | satisf. | 0.000 | 1.000 | 58 | satisf. | satisf. | 0.000 | 1.000 |
| 59 | satisf. | satisf. | 0.000 | 1.000 | 60 | satisf. | satisf. | 0.000 | 1.000 |
| 61 | satisf. | satisf. | 0.000 | 1.000 | 62 | satisf. | satisf. | 0.000 | 1.000 |
| 63 * | not sa. | satisf. | 0.079 | 0.921 | 64 | satisf. | satisf. | 0.000 | 1.000 |
| 65 | satisf. | satisf. | 0.000 | 1.000 | 66 | satisf. | satisf. | 0.000 | 1.000 |
| 67 | satisf. | satisf. | 0.000 | 1.000 | 68 | satisf. | satisf. | 0.000 | 1.000 |
| 69 | satisf. | satisf. | 0.002 | 0.998 | 70 | satisf. | satisf. | 0.141 | 0.859 |
| 71 | satisf. | satisf. | 0.000 | 1.000 | 72 | satisf. | satisf. | 0.000 | 1.000 |
| 73 | satisf. | satisf. | 0.000 | 0.999 | 74 | satisf. | satisf. | 0.012 | 0.988 |
| 75 | not sa. | not sa | 0.743 | 0.257 | 76 | satisf. | satisf. | 0.000 | 1.000 |
| 77 | not sa. | not sa. | 0.967 | 0.033 | 78 | satisf. | satisf. | 0.000 | 1.000 |
| 79 | satisf. | satisf. | 0.000 | 1.000 | 80 | satisf. | satisfy. | 0.000 | 1.000 |

*denotes a data row that was classified incorrectly.

*Table 5.* Agreement matrix for validating network using Moores Mill data file.

| | *ACTUAL "NOT SATISFIED"* | *ACTUAL "SATISFIED"* | *TOTAL* |
|---|---|---|---|
| Classified as "NOT SATISFIED" | 6 | 0 | 6 |
| Classified as "SATISFIED" | 4 | 70 | 74 |
| TOTAL | 10 | 70 | 80 |
| True-Positive Ratio | 0.60 | 1.00 | n/a |
| False-Positive Ratio | 0.00 | 0.40 | n/a |
| True-Negative Ratio | 1.00 | 0.60 | n/a |
| False-Negative Ratio | 0.40 | 0.00 | n/a |
| Sensitivity | 60.00% | 100.00% | n/a |
| Specificity | 100.00% | 60.00% | n/a |

actual NOT SATISFIED classification is 60% or .6 (false-positive), which means that there is a 60% probability that the computer will detect that the resident is not satisfied.

The ratio values and the percentages for sensitivity for Actual "Satisfied" and specificity for Actual "Not Satisfied" are the same for Tables 3 and 5. However, the network misclassified 4 data rows that were actually NOT SATISFIED but classified as SATISFIED which explains the 60% for specificity. Since the network was not as efficient in detecting Not Satisfied cases, the value of the true-positive ratio and the sensitivity decreased.

## 7.3. ROC for Validating Neural Network

Figure 10 and Figure 11 represent the ROC curves for the validation data for the network model. As mentioned in section 6.3, the circle plotted on the curve shows the intersection of the true-positive and the false-positive ratio on the y-axis for each classification category, and converts continuous probabilities to binary classifications for the trained network. The area under the curve represents how well the network is performing. A value close to 1 means that the network is discriminating very well between the different output categories. The area under the curves in Figures 10 and 11 is 0.9307. This implies that the overall effectiveness of the network is in discriminating between different output categories when validating the trained network is 93.07%.

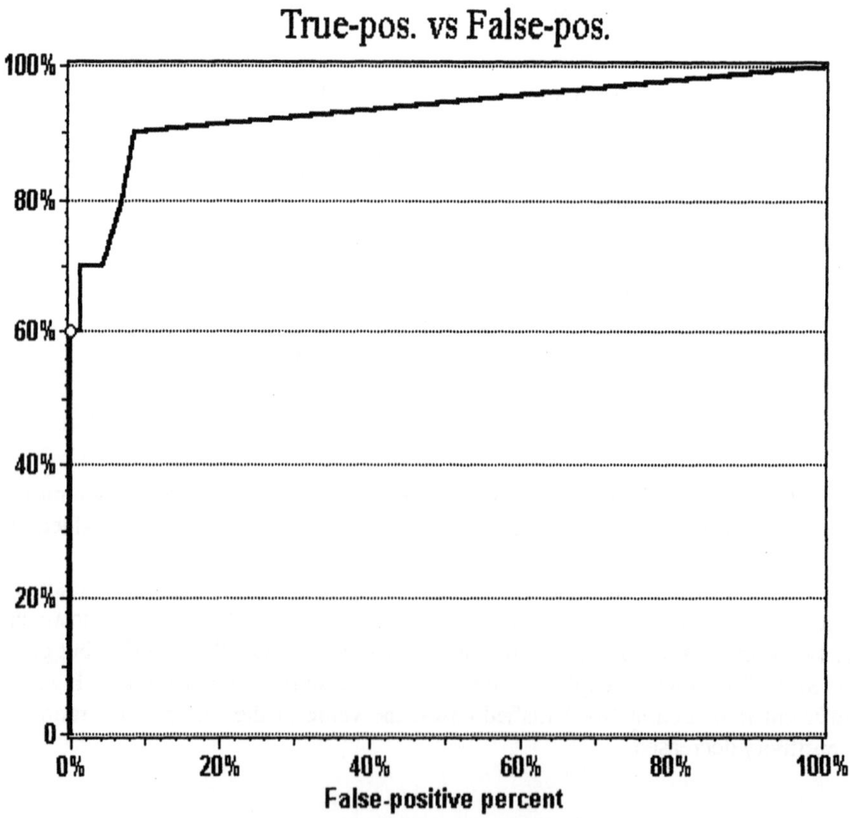


Figure 10. ROC for NOT SATISFIED classification test data

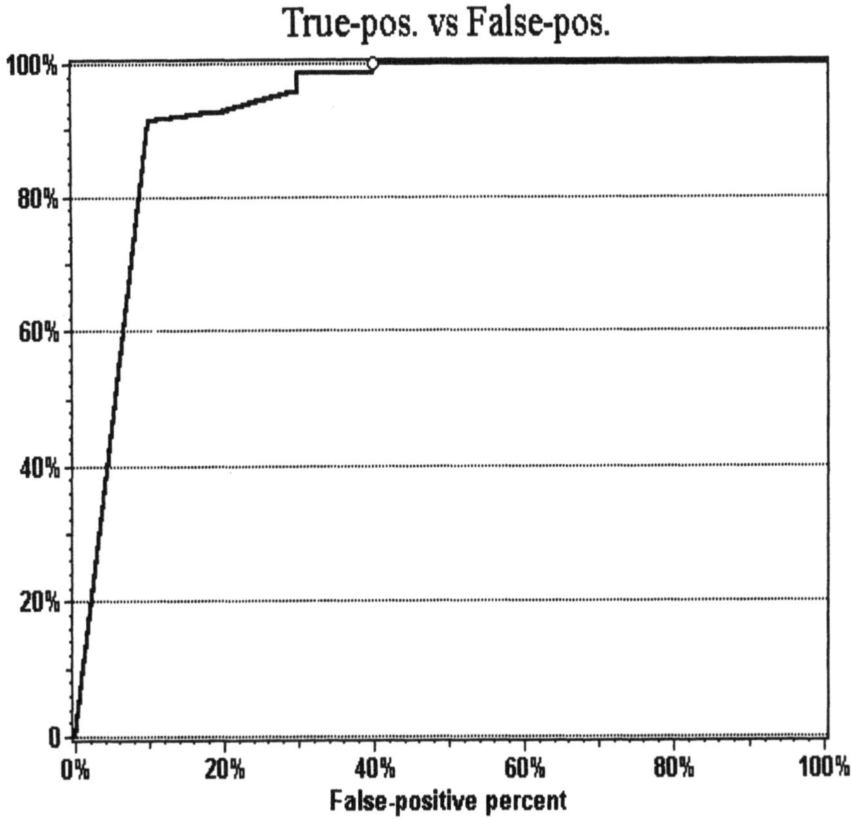


Figure 11. ROC for SATISFIED classification test set

# 8. CONCLUSIONS

In essence, neural networks are mathematical constructs that emulate the processes people use to recognize patterns, learn tasks, and solve problems. Neural networks are usually characterized in terms of the number and types of connections between individual processing elements, called neurons, and the learning rules used when data is presented to the network. Every neuron has a transfer function, typically non-linear, that generates a single output value from all of the input values that are applied to the neuron. Every connection has a weight that is applied to the input value associated with the connection. A particular organization of neurons and connections is often referred to as a neural network architecture. The power of neural networks comes from their ability to learn from experience (that is, from historical data collected in some problem domain). A neural network learns how to identify patterns by adjusting its weights in response to data input. The learning that occurs in a neural network can be supervised or unsupervised. With supervised learning, every training sample has an associated known output value. The difference between the known output value and the neural network output value is used during training to adjust the connection weights in the network..

This research developed a residential satisfaction decision support system that can assist owners in making decisions that will meet their residents' needs. The system is based on neural networks. Residential satisfaction was investigated at two affordable housing multifamily rental properties located in Atlanta, Georgia named Defoors Ferry Manor and Moores Mill. Nonprofit housing developers, Atlanta Mutual Housing Association (AMHA) and Atlanta Neighborhood Development Partnerships (ANDP), respectively own Defoors Ferry Manor and Moores Mill

The neural network was trained using Defoors Ferry Manor data, and it took 49 seconds to train the network. Seventy-nine hidden neurons were trained. The neural network was applied to 99 data rows used to train the network. Ninety-seven of those rows were classified correctly and 2 rows were classified incorrectly. The ROC (Receiver Operating Characteristic) graph showed the efficiency of the network, and it was concluded that the network was 97.40% effective in making correct classifications.

The network was trained using data from Defoors trained data set; afterwards, the network was validated by running the network on Moores Mill test data and observing how efficient the network was in discriminating between different output categories. The Moores Mill test dataset has the same input variables and output variable as Defoors. There were 80 data rows in the Moores Mill train dataset. Out of the 80 data rows, 4 rows were classified incorrectly. When the network was applied to 80 rows of the data, there were 10 cases where residents were "NOT SATISFIED"; but the network classified 4 of those cases as "SATISFIED".

The statistics related to the network's performance were that there was a 100% chance that the network will correctly predict a resident is satisfied. On the other hand, the specificity of the network for the actual SATISFIED classification was 60%, which means that there is a 60% chance that the computer will detect when the resident is not satisfied. The network's overall effectiveness in discriminating between different output categories when validating the network was 93.07%.

## NOTES

1. Category responses are 1=will recommend, 2=will not recommend, and 3=do not know.
2. Category responses are 1=better off than before, 2=worse off than before, and 3=about the same as before.
3. Category responses are 1=very unsafe, 2=somewhat unsafe, 3=somewhat safe, and 4=very safe.

# REFERENCES

Ackley, D. H., Hinton, G. E., and Sejnowski, T. J. (1985). "A Learning Algorithm For Boltzmann Machines." Cognitive Sciences, 9, 147-169.

Anderson, J. A., Silverstein, J. W., Ritz, S. A., and Jones, R. S. (1977). "Distinctive Features, Categorical Perception, and Probability Learning: Some Appplications Of A Neural Model." Psychological Review, 84, 412-451.

Brandt, R. D., Wang, Y., Laub, A. J., and Mitra, S. K. (1988). "Alternative Networks For Solving And Traveling Salesman Problem And The List-Matching Problem." Proceedings International Conference on Neural Networks, 2, 333-340.

Broomhead, D. S., and Lowe, D. (1988). "Multivariable Function Interpolation And Adaptive Networks." Complex Systems, 2, 321-355.

Carpenter, G. A., and Grossberg, S. (1988). "The ART of AdaptivePattern Recognition By A Self-Organizing Neural Network." IEEE Computer, 21, 77-88.

Durbin, R., and Willshaw, D. (1987). "An Analogue Approach To The Travelling Salesman Problem Using An Elastic Net Method." Nature, 326, 689-691.

Favata, F., and Walker, R. (1991). "A Study Of The Application Of Kohonen-Type Neural Networks To The Travelling Salesman Problem." Biological Cybernetics, 64, 463-468.

Fukushima, K. (1980). "Neocognitron: A Self-Organizing Neural Network Model For A Mechanism Of Pattern Recognition Uneffected By Shift In Position." Biological Cybernetics, 36, 193-202.

Goldstein, M. (1990). "Self-Organizing Feature Maps For The Multiple Traveling Salesman Problem (MTSP)." Proceedings IEEE International Conference on Neural Networks, 258-261.

Gupta, J. N. D., and Sexton, R. S. (1999). "Comparing Backpropagation With A Genetic Algorithm For Neural Network Training." Omega, 27, 679-84.

Hartigan, J. A. (1975). Clustering Algorithms, Wiley, New York.

Hegde, S., Sweet, J., and Levy, W. (1988). "Determination Of Parameters In A Hopfield/Tank Computational Network." Proceedings IEEE International Conference On Neural Networks, 291-298.

Hopfield, J. J. (1982). "Neural Networks And Physical Systems With Emergent Collective Computational Abilities." Proceedings of the National Academy of Sciences of the USA, 79, 2554-8.

Hopfield, J. J., and Tank, D. W. (1985). "Neural Computation Of Decisions In Optimization Problems." Biological Cybernetics, 52, 141-152.

Jacobs, R. A., and Jordon, M. I. (1991). "A Competitive Modular Connectionisst Architecture." Neural Information Processing Systems 3, R. P. Lippma, ed., Morgan Kaufmann, San Mateo, CA, 733-767.

Kohonen, T. (1982). "Self-Organized Formation Of Topologically Correct Feature Maps." Biological Cybernetics, 43, 59-69.

Kohonen, T. (1988). Self-Organisation and Associative Memory, Springer, New York.

Lai, W. K., and Coghill, G. G. (1992). "Genetic Breeding Of Control Parameters For The Hopfield/Tank Neural Net." Proceedings International Joint Conference on Neural Networks, 618-623.

Montana, D. J. (1995). "Neural Netowrk Weight Selection Using Genetic Algorithms." Intelligent Hybrid Systems, S. Goonatilake and S. Khebbal, eds., Wiley, Chichester, 85-104.

Sexton, R. S., Gupta, J. N. D., Smith, B. N., and Montagno, R. V. (1998). "Neural Network Training Via Genetic Algorithm And Backpropagation: An Empirical Comparison." Working paper, Department of Management, Ball State University, Muncie, Indiana.

Smith, K. A., and Gupta, J. N. D. (2000). "Neural Networks in Business: Techniques and Applications for the Operations Researcher." Computers & Operations Research, 27, 1023-1044.

Smith, K. A., Palaniswami, M., and Krishnamoorthy, M. (1996). "Traditional Heuristic Versus Hopfield Neural Network Approaches To A Car Sequencing Problem." European Journal of Operational Research, 93, 300-316.

Van Den Bout, D. E., and Miller III, T. K. (1989). "Improving The Performance Of The Hopfield-Tank Neural Network Through Normalization and Annealing." Biological Cybernetics, 62, 129-139.

Wong, B. K., Bodnovich, T. A., and Selvi, Y. (1997). "Neural Network Applications In Business: A Review And Analysis Of The Literature (1988-1995)." Decision Support Systems, 19, 301-320.

Zurada, J. M. (1992). An Introduction To Artificial Neural Systems, West Publishing, St. Paul.

# Chapter 11

# EVACUATION SIMULATION IN UNDERGROUND MALL BY ARTIFICIAL LIFE TECHNOLOGY

**Hitoshi Furuta and Masahiro Yasui**

## 1. INTRODUCTION

In recent years, underground malls are popular and form a part of downtown, because ground has no available spaces in Japan [1]. Since the underground malls have complicated configuration and connections, it is difficult for visitors to evacuate in the event of a disaster. Therefore, it is necessary to develop a disaster prevention measure to improve the safety of people using the underground malls.
Then, it is inevitable to predict how people behave for the anxiety and confusion during the disaster and to grasp the total action during the disaster. In this study, a new simulation system is developed to consider the evacuation behavior in the underground malls during the disaster. So far, a lot of methods have been proposed to the simulation of evacuation, whose models are based upon many and complex factors [2,3,4,5,6]. For more accuracy, it is necessary to take into account human psychological factors, visibility, and so on. However, when those models are applied to the real cases, some problems may arise on calculation time and accuracy.
A new simulation model is proposed, which can express complex human behavior by a simple model that defines actions probabilistically, by introducing artificial life technology. If many factors are considered in the simulation, its implementation requires a lot of computational time.

## 2. ARTIFICIAL LIFE TECHNOLOGY

### 2.1 History of artificial life technology

In September, 1987, artificial life workshop was held at Los Alamos, New Mexico, US. As the leadership of C. G. Langton, artificial life technology has been developed so far.

In 1990, Tierra was developed by T. Ray, which was quite important result in the initial stage of artificial life researches [7]. (Figure 1) J. L. Casti of the Santa Fe research institute said that January 4, 1990 is the day which should be kept in mind, because a life that is not organized by carbon was born in computer for the first time ever.

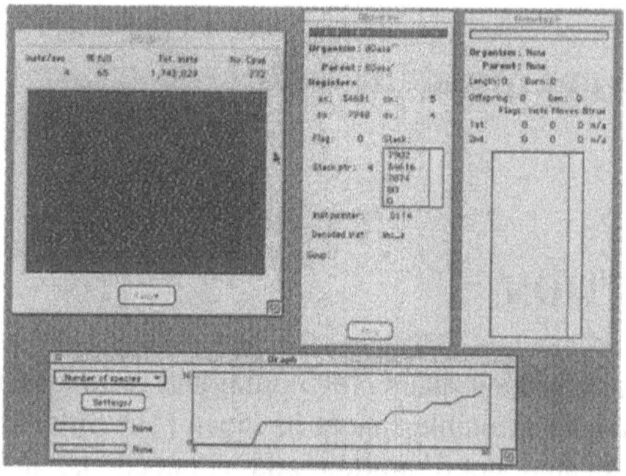

Figure 1. Tierra model by T. Ray

## 2.2 Artificial life technology

Characteristics of living beings are growth, self-duplication, metabolism, environmental adaptation, evolution and so on [8,9,10,11]. From this point of view, it can be considered that actions do not have so many varieties in the human decision. Therefore, the variety in human decisions can be imitated by selecting several possible actions. In order to realize this function, it is useful to apply the generic technology that has been completed through artificial life.

## 2.3 Emergence

The concept of emergence has been considered in the philosophical field since a long time ago [11,12]. In artificial life, it is supposed that emergence is universally involved in biological phenomenon such as birth of a life, ontogenesis, and evolution. It is considered that emergence influences many phenomena such as birth of heart, sociogenetic, economy, evaluation of culture and so on.

In order to generate "emergence", hierarchical structures of functions are needed [5,6,7]. Each element in the lower layer has only a simple relation with each other, but it does not have a relation with all elements. Namely, each element in the lower layer has only a relation with the neighborhoods. As a result, a general order in the upper layer is formed (bottom-up). The order formed in the upper layer becomes the boundary condition to the actions in the lower layer. Therefore, the nonlinear feedback between the upper and lower hierarchies is formed and causes a complicated behavior. This is called as "emergence". Concept of "emergence" is shown in Figure 2.

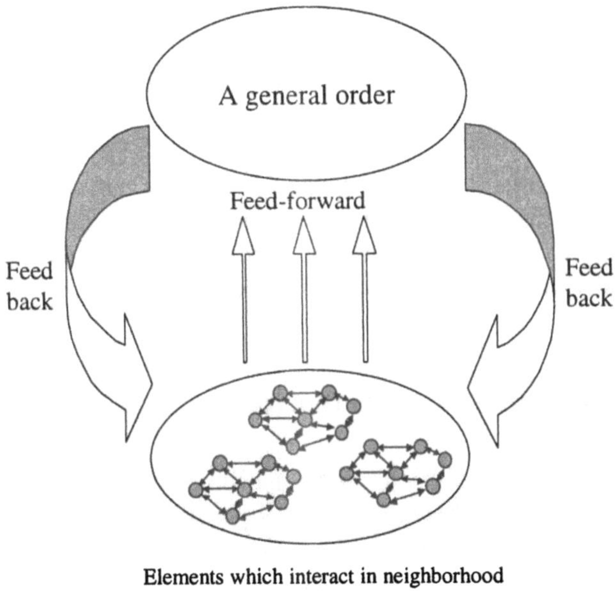


Figure 2. Concept of "emergence"

# 3. HUMAN BEHAVIOR DURING DISASTERS

## 3.1 Human actions during disasters

Human behavior can be roughly divided into two typed such as leader-action and following-action.

*Table 1*. Difference of behavior during disaster

| | leader-action (active) | following-action (passive) |
|---|---|---|
| man | 59.8% | 40.2% |
| woman | 34.0% | 66.0% |

From Table 1, it is obtained that 46.9% of people take leader-action and 53.1% of people take following-action as a whole.

The difference between leader and following actions is described below:.

■ Leader-action.

● People acting as the leader can recognize paths to exits and go to an emergency exit, and inform it to human beings surrounding.

■ Following-action.

● People following the human beings around them, because they do not recognize paths to exits.

Also, it may be possible that human being fall in panic during disasters. Then it is imagined that there are struggles to survive even if it shoves others aside and people may behave in an irrational way such that occasionally increase own or others' risk. However, there is no case of panic in actual disaster situations.

## 3.2 Rate of recognizing paths to exits

The rate of recognizing paths to exits is different between weekdays and weekends. On weekdays, there are many people of using underground mall to commute, whereas on weekends there are people who often use it for shopping or eating. Table 2 presents the results of questionnaire about whether they know exits or not. Table 2 shows that 73.3% of people recognize exits on weekdays and 63.2% of people recognize the exits on weekends as a whole.

*Table 2.* The rate of recognizing exit

| | recognize exit | | not recognize exit | | uncertainty |
|---|---|---|---|---|---|
| | immediate answer | answer after a while | immediate answer | answer after a while | |
| Weekdays | 73.3% | 9.4% | 10.7% | 5.5% | 1.2% |
| Weekends | 63.2% | 10.0% | 19.5% | 6.3% | 1.0% |

When disasters occur, people do not necessarily recognize the nearest exit. If they can not find any near exit in a short time, they tend to evacuate to not near exit but known exit. On the other hand, even if they recognize any exit, they may act depending on the surrounding situation.

## 3.3 Safety recognition in Osaka underground mall

Woman has a tendency to consider that underground mall is less safe than man does. Table 3 presents the results of questionnaire about recognition of safety for weekdays and weekends.

*Table 3.* The rate of recognizing safety in Osaka underground mall

| | think safe | think unsafe | no idea | uncertainty |
|---|---|---|---|---|
| Weekdays | 14.0% | 75.9% | 9.8% | 0.3% |
| Weekends | 15.8% | 72.6% | 11.3% | 0.3% |

## 3.4 Emergence in evacuation

During disasters, human being differs in the way of recognition of the disasters in an underground mall. Moreover, the transition of information is also different. However, people become a crowd and evacuate as a whole regardless of those differences. The crowd is a large number of people who act together. The crowd can be divided into the following three groups:

- The crowd tends to force themselves to the corner by the disaster
- The crowd who can appropriately correspond and judge to the disaster
- The crowd who is ignorant or indifferent to the disaster

A representative factor forming a crowd is the interaction of each human action. The interaction of each human being is realized through the information from the environment. As the information from the environment, leading light (sign) and information from the crowd can be considered. Namely, each human action influences on the crowd and the crowd influences on each human action. Thus, each human being and the crowd can possibly show "emergence" due to the interaction among them. Concept of emergence in evacuation is shown in Figure 3.

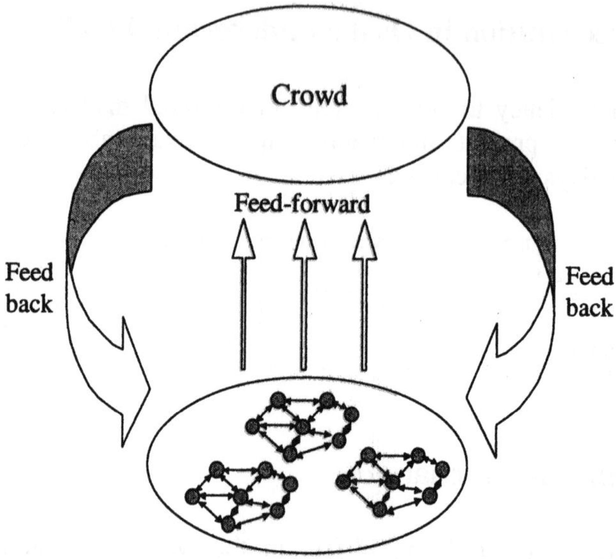


Human interaction in neighborhood

Figure 3. Concept of emergence during evacuation

## 3.5 Occurrence of second disaster

While recognizing some disaster, there are many people who want to evacuate immediately from the underground mall, because they understand it is dangerous. It is expected that the neighborhood of exits will be very congested. Therefore, it is likely that there may occur such second disasters as domino phenomenon and avalanche of the crowd. While domino effect is similar to avalanche of the crowd, they are different phenomena. Domino effect is the expansions to a line from back to front so that human beings behind force human beings ahead down. Avalanche of the crowd is the expansion of the crowd to multi-directions with the shape of lump from front to back so that human beings behind and left or right are involved in the fall of human beings ahead. Avalanche can occur when unstable balance is maintained by collisions of human beings. The difference between domino effect and avalanche of the crowd is shown in Table 4.

*Table 4.* Comparison domino effect and avalanche of the crowd

| | domino effect | avalanche of crowd |
|---|---|---|
| density to occur | even if density of population is 3 to 5 people/㎡ | if density of population is more than 10 people/㎡ |
| effect of pressure | on tumbling | before tumbling |
| direction of tumbling | from back to front | from front to back |
| shape of tumbling | a line | the shape of lump |

# 4. APPLICATION EXAMPLE

## 4.1 Model of human being

It is assumed that each human being has the following characteristics:

- It recognizes paths to exits or not

And it has the following attitudes:

- Heading direction
- Intensity to the heading direction

Moving of human being is simulated by selecting the moving direction, intensity and speed of moving. By the difference of moving speed, the difference of old and young persons can be identified. Each model decides the direction by choosing a direction. After the heading direction is decided, it decides a movement cell. The model which recognizes paths to the exit heads directly for the exit. On the other hand, the model which does not recognize paths to the exit decides the heading direction probabilistically from the following factors:

- The direction that the human being in the neighborhood heads for
- The intensity to the present direction
- The information from the human being in the neighborhood
- The leading light

The direction that a lot of human beings in the neighborhood go is sets up to be chosen highly. It is probabilistically decided as counting the number of directions chosen by neighbor human beings and the intensity to the present direction is strengthen.
For example, consider the following case presented in Table 5.

*Table 5.* Example of choosing direction

| | | |
|---|---|---|
| direction of heading for | | south |
| the intensity to the heading direction | | 3 |
| the number of human in the neighborhood | | 10 |
| the number of the heading direction that human beings in the neighborhood choose | north | 4 |
| | east | 3 |
| | south | 2 |
| | west | 1 |

For this case, probability of choosing each direction is assumed to be.

$$\text{north} : \text{east} : \text{south} : \text{west} = 4 : 3 : 2+3 : 1$$
$$= 4 : 3 : 5 : 1 \qquad (1)$$

Thus, south is chosen because it has the highest probability. When direction is chosen within the region of recognizing leading light, it is assumed that the direction to the leading light is selected with high probability. This is realized by adding a fixed value to the direction to the leading light.
For example, if the adding value is set up to be 5, the ratios become to be.

north : east : south : west = 4 : 3+5 : 5 : 1
= 4 : 8 : 5 : 1 (2)
Moving direction is decided by the ratios given in Eg.2. In this case, since probability to the east is the highest, many people select this direction and therefore it causes the emergence (i.e., moving to the east as a whole) easily.
If people who do not recognize paths to the exit are informed the paths to the exit from human beings who recognize, they tend to choose the same direction according to the information. When speed of the information is fast or the purveyors arrive at an exit, the simulation is implemented in the same way as the case without the information regarding the exit. When people approach to the exit without recognizing paths to the exit in the near region to the exit, they are given a fixed probability to the present direction.
Then, it is considered that emergence appears with ease, as the crowd becomes larger. A flowchart of choosing a direction without recognizing paths to exits is shown in Figure 4.

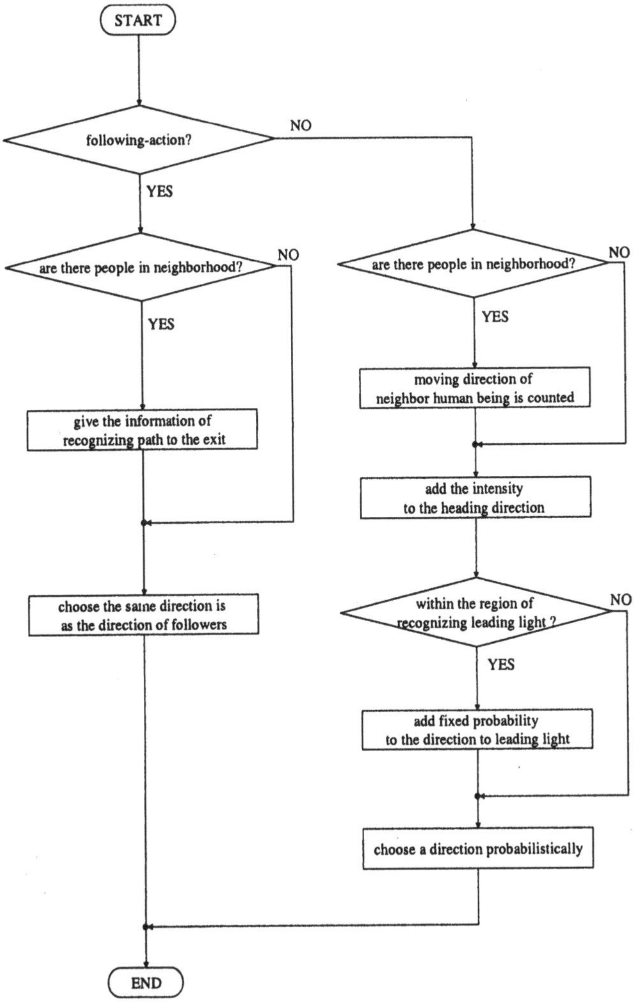


Figure 4. Flowchart to choose a direction by people without recognizing paths to exits

## 4.2 Realization of recognizing paths to exits

Intersections and exits are set as nodes, and passages are set as links. Each node has each identification number given as nearer exit has higher values and exit has the highest value. If a human being recognizes paths to an exit, the probability of choosing the direction to the exit is assumed to be high. However, he/she does not always select the shortest path, even if he/she knows the direction to the exit. This fact can be considered in the simulation, because many trials are implemented in the simulation. The flowchart is presented in Figure 5, in which people without recognizing paths to exits select the moving direction.

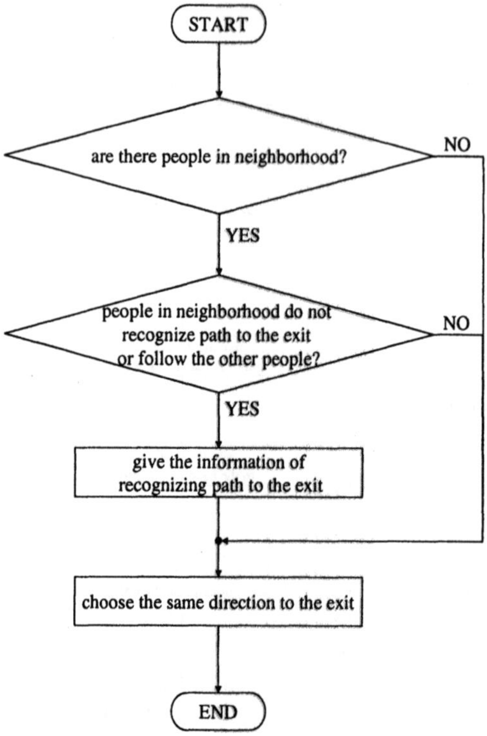


Figure 5. Flowchart to choose a direction by people recognizing paths to exits

## 4.3 Decision of movement cell

Characteristics of human actions are as follows, when human beings are moving:

- Human being tries to avoid physical contacts to other people.
- Human being tries to make axis of shoulder square to the direction of movement.
- Human being tries to go straight ahead unless external force is given.
- Human being tries to avoid a contact with wall.

It is difficult to develop model that satisfy all the above requirements exactly. Even if an accurate model can be accomplished, there arises a problem of enormous computation time. In this study, it is attempted to develop a simple model that provides almost the same result by introducing artificial life technology.

The model is based on the assumptions that the heading direction is considered as the base and the next movement is selected from six actions that three cell in front, two cells in right and left, and stay in the same cell. In

Figure 6, a human being is in the center cell and white arrow means the heading direction and the cells with black arrow indicates movable cells at the next step. Then, the following probabilities are assumed:

straight : sidewinder : right and left : stay = 500 : 100 : 20 : 1 (3)

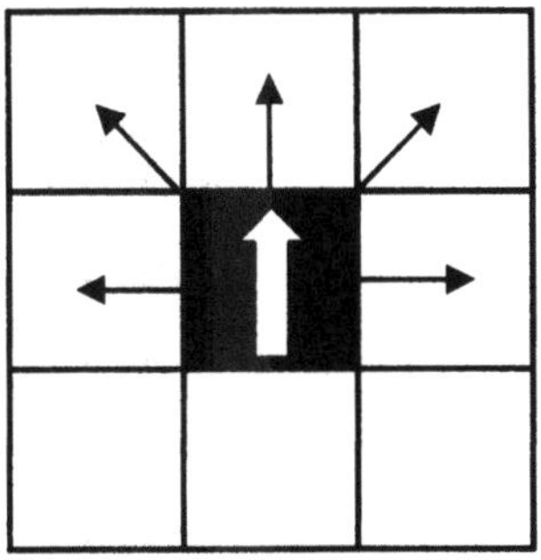

Figure 6. Movable cell

When moving against a wall, the following ratios are applied.

straight : sidewind : right and left : stay = 100 : 500 : 20 : 1 (4)

## 4.4 Verification of proposed model

A virtual underground mall shown in Figure 7 is considered to examine the effectiveness of the proposed model.

Figure 7. A virtual underground mall

It is assumed that the number of human beings is 200 and they are generated at random, and the ratio of human beings who recognize paths to exits is 30% or 60%. The execution result for the ratios of 30% and 60% are presented in Figure 8 and Figure 9, respectively.

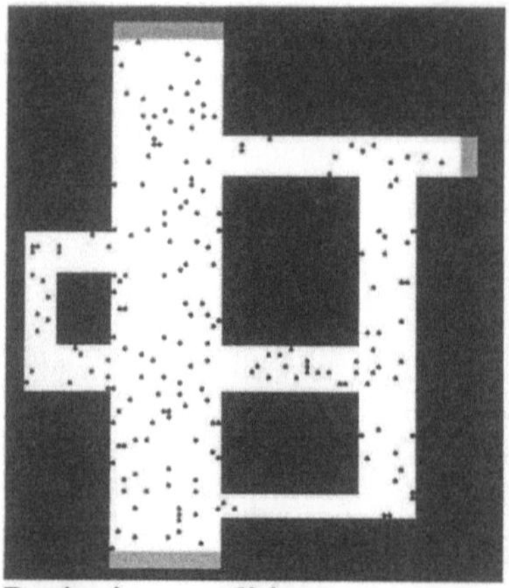

Beginning condition
Evacuation rate : 0%

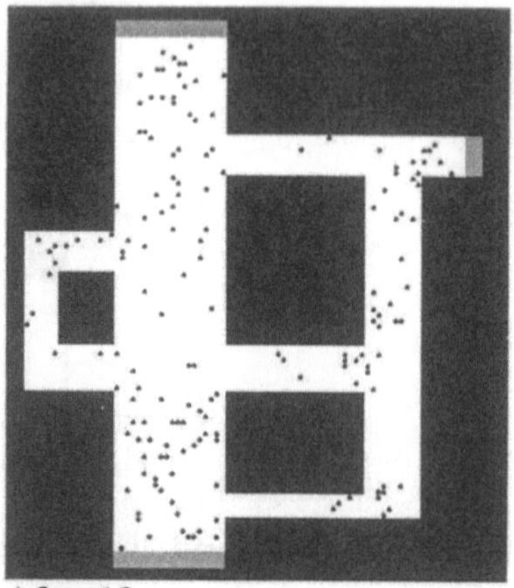

After 10 steps
Evacuation rate : 17%

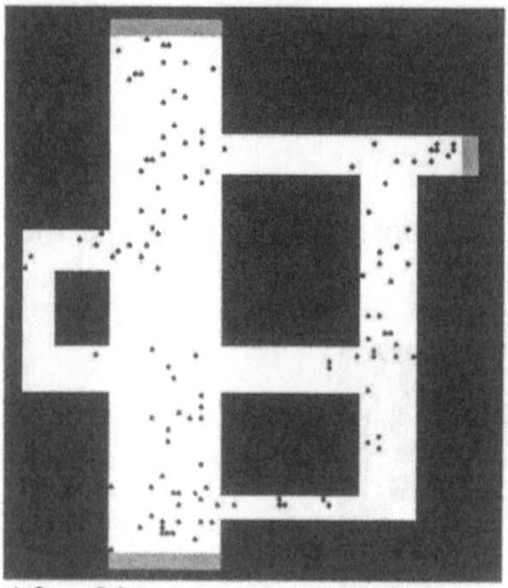

After 20 steps
Evacuation rate : 39%

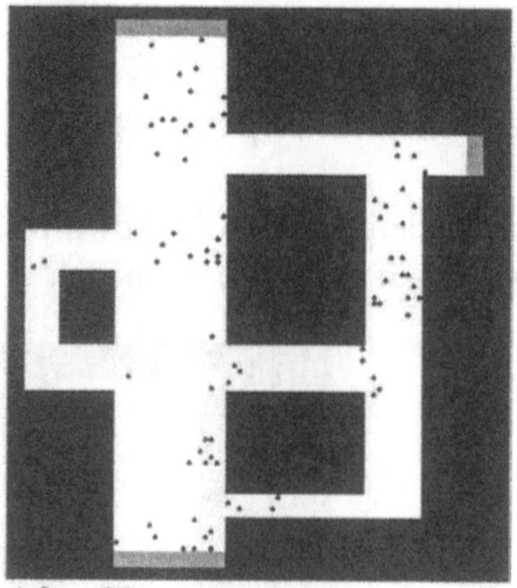

After 30 steps
Evacuation rate : 59%

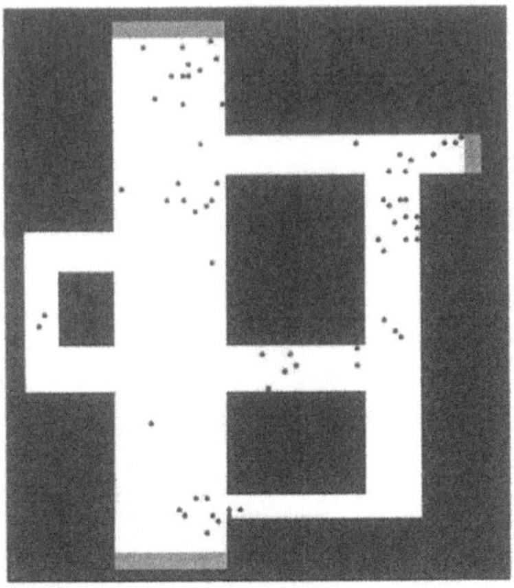

After 40 steps
Evacuation rate : 67%

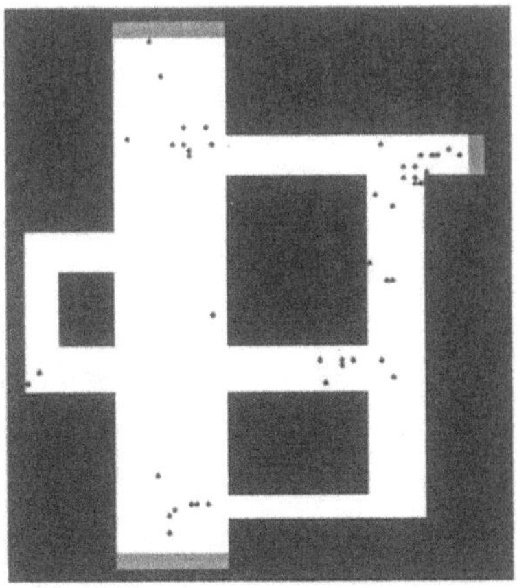

After 50 steps
Evacuation rate : 77%

*Figure 8.* Changes of evacuation process for recognizing rate 30%

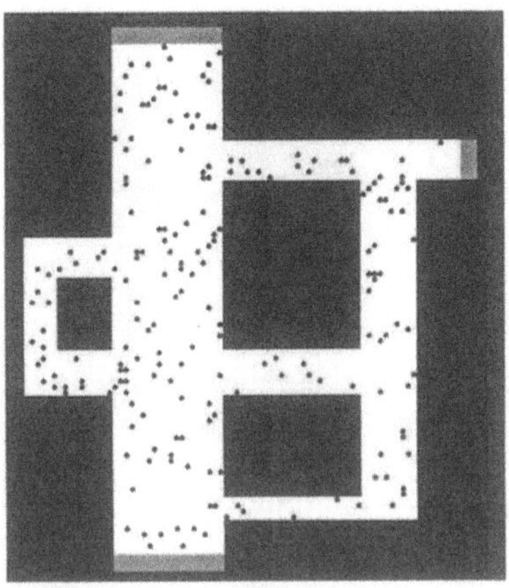

Beginning condition
Evacuation rate : 0%

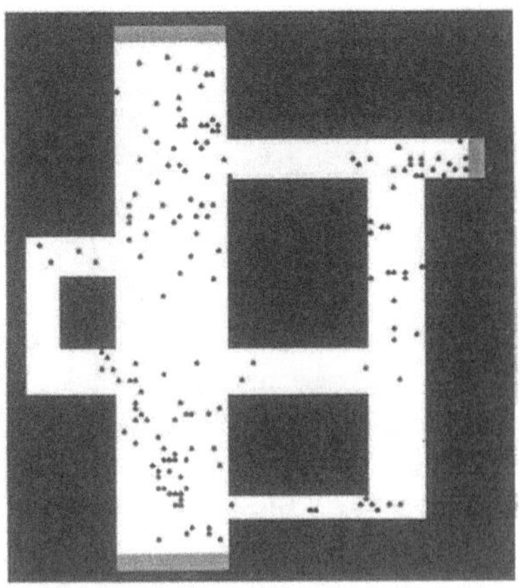

After 10 steps
Evacuation rate : 19%

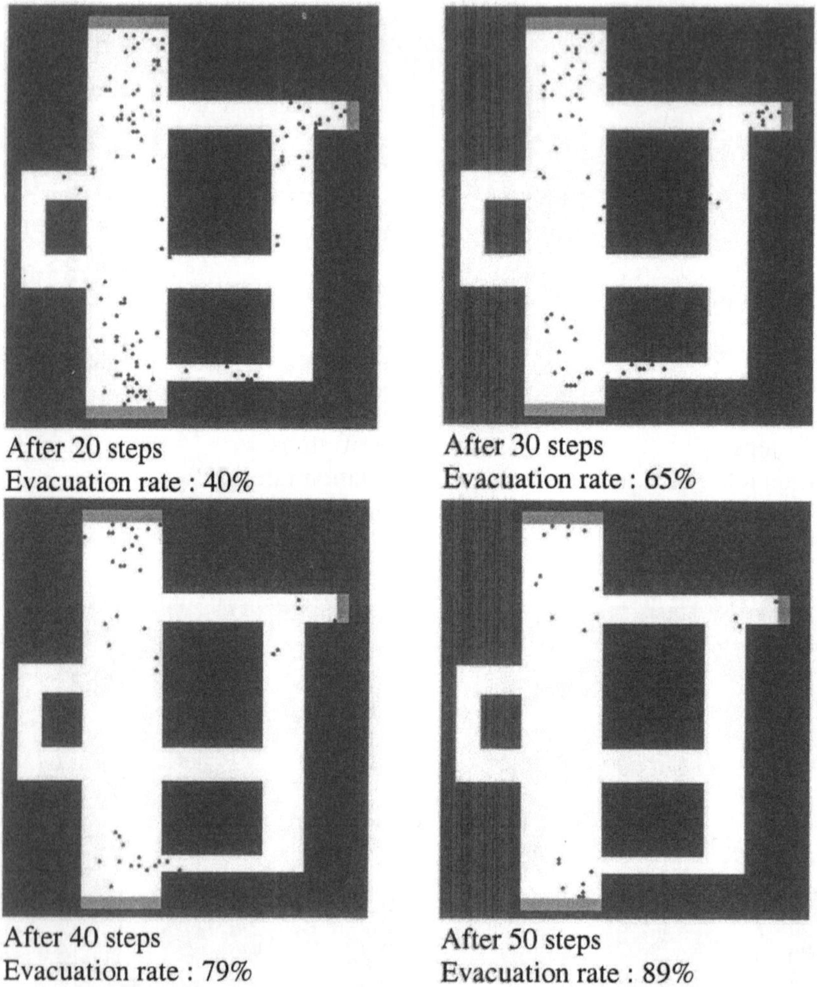


Figure 9. Changes of evacuation process for recognizing rate 60%

From these figures, it is seen that human beings without recognizing paths to exits form a stream of crowd. This implies that all the people do not recognize the path to the exit. Namely, the explicit information of the path to the exit is not always provided in this model. The form of crowd is created by the interaction in the neighborhood.

Paying attention to the evacuation rate, evacuation rate is not different until 20 steps, even though there are some differences in the rate of recognizing the exits. However, evacuation rate becomes different after 30 steps. As steps proceed, the difference of evacuation rate becomes larger.

On the other hand, it is possible to examine the state of evacuation during the disaster by using this model. For example, it is possible to investigate the change of the time to complete evacuation by the change of rate recognizing paths to the exit. Thus, the number of steps needed for 80% completion of evacuation is examined by changing the rate of recognizing paths to the exit. The average of examination results is shown in Figure 10.

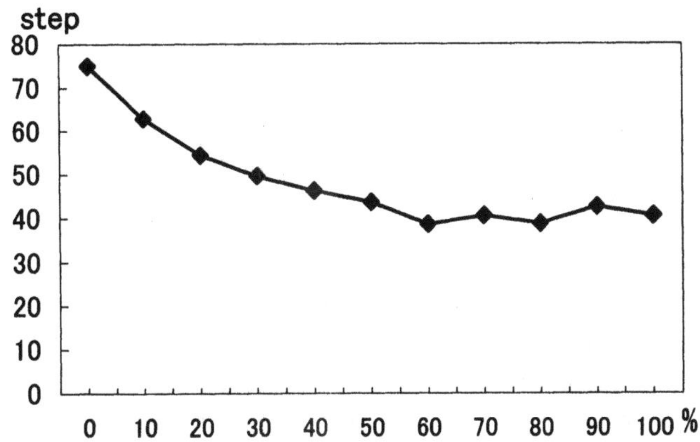


Figure 10. Steps needed for 80% completion of evacuation

From Figure 10, it is obtained that if the recognizing rate is 30%, steps needed for 80% evacuation are approximately 50 and if the recognizing rate is 60%, steps needed for 80% completion of evacuation are approximately 40. However, the difference diminishes when the recognizing rate becomes more than 60%.

# 5. CONCLUSIONS

When considering complex factors on human decision, it needs a lot of load and time. It is almost impossible to model each human being individually. In this study, an attempt was made to develop a simple simulation model by using the concept of probability. Although the model is simple, emergence can appear by the interaction of each model. Therefore, it is possible to express complex actions of human decisions by probability and the simulation model proposed in this study is effectiveness.

However, there still remain many problems to be overcome in the future. While an emergence would appear, it does not be sufficiently proven to be a true emergence. It is necessary to implement many simulations for various cases. .

In this study, human actions are defined n terms of probability. The probability is subjectively decided so that it needs to investigate the human behaviors experimentally. Moreover, it is necessary to examine the applicability of the proposed model for a real underground mall and many various environments. Also, disaster should be defined in a more detail way; namely, the characteristics of disasters should be identified. Fire or earthquakes are possible disasters to be considered.

# REFERENCES

[1] Kato, Yasuyoshi, *Cellular Automaton method*, MORIKITA Press, Oct. 1998 (in Japanese)

[2] Burks, Arthur. W. ed., *Essays on Cellular Automata*, Urbama: Illinoi Univ. Press, 1970

[3] Bandini, Stefania and Mauri, Giancarlo, *Proceedings of the Second Conference on Cellular Automata for Research and Industry*, Springer-Verlag, 1997

[4] Ilachinski, Andrew, *Cellular Automata*, World Scientific 2001

[5] Kato, Yasuyoshi, *Cellular Automaton method*, MORIKITA Press, Oct. 1998 (in Japanese)

[6] Weiss, Gerhard, *Multiagent Systems*, MIT Press

[7] http://www2.create.human.nagoya-u.ac.jp/~ari/stuff/10years.html

[8] Shiraki, Wataru, *Application of Artificial Life Technology to Engineering*, Science and Technology Press, 2003 (in Japanese)

[9] Weisbuch, Gerard, *Complex Systems Dynamics*, Addison Wesley, 1991

[10] Tsutomu HOSHINO, *Dreams and Troubles of Artificial Life*, SYOUKABOU, Aug. 1994

[11] Arita, Ryuya, *Artificial Life*, Science and Technology Press, Jan 2000 (in Japanese)

[12] Epstein, Joshua and Axtell, Robert, *Growing Artificial Societies – Social Science from the Bottom Up*, MIT Press, 1996

# Chapter 12

# EPISTEMIC UNCERTAINTY AND THE MANAGEMENT OF HIGH RISK EXPOSURES

**Mark Jablonowski**

## 1. INTRODUCTION

Risk assessments involve establishing the probability of adverse consequences. It is these assessments that guide the risk management process. Traditionally, risk assessment has relied on estimation of precise probabilities from data. These numbers then serve as the basis for various well-defined risk financing, loss control and loss prevention decisions.

In the "real world", risk assessments are subject to considerable pitfalls. Most of these result from the fact that real world complexities and dynamics introduce irreducible knowledge imperfections. We will refer to these generally as epistemic uncertainties, as distinct from the variability introduced by randomness. Limited data availability results in the inability to specify probability distribution exactly. This drawback, related fundamentally to the age-old "problem of induction", arises from the fact that many theoretical distributions can be plausibly fit to limited data. From a decision standpoint, the problem becomes, "which to choose?".

The problem is illustrated in Figure 1. The probability and loss characteristics of risk are presented on a two-dimensional graph, or risk map. Risk assessment takes the form of trying to identify the probability distributions that relate possible losses to their probability of occurrence (usually, on an annualized basis). Fitting distributions to data is relatively straightforward when data is readily available. This is usually the case with smaller, more "frequent" events. Once the data on which to base the fit becomes scarce, as they invariably do as losses become sufficiently large, various theoretical probability distributions become (more or less) plausible candidates. This is indicated by the divergence of candidate theoretical distributions.

The risk manager finds him or herself in the uncomfortable position of having to make some of our most important decisions - those involving low probability/ high consequence (i.e., "high risk") exposures - with little (or no) information. The problem for the risk management decision maker becomes how to incorporate these uncertainties into the decision-making process for dealing with high risk exposures (Jablonowski, 2000, 2002).

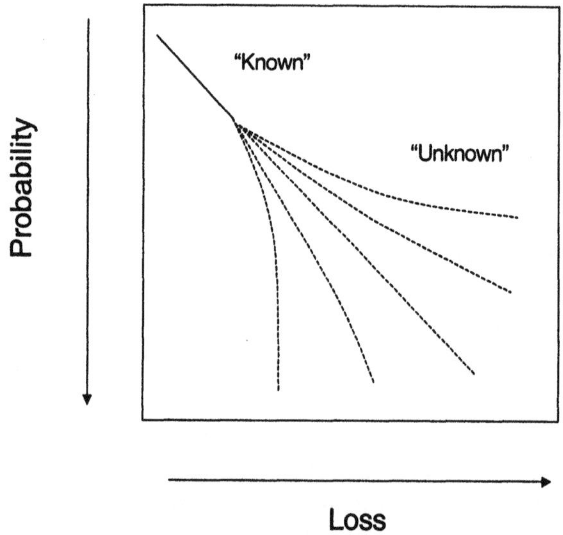


Figure 1. Risk Assessment Under Epistemic Uncertainty (Imperfect Knowledge)

We will review here an exploratory approach to identifying areas of uncertainty in risk assessments. Describing, as best we can, the degree of epistemic uncertainty involved is the key first step to proceeding with some sort of decision strategy with respect to high risk exposures. We then proceed to discuss several potential decision criteria for decision in high risk situations.

## 2. CIRCUMSCRIBING EPISTEMIC UNCERTAINTY

The epistemic uncertainty involved in various risk estimates is separate and distinct from uncertainty due to randomness. As a result, the standard tools of statistical analysis, such as the development of probabilistic confidence intervals, do not apply. Rather, we need to approach epistemic uncertainty from the standpoint of possibility: What are the possible potentials for loss within the exposure mechanism given? This "possibilistic" approach is used to augment analysis in terms of the probability and consequence characteristics of loss when assessing risk. Recognizing, and measuring (as best we can) this uncertainty is the essential first step to understanding high risk exposures and what we can (and can not) do about them.

## 2.1 Exploratory Modeling

Assessing high risk exposures on a possibilistic basis depends on our ability to enumerate, as best we can, plausible alternative risk scenarios. Each alternative represents a possibility that we must consider in our final decision as to how to deal with that environment. Assessment of the extent of epistemic uncertainty in this fashion is the basis of exploratory modeling (Bankes, 1993). Exploratory modeling recognizes that under epistemic uncertainty there may be a collection, or ensemble, of models consistent with the data under study. We do not have sufficient knowledge to declare that one, exact model represents the "true" model. Exploration, therefore, results in the specification of multiple plausible scenarios consistent with the data (or perhaps rather, lack thereof). No attempt is made to try to summarize or otherwise combine the models into one single "best" model. Rather, the plurality of models is left for further consideration in the decision process.

This process is distinct from those applied when we believe that the underlying model parameters are random variates. In that case, we would turn to statistical methods, such as linear regression, to specify the underlying stochastic model. While exploration does not exclude stochastic models, it does not limit itself to uncertainties which result from randomness. As is obvious from the study of risk, models of epistemic and probabilistic uncertain can coexist.

Exploration also varies from sensitivity analysis. In sensitivity analysis, as properly defined, the values of the underlying model parameters remain static (i.e., known). The variables are changed systematically, and the behavior of the model noted. This assumes knowledge of the model. What is unknown, to the analyst at least, is the behavior of the model "output(s)" under perturbations of its various "inputs". This behavior is specified in the derivatives (or partial derivatives) of the model, which it turn requires the model parameters be known. As the complexity of the model increases, we may need to determine sensitivity numerically. Numerical methods that attempt to identify the parameters of complex models (e.g., steepest ascent, and related response surface methodologies) are, once again, distinct from the exploratory methods we suggest here in that they assume the system is precisely "knowable" (albeit in some intractable analytical form).

The methods of generating exploratory scenarios vary. More often than not, they owe more to the process of discovery than to the analytics of verification and validation. In exploratory modeling we seek not so much to discover the unknown, as to discover how much we don't know. Discovery is very much a creative process, not just an analytical one (Kanatarovich, 1993). While sampling and search through prospective domains of exploration can be systematized using a variety of analytical methods, fleshing out those domains remains very much an ad hoc matter. The means are often suited to the challenges at hand. As such, many exploratory techniques are developed as part of the application process itself.

Heretofore, perhaps the widest application of exploratory models as we have defined them here has been their use in planning and long range forecasting, both on a organizational and social level. This scenario based approach was originally developed based on a need to explicitly recognize the affects of epistemic uncertainty on model building in the planning process (Wack, 1985). While usually based on narrative scenarios, scenario-based planning often incorporates mathematical models as well (e.g., various population growth models). A large and useful body of techniques for scenario generation and application has emerged from such applications (Lempert etal., 2003). These include the development of structured methods for the elicitation of expert opinion, and robust simulation modeling techniques.

To aid in the exploration process, computers are often utilized. Heuristic methods facilitated by the use of modern electronic computers include interactive visualizations and multiple simulations. In fact, realizing the full power of exploration requires substantial computational power. With expanded computational capability readily available today, due to the increased accessibility of powerful "desktop" computing environments, the idea of exploration is more attractive now than ever before.

Even with the availability of large amounts of computing power at our disposal, exploration may seem at first glance as an effort cursed by the dimensionality of the very models it attempts to explore. Exploration within an admittedly uncertain environment immediately calls to mind a search space of immense proportions. The key to effective exploration, however, lies in its ability to circumscribe uncertainty. Developing plausible ways to do so is essential to the process of exploration. In fact tractable strategies exist, and continue to be developed. Consider for example bounding the slope, m, of a simple linear equation of the form y=mx+b. Plausible bounds may be formulated rather simply using intervals. Interval calculations prove to be a computationally efficient method than simulation for handling the ensemble that results.

Consider now the expansion of the exploration to an unknown model form. Let's say that all we know is that the function is continuously differentiable, bounded, and monotonic. The ensemble of possibilities is huge. However, we may still be able to systemize search among a more limited landscape by selectively searching through various functional forms (e.g., polynomials of arbitrary order) (Bankes, 1993). Again, as in the process of discovery, our effective "search space" may be quite large. Yet, selective guidance, based perhaps on past inductive successes, permits progress in this complex area. Carefully crafted sampling and search are common to success in both exploration and discovery.

As Bankes notes, the results of an exploratory analysis will not typically be mathematically rigorous., but rather present "an imperfect image of the complete ensemble..." (Bankes, 1993, p. 443). And further, "Given a fixed analytic budget (in dollars, people, or time), the analysis must provide the most useful results possible based on what we know about the problem at hand." (Bankes, ibid). The question is if, or to what extent, the results are useful to the problem at hand. The process must ultimately be directed by the questions we seek answers to. In this way, the search space itself can often be made more tractable. For example, in the analysis of risk, we are concerned only with the results of actions that cause losses to the entity (as opposed to gains). Exploration of candidate probability distributions and their properties may therefore be limited to examination of the negative semi-variance, for instance. Exploration may be further limited to the extent our decisions about risk depend on thresholds. Does the possibility exist (based on the exploration) that we may exceed some critical threshold? Exploration exposes epistemic uncertainty. The degree to which we are able to determine the extent this uncertainty permeates our models is a matter of our own ingenuity.

When making high risk decisions, the risk manager faces considerable uncertainty about the "true" probability distribution of losses. Rather than ask "which to choose?" in the face of uncertainty, the exploratory modeler asks, "why choose?". To the extent that uncertainty can be properly circumscribed, and this information carried forward to the decision phase, we have preserved valuable information that can affect the decision process.

## 2.2 A Natural Measure of Uncertainty

Certain intuitive concepts fall into place under the exploratory approach. For one thing, the divergence of opinion often seen among "knowledgeable" experts is accommodated. This divergence does not suggest any of the experts are wrong, but merely that there are multiple plausible candidates for the "true" distribution. Averaging the results often hides this divergence. We lose valuable information about uncertainty. Under exploratory modeling, methods that embrace the variety of expert's opinions are encouraged. Methods for the elicitation of expert knowledge, such as the so-called Delphi technique and other types of formalized "brainstorming", attempt to preserve divergences of opinion rather than artificially suppress them.

When we have perfect knowledge of the environment, we can specify models exactly. Under complete ignorance, all models are essentially "possible". The knowledge we do have, however imperfect, constrains the possibilities between these two extremes. As a result, the divergence of the ensemble of explorations provides a natural measure of uncertainty due to knowledge imperfection: The wider the spread of the estimates, the greater the uncertainty. This natural measure of epistemic uncertainty can be used to guide further exploration, and also becomes part of the subsequent decision making process.

We can formalize the measures using absolute or relative interval measures, or, to the extent that the exploratory ensemble consists of distinct elements, set cardinality. These intervals are not developed from data, but are rather a response to the data available. Ultimately, they are judged instrumentally, by how well they permit us to reason about a complex and uncertain world.

## 2.3 Relation to Uncertainty Logics

While the uncertainty we encounter in most real-world risk assessments extends beyond that recognized by the theory of probability, its features can still be captured in robust theoretical models. Primary among these is the logic of fuzzy sets. Fuzzy sets are a generalization of interval-valued sets that allow for the possibility of various outcomes in the face of imperfect knowledge (Zimmermann, 1991). They are used to model imprecision. Basically, a fuzzy set defines a spectrum of possibilities, using numerical expression.

In terms of Figure 1 above, we could view the extreme distributions on the chart as defining the bounds of our uncertainty interval. Notice that as losses get larger (and data gets more scarce), the interval of uncertainty expands. In this way, exploratory modeling provides a direct empirical link to the creation of such intervals. To the extent that certain distributions within the envelope can be assigned different degrees of "credibility", we have defined a graded interval of possibility known as a fuzzy membership function.

Indeed, by purely intuitive criteria, we see that the projections of multiple plausible models, such as we have shown in Figure 1, impart a visual fuzziness to the model building process. This visual diffusion is a good analogy for the type of uncertainty we feel. The true answer lies within a haze of uncertainty represented by multiple alternatives. These intuitive criteria can be formalized by giving the visual spread mathematical meaning. This can be done using simple intervals, or "nested" intervals graded by level of confidence.

Exploration and the development of formal models using intervals values and fuzzy logics are intimately connected. Together they form the basis of a strategy for thinking about epistemic uncertainty.

## 2.4 Examples from Risk Modeling

The question remains as to how we can practically incorporate exploratory models into the risk assessment process. Critical to the endeavor is our ability to generate candidate models efficiently. Intuition has been a mainstay of such approaches. A more defined and controllable methodology entails the development of multiple plausible models within the same, formal modeling framework. This approach has only recently gained practicality with the advent of very powerful, yet accessible, computers. We will detail two simple, real world examples of the exploratory modeling environment will help illuminate some of its features. Both are from the field of risk assessment and management at the organizational level. The uncertainty inherent in risk estimates has long been recognized at the level which the assessment of very low probabilities associated with very high stakes outcomes presents grand challenges to society (such as "global warming"). Observation suggests that significant uncertainties enter at far lower probabilities. As a result, the frontier for application of exploration to uncertain models is much wider than currently assumed.

### 2.4.1 Exploring an Actuarial Simulation

We consider first a large manufacturing firm with significant exposure to public liability from both its operations and products. In order to gain greater insight into the properties of a firm's exposures in this area, for the purpose of insurance purchasing and loss prevention, actuarial studies are often commissioned. The results of such studies usually include a probability distribution of the probabilities of exceeding some annual aggregate loss (more technically, a complementary cumulative distribution function, or CCDF).

In our case study, it was decided that a fairly straightforward actuarial approach of fitting separate loss frequency and distributions to actual loss data would be appropriate. The distributions were combined into an aggregate distribution using Monte Carlo simulation. The result was a parameterized model that could be used to test various risk management alternatives. For example, various insurance retentions ("deductibles") could be applied to the simulation, and the results noted. If the firm had its own "captive" insurance subsidiary, the distribution would provide crucial information on the probability of exceeding various financial thresholds.

Figure 2 shows the result of the simulation process. Based on initial, best guess estimates of parameters based on data fitting, the result is shown in the figure as the dashed line. All simulations, including mathematical manipulations were done using a common spreadsheet computer program on a "desk top" personal computer.

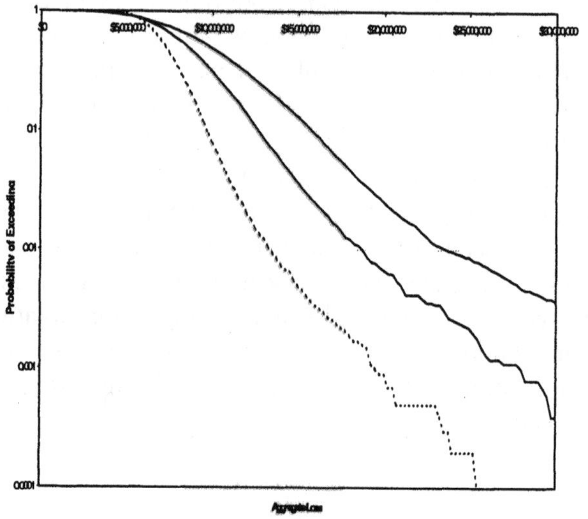


Figure 2. Exploring an Actuarial Simulation

Given data limitations and other areas of uncertainty, it was decided to explore further models. Of primary concern was the tradeoff between severity and frequency. Upon review of the original model, concern surfaced as to whether the aggregate loses were in fact more "frequency driven", or more based on severity. Given some plausible assumptions, it was suspected that the contribution of frequency of loss in the initial estimate might have been overstated. Conversely, the severity, or potential size of loss, may have been understated, based on expert understanding of the mechanics of loss in this environment. Two more plausible simulations were run using the new models. These are shown in the figure as an additional two solid lines.

The results of this exploration were telling. While all three models provide similar results to an aggregate annual loss of $5,000,000, they start to diverge significantly above that. At $10,000,000 of losses a year, the divergence in probabilities is pronounced. Such uncertainties would certainly have an effect on, say, the level of exposure to hold on the firm's own account (vs. commercial insurance). Based on the analysis, appropriate actions could be taken. For example, increased caution (i.e., regret minimization) may be reflected in a more conservative retention/ captive usage decision. What we forego is the possibility of additional tangible cost savings due to reduced reliance on commercial insurance. What we gain is protection against decisions that could cause us considerable financial instabilities down the road.

Note that though similar in appearance, the results of exploratory modeling are far different from the calculation of statistical confidence intervals. What we attempt to capture here is uncertainty due to knowledge imperfection, not randomness in the sampling process. In addition, the common interpretation of such intervals, as a "probability of a probability", becomes problematic in the face of uncertainty that is not properly treated as a form of randomness. Practically, propagation of this uncertainty via the probability calculus leads to different results than propagation via the calculus of intervals (Cooper, 1994).

### 2.4.2 Investment in Loss Prevention

The investment in loss prevention activities remains one of the most critical decisions in the risk management process. Economic analysis and justification of loss prevention expenditures proceeds, classically, as a comparison of discounted costs and benefits, appropriately weighted by probabilities. Very simply, expected benefit, E(B), of initiating protection may be calculated as,

Where, $$\sum_{t=1}^{T} (p-p^*)(L)/(1+r)^t$$

p = Annual probability of loss without the protective measure
p* = Annual probability of loss with the protective measure
L = Loss reduction for mitigation
r = annual rate of return, or discount rate
T = Useful life of the protective measure

When expected benefit exceed expected cost, we institute protection. Where expected cost is greater that expected benefit, we do not (Kunreuther, 2000). This simple, intuitive measure functions fine when probabilities are known. When epistemic uncertainty enters, the commensurate complications are introduced. When uncertainty as to post-prevention, pre-prevention or both, exists, we must account for it in the decision process. We may examine the degree of epistemic uncertainty present by performing an exploratory analysis.

Consider the individual firm's decision to install sprinkler protection at a major production facility. We assume the initial cost of the protection measure is $80,000, with a useful life of 20 years. The annual probability of a complete loss of the facility due to fire is assumed to be 015. With the proposed sprinkler protection the probability of loss drops to .0001. The potential reduction in loss ("loss expectancy") is $1,000,000. An annual discount rate of 6 percent is assumed. The net expected stream of discounted benefits equals $99,361. Since the long-run expected benefit is greater than cost, the protective measure should be installed.

Obviously this decision depends critically on our ability to assess pre- and post loss probabilities. The problem is that this precision may not be available in the "real world". We rarely know enough about the mechanisms and statistics of low probability/ high consequence losses, or the effectiveness of protective measures in "field" applications, to be able to make such accurate assessments.

Exploration of this model would involve multiple possibilities for the probabilities of loss and effectiveness of protection. To explore the model in our example, an electronic spreadsheet was developed that allowed manipulation of the expected cost/ benefit model using different possible probabilities. Rather than deriving a precise measure of expected benefits, the results are shown with various degrees of possibility. These possibilities were "weighted" by degree of possibility by a group of experts. The degree is shown as a fuzzy membership in the set of all possibilities, along a scale of 0 to 1. The higher the membership, the greater the possibility an expected benefit belongs to the set of potential benefits. The results are shown in Figure 3.

Note that a range of benefits are shown, with various degrees of possibility. The result of the traditional analysis ( in this case, an expected benefit of $99361.) is shown as a high possibility. But other expected values are possible as well, both above and below the expected cost of the measure. The possibility of "1" for a range of values between 0 and $5,000 means that all values in that range are strong possibilities, all equally "good". Note also that in this case the experts suggest there is a distinct possibility ("1") that the protective measure will have no effect on reducing the probability of loss at all. With the installation at hand, this possibility was suggested by the fact that plant construction factors (i.e, flammable building materials) could overwhelm the ability of the proposed system to respond effectively. This combined with the uncertainty engendered by a unique sprinkler design needed to accommodate an intricate variety of manufacturing processes. Note further that this "unsureness" was not simply a matter of statistical reliability of, say, the water supply. Such reliability, however, adds another complicating factor which further deepens the epistemological uncertainty involved in the assessment.

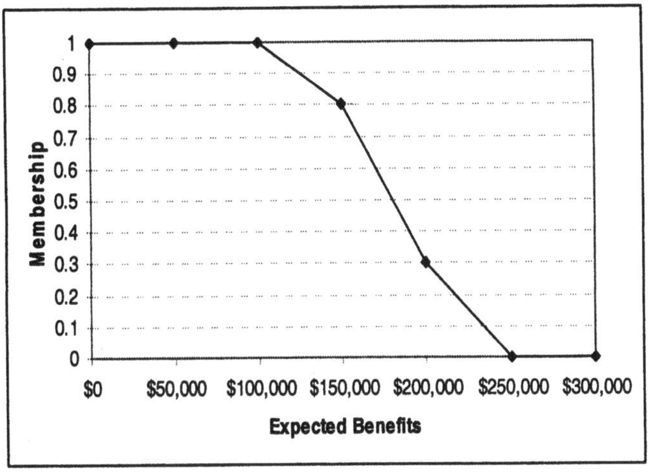


Figure 3. Results of Exploratory Modeling of Loss Prevention Measures

Were greater knowledge of losses and prevention mechanisms exist, perhaps via more controlled conditions and a more complete level of statistical knowledge of function (and malfunction), the fuzzy membership function would surround our "best guess" estimate more tightly. In these cases, and approximate decision can be made with some degree of confidence. That approximate yet supportable decisions are made in such cases is undeniable. Yet there are considerable real-world situations were uncertainty results in wide, and quite problematic, intervals of possibility. This is what makes many of these decisions so difficult. These conditions must be addressed in the decision process.

## 3. FROM EXPLORATION TO DECISION

The exploratory approach exposes epistemic uncertainties that require specialized treatment within the decision process. Care must be taken in applying such criteria lest the value of exploration be lost. Key to the exploratory approach is representing as much of what we don't know as far into the decision process as we can. We examine some possible criteria for decisions under epistemic uncertainty. Starting with simple extensions of the expected value criterion, we move to criteria that more accurately reflect limitations on knowledge when data becomes scarce.

## 3.1 Extended Expected Value Criteria

When probabilities and consequences can be measured with relative accuracy, decisions may be based on expected values (probability × loss) (Kammen and Hassenzahl, 1999). Expected value calculations can be easily extended to exploratory results. If we consider the upper (u) and lower (l) bounds of an exploration as constituting an interval, interval valued expected values could be easily computed using the "interval average", (u+l)/2. Let's say exploration suggests an interval of [.01,.10] as bounding the probability estimate of a loss of $1,000,000. The point estimate of probability based on our interval average would be ( .01+.10)/2, or .055. The expected value of loss n this case is (.055 × $1,000,000) = $55,000. This number could be used directly for expected cost/ benefits comparisons, and the like. Similar approaches are available when bounds are based on fuzzy membership functions rather than pure intervals.

One problem with direct extension of expected value in this fashion is that we lose valuable uncertainty information gained through the exercise of exploratory modeling. In the case of interval estimates, for example, the interval will once again be reduced to a single, "best guess" estimate. In our simple example above, we do not distinguish between a decision based on the interval average of .055 and a precise probability of .055. Yet the fact that the former measurement involved greater uncertainty might certainly be relevant to our decision process. A method that preserves much (but not all) of the uncertainty information gained through the exploratory exercise is the consideration of regret. Regret may be defined most simply as the difference between the actual decision outcome and that which would be optimal under the circumstances. Introduction of regret recognizes that uncertainty, specifically that regarding the "down side" of the estimate, can have an affect on the decision process. Applied to the results of an exploration, regret minimization would suggest that we take the most conservative bound developed through our exercise as the basis for our expected value calculations. In the actuarial exploration discussed above (Section 2.3.1), for example, we would use the outer most plausible bound of the exploration to perform the calculations. These would then be utilized for traditional analysis based on expected values, such as the computation of insurance premiums.

## 3.2 Minimax and Precaution

When exploration indicates that the probabilities of outcomes under study are sufficiently "unknown", we may choose to abandon the probability dimension as a guide to action. Decisions based soley on the consequence dimension of risk suggest a minimax approach to decision. Under minimax, we choose those actions that minimize the maximum possible loss, regardless of probabilities. The concept of regret avoidance remains central, but now on an absolute basis. No attempt is made to introduce relativities via probability weightings of any sort (e.g., expected values).

In terms of risk management via loss prevention and mitigation, this approach is maximally conservative with respect to risk. The decision maker is theoretically willing to spend up to the amount of loss to prevent the loss. That is, as long as the difference between worst case outcome and cost of prevention/ mitigation is positive, we undertake the preventive measure.

The minimax also forms the basis of precautionary approaches to decision, which suggest that when the consequences are high, and probabilities uncertain, we choose a maximally conservative approach with respect to the risk. This "precautionary principle" is being used more frequently to guide in the governmental regulation of technological activities that have the potential for widespread harm (Raffensberger etal, 1999). Consider the case of "global warming". There is scientific evidence that points to the fact that the mean temperature of the earth is rising. The supposed link to human activity is the production of so called green house gases due to increasing industrialization. The link between global warming and industrial production is, however, tenuous. That is, considerable uncertainty surrounds the probabilistic assessment of the risk of this industrial byproduct. Nonetheless, given the catastrophic consequences of this environmental trend, precaution suggests we apply a minimax approach to the regulation of the production of green house gases. This means spending on their curtailment, or baring the effectiveness of any genuine protective measures (see Section 2.3.2), we avoid the activity altogether.

In the application of the minimax criteria, exploration seeks to determine if, when consequences get serious enough, their probabilities are sufficiently "unknown". If so, we abandon the probability criteria as desiderata, and concentrate solely on the loss dimension: If consequences are sufficiently bad, we take the appropriate action to avoid these consequences.

# 4. CONCLUSIONS

Risk management under conditions of uncertainty brought about by knowledge imperfection presents a considerable challenge. The existence of this form of uncertainty, as distinct from variability introduced by randomness, is recognized in the formal logics of intervals and fuzzy sets. Recognition of this form of uncertainty is especially important in the analysis of low probability/ high consequence (high risk) exposures. The consequences of "wrong" decisions in the context of high risk exposures are extreme and quite possibly irreversible.

Exploratory modeling refers to the identification of epistemic uncertainty via the identification of possible models consistent (i.e., bounded by) the information at hand. No attempt is made to eliminate, or even summarize this data, lest valuable information about this uncertainty be lost. Exploration takes the uncertain landscape of risk as it is, and seeks to mold decision around this landscape. In doing so, we manipulate decision criteria to best suit our goals, and not the factual basis for the decision. The result is decisions made on a more realistic, and therefore (hopefully), better basis.

Epistemic uncertainty, once identified via the process of exploration, can be expressed using intervals or fuzzy membership functions. Extensions of expected value criteria to the domain of intervals and fuzzy memberships of the associated probability distributions is often suggested. Calculating expected values based on interval (or fuzzy) averages can result in considerable loss of information gained during the exploratory exercise. By including considerations of regret, we choose among the results of our exploration so as to act with appropriate conservatism with respect to those uncertainties present. When probabilities are sufficiently unknown, we turn to minimax and precautionary criteria for dealing with high risk environments. These criteria emphasize avoidance of the "worst case" in terms of consequences, in effect ignoring probabilities altogether.

When dealing with epistemic uncertainty, we must adjust our decision criteria to suit the degree of knowledge imperfection. Exploratory modeling becomes the essential first step in defining the extent of these knowledge imperfections.

# REFERENCES

Bankes, S., "Exploratory Modeling for Policy Analysis", Operations Research, May-June, 1993.

Cooper, J. A., "Fuzzy Algebra in Risk Assessment", SERA: Safety Engineering and Risk Analysis (ASME), 1994.

Jablonowski, M., "Dealing With Danger: A Fuzzy Approach", in Thomas Whalen, ed., Proceedings of the 19th International Conference of the North American Fuzzy Information Processing Society (NAFIPS 2000), IEEE Press, 2000.

Jablonowski, M., "Why Risk Analyses Fail", CPCU Journal (Society of Chartered Property and Casualty Underwriters), Winter 2000.

Jablonowski, M., "Are Formal Risk Assessments of Any Use?", Risk Management, August, 2002.

Kammen, D. M. and D. M. Hassenzahl, Should We Risk It? Exploring Environmental, Health and Technological Problem Solving, Princeton University Press, 1999.

Kantorovich, A., Scientific Discovery: Logic and Tinkering, State University of New York Press, 1993.

Kunreuther, H., "Decision Making for Protective Measures", in Extreme Events: Developing A Reasearch Agenda for the 21st Century (National Science Foundation, 2000)

Lempert, R. J., S.W. Popper and S. C. Bankes, Shaping the Next One Hundred Years: New Methods in Quantitative, Long-Term Policy Analysis, Rand, 2003.

Raffensberger, C., J. Tickner and W. Jackson, eds., Protecting Public Health and the Environment: Implementing the Precautionary Principle, Island Press, 1999.

Wack, P., "Scenarios: Uncharted Waters Ahead", Harvard Business Review, September-October, 1985.

Zimmermann, H.-J., Fuzzy Set Theory and its Applications, Kluwer Academic Publishers, Boston, 1991.

# Chapter 13

# EXPERIMENT WITH A HIERARCHICAL TEXT CATEGORIZATION METHOD ON WIPO PATENT COLLECTIONS

**Domonkos Tikk, György Biró, and Jae Dong Yang**

## 1. INTRODUCTION

The immense and exponentially growth in the number of electronic documents stored on the internet, corporate intranets and data warehouses necessitates powerful algorithms and tools that are able to deal with data of such quantity. An obvious way to handle the vast number of documents is organizing them into category systems. Category systems are usually hierarchic (called taxonomy) because that offers straightforward way to find and browse data at arbitrary refinement. E.g. documents on large internet directories, such as Yahoo! and Google, are categorized into taxonomy. This storage technique requires efficient automatic categorization methods as manual text categorization is no longer amenable in that size, requiring a vast amount of time and cost.

The purpose in automatic text categorization (TC) is to assign a document to appropriate category/ies (or topic) being selected from a predefined set of categories. Originally, research in TC addressed the binary problem, where a document is either relevant or not w.r.t. a given category. In real-world situation, however, the great variety of different sources and hence categories usually poses multi-class classification problem, where a document belongs to exactly one category selected from a predefined set Baker and McCallum, 1998; Weiss et al., 1999; Wiener et al., 1993; Yang, 1999. Even more general is the case of multi-label problem, where a document can be classified into more than one category. While binary and multi-class problems were investigated extensively Sebastiani, 2002, multi-label problems have received much less attention Aas and Eikvil, 1999.

As the number of topics becomes larger, multi-class categorizers face the problem of complexity that may incur rapid increase of time and storage, and compromise the perspicuity of categorized subject domain. A common way to manage complexity is using a hierarchy (in this paper we restrict our investigation to tree structured hierarchies), and text is no exception Chakrabarti et al., 1998. Internet directories (see e.g. Yahoo; http://www.yahoo.com) and large on-line databases are often organized in hierarchy.

Patent databases are typically such where the use of a hierarchical category system is a necessity. Patents cover a very wide area of topics, and each field can be further divided into subtopics, until a reasonable level of specialization is reached. The International Patent Classification (IPC) is a standard taxonomy developed and administered by WIPO (World Intellectual Property Organization) for classifying patents and patent applications. The use of patent documents and IPC for research into automated categorization is interesting for the following reasons Fall et al., 2002:

1 IPC covers a huge range of topics and uses a diverse technical and scientific vocabulary.

2 IPC is a complex, hierarchical taxonomy, where over 40 million documents have been classified worldwide. The number of documents classified each year is rising fast.

3 Domain experts in national patent offices currently classify patent documents fully manually. These experts have an intimate knowledge of the IPC system.

4 Patent documents are often available in several languages. Professional translators have already performed large numbers of translations manually.

As a courtesy of WIPO, we could experiment with the WIPO-alpha English and WIPO-de German patent databases issued in late 2002 and early 2003, respectively. (Collections are available after registration at http://www.wipo.int/ibis/datasets/index.html.) WIPO-alpha is a large collection (3 GB) of about 75000 XML documents distributed over 5000 categories in four levels (the top four levels of IPC); WIPO-de is an even larger collection of about 110000 XML documents defined on the same taxonomy (IPC). At WIPO, they experimented with several text categorization technique on the WIPO-alpha collection, see Fall et al., 2003a.

Our primary purpose with this database is to analyze the applicability of our algorithm, having been tested successfully on smaller corpora (see Tikk and Biró, 2003; Tikk et al., 2003), on a very large real-world collection in terms of efficiency and feasibility (time and space requirements).

The paper is organized as follows. Section 2 gives an overview on UFEX and the major features implemented in HITEC. Section 3 reports on our experiences on WIPO collections. The conclusion is drawn in Section 4.

## 2. THE CLASSIFIER

UFEX (Universal Feature Extractor) method aims at determining relevant characteristics of a set of categories based on training entities. It is particularly optimized to handle hierarchically organized category structures. The nature of the training entities is independent from UFEX as it applies an internal representation form, therefore it is able to work on arbitrary kind of data (e.g. text, image, numerical measurements) that can be described by numerical vectors of features. The basic idea of UFEX is described in details in Tikk et al., 2003. For simplicity, in the next we will use the TC-specific notations. Here we remark again that, nevertheless, UFEX is designed to be able to process arbitrary numerical data.

The core idea of UFEX is an iterative learning module that gradually trains the classifier to recognize constitutive characteristics of categories and hence to discriminate typical documents belonging to different categories.

Characteristics of categories are captured by typical terms occurring frequently in documents of the corresponding categories. We represent categories by weight vectors, called category descriptors (or simply descriptors), where an element of this vector refers importance of a term (typically word) discriminating the given category from others. The training algorithm of UFEX sets and maintains category descriptors in a way that allows the classifier to be able to categorize documents with high accuracy in the appropriate category. The training starts with zero descriptors.

We now briefly describe the training procedure. First, when classifying a training document we compare it with category descriptors and assign the document to the category of the most similar descriptor. When this procedure fails finding correct category we raise the weight of such features in category descriptors that appear also in the given document. If a document is assigned to a category incorrectly, we lower the weight of such features in descriptors that appear in the document. We tune category descriptors by finding the optimal weights for each feature in each category descriptor by this awarding–penalizing method. The training algorithm is executed iteratively and ends when the performance of the classifier cannot be further improved significantly. See the block diagram of Figure 1 for an overview and details in Subsection 2.2. about the training algorithm. For test documents the classifier works in one pass by omitting the feedback cycle.

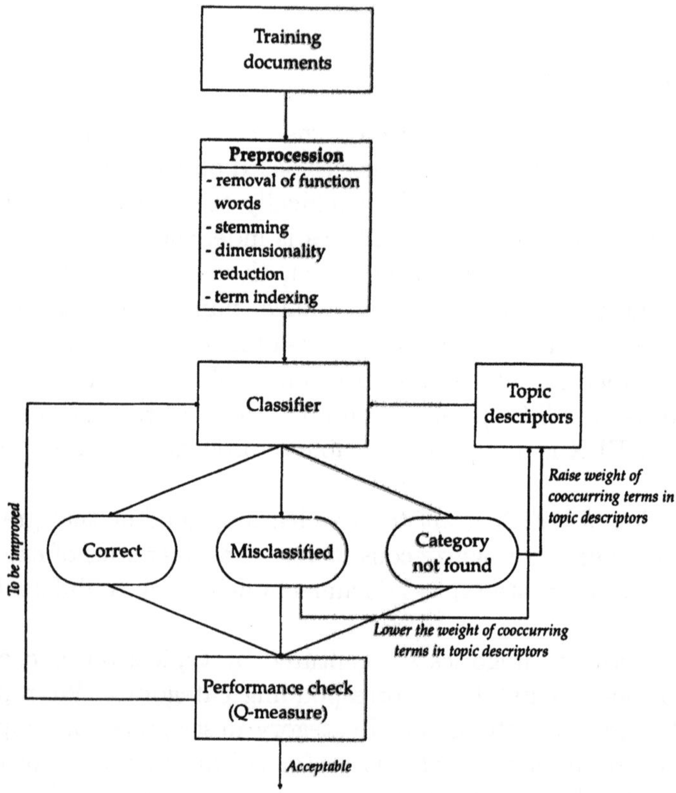


*Figure 1.* The flowchart of the training algorithm of UFEX

The rest of this section is organized as follows. Subsection 2.1. describes the topic hierarchy, vector space model and descriptors. Subsection 2.2. presents classification and the training method.

## 2.1. Notations

Let $C$ be the fixed finite set of categories organized in a topic hierarchy. In this paper, we deal with tree structured topic hierarchies, and we do not allow multiple parenthood unlike in our previous work Tikk et al., 2003.

Let $\mathcal{D}$ be a set of text documents and $d \in \mathcal{D}$ an arbitrary element of $\mathcal{D}$. In general, documents are pre-classified under the categories of $\mathcal{C}$, in our case into leaf categories. We differentiate training, $d \in \mathcal{D}_{\text{Train}}$, and test documents, $d \in \mathcal{D}_{\text{Test}}$, where $\mathcal{D}_{\text{Train}} \cap \mathcal{D}_{\text{Test}} = \emptyset$, and $\mathcal{D}_{\text{Train}} \cup \mathcal{D}_{\text{Test}} = \mathcal{D}$. Training documents are used to inductively construct the classifier. Test documents are used to test the performance of the classifier. Test documents do not participate in the construction of the classifier in any way.

Each document $d_j \in \mathcal{D}$ is classified into a leaf category of the hierarchy. No document belongs to non-leaf categories. We assume that a parent category owns the documents if its child categories, i.e., each document belongs to a topic path containing the nodes (representing categories) from the root to a leaf. Formally,

$$\text{topic}(d_j) = \{c_1, \ldots, c_q \in \mathcal{C}\} \tag{1}$$

determines the set of topics $d_j$ belongs to along the topic path from the highest to the deepest. Note that the root is not administrated in the topic set, as it owns all documents. $c_q$ denotes leaf-category, and the index refers to the depth of the category.

Texts cannot be directly interpreted by a classifier. Because of this, an indexing procedure that maps a text $d$ into a compact representation of its content needs to be uniformly applied to all documents (training and test). We apply the usual vector space model, where a document $d_j$ is represented by a vector of term weights

$$d_j = (w_{1j}, \ldots, w_{|\mathcal{T}|j}), \tag{2}$$

where $\mathcal{T}$ is the set of terms that occurs at least once in the training documents $\mathcal{D}_{\text{Train}}$, and $0 \leq w_{kj} \leq 1$ represents the relevance of $k$th term to the characterization of the document $d$. Before indexing the documents function words (i.e. articles, prepositions, conjunctions, etc.) are removed, and stemming (grouping words that share the same morphological root) is performed on $\mathcal{T}$.

We experimented with tf×idf (3) and entropy (4) weighting Salton and McGill, 1983:

$$w_{kj} = o_{kj} \cdot \log\left(\frac{N}{n_k}\right), \tag{3}$$

$$w_{kj} = \log(o_{kj}+1)\left(1+\frac{1}{\log N}\sum_{i=1}^{N}\left[\frac{f_{ki}}{n_k}\log\left(\frac{f_{ki}}{n_k}\right)\right]\right). \tag{4}$$

Here $o_{kj}$ is the occurrence of the $k$th term in $d_j$; $n_k$ is the number of documents for which $k$th term occurs at least once; $N = |\mathcal{D}_{\text{Train}}|$. Term vectors (2) are normalized before training.

We characterize categories analogously as documents. To each category is assigned a vector of descriptor term weights

$$\mathrm{descr}(c_i) = \langle v_{1i}, \ldots, v_{|T|i} \rangle, \quad c_i \in \mathcal{C} \tag{5}$$

where weights $0 \leq v_{1i} \leq 1$ are set during training. All weights are initialized as 0. The descriptor of a category can be interpreted as the prototype of a document belonging to it.

## 2.2. Classification and training

### 2.2.1. Classification

When classifying a document $d \in \mathcal{D}$ the term vector representing $d$ (2) is compared to topic descriptors (5). The vector of $d$ is matched against a set of descriptors and based on the result the classifier selects (normally) a unique category.

The classification method works downward in the topic hierarchy level by level. First, it determines the best among the top level categories. Then its children categories are considered and the most likely one is selected. Considered categories are always siblings linked under the winner category of the previous level. Classification ends when a leaf category is found. This, in fact, is a greedy algorithm where the best category is selected based on a conformity measure defined next.

Let us assume that we have to select from $m$ categories at an arbitrary stage of the classification of document $d_j$: $c_1, \ldots, c_m \in \mathcal{C}$. Then we calculate the conformity of term vector of $d_j$ and each topic descriptors $\mathrm{descr}(c_1), \ldots, \mathrm{descr}(c_m)$, and select that category that gives the highest conformity measure. We applied the unnormalized cosine measure that calculates this value as a function $f$ of the sum of products of document and descriptor term weights:

$$\mathrm{conf}(d_j, \mathrm{descr}(c_i)) = f\left(\sum_{k=1}^{|T|} w_{kj} \cdot v_{ki}\right), \tag{6}$$

where $f : \mathbb{R} \to [0,1]$ is an arbitrary smoothing function with $\lim_{x\to 0} f(x) = 0$ and $\lim_{x\to\infty} f(x) = 1$. The smoothing function is applied (analogously as in control theory) to alleviate the oscillating behavior of training.

Summarizing, we give the pseudo-code of UFEX's classification algorithm for a document $d_j$. It starts from the root category.

Step 1: Calculate (6) for all $m$ sibling categories of the given level.

Step 2: Select the category that has the highest conformity measure with $d_j$: $c_{\mathrm{best}}$.

Step 3: If $c_{\text{best}}$ is a leaf-category then stop; otherwise go to Step 1.

McCallum McCallum et al., 1998 criticized the greedy topic selection method because it requires high accuracy at internal (non-leaf) nodes. In order to alleviate partly the risk of a high level misclassification, we control the selection of the best category by a minimum conformity parameter $\text{conf}_{\text{min}} \in [0, 1]$, i.e. the greedy selection algorithm continues when

$$\text{conf}(d_j, \text{descr}(c_{\text{best}})) \geq \text{conf}_{\text{min}} \tag{7}$$

satisfied, where $c_{\text{best}}$ is the best category at the given level. This means that we stop in Step 2 if the best category does not satisfy the minimum conformity condition of (7).

Another type of problem occurs when there are several categories having approximately the same conformity with $d_j$ as $c_{\text{best}}$. In such a case it is reasonable to consider a set of categories as the best ones, and continue the selection method among their children. Formally, we can set a parameter $\text{conf}_{\text{relax}} \in [0, 1)$, typically around 0.9 and select in Step 2 a set of categories satisfying:

$$C_j = \{c | \text{descr}(c_{\text{best}}) - \text{descr}(c) \leq \text{conf}_{\text{relax}}\}.$$

### 2.2.2. Training

In order to improve the effectiveness of classification, we apply supervised iterative learning, i.e. we check the correctness of the selected categories for training documents and if necessary, we modify term weights in category descriptors. Term weights are modified when a document is classified incorrectly.

The classifier can commit two kinds of error: it can misclassify a document $d_j$ into $c_i$, and usually simultaneously, it cannot determine the correct category of $d_j$. The following training algorithm of UFEX aims at minimizing both types of error. We scan all the decisions made by the classifier and process as follows.

For each considered category $c_i$ at a given level we accumulate a vector $\delta(c_i) = \langle \delta(v_{1i}), \ldots, \delta(v_{\mathcal{T}i}) \rangle$ where

$$\delta(v_{ki}) = \alpha(\text{conf}_{\text{req}} - \text{conf}(d_j, \text{descr}(c_i))) \cdot w_{ij}, \quad 1 \leq k \leq \mathcal{T} \tag{8}$$

where $\text{conf}_{\text{req}} = 1$ when $c_i \in \text{topic}(d_j)$, 0 otherwise. Here $\alpha \geq 0 \in \mathbb{R}$ is the learning rate. The category descriptor weight $v_{ki}$ is updated as $v_{ki} + \delta(v_{ki})$, $1 \leq k \leq \mathcal{T}$, whenever category $c_i$ takes part in an erroneous classification. If $d_j$ is misclassified into $c_i$ then $(\text{conf}_{\text{req}} - \text{conf}(d_j, \text{descr}(c_i)))$ is negative, hence the weight of co-occurring terms in $d_j$ and $c_i$ are reduced in the category descriptor of $c_i$. In the other case, if $c_i$ is the correct but unselected category of $d_j$, then $(\text{conf}_{\text{req}} - \text{conf}(d_j, \text{descr}(c_i)))$ is positive, thus the weight of co-occurring terms in $d_j$ and $c_i$ are increased in the category descriptor of $c_i$.

Summarizing, we give the pseudo-code of the training algorithm of UFEX.

Step 1 Calculate for each category $\delta(c)$ defined in (8).

Step 2 Each category descriptor $\text{descr}(c)$ is updated by $\text{descr}(c) - \delta(c)$.

1 When $c$ is incorrectly selected then $\text{conf}_{\text{req}}(c) = 0$ and $\delta(c)$ is negative, hence the weight of co-occurring terms in $c$ and $d_j$ are decreased that incurred the incorrect selection.

2 When the correct $c$ is not found then $\text{conf}_{\text{req}}(c) = 1$ and $\delta(c)$ is positive, therefore the weight of co-occurring terms in $c$ and $d_j$ are increased to force the correct selection.

Step 3 Repeat Step 1 and Step 2 for all documents in the training set.

Step 4 If the terminal condition is satisfied then stop; otherwise repeat Step 1–Step 3.

We also experimented with a more sophisticated weight setting method where the previous momentum of the weight modifier is also taken into account in the determination of the current weight modifier. Let $\delta^{(n)}(v_{ki})$ be the weight modifier in the $n$th training cycle, and $\delta^{(0)}(v_{ki}) = 0$ for all $1 \leq k \leq \mathcal{T}$. Then the weight modifier of the next training cycle is $\delta^{(n+1)}(c_i) = \langle \delta^{(n+1)}(v_{1i}), \ldots, \delta^{(n+1)}(v_{\mathcal{T}i}) \rangle$, and its elements are calculated as

$$\begin{aligned} \delta^{(n+1)}(v_{ki}) = \alpha \cdot (\text{conf}_{\text{req}} - \text{conf}(d_j, \text{descr}(c_i))) \cdot w_{ij} \\ + \delta^{(n)}(v_{ki}) \cdot \beta \end{aligned} \tag{9}$$

where $\beta \in [0, 1]$ is the momentum coefficient. The value of $\alpha$ and $\beta$ can be uniform for all categories, or can depend on the level of the category. We experienced that at a lower value, typically 0.05..0.2 is better if the number of training documents is plentiful, i.e. higher in the hierarchy, and a higher value is favorable when only a few training documents are available for the given category, i.e. at leaf categories.

This modification changes the pseudo-code of the training algorithm as

- In an initial step (Step 0) $\delta^{(0)}(v_{ki}) = 0$ are set.
- In Step 1 we calculate $\delta(c)$ as defined in (9)
- In Step 4, we increase the training cycle counter by 1.

The number of nonzero weights in category descriptors increases as the training algorithm operates. In order to avoid their proliferation, we propose to set descriptor term weights to zero under a certain threshold.

The training cycle is repeated until the given maximal iteration has not been finished or the performance of the classifier reaches a quasi-maximal value. We use the following optimization (or quality) function (introduced in Tikk et al., 2003) to measure inter-training effectiveness of the classifier for a document *d*:

$$Q(d) = \frac{\#(\text{correctly found topics of } d)}{\#(\text{total topics of } d)} \cdot \frac{1}{1 + \#(\text{incorrectly found topics of } d)}.$$

The overall $Q$ is calculated as average of $Q(d)$ values:

$$\overline{Q} = \frac{\sum_{d \in \mathcal{D}_{\text{Train}}} Q(d)}{|\mathcal{D}_{\text{Train}}|} \tag{10}$$

The quality measure $\overline{Q}$ is more sensible to small changes in the effectiveness of the classifier than, e.g., $F$-measure van Rijsbergen, 1979 that we use to qualify the final performance of the classifier (see Section 3). Hence, it is more suitable for inter-training utilization. By setting a maximum variance value $\text{var}_{\text{max}}$ (typically 0.95..1.00) we stop training when actual $\overline{Q}$ drops below the $\text{var}_{\text{max}} \cdot \overline{Q}^{\text{best}}$, where $\overline{Q}^{\text{best}}$ is the best $\overline{Q}$ achieved so far during training.

## 3. EXPERIMENTS ON WIPO PATENT COLLECTIONS

### 3.1. The document collections

WIPO offers two patent document collections for research: WIPO-alpha that consists of 3 GB English patent documents (in total about 75000 documents) and WIPO-de collection that contains German patent applications (in total about 110000 documents). The documents are in XML format. Collection are provided as two sub-collection of a training set of 46324 (84822) English (German) documents, and a test set of 28926 (26006) English (German) documents, respectively. Documents are assigned one main category, and can be also linked to several other categories. The indexers used the top four levels of IPC taxonomy (termed: section, class, subclass, and main group; top-down) when attributing IPC codes to documents.

Training collections consist of documents roughly evenly spread across the IPC main groups, subject to the restriction that each subclass contains between 20 and 2000 documents. Test collections consist of documents distributed roughly according to the frequency of a typical year's patent applications, subject to the restriction that each subclass contains between 10 and 1000 documents. All documents in test collections also have attributed IPC symbols, so there is no blind data.

Each document includes a title, a list of inventors, a list of applicant companies or individuals, an abstract, a claims section, and a long description. These information are store in separate XML fields. Detailed descriptions about the collections can be found in Fall et al., 2002 and Fall et al., 2003b.

## 3.2. Performance measures

We have adopted three heuristic evaluation measures for categorization success proposed by the provider of the WIPO collections Fall et al., 2002. Let us suppose that the method returns an ordered list of predicted IPC codes, where the order is determined by the confidence level (see (6)). Then we can define the following measures (see Figure 2):

1 Top prediction (briefly: Top) The top category predicted by the classifier is compared with the main IPC class, shown as [mc] in Figure 2.

2 Three guesses (Top 3) The top three categories predicted by the classifier are compared with the main IPC class. If a single match is found, the categorization is deemed successful. This measure is adapted to evaluating categorization assistance, where a user ultimately makes the decision. In this case, it is tolerable that the correct guess appears second or third in the list of suggestions.

3 All classes (Any) We compare the top prediction of the classifier with all classes associated with the document, in the main IPC symbol and in additional IPC symbols, shown as (ic) in Figure 2. If a single match is found, the categorization is deemed successful.

Although in Fall et al., 2002 it is suggested to use these measures solely on IPC class level, in our experiments we also use them on the lower subclass and main group levels.

## 3.3. Dimensionality reduction

When dealing with a huge document collection, the large number of terms, $|\mathcal{T}|$, can cause problem in document processing, indexing, and also in category induction. Therefore, before indexing and category induction many authors apply a pass of dimensionality reduction (DR) to

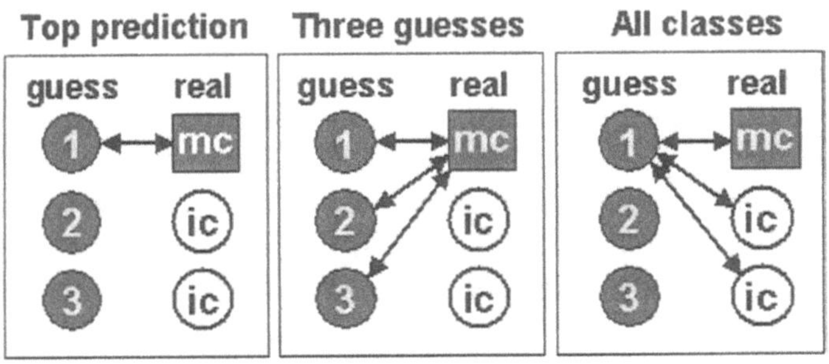


*Figure 2.* Explanation to the three evaluation measures Top, Top 3, Any Fall et al., 2002

reduce the size of $|\mathcal{T}|$ to $|\mathcal{T}'| \ll |\mathcal{T}|$ Sebastiani, 2002. Beside that it can speed up the categorization, papers also reported that it can increase the performance of the classifier with a few percent, if only a certain subset of terms are used to represent documents (see e.g. Koller and Sahami, 1997; Wibovo and Williams, 2002).

In our previous experiments Tikk et al., 2003 we also found that performance can be increased slightly (less than 1%) if rare terms are disregarded, but the effect of DR on time efficiency is more significant. We reduced $|\mathcal{T}|$ by disregarding terms that either occur less than $\text{min}_{\text{occur}}$ times, or occur more often than a certain threshold in $\mathcal{D}_{\text{Train}}$, i.e. if $n_k/|\mathcal{D}_{\text{Train}}| \geq \text{max}_{\text{freq}}$. By the former process we disregard words that are not significant in the classification, while by the later process we ignore words that are not discriminative enough between categories. The typical values are $\text{min}_{\text{occur}} \in [1..10]$ and $\text{max}_{\text{freq}} \in [0.05..1.0]$.

The construction of patent documents provides another way of DR as well. One may select certain XML fields as the basis of the term set (dictionary), and index the other parts of the documents using this dictionary. E.g., the long description part can be ignored for dictionary creation because it may contain lot of dummy words.

## 3.4. Results

We present the results obtained by HITEC on WIPO-alpha collection by means of a series of figures (Figure 3–7) and a summarizing table (Table 1). We differentiated results based on confidence level. Here 0.0 means that all guesses are considered, while 0.8 means that only those decision are considered where the confidence level is not less than 0.8. Obviously, the higher is the confidence level, the lower is the number of considered documents. The figures show all the three performance measures at class, subclass and main group levels by increasing confidence levels of 0.1 step. Table 1 compares some significant values of each parameter setting.

The best results have been achieved after 7 iterations when only XML fields of inventors, applicants, title, abstract and claims are used for dictionary creation ("iptac" setting); entropy weighting (4) is used; $\min_{\text{occur}} = 2$ and $\max_{\text{freq}} = 0.25$. (Figures 3). The other settings, e.g. "ipta" that appears in Table 1, are modification of this one. We denote there only the modified values.

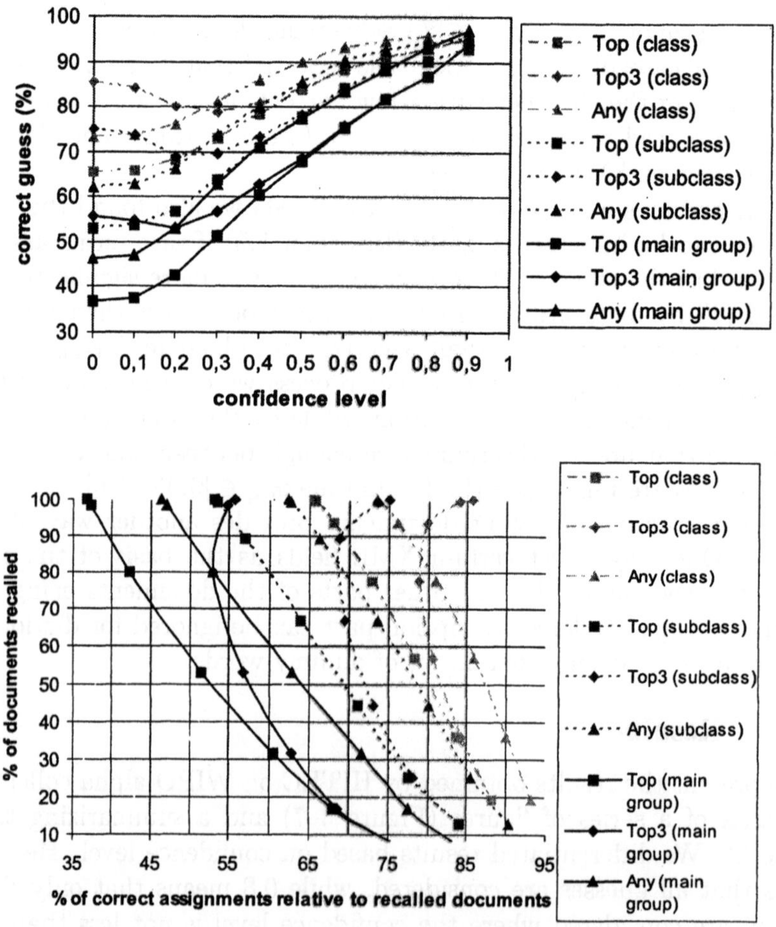


*Figure 3.* Setting "iptac". Precision by confidence levels. b) Comparisons of precisions, extrapolated to 100% recall

The next setting delivered very similar results as "iptac"; obtained when claims fields is disregarded for dictionary creation ("ipta" setting). See Figure 4.

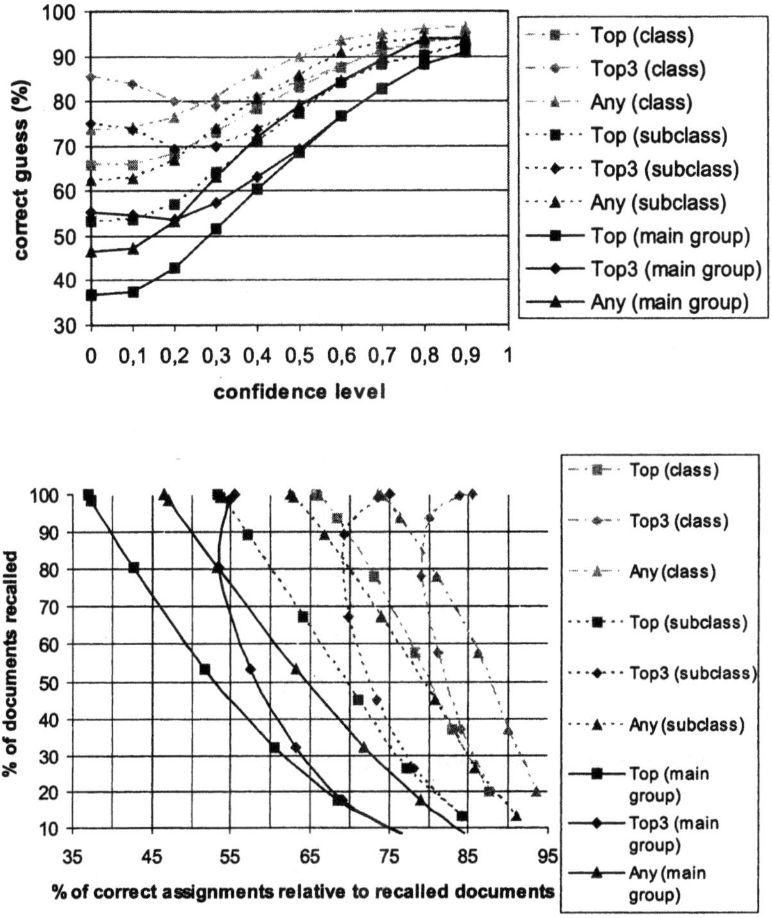


*Figure 4.* Setting "ipta" with entropy weighting and $min_{occur} = 2$ and $max_{freq} = 0.25$. a) Precision by confidence levels. b) Comparisons of precisions, extrapolated to 100% recall

We investigated the effect if only the main category of each patent document is used for training. This experiment was suggested by the developers of the collection Fall et al., 2002, and can be argued that this selection makes ambiguous training documents (having more topics) unique for training purpose. The obtained result did not support this hypothesis, the obtained results were inferior than the ones with regular setting (Figure 5).

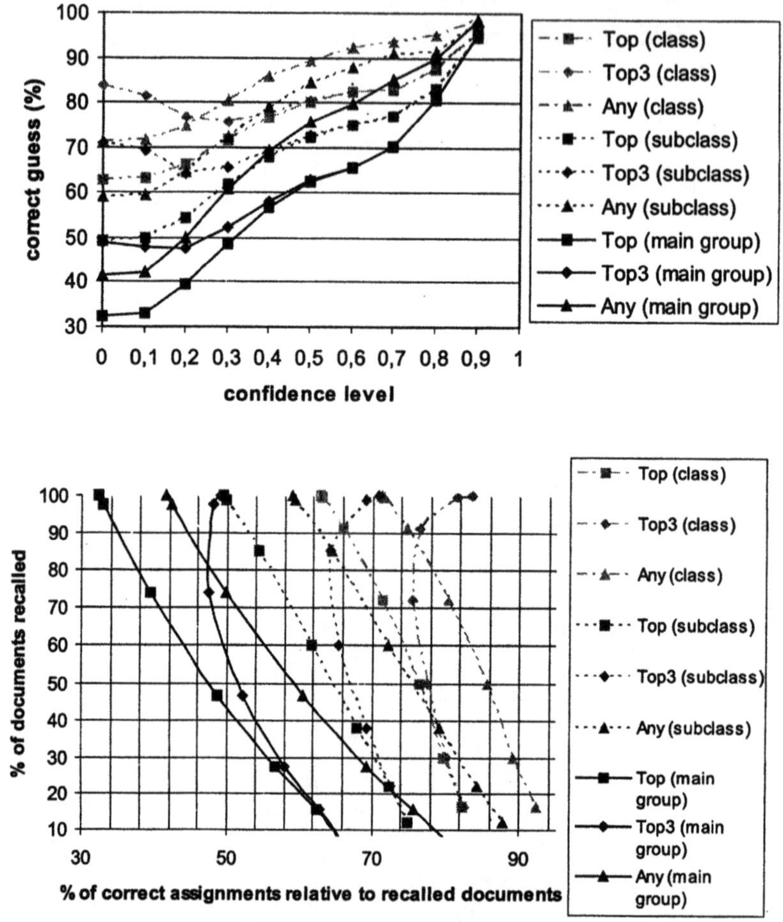


*Figure 5.* Setting: only main categories used for training. Precision by confidence levels b) Comparisons of precisions, extrapolated to 100% recall

We also investigated the use of tf×idf weighting (3). The obtained results are considerably worse then the ones by entropy weighting, but the time requirement for indexing the collection is decreased by about 30%, because tf×idf weighting requires one pass less for indexing. See Figure 6. The inferiority of the results are also due to the high $\max_{var}$ parameter that is 0.5 here, while 0.01 with other setting. Consequently, the number of documents taken into account at high consistency levels is significantly higher.

The next experiments were obtained when semantic information were propagated back to the learning phase. This modification takes into

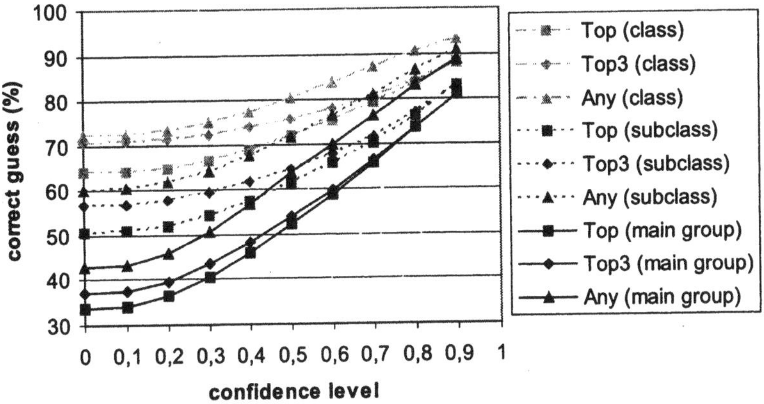


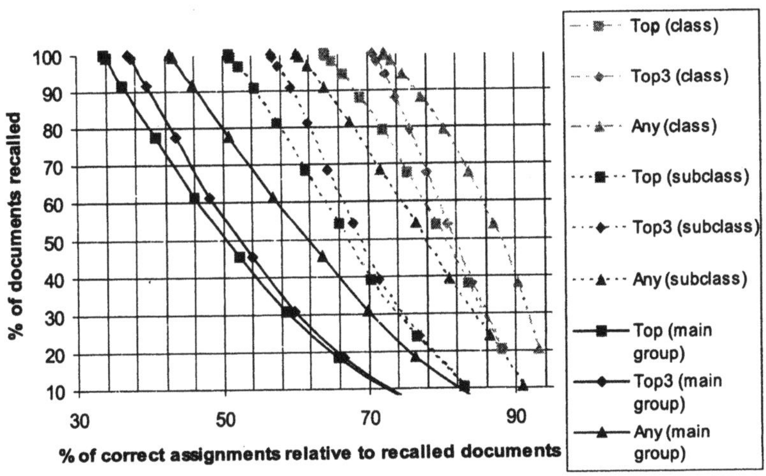


*Figure 6.* Use of tf×idf weighting. a) Precision by confidence levels. b) Comparisons of precisions, extrapolated to 100% recall

account the location of the clue word in a sentence and the location of an important sentence in a paragraph. Cumulating these information we can determine areas in the text that are more important than others. This modification has great effect on the results at low confidence levels since it increases certain performance measure values by more than 3%. See Figure 7 and Table 1.

Table 1 also contains reference results from Fall et al., 2003a for IPC class and subclass levels. The referred paper does not contain results for main group level. We assumed that the results of Fall et al., 2003a refer to the 0.0 confidence level (the most difficult setting), although it is

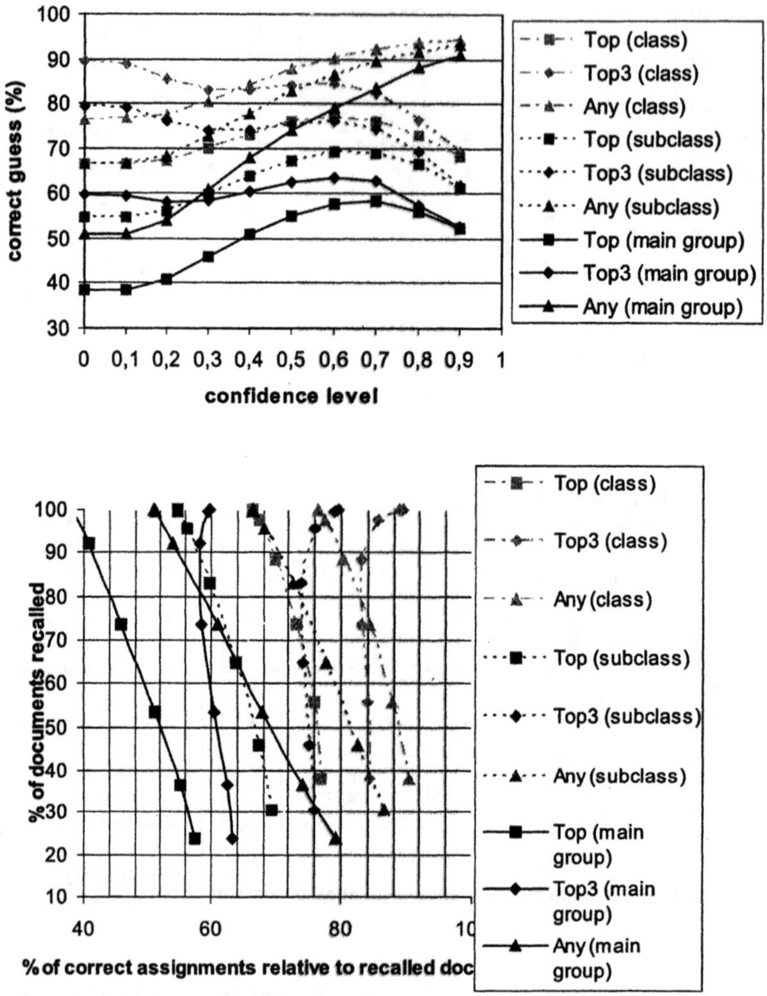


*Figure 7.* Using semantic information. a) Precision by confidence levels. b) Comparisons of precisions, extrapolated to 100% recall

not indicated explicitly. One can observe that HITEC outperforms the best technique experimented with in Fall et al., 2003a by at least 10.41% at each level and performance measure, and the difference increases at deeper IPC levels. We indicated the difference between our best results and of Fall et al., 2003a also in the table.

It worth to note that the graph of Top 3 measure is dropping when the consistency level increases. The reason is that at lower consistency level more categories are returned, and based on the results, in some case

*Table 1.* Summary of results on WIPO-alpha patent collection (Abbreviation of method names: NB – Naïve Bayes, SVM – Support Vector Machine, $k$-NN – $k$ Nearest Neighbors)

| Evaluation measure | Setting | IPC/conf. level cl./0.0 | cl./0.8 | s.cl./0.0 | m.g./0.0 |
|---|---|---|---|---|---|
| Top | ipta | 65.75 | 92.93 | 53.25 | 36.89 |
| | iptac | 65.50 | 92.93 | 53.14 | 36.78 |
| | main | 62.81 | 84.37 | 49.41 | 32.28 |
| | tfidf | 64.04 | 83.56 | 50.76 | 33.75 |
| | semantic | 66.41 | 72.86 | 54.63 | 38.38 |
| | best of WIPO paper | 55.00 NB,SVM | – | 41.00 SVM | – |
| | difference | 11.41 | – | 13.63 | |
| Top3 | ipta | 85.56 | 92.93 | 75.05 | 55.44 |
| | iptac | 85.61 | 92.93 | 75.00 | 55.58 |
| | main | 83.61 | 84.37 | 70.89 | 48.98 |
| | tfidf | 70.71 | 84.11 | 56.57 | 37.05 |
| | semantic | 89.41 | 76.45 | 79.48 | 59.64 |
| | best of WIPO paper | 79.00 NB | – | 62.00 $k$-NN | – |
| | difference | 10.41 | – | 17.48 | |
| Any | ipta | 73.68 | 95.83 | 62.45 | 46.46 |
| | iptac | 73.41 | 95.64 | 62.28 | 46.38 |
| | main | 71.32 | 94.97 | 58.97 | 41.51 |
| | tfidf | 72.22 | 90.38 | 60.18 | 42.71 |
| | semantic | 76.46 | 93.48 | 66.36 | 50.90 |
| | best of WIPO paper | 63.00 NB | – | 48.00 SVM | – |
| | difference | 13.46 | – | 18.36 | |

the one returned with lower consistency can be the correct one, but that is left out at higher consistency levels. When the consistency level goes higher, much fewer documents are considered, and then the Top 3 values increases again. (This arguing does not apply for the tfidf setting with high variance, because there the number of inferred documents is low even at low consistency levels.)

One can observe that the relationship between evaluation measures is Top $<$ Any $<$ Top 3 except when tf$\times$idf weighting scheme is applied: then Any gives the best values. Naturally, the lower we go in the taxonomy the more imprecise predictions are. At IPC class level the Top 3 measure of HITEC at the lowest confidence level (i.e. when basically all documents are considered) attains 89.41% with semantic analyzer, which is a quite significant result. This value hints that the algorithm can be used for large document corpora in real-world applications. Because of the very large taxonomy and range of documents the results on the main group level seems to be quite weak. However, if we consider that human experts can do this categorization work with about 64% accuracy then this result turns out to be much more significant.

Table 2 shows our experiments on the German patent collection WIPO-de. The table presents selected results with the best setting (using semantic information). The results are comparable with the ones achieved for English patent documents. Consequently, we can conclude that the high performance of HITEC is practically independent from the language of document corpus and HITEC can be used generally for document classification tasks.

*Table 2.* Summary of results on WIPO-de patent collection

| Evaluation measure | IPC/conf. level class/0.0 | class/0.8 | subclass/0.0 | main group/0.0 |
|---|---|---|---|---|
| Top | 65.02 | 86.95 | 55.37 | 37.93 |
| Top3 | 87.14 | 89.00 | 77.61 | 57.34 |
| Any | 75.04 | 96.95 | 66.88 | 50.79 |

Let us shortly remark the time efficiency of the method. Our experiments were executed on a regular PC (Linux OS, 2 GHz processor, 1 GB RAM). The indexing of the entire train collection took around one hour with entropy weighting and just over 40 minutes with tf×idf weighting. The training algorithm (7 iterations) required about 2 hours with each settings. If more iterations were done the results did not improved significantly.

# 4. CONCLUSION

We presented HITEC, an automated text classifier and its application categorize to English and German patent collections of WIPO under the IPC taxonomy. IPC covers all areas of technology and is currently used by the industrial property offices of many countries. Patent classification is indispensable for the retrieval of patent documents in the search for prior art. Such retrieval is crucial to patent-issuing authorities, potential inventors, research and development units, and others concerned with the application or development of technology. An efficient automated patent classifier is crucial component in providing an automated classification assistance system for categorizing patent applications in the IPC, that is a main aim at WIPO Fall et al., 2002. HITEC can be a prominent candidate for this purpose.

## Acknowledgments

This work was supported by Korea Science and Engineering Foundation (KOSEF) Grant No. R05-2003-000-11986-0.

# REFERENCES

Aas, L. and Eikvil, L. (1999). Text categorisation: A survey. Raport NR 941, Norwegian Computing Center.

Baker, K. D. and McCallum, A. K. (1998). Distributional clustering of words for text classification. In Proc. of the 21th Annual Int. ACM SIGIR Conference on Research and Development in Information Retrieval (SIGIR'98), pages 96–103, Melbourne, Australia.

Chakrabarti, S., Dom, B., Agrawal, R., and Raghavan, P. (1998). Scalable feature selection, classification and signature generation for organizing large text databases into hierarchical topic taxonomies. The VLDB Journal, 7(3):163–178.

Fall, C. J., Törcsvári, A., Benzineb, K., and Karetka, G. (2003a). Automated categorization in the international patent classification. ACM SIGIR Forum archive, 37(1):10–25.

Fall, C. J., Törcsvári, A., Fievét, P., and Karetka, G. (2003b). Additional readme information for WIPO-de autocategorization data set. http://www.wipo.int/ibis/datasets/wipo-de-readme.html.

Fall, C. J., Törcsvári, A., and Karetka, G. (2002). Readme information for WIPO-alpha autocategorization training set. http://www.wipo.int/ibis/datasets/wipo-alpha-readme.html.

Koller, D. and Sahami, M. (1997). Hierarchically classifying documents using a very few words. In International Conference on Machine Learning, volume 14, San Mateo, CA. Morgan-Kaufmann.

McCallum, A., Rosenfeld, R., Mitchell, T., and Ng, A. (1998). Improving text classification by shrinkage in a hierarchy of classes. In Proc. of ICML-98. http://www-2.cs.cmu.edu/~mccallum/papers/hier-icml98.ps.gz.

Salton, G. and McGill, M. J. (1983). An Introduction to Modern Information Retrieval. McGraw-Hill.

Sebastiani, F. (2002). Machine learning in automated text categorization. ACM Computing Surveys, 34(1):1–47.

Tikk, D. and Biró, G. (2003). Experiments with multilabel text classifier on the Reuters collection. In International Conference on Computational Cybernetics (ICCC03), pages 33–38, Siófok, Hungary.

Tikk, D., Yang, J. D., and Bang, S. L. (2003). Hierarchical text categorization using fuzzy relational thesaurus. Kybernetika, 39(5):583–600.

van Rijsbergen, C. J. (1979). Information Retrieval. Butterworths, London, 2nd edition. http://www.dcs.gla.ac.uk/Keith.

Weiss, S. M., Apte, C., Damerau, F. J., Johnson, D. E., Oles, F. J., Goetz, T., and Hampp, T. (1999). Maximizing text-mining performance. IEEE Intelligent Systems, 14(4):2–8.

Wibovo, W. and Williams, H. E. (2002). Simple and accurate feature selection for hierarchical categorisation. In Proc. of the 2002 ACM symposium on Document engineering, pages 111–118, McLean, Virginia, USA.

Wiener, E., Pedersen, J. O., and Weigend, A. S. (1993). A neural network approach to topic spotting. In Proc. of the 4th Annual Symposium on Document Analysis and Information Retrieval, pages 22–34.

Yang, Y. (1999). An evaluation of statistical approaches to text categorization. Information Retrieval, 1(1–2):69–90. http://citeseer.nj.nec.com/yang97evaluation.html.

# Chapter 14

# STUDY OF TRANSPORTATION AND UNCERTAINTY

**Shinya Kikuchi**

## 1. INTRODUCTION

An increasing number of engineers and planners advocate the proper treatment of uncertainty in the analysis of transportation. This trend is certainly in the right direction, when one realizes the abundance of uncertainty, in the data, in the knowledge, in the dynamics in the demographic, economic and social trends, and technology developments. How to present what is known and what is not known (or not sure about) clearly is the bedrock of the scientific approach. In the practice of transportation engineering and planning, however, making distinction between the two, what is known what is not known, is often smeared by too much uncertainty.

In view of the recent development in the theory of uncertainty in systems science, the greater public demand for accountability in the planning process, and the greater degree of complexity in the transportation issues, this short paper discusses the nature of uncertainty in the analysis of transportation engineering and planning. The intent of the paper is to promote discussions on the diversity of uncertainty types and the appropriate formalisms to deal with it.

## 2. SCOPE AND NATURE OF ANALYSIS, AND UNCERTAINTY

**The Scope**

Transportation facilities and services are public works whose objective is to achieve what is good for the individuals and for the society over many years. Today's transportation engineering and planning places emphasis on how transportation facilities impact the socio-economic systems (not just solving congestion) over the long run. This means that the scope of transportation engineering expands as new societal issues emerge: from mobility, to environment, energy, social equity, public health, livability, and to national security.

As suggested by Dickey (1983), the issues that transportation study deals with are three types: problems affecting transportation, problems of transportation, and problems affected by transportation. The 3C principle in the federal transportation regulation, a coordinated, cooperative, and continuing approach, reaffirms the view that the scope of transportation is unbounded and dynamic.

Because the issues of transportation change with time and space, uncertainty and risk are associated with every decision on investment, policy, and technology development. For example, decisions on ITS, perhaps the most popular planning activity today, face enormous uncertainty in its long-term impacts on the society including investment, privacy, government's accountability, security, and cost responsibility, and many unforeseen effects. Thus, the scope of transportation analysis is in fact bound by uncertainty.

**The Nature of Transportation Analysis**

Underlying in transportation analysis is the aspects of human, both as the user and non-user of the facilities and services. A human, individually and collectively, has feeling, value, and desire, which are not only difficult to describe and measure, but also change over time and space. This feature set transportation analysis apart from other engineering fields.

Other engineering disciplines study the properties of objects and their behaviors in order to develop a product or system that satisfies a set of well-defined objectives, e.g. minimum cost and maximum safety. For example, in structural engineering, engineers are basically interested in one aspect, to know how the material will react to the stress, and design a structure so that it does not fail. Failure is clearly defined.

In transportation, in contrast, we want to know how people and society react to a set of stimuli including changes in infrastructure, land use pattern, regulations, demographics. Most cases, however, we do not know exactly what success is and what failure of the decision until many years later is. Further, experiments are not possible. The solution is usually tailored to the case-by-case unique local conditions. Because of the human factor and the changing nature of the scope, the transportation issues are complex with many factors interacting in a complicated manner.

The nature of transportation study is characterized by the following, (1) human factor (individually and collectively) affects the system behavior; (2)

objectives are many, and not well-defined, (3) many elements, transportation and non-transportation, are involved in a complex manner, (4) the system performance changes dynamically over time with the changes in the socio-economic system, and (5) the solutions, if found, are local, not universal.

## Uncertainty and Information

Given the scope and the nature of transportation study above, the issue of uncertainty is inseparable to the study of transportation. Uncertainty is the state of lack of information and, as a result, it is difficult to make decisions; as such, uncertainty and information are dual. The more information is gained, the less uncertain the situation becomes. Collecting information is a crucial activity of transportation. This means obtaining the data, developing knowledge base, defining clear goals, and communicating the ideas. How to measure the usefulness of information, how to decide on the level of details, how to combine different types of information to create new knowledge, and how to control propagation of uncertainty along the analysis path, are some of the tasks that should accompany in any analysis of transportation.

Information takes different forms. Some information is statistical, which allows treatment by probability theory. However, the majority is in the form of perception and linguistic form. How to incorporate such non-statistical information in the analysis process has been the challenge. This is one of the topics to be discussed in this paper.

Despite the enormous uncertainty in transportation analysis, the subject of information and uncertainty has not been seriously dealt with in a quantitative manner. Engineers and planners have avoided facing uncertainty head-on, rather many times the issue was veiled by self-serving treatment. Uncertainty is considered something undesirable and uncomfortable; and hence, it is removed by making assumptions, or at best, by conducting a sensitivity analysis. Even though the uncertainty in the initial data may be treated with the statistical analysis, along the analysis steps, uncertainty vanishes, and the final outcome is often presented with certainty. Kenneth Boulding (1974) states, "An important source of bad decision is illusion of certainty." This is also convenient, since politicians, the final decision-maker, are reluctant to hear about uncertainty about the consequences of the decision.

# 3. FOUR ACTIVITIES OF TRANSPORTATION ANALYSIS: PREDICTION, DIAGNOSIS, ABDUCTION, AND CONTROL

Most of the analytical problems that we deal with fit in one of the following four classes. They are to predict, to diagnose, to abduct, and to control. To conduct any of these classes of activities, necessary elements are, input or data, knowledge base or model, output or result, and objectives or goals. The relationships among them are presented in Figure 1.

**Prediction** is the activity to predict the outcome or results, based on the input and the knowledge base. This is similar to finding the value of y given x in y= f(x), where f(x) is the knowledge base. Prediction is perhaps the most common activity performed in transportation, such as travel demand forecasting, analysis of capacity and system performance when the initial condition is given. If the input is not certain, then uncertainty propagates along the analysis steps so that the degree of uncertainty of the output may become greater than that of the input.

**Diagnosis** means finding the cause or input, based on the outcome and the knowledge base. This is similar to solving an equation, say $2x^2 + x + 4 = 6$ for x, where the outcome is known and the knowledge (model) is also known. Diagnosis is a much more difficult problem than prediction. Usually, more than one solution exists (as seen in even the simple problem above). Yet, diagnosis is an increasing important subject in transportation, because accountability and potential of litigation as a result of accidents and any negative outcomes are today's pressing issue.

**Abduction** means fine-tuning the knowledge-base, based on the input and output. An example is a regression analysis, which is to find a relationship from the data on input and output. The back-propagation neural network is another example of knowledge building from input and output. Calibration of model parameters can be in this category also.

**Control** means regulating the input (and the knowledge base) in order to achieve a goal or to match the output to a target. If the output and target do not match, then the input is adjusted, or sometimes the parameters of the model are adjusted. Control can be performed by trial and error or by mathematical programming. The former may be the case of solving a problem by iteration. The latter may be optimization using mathematical programming. Control can also mean regulation and calibration, to regulate the flow of traffic, traffic signal, or land use. Many of ITS strategies are in this category.

In these activities, uncertainty is inseparable, in the data, in the knowledge base, and in the goals. We need to look at our limits in the ability to predict, diagnose, abduct, and control in the face of uncertainty.

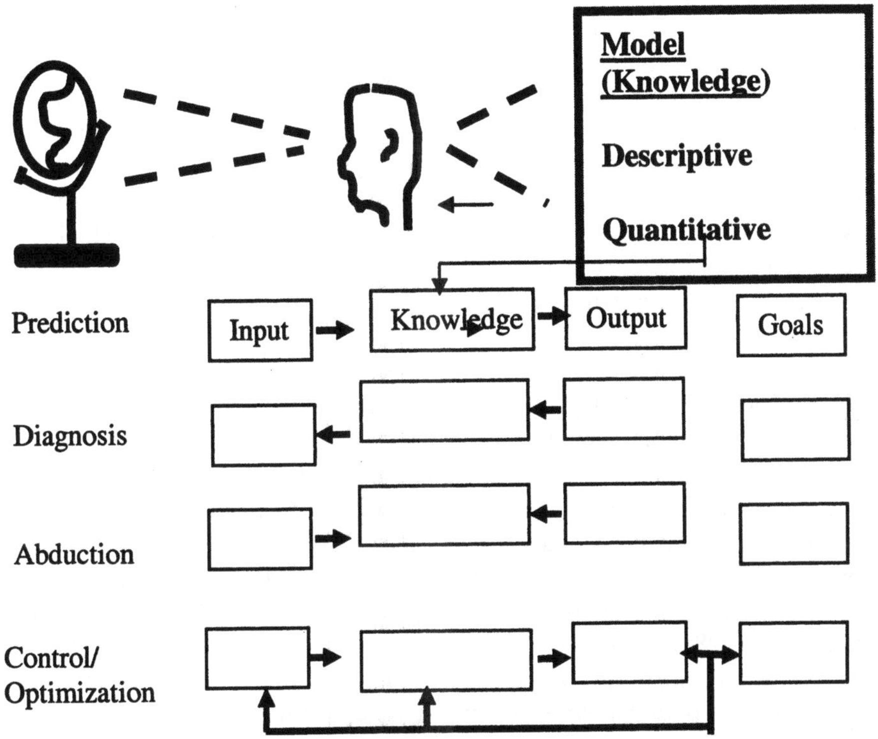


Figure 1 Four Activities of Transportation analysis

## 5. MODELING UNDER UNCERTAINTY

Let us now focus on the presence of uncertainty and its effects in the analysis process. Figure 2 shows the analysis chain consisting of observation of phenomena, model building, analyst's interpretation, and application.

**Uncertainty in the Observation of Phenomena**. The analyst observes the phenomena, collects data, and transforms it to information, which becomes the input to model building. The data that is collected may be in the form of: numerical, descriptive, or illustrative. Uncertainty in the data may be inherent to the phenomenon itself (e.g., randomness), in the measurement, or a by-product of information transmission. Regardless of the sources, these uncertainties undoubtedly create bias and perception in the mind of the analyst.

**The Model Framework.** A model is a pair of spectacles through which the analyst views the phenomena and captures its essence. Basically, any model represents the connection between cause and effect (or stimulus-response). A model may take a functional form, a rule base ("if ... then..... ,"), or a combination of the two: The mathematics in the model, however, must adhere to the axioms on which the mathematical theory is based. In the case of models that represent uncertainty, mathematics may be probability theory, possibility theory, fuzzy set theory, or other theories of evidence. Use of any of these frameworks depends on the nature of the phenomena for which the analyst wishes to model, and also depends on the purpose of application. In some cases, a deterministic assumption may be sufficient for a particular application.

**Analyst's Interpretation of the Model.** Each analyst interprets the results of the model differently depending on the context and his/her bias. Subjectivity enters in interpretation always, particularly, when uncertainty is involved. A precise value obtained from the model may be interpreted as approximate values. For example, if an analyst hears that the capacity of a section of a highway is 2456 veh/hr, he/she interprets as a rounded value near it, say, 2,500 veh/hr. Another example, if a travel demand model provides the future volume to be 28,673veh/day, it may be perceived as 29,000veh/day. A subjective filter exists in the analyst's mind, and it works differently, conservatively or optimistically, depending on application.

**Application of the Model Results to Decisions.** The model and the results are then applied to prediction, diagnosis, abduction, control/optimization, leading to decisions on the design, investment, and policy. The precision for the parameters of design is controlled by the purpose of application and objectives. For example, in the case of predicting the traffic volume, one requires different levels of accuracy when determining the thickness of pavement and when determining the number of lanes. The precision level is not only determined by the application but also it is limited by the degree of precision of the initial data, measurement, model accuracy. For example, the travel demand forecasting that is extremely in detail in the model choice step but based on crude data on trip generation may be of little value from the standpoint of accuracy.

In summary, to deal with uncertainty, one needs to able to answer the following question.

When observing the phenomena and collecting data – Is uncertainty inherent in the phenomena, observation, measurement, or subjective bias? What causes uncertainty?

When formulating the model – Are the mathematical representation and the types of uncertainty observed consistent with the uncertainty in the observation of the phenomena?

When interpreting the result of the model – How much bias does an analyst have when interpreting the model result and application?

When applying the model result - Is the level of accuracy in the result sufficient for the problem, too much or too little? Can designer understand the uncertainty involved and know how to reflect it in the design decision?

In summary, the main concern is the amount of uncertainty in the outcome and how much it ultimately affects the consequences of the decision.

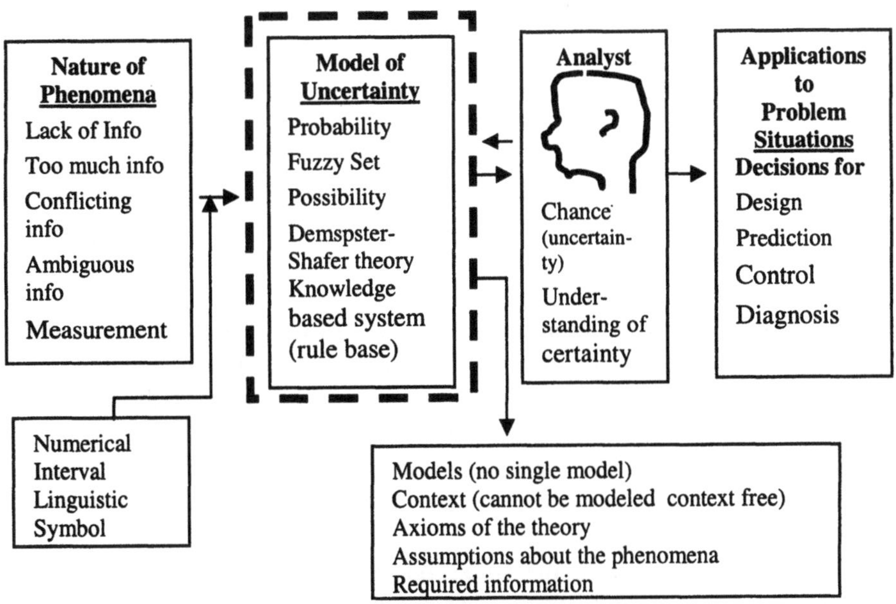


Figure 2 Process of modeling in the environment of Uncertainty

## 6. TWO CASES OF UNCERTAINTY

Formally, uncertainty needs to be looked at from the standpoint of truth or false of a proposition in the face of evidence. Uncertainty about the truth of a proposition "x is A" is associated with two conditions as shown below, where x and A are sets.

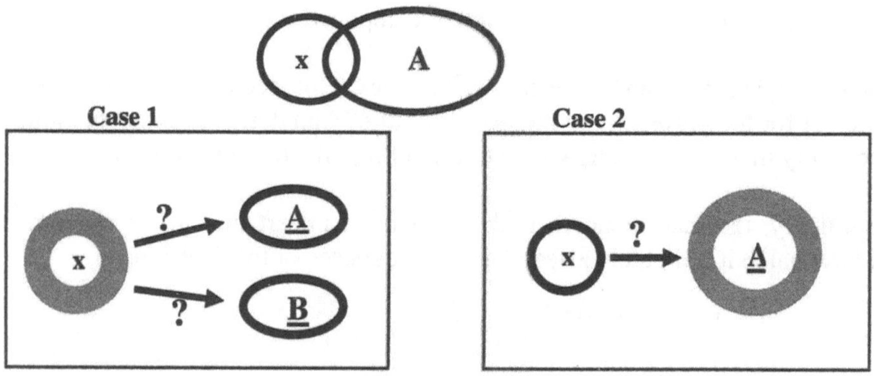


Case 1: x is not clear. In other words, information about x is incomplete, thus "x is A" cannot be asserted with certainty. This type of uncertainty is called ambiguity.

Case 2: A is not well defined. In other words, boundaries of the definition of A are not clear, thus "x is A" cannot be asserted with certainty. This type of uncertainty is called vagueness.

Until recently distinction between two cases was not clear, and the theory of probability has been used for any type of analysis involving uncertainty. Understanding the difference between these two cases clarifies the use of proper mathematical tool for different situations. George Klir, in his various books, Klir, et al.(1995,1999,2000) has established the distinction between these two cases, and has developed the framework of theory of uncertainty .

Case 1, the truth of "x is A" is measured the weight of evidence pointing to A. Thus, to know the mechanism that generates different outcomes needs to be understood, and organized into proper form of evidence. While we do not go in details in this paper, the evidential patterns are now categorized into three distinct types, probability distribution, possibility distribution, and belief function of Dempster-Shafer theory. Each of these three distributions forms a distinct well-established mathematical formalism of measure theory. In the following two common distributions, probability and possibility, are discussed briefly. These distributions are perhaps most relevant for transportation analysis.

The probability distribution is one of evidential patterns in which each piece of evidence points to well-defined set (outcome) exclusively; hence, each piece of evidence is conflicting. The causal mechanism is not known, and thus, the outcome is random (to the eyes of the analyst), such as the case of rolling a die. The truth of "x is A" is only measured by weighing the evidence or frequency pointing to A. This is usually the case of observation of random phenomena. Probability theory measures propensity of occurrence of events.

Possibility distribution, on the other hand, is an evidential pattern in which each piece of evidence points to nested sets; thus, the body of evidence is not conflicting but agreeing one another. Each piece of evidence differs only by the degree of agreement. Perception of time or cost, or the concept of approximate value, generally follows the possibility distribution. The truth is measured by the degree of agreement. In this case, because the evidence is not specific to a set, the optimistic and pessimistic ways of weighing the evidence exist. The former is called the possibility measure, and the latter, the necessity measure. Possibility theory measures the strength of disposition. Incidentally, the belief function of Dempster-Shafer theory represents the generalized evidential pattern in which probability and possibility distributions subsumed.

**Case 2** arises as a result of the unclear definition of the set; the truth of x is A cannot be asserted due to the vagueness of the word or image of A. Fuzzy set represents this type of uncertainty associated with language. This representation has been found to be very useful in modeling human behavioral pattern, because most of human decision and behavior is based on language based command.

At the human level decisions, either as an individual or as a group, reasoning is usually conducted in language. It is inevitable that vagueness accompanies the language-based communication. Although language conveys the nuance and sensitivity of context dependency much more effectively than the numerical/mathematical expression, it is difficult to formalize and to model the process. Fuzzy set theory is the formalism that allows modeling of language based reasoning and representation of the perceived condition.

**Mathematical Theories of Uncertainty Treatment**

Figure 3 summarizes the types of mathematical framework when analyzing the truth of "x is A," as a function of the nature of information about x and the character of set A. Three frameworks are presented. They are probability theory, possibility theory and Dempster-Shafer theory.

Probability theory is applicable when the information about x is statistical and set A is crisp set. Probability theory, however, can be applied to the situation when A is a fuzzy set, under a certain condition, (when the membership function of set A and the membership function of set "not A" are defined as a perfect complement).

Possibility theory is applicable when the information about x is perceptive (or possibility distribution), and set A can be either crisp or fuzzy set. In this case, the specific measures are possibility and necessity, representing optimistic and conservative view of disposition, depending on how to weigh the evidence toward A.

In addition, when the information is both probabilistic and also possibilistic, then Dempster-Shafer theory is appropriate. This is a comprehensive framework in which both possibility and probability theories are subsumed. Under this theory, two measures, belief and plausibility measures, are used to represent the optimistic and conservative dispositions.

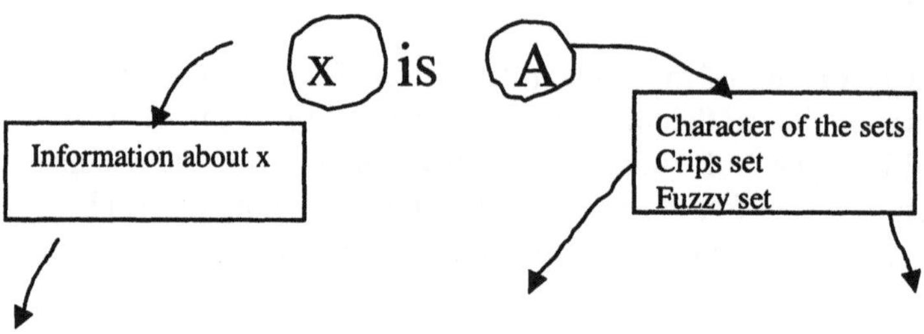


| | | Nature of set A | |
|---|---|---|---|
| | | Crisp set | Fuzzy set |
| Type of Information about x | Probabilistic (Info points to conflicting set) | Probability theory<br>Probability | Probability theory<br>Probability |
| | Possibilistic (Info points to nested sets) | Possibility theory<br>Possibility measure<br>Necessity measure | Possibility theory<br>Possibility measure<br>Necessity measure |
| | Combined (Info points to combination of conflicting and nested sets) | Dempster-Shafer theory<br>Belief measure<br>Plausibility measure | Demspter-Shafer theory<br>Belief measure<br>Plausibility measure |

Figure 3. Different theories of uncertainty in the context of the truth of x is A

# 7. TRANSPORTATION ANALYSIS CONDUCIVE TO UNCERTAINTY TREATMENT

It is clear from the above that uncertainty refers to the situations in which determining the truth is difficult, (1) due to the lack of evidence and (2) due to the lack of clear definition of words. For the former situation, the evidential pattern that is consistent with the observed phenomena needs to be established, and they are typically probability distribution or possibility distribution. These distributions are handled in the domain of *measure theory*, which measure the truth of a proposition. This section presents the types of transportation subjects for which these frameworks, measure theory and fuzzy set theory, are suited.

**Subjects Suited for Analysis Using Measure Theory**

Measure theory is relevant when evaluating the truth of a proposition "x is A," a situation when x is classified to A. In the transportation problem, a number of problems are in this category. Following are the problems often encountered in transportation. x may be any current or projected condition, and A is a particular domain or a set of interest. Decision depends on the truth of this proposition. In the following we list typical analysis situations.

Classifying a situation into one of the predetermined classes, e.g.,
Assigning the current (or future) traffic condition to one of level of service categories.

Comparing the estimated transit ridership with a threshold ridership value that justifies the investment.

Comparing two quantities for their ranking, or comparing a value with a reference value, e.g.,
Selecting an alternatives based on comparison of the utilities of each alternative.

Examining feasibility of arrival by comparing the estimated arrival time with the desired arrival time.

Determining preference
Setting ranking among available alternatives.

In dealing with these problems, either probability theory or possibility theory is applicable depending on the nature of evidential pattern of x and the definition of set, A. Probability theory is applicable when x, the predicted condition, is known or assumed to be random, and also when A is well defined. Thus, the uncertainty of "x is A" is expressed in probability and it is interpreted as propensity of occurrence.

Possibility theory, on the other hand, is useful when the information about x is approximate, such that evidence is given in a set of ranges. Imagine an experiment in which each observation (experiment) yields a range (rather than a specific value), and the size of the range is different for every observation. This is similar to asking individuals in a group about acceptability of transit fare and each answers a different range. In this situation, probability theory cannot be applied because each piece of evidence is not pointing to well-defined sets. Under this evidential pattern, the truth of "x is A," is measured two ways, one, weighting any positive evidence pointing to A (which may also point to not A): two, weighing evidence that exclusively pointing to A. The former is called the possibility measure, and the latter is called the necessity measure.

## Subject Suited for Fuzzy Set Representation

The nature of uncertainty that is related to vagueness and approximation is treated by fuzzy set. The following shows the types of notion that is suited for treatment by fuzzy set.

Notion of desire, e.g.,

Desired departure time and desired arrival time
Desired design value
Objective and goals

Notion of satisfaction and acceptability (vague threshold values), e.g.,

Satisfactory cost, acceptable cost, willing value for payment,
Acceptable level of error,
Acceptable delay,
Acceptable air pollution level.
Stated preference

Perception and quantities based on memory, e.g.,

Past travel time, distance, appearance and condition.

Descriptive condition, e.g.,

Traffic congestion - bad traffic and good traffic condition,
Comfort, safety, level of service.

Imprecise Values - hard to measure or hard to summarize, e.g.,
Sight distance, reaction time, capacity of roadway, value of time, cushion value in design.

Possibility
Travel time, capacity of roadway or a system

Similarity

Among the above examples, one's estimated travel time by automobile between two points is a typical case of possibility. Although the statistically based information about past travel times may exist, because one is able to control the travel time somewhat, by driving fast or slowly, the travel time *estimated by the traveler before travel* is a set of values which are "possible" to achieve by the traveler and differ among the individuals. Another case is the capacity of a roadway, whose precise value is perhaps impossible to determine; it is understood as the approximate value, which the roadway can "possibly" handle. To some extent, the forecast value of travel demand is in this category.

# 8. COMMENTS ON SPECIFIC TRANSPORTATION ANALYSIS ACTIVITIES

## Comparison and Classification involving Approximate Values

Comparing values, for example, cost vs. benefits, expected outcome vs. target, and utilities of an alternative vs. another, is a common problem during any evaluation process. Classifying a given situation into one of predetermined classes also falls into the same class of problem. An example is determining the traffic condition with a threshold value of the level of service. These situations examine whether a value belongs to another set or not. In determining the truth of M>N, the problem is to find the truth that M belongs to a set of numbers greater than or equal to N.

When the values to be compared are not exact value, no clear answer can be given. The traditional approach has been to assume that the values are random, and assume a probability distribution for each value, and then to state one value is greater than the other in probability. A typical example of this approach is the stochastic choice model when utilities are compared. Yet, the utility of an alternative, to the mind of the decision maker, is not a single value, perhaps, it is an approximate number.

The values to be compared may not always be a random nature; the values may be an inherently approximate number due to perception of the analyst. Consider the case of comparing the travel times of several available routes with each travel time known to him in approximate number with a range. This is the case of comparing fuzzy numbers associated with the perception of the time. This case each number follows a possibility distribution and the answer that one value is greater than the other is given in possibility, hence, optimistic and conservative answers.

Transportation planning community for many years has concentrated on improving the accuracy of forecast traffic volume. One of the premises of transportation analysis is that transport activity and economic activity are linked. It is known that forecast economic activity into the next 20 year is impossible task. Nevertheless, the transportation planners have continued to "perfect" the long term forecasting models. The purpose of forecast is to examine whether the future conditions warrant or justifies a certain action today. Thus the forecasting issue must be linked to the issue of comparing numbers. The forecast volume is not a random value. It is a possible value. It is based on various assumptions and the initial random nature of traffic volume of today has transformed into a possible value. Thus, the comparison should be conducted in the manner consistent with the possibilistic framework.

## Computation and Reasoning

Arithmetic operations of numbers are basic to all analyses. When the values are uncertain, having a range, randomness or approximate nature, then computation naturally yields not only cumbersome but also uncertainty propagates in the computation. Usually as more approximate numbers are manipulated the more uncertainty the outcome becomes. The compound effects of different models when they are chained are significant. One needs to conserve uncertainty, and also at the same time, one needs to control propagation of it in order for the outcome of the computation to be meaningful.

A typical example is the computations involved in the four-step travel demand forecasting process. Each step contains uncertainty starting from the trip generation and attraction. If the uncertainty is not masked by assumption of single number and definite equation, e.g., the gravity model, uncertainty propagates and the travel forecast must have a wide variation. (which is actually the case). In many cases, planners are interested in the range rather than a specific value. This means that the analyst should strive to conserve uncertainty.

Increased uncertainty means weakened strength of reasoning. This affects the credibility of the planner, ironically presenting uncertainty is more honest presentation. Reasoning process is affected by the propagation of uncertainty.

Our reasoning process, however, is based on incomplete data, incomplete knowledge of the phenomena, incomplete understanding of causalities, association, incomplete understanding of our goals, and or what we want to achieve in the long run. Analyst makes the case to the decision maker as to recommendation. Especially, in transportation planning, the reasoning process is based on a chain of reasons. For each reasoning step, a model is used. Since each model contains uncertainty, or information loss or information addition occurs. Such losses and additions should be minimized.

## Composite Picture and Presentation

Transportation analysis involves getting the composite image of a situation. Many attributes exist and they form a composite image of a situation. Let us think the case of the definition of level of service. It is a driver perceived traffic condition according to HCM. Many attributes of LOS are not independent and the driver forms an image of the condition, which is hard to describe. The weights among the attributes are not clear, and definitely they are not additive because the attributes are not independent. A composite image is formed by set operations, e.g., union and intersection; however, the operations should not be a clear-cut binary operations, but rather emulate the sensitivity of language based operations, which is fuzzy set operations strive to achieve.

Presentation is an aspect that is emphasized greatly in transportation planning today. Technology enables us to create an image that allows the audience to make judgement rather than the planner describing the situation. The traffic flow simulation has become a popular tool to present the situation graphically and allows the audience to make the judgement. Such illustrations tend to impress the audience.

In order to make the credible conclusions, however, we must understand the entire picture of how much we know and how much we do not know, and what more needs to be done to be certain. It is important that the public should not be swayed by the pretty image of the simulation and not being told about the uncertainty hidden in the output. It is important to present what is known and what is not known clearly.

# 9. CONCLUSION

Transportation affects our lives in innumerable ways. It controls the way of life, economy, environment, and the livability. It fosters technological innovations and affects commerce. It alters pattern of human settlement. Studying transportation means understanding how engineering, humanity, economy, politics, and nature work together in a dynamic manner, and to contribute to the creation of a better living condition. This paper advocates incorporating uncertainty in the analysis of transportation in a serious manner, and selecting proper framework when handling uncertainty and presenting it.

How to deal with uncertainty is one of critical challenges, because it dictates the strength and weakness of logic and reasoning process. The lack of ability to deal with uncertainty affects the credibility of the profession. This may be a reason that the position of transportation has been ambiguous in terms of its membership in the scientific community. Integrity of transportation engineering and planning lies how uncertainty is being treated and represented.

Every field of science and engineering has a set of founding rules and principles. In the case of fluid mechanics, it is the Euler's equation; in structural engineering, it is the Newton's laws, etc. Given the nature of study, what are the governing principles in transportation analysis? We may borrow some from physical and economic principles, but the principles that deal with the complex relationships between transportation and the socio-economic issue are not yet available. Theories of uncertainty should find the niche in this area and the fundamental principles of study of transportation may lie in treatment of uncertainty.

# REFERENCES


Boulding, Kenneth E. (1974), Reflection on Planning: the Value of Uncertainty, *Technology Review*, October/November 1974 p.8.

Dickey, John W.,(1983) *Metropolitan Transportation Planning*, Second Edition, McGraw Hill, p. 39.

Dubois Didier, and Henri Prade, (1980) *Fuzzy Sets and Systems: Theory and Applications*, Academic Press, San Diego, CA

Kikuchi, Shinya, (1993) "The Fuzzy Set Theory Approach to Transportation Problems" *Proceedings of Japan Society of Civil Engineers*, Vol. 458, pp 1-6.

Kikuchi, S. and M. Pursula, (1997) *Treatment of Uncertainty in Study of Transportation: Fuzzy Set Theory and Evidence Theory*, Vol. 124, No. 1, *Journal of Transportation Engineering*, American Society of Civil Engineers, pp1-8.

Klir, George and Bo Yuan, (1995), *Fuzzy Sets and Fuzzy Logic, Theory and Applications*, Prentice Hall, 1995.

Klir George, (2000), Fuzzy Sets: an Overview of Fundamentals, Applications, and personal View, Beijing University Press.

Klir, George and Mark Wierman, (1999), *Uncertainty Based Information, Elements of Generalized Information Theory*, Phisica-Verlag, Heiderberg, New York.

Yager, Ronald, and Dimitar P. Filev, (1994) Essentials of Fuzzy Modeling and Control, John Wiley and Sons, Inc.

# Chapter 15

# MULTI AGENT SYSTEMS APPROACH TO PARKING FACILITIES MANAGEMENT

**Mauro Dell'Orco and Dušan Teodorović**

## 1. INTRODUCTION

The number of trips by private cars has significantly increased in recent decades in many cities. At the same time, parking capacities have not kept up with this increase in urban travel demand. Streets in many cities are overloaded. Everyday, a significant percentage of drivers in single-occupancy vehicles are searching for a parking space. Additionally, less experienced drivers or out-of-towners further contribute to the increase of traffic congestion. This complex situation results in increased travel times and number of stops, unexpected delays, greater travel costs, inconvenience to drivers and passengers, increased air pollution and noise level, and increased number of traffic accidents.

Expanding parking capacities is extremely costly, and sometimes environmentally damaging. Planners, engineers, economists, and city authorities introduced the concept of "congestion pricing" in an attempt to reduce fast growing traffic congestion. Congestion pricing assumes introducing different fees for streets/roads/zones/parking facilities usage. Various fees or tolls that vary with a location in the network, time of a day and/or level of traffic congestion could be proposed. In other words, drivers should pay for using specific road, corridor, bridge, parking facility, or for entering particular area during some time periods. It seems, that the basic economic concepts of supply and demand should be more utilized when solving complex urban traffic congestion problems. The basic idea behind the concept of congestion pricing is to force drivers to travel and use transportation facilities more during off-peak hours and less during peak hours, as well as to increase usage of underutilized transportation facilities. Successfully planned and implemented congestion pricing can have as consequences significant toll revenue, and drivers' responses in parking facilities used, departure time and destination traveled. This can result in decreased total number of vehicle trips, decreased total number of vehicle trips during peak periods, increased number of vehicle trips during off-peak periods, increase in ridesharing, greater number of passengers in public transit, and in some cases increased cycling, and walking.

Parking facilities management has significant influence on the level of traffic congestion. In this paper the agent-based model for parking facilities management has been developed. The Agent based model developed in this paper that represents the "Bottom-up approach" to problem solving is appropriate tool that can help us to better understand complex nature of urban traffic congestion. In this way, it will become easier to predict and/or control the overall performance of a complex urban traffic system.

The paper is organized in the following way. A statement of the problem is given in Section 2. A Multi Agent Systems approach to the problem of parking facilities management is described in Section 3. The results obtained by using our proposed model in the case study of Bari are given in Section 4. Section 5 contains conclusions and directions for further research.

## 2. PARKING FACILITIES MANAGEMENT: THE EVOLUTION OF UNPLANNED COORDINATION

Urban traffic congestion problems show a complex behavioral pattern. Like emergent phenomena, traffic congestion is frequently unpredictable and even sometimes counterintuitive. It is very difficult, if not impossible, to explain explicitly the relationship between traffic authorities' actions and individual (drivers) behavior. The phenomenon cannot be successfully analyzed and explained through analytical models. The only way for analyzing this emergent phenomenon is the development of simulation models that can simulate behavior of every agent. Agent-based modeling is an approach based on the idea that a system is composed of a decentralized individual "agents" and that each agent interacts with other agents according to the localized knowledge.

In our case, the interacting agents might be drivers, parking authorities, law enforcement and city government. If a kind of "central planner", responsible for minimization of the total urban traffic congestion, existed, all participants in urban traffic (drivers and parking authorities) would be obliged to strictly follow his orders. In real life this situation does not exist. Because of that, it is extremely important to explore how unplanned coordination evolves under different parking pricing strategies and different levels of parking enforcement, and whether it produces similar results like the global coordination planned by the "central planner", whose main objective is the minimization of the total urban traffic congestion. In our city parking system, we assume that the following agents are present: (a) drivers; (b) parking authorities; (c) law enforcement. Through our proposed model, we study the evolution of unplanned coordination among independent agents in a "market selection game" in the case of different parking pricing strategies and different levels of parking enforcement.

The model developed in this paper is an Agent-based model. In the city we studied, each part (drivers and parking authorities) acts based on its local knowledge and competes with other parts. Agents that represent parking authorities and law enforcement can increase the capacity, or significantly change parking fee policy. Agents that represent drivers learn all the time, and change their chosen parking facilities. Through the aggregation of individual drivers, parking authorities and law enforcement behavior, the overall picture of the city parking system emerges.

We consider the situation in which few different parking facilities operate in urban traffic network and "compete" among themselves. The parking facilities define parking supply that is characterized by facility locations, working hours, and parking fees. Drivers adjust their parking plans to the proposed parking supplies. This results in the creation of drivers' itineraries through the network.

Parking facilities compete with each other for "drivers market share", that represents the percentage of the total number of drivers that facility can attract, and highly depends on the proposed parking facility supply. Parking "competition" we consider is a kind of iterated game: at the end of each iteration, any parking facility can increase its profit by changing operating strategy that includes adjusting working hours and changing parking fees.

Parking fees can vary day-to-day or within day. In the latter case, parking fee at peak hours would increase substantially compared with parking fee at off-peak hours. In this paper updating of parking fee is carried out only day-to-day.

To analyze parking strategies, we introduce a non-cooperative evolutionary model. In other words, the proposed model does not consider agents' deliberate cooperative behavior, and describes the changes over time of the agents' behavior. The agent's fitness is based on the success each agent has in playing the game. Agents follow particular strategy for a certain period of time. They are free to change the followed strategy at the end of each iteration.

Let us assume that every network user chooses the parking facility based on perceived parking searching time, parking fee, number of previous rejections from that parking facility, etc. Perceived parking search time, or perceived parking fee is, very often, fuzzy. In other words, when subjectively estimating parking search time, expressions are used such as "it takes about 15 minutes." It is rarely if ever heard that searching time is 13 minutes and 45 seconds. The claim that parking searching time is "about 15 minutes" is the result of a subjective feeling, an individual's subjective estimate. This is not the result of any measuring or the realization of a random variable representing parking searching time. If we were to record parking searching time over a longer period of time, we would receive a series of different values, each representing one realization of the random variable that represents parking searching time. When we subjectively estimate parking searching time, we do not have information regarding the probability density function of search time; rather, we base our estimate on experience and intuition. Network users perceive also certain parking fee as “expensive”, “reasonable”, “not so expensive”, etc. Users have a specific preference regarding the choice of each of the possible parking options. This preference can be “stronger” “medium” or “weaker.” Let us introduce into the discussion a preference index that can take values from the interval of 0 to 1. When the user has an absolute preference for a specific parking facility, we consider the preference index to be equal to 1. This preference index decreases along with a decrease in the strength of the preference. Obviously, perceived parking searching times, perceived parking fees, and the strength of the user's preference can be expressed by fuzzy sets such as “very short travel time”, “short travel time”, “expensive road fee”, “acceptable road fee”, “very strong preference,” “strong preference,” etc (Figure 1).

Our agents use approximate reasoning in decision-making process. In other words, agents’ parking facility choice decisions are made based on a set of *if....then* rules, in which antecedents and conclusions are based on the attributes of alternatives. In this paper, we considered the following attributes:

- Lot searching time;
- Location of parking facility;
- Number of previous rejection from each facility;
- Duration of stay;
- Parking fee;

– Level of parking enforcement (for illegal parking).

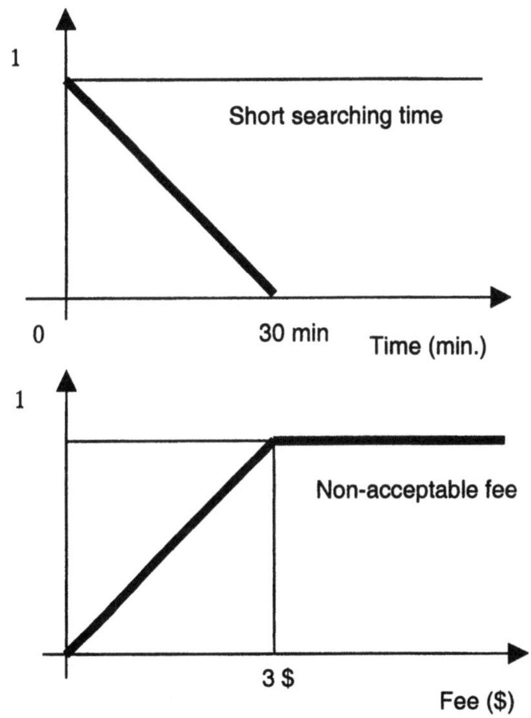


Figure 1 – Examples of perceived parking searching times, and perceived parking costs, expressed by fuzzy sets

# 3. MATCHING TRANSPORTATION SUPPLY AND TRANSPORTATION DEMAND UNDER DYNAMIC PARKING PRICING: MULTI AGENT SYSTEMS APPROACH

In order to describe complex process of matching transportation supply and transportation demand under dynamic parking pricing, we use relatively new computational paradigm - Multi Agent Systems. We propose three different types of agents: (a) drivers; (b) parking facilities; (c) government of city. We assume that agents' behavior highly depends of the current traffic situation. We also assume that agents have capability to recognize various situations, make rational decisions, and learn from experience. Every agent has full autonomy in decision-making process. This means that every agent-driver has full autonomy in choosing the parking facility and departure time. At the same time, every agent-parking facility has full autonomy in applying certain parking fee during certain time interval, and city government Practically, agent-drivers and agent-parking facility "negotiate" all the time for using specific parking facilities during specific time intervals. Obviously, agent-drivers and agent-parking facilities frequently have different goals and through the "negotiation", they try to find compromise solution. In this way, parking facility occupancies is the result of many independent decisions made by individual agents.

When making decisions, agent-drivers use experience and intuition. In order to have fair negotiation between agent-drivers, and agent-parking facilities, we assume that agent-parking facilities also use experience and intuition in decision-making process. When describing different decisions made at various stages of a process, human beings prefer to use qualitative expressions instead of quantitative ones. We assume the same for agent-parking facilities. In this way, the strategies of the agents can be formulated in terms of numerous descriptive rules. The qualitative or fuzzy nature of the human way of deciding has encouraged us to make an attempt at developing fuzzy systems that would control processes of matching parking supply and transportation parking demand.

## 3.1. Fuzzy Rule Base for Parking Facility Choice

The type of the planned activity, time of day, day of the week, current congestion of a particular routes, knowledge of a city streets and parking fees, and potentially available parking places have significant influence on chosen route between origin and destination. Every parking facility choice is composed of a set of vague rules. It is sometimes difficult to describe these rules explicitly. Drivers make their parking facility choice after comparing the characteristics of the alternative parking facilities (charged parking facilities, free parking facilities, and illegal parking “facilities”). We assume that users make theirs parking-choice decisions based on distance from the parking facility, perceived parking search time, perceived parking cost, and based on experience from the past.

Users' perceived parking search times and parking fees could be represented by corresponding fuzzy sets. We also assume that the user has a certain preference for the choice of a certain parking facility. This preference can be “stronger” or “weaker.” Moreover, could be represented by fuzzy sets.

Only few papers used fuzzy logic in modeling urban route choice: see for example [0], [0], [0], [0], [0], [0], [0], [0]. Even less papers have previously used Fuzzy logic to model the parking facility choice [0].

When choosing parking facility any driver can choose charged parking facility, free parking facility (if any), or he/she can decide to park illegally (illegal parking “facility”).

Since the fine for illegal parking can be considered like a “very very high parking fee”, and the parking fee is zero for free-of-charge parking, in this paper all possible parking options have been considered like charged parking with different fees. In this way, we could propose a unique fuzzy logic rule base for all possible parking options. The fuzzy rule basis that we propose is the following:

Rule 1:

**If** parking facility is CLOSE and number of previous rejections is LOW and duration of stay is LONG and parking fee is LOW

**Then** preference to choose parking facility is VERY HIGH

else

Rule 2:

**If** parking facility is CLOSE and number of previous rejections is LOW and duration of stay is SHORT and parking fee is LOW

**Then** preference to choose parking facility is HIGH

else

Rule 3:

**If** lot-searching time is SHORT and parking facility is CLOSE and number of previous rejections is LOW and parking fee is MEDIUM

**Then** preference to choose parking facility is HIGH

else

Rule 4:

**If** parking facility is CLOSE and number of previous rejections is HIGH and duration of stay is LONG and parking fee is LOW
**Then** preference to choose parking facility is MEDIUM

else
Rule 5:
**If** parking facility is FAR and number of previous rejections is LOW and parking fee is LOW
**Then** preference to choose parking facility is MEDIUM

else
Rule 6:
**If** lot-searching time is LONG and parking facility is CLOSE and parking fee is MEDIUM
**Then** preference to choose parking facility is MEDIUM

else
Rule 7:
**If** duration of stay is SHORT and parking fee is VERY VERY HIGH and enforcement is WEAK
**Then** preference to choose parking facility is MEDIUM

else
Rule 8:
**If** parking facility is CLOSE and number of previous rejections is HIGH and duration of stay is SHORT and parking fee is HIGH
**Then** preference to choose parking facility is LOW

else
Rule 9:
**If** parking facility is FAR
**Then** preference to choose parking facility is LOW

else
Rule 10:
**If** lot searching time is LONG and parking facility is FAR
**Then** preference to choose parking facility is VERY LOW

else
Rule 11:
**If** duration of stay is LONG and parking fee is VERY VERY HIGH and enforcement is STRONG
**Then** preference to choose parking facility is VERY LOW

Rule 12:
**If** lot searching time is SHORT and parking facility is FAR and number of previous rejections is HIGH and duration of stay is

SHORT and parking fee is HIGH
**Then** preference to choose parking facility is VERY LOW

The proposed approximate reasoning algorithms takes into account driver's perception of the parking facility location, duration of the parking search time, duration of stay, driver's perception of the requested parking fee, as well as gained experience in using specific parking facility.

## 3.2. Fuzzy rule base for updating the parking fee

We already mentioned that in order to have "fair negotiation" between agent-drivers, and agent-parking facilities we assume that agent-parking facilities also use experience and intuition in decision-making process. When describing different decisions made at various stages of a process, human beings prefer to use qualitative expressions instead of quantitative ones. We assume the same for agent-parking facilities. In this way, the strategies of the agents can be formulated in terms of numerous descriptive rules.

We assume that the following very simple fuzzy rule base can appropriately describe agent-parking facility behavior.

Rule 1:
**If** demand for parking is LOW
**Then** parking fee is LOW
Else
Rule 2:
**If** demand for parking is MEDIUM
**Then** parking fee is MEDIUM
Else
Rule 3:
**If** demand for parking is HIGH
**Then** parking fee is HIGH

## 3.3. Fuzzy rule base for updating the enforcement

As well as agent-drivers, and agent-parking facilities there is in our scenario another competing Agent - the city government. It organizes the enforcement, trying to allocate in optimal way the resources among the different demands. To this aim, the experience of previous days is useful in decision process.

In this case, the following fuzzy rule base can describe city government behavior:

Rule 1:

**If** demand for illegal parking is LOW
**Then** enforcement is WEAK

Else

Rule 2:

**If** demand for illegal parking is MEDIUM
**Then** enforcement is MEDIUM

Rule 3:

**If** demand for illegal parking is HIGH
**Then** enforcement is STRONG

Usually, there are no sudden daily changes in parking occupancy. In other words, the expectation is that traffic volumes vary smoothly over days. The proposed approximate reasoning algorithm also shows that parking facility charges vary smoothly over time. Driver entering the parking facility that has high demand should pay more because he/she is privileged to use relatively scarce resource.

## 4. THE CASE STUDY

The proposed model has been applied to the case of Bari, a medium-sized town of Southern Italy. In Figure 2, a simplified scheme of locations and capacities of charged parking in the Bari CBD is shown.

For sake of simplicity, we assume that:

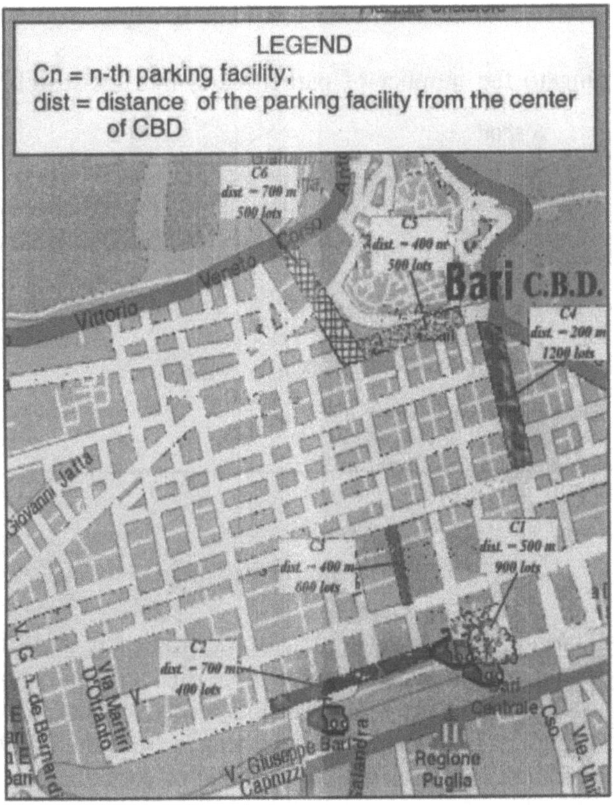


Figure 2 – Location and capacities of charged parking in Bari CBD

there are two types of demand - long stay and short stay (Figure 3). From a field survey, it came out that the first one is for work purposes, and takes place from 6 to 9 a.m. with an approximately sinusoidal distribution having a maximum of 97 cars/min. The second one is for other purposes (shopping, small business, etc…), and takes place from 8:30 to 11:30 a.m. with a constant value of 0.7 cars/min. The total demand from 6:00 to 11:30 is 8500 cars;

- at first, there are 6 charged parking facilities;
- free parking along the streets is allowed everywhere in CBD, and inner lots are occupied first;
- illegal parking is used to reach the closest proximity of destination, and it is influenced by level of enforcement;
- at the beginning, level of enforcement is weak.

Examples of membership functions are represented in Figures 3 and 4. The values of attributes used in the simulation can be:

- constant for each facility, like charged parking distances from CBD center;
- variable according to the number of parked cars, like lot searching time or

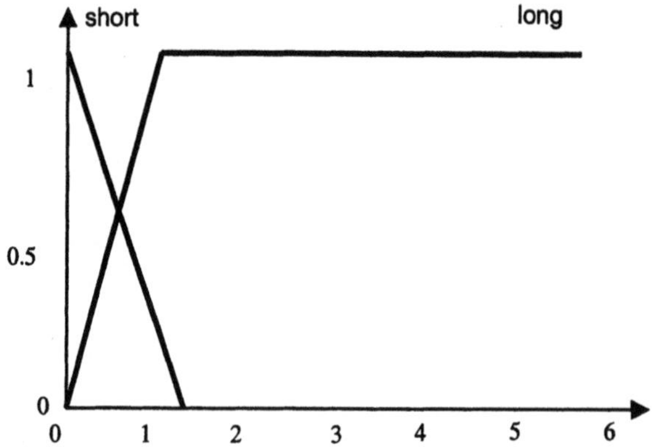


Figure 3 – Duration of stay (hrs)

distance of free parking from CBD center. In the first case, it is obvious that the greater the number of parked cars is, the more difficult finding a lot. In the second case, variability is in the sense that lots closer to the CBD center are supposed to be occupied first, therefore the distance from the CBD center, at which lots can be found, increases with the number of parked cars;

- random, like duration of stay.

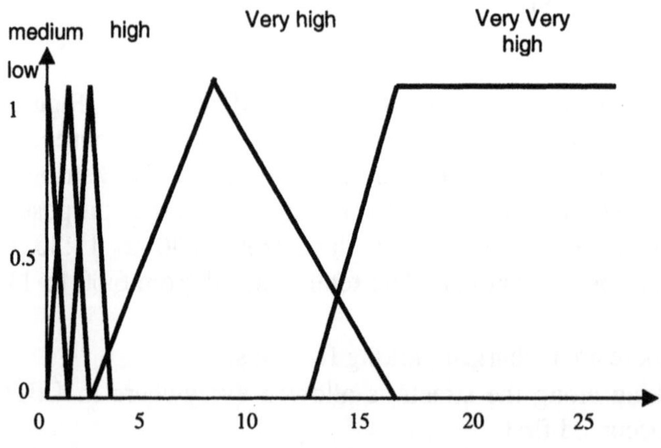


Figure 4 – fee ($)

In Figures 5, 6 and 7, lot-searching time versus number of parked cars, probability density function for long stay, and distance from CBD center versus number of parked cars are reproduced.

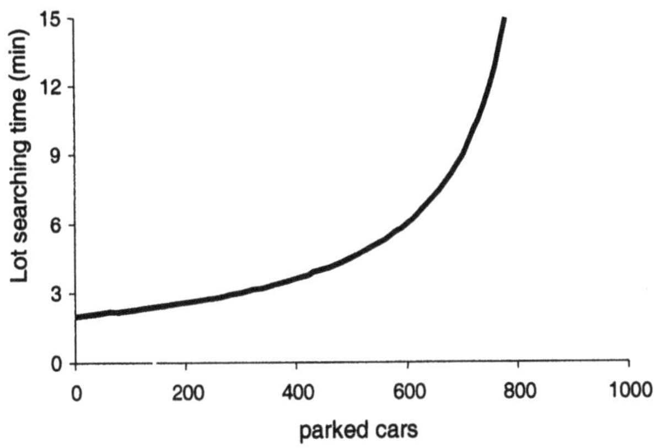


Figure 5 – LST vs number of parked cars

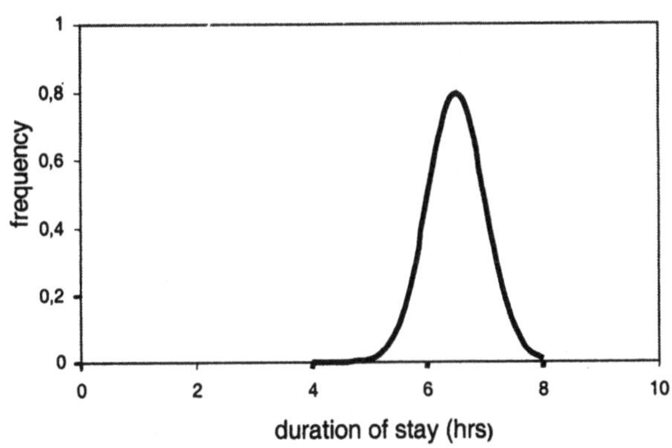


Figure 6 – duration of stay probability density function

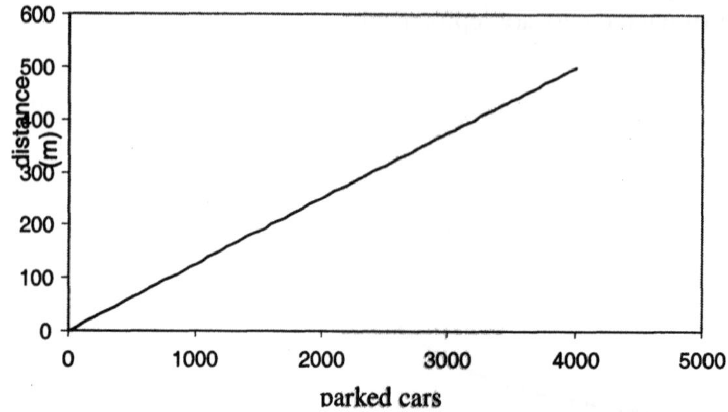


Figure 7 – distance of free parking facilities from CBD center

According to the time interval, we find the number of cars demanding service and draw the duration of stay, both for long and short-stay group. Then, the search of a parking lot is simulated applying the proposed model; in this way, we obtain the parking fees for each facility and the level of enforcement. Fees and enforcement are updated on a daily basis

These outcomes are shown in Figures 8 and 9, while Figure 10 reproduces the occupancy of facilities. If we hypothesize that a new charged parking facility is built in an intermediate location, the model allows us calculating the changes in fees. The new fees and enforcement are reported in Figure 11. It can be noted that the final level of enforcement is slightly lower than the former one, while the level of fee for the new parking facility is a low intermediate one, possibly due to its location.

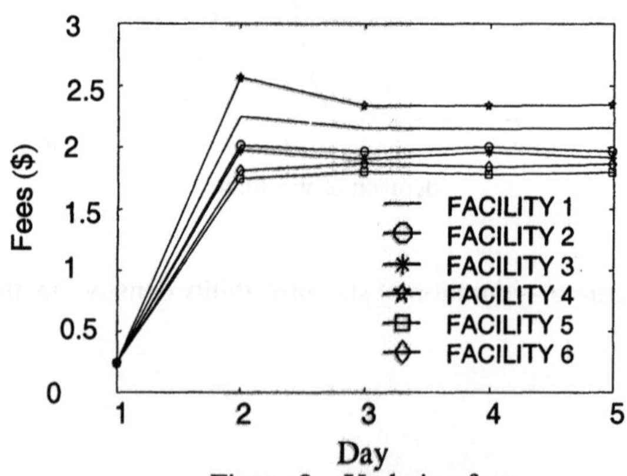


Figure 8 – Updating fees

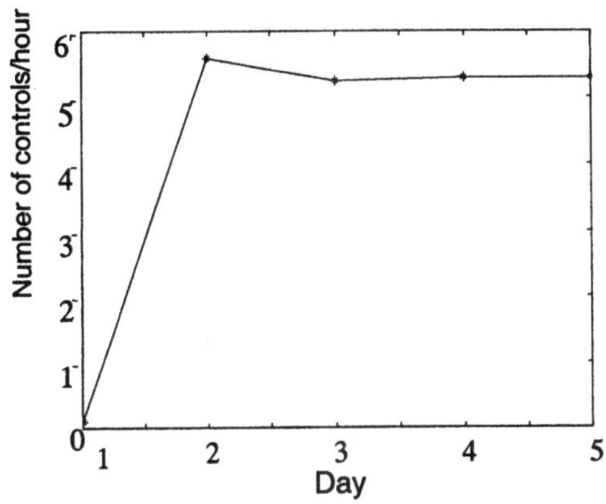


Figure 9 – Updating enforcement

Similar analysis could be easily made in the case of closing some of the existing parking facilities, or in the cases of significant increase in current

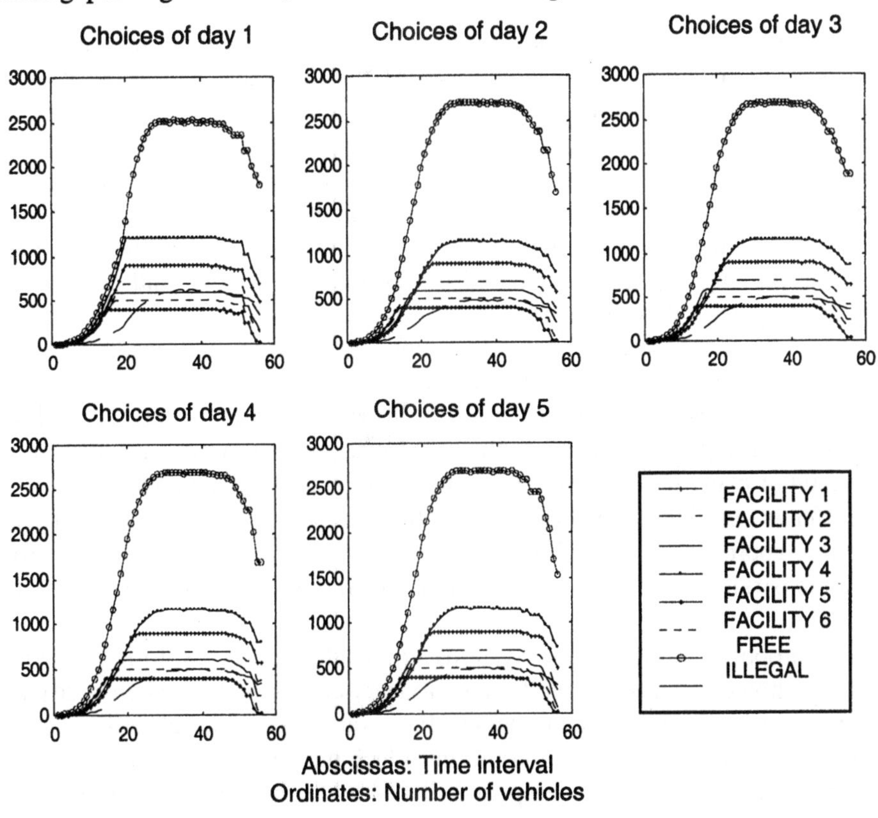


Figure 10 – Facilities occupancy

parking fees.

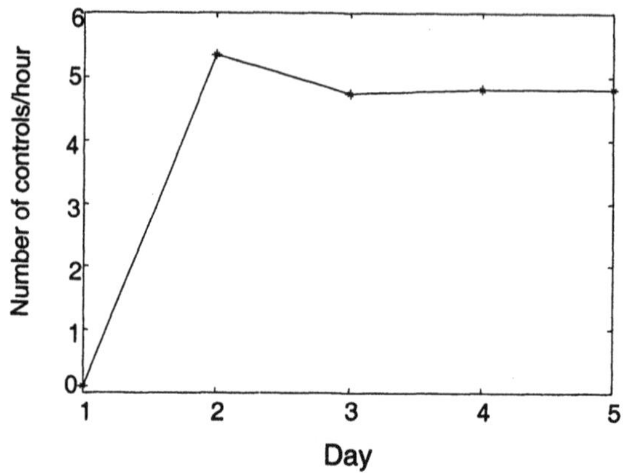


Figure 11 - Updating enforcement

# 5. CONCLUSIONS

We considered the existence of dynamic parking pricing in the urban transportation network. Considered time unit was one day. In other words, we updated fees and enforcement on a daily basis. In future applications, parking fees could be also updated on hourly basis. Proposed agent based model has capabilities to show flows propagation towards parking facilities in the urban traffic transportation network, and dynamic parking facility occupancy, as well as to calculate parking search times. Our model considers that each part (drivers, parking facilities) act based on its local knowledge and cooperates and/or competes with other parts.

Model developed allows agents that represent parking facilities to increase the capacity, or to significantly change parking -pricing policy, while the agents that represent drivers learn all the time, change their parking facility choices, and departure times. In our approach, parking facility occupancy is the result of many independent decisions made by individual agents.

Different pricing strategies should be the part of the comprehensive solution approach to complex traffic congestion problems. Undoubtedly, dynamic parking pricing represents one of the important demand management strategies.

The proposed Multi Agent Systems approach enables exploring various parking choice models, as well as various pricing schemes when studying complex parking pricing phenomenon. For example, traffic authorities, local governments and private sector could introduce higher parking tariffs for solo drivers, or they could provide special discounts to vanpoolers. In this paper we have tried to explore different parking pricing schemes, as well as to study the extent of previously gained driver's experience, and how does it affect parking facility choice.

Obviously, parking pricing should be carefully studied in the context of the considered city area (suburb, down-town, residential, commercial, retail use areas). The main role of any dynamic parking pricing would be in reducing the total number of vehicle trips, and in shifting commuters to alternative modes of transportation.

# REFERENCES

Akiyama, T., Tsuboi, H., Description of route choice behaviour by multi-stage fuzzy reasoning. Paper presented at the Highways to the Next Century Conference, Hong Kong, (1996).

Bonabeau, E., Dorigo, M., Theraulaz, G., *Swarm Intelligence*, Oxford University Press, Oxford, 1999.

Bond, A. H., Gasser, L., *Readings in Distributed Artificial Intelligence*, Morgan Kaufmann, (1988).

Dell'Orco, M., Ottomanelli, M., Sassanelli, D., Modeling Uncertainty in Parking Choice Behaviour, 82nd Annual Meeting of the Transportation Research Board – Washington DC (2002)

Dean, J., Gumerman, G., Epstein, J., Axtell, R., Swedlund, A., Parker, M., McCarroll, S., "Understanding Anasazi Culture Change through Agent-Based Modeling", in Kohler, J., and G. Gumerman (Eds.): *Dynamics in Human and Primate Societies*, Santa Fe Institute, (2000).

Hawas, Y.E.; Developing fuzzy route choice models using neural nets, Proceedings of the IEEE Intelligent Vehicle Symposium, (Volume 1), 71 –76, (2002).

Henn, V., Fuzzy route choice model for traffic assignment, Fuzzy Sets and Systems, 116, 103-118, (2000).

Kikuchi, S., Rhee, J., Teodorović, D., Applicability of An Agent-Based Modeling Concept to Modeling of Transportation Phenomena, YUJOR , 12 , 141-156, (2002).

Lotan, T ,Modeling discrete choice behavior based on explicit information integration and its application to the route choice problem, IEEE Transactions on Systems, Man and Cybernetics, Part A, 28, 100 –114, (1998).

Lotan, T., Koutsopoulos, H., Route choice in the presence of information using concepts from fuzzy control and approximate reasoning, Transportation Planning and Technology, 17, 113-126, (1993).

Lotan, T., Koutsopoulos, H., Models for route choice behaviour in the presence of information using concepts from fuzzy set theory and approximate reasoning, Transportation, 20, 129-155, (1993).

Lučić, P., Teodorović, D., Transportation Modeling: An Artificial Life Approach, Proceedings of the 14th IEEE "International Conference on Tools with Artificial Intelligence", 216-223, November 4-6, 2002 Washington D.C.

Lučić, P., Teodorović,D., Computing with Bees: Attacking Complex Transportation Engineering Problems, *International Journal on Artificial Intelligence Tools*, 12, 375-394, (2003).

Peeta, S., Pasupathy, R., Analyzing Traffic System Evolution Using Multi-Agent Simulation, 80th Annual Meeting of Transportation Research Board (CD-ROM), (2001)

Portugali, J., Benenson,I., Human Agents between Local and Global Forces in a Self-organizing City in Schweitzer, F. (Ed.): *Self-organization of Complex Structures*, Gordon and Breach Science Publishers, (1997).

Resnick, M., *Turtles, Termites, and Traffic Jams: Explorations in Massively Parallel Microworlds.* The MIT Press, (1994).

Shaout, A.; Cherri, A.; Cotner, J., Fuzzy route choice: an enhancement to future automotive navigation systems, 1993., Proceedings of the 36th Midwest IEEE Symposium on Circuits and Systems, (Volume 1), 212 –215, (1993).

Teodorović, D., Fuzzy Logic Systems for Transportation Engineering: The State of the Art, Transportation Research, 33A, 337-364, (1999).

Teodorović, D., Invited Review: Fuzzy Sets Theory Applications in Traffic and Transportation, European Journal of Operational Research, 74, 379 - 390, (1994)

Teodorović, D., Kalić, M., A Fuzzy Route Choice Model for Air Transportation Networks, Transportation Planning and Technology , 19, 109-119, (1995).

Teodorović, D., Kikuchi, S., Transportation Route Choice Model Using Fuzzy Inference Technique, Proceedings of ISUMA 90 – Proceedings of the First International Symposium on Uncertainty Modeling and Analysis,140 - 145, IEEE Computer Society Press, College Park, U.S.A., December, (1990).

Teodorović, D., Vukadinović, K., *Traffic Control and Transport Planning: A Fuzzy Sets and Neural Networks Approach,* Kluwer Academic Publishers, Boston/Dordrecht/London, (1998).

Teodorović, D., Vukanovic, S., Obradovic, K., Modeling Route Choice With Advanced Traveler Information by Fuzzy Logic, Transportation Planning and Technology, 22, 1-25, (1998).

Vythoulkas, P.C., Koutsopoulos, H.N., Modeling discrete choice behaviour using concepts from fuzzy set theory, approximate reasoning and neural networks, Research Report TSU 817, ESRC Transport Studies Unit, University of Oxford, Oxford, (1994).

Wahle, J., Schreckenberg, M., A Multi-Agent System for on-line Simulations based on Real-World Traffic Data, in Proceedings of the IEEE Hawaii International Conference on System Science, (2001).

Wooldridge, M., Jennings, N., Intelligence Agents: Theory and Practice, The Knowledge Engineering Review, 10, 115-152, (1995).

Yu; J.W., Peeta, S., A hybrid model for en-route driver behavior under real-time information, Proceedings of The IEEE 5th International Conference on Intelligent Transportation Systems, 301-306, (2002).

## Chapter 16

# MODELING TRANSPORTATION CHOICE THROUGH UTILITY-BASED MULTI-LAYER FEEDFORWARD NETWORKS

**Giulio Erberto Cantarella and Stefano de Luca**

## 1. INTRODUCTION

McCulloch & Pitts (1943) devised Artificial Neural Networks (McCulloch & Pitts, 1943) as a model (a very crude one indeed) of the neural tissue which made up the neural system of any animal (from brainless insects to human beings). More recently, ANNs have been considered just very flexible analytical functions, without any further reference to the initial biological background. As such an ANN is no more than a vector-valued function, $\mathbf{y} = \varphi(\mathbf{x})$, defined by a serial-parallel composition of several sub-modeling real-valued functions, that is a Parallel Distributed Processing (PDP) model made-up by processing units (*PUs*) (often called neurons from the biological metaphor) connected so that input values, $\mathbf{x}$, are forwarded through intermediate (hidden) *PUs* to the output *PUs*, which provide the output values, $\mathbf{y}$.

Among the several proposed types, Multi-Layer FeedForward Networks (MLFFNs) are the most used as a powerful tool for regression and classification analysis. In a MLFFN all *PUs* are grouped into layers, and the layers are sequentially ordered from the input one to the output one so that each *PU* is only connected with all those in the upstream layer (if any) and in the downstream layer (if any) but not with those in the same layer or in other layers. MLFFNs are black box models useful when no satisfactory theoretical paradigm is available, but their parameters may hardly be given a clear interpretation.

Recently MLFFNs have been applied to travel demand analysis, say to analyze how user socio-economic characteristics and level-of-service provided by transportation supply affect demand flows. The first papers (Reggiani e Tritapepe, 1998; Nijkamp et al., 1996; Schintler and Olurotimi, 1998; Shmueli et al., 1998; Shmueli, 1996; Mozolin et al., 2000) addressed different demand analysis issues (trip generation, trip distribution and modal split) with satisfactory results. On the other hand all these papers have been calibrated[1] on aggregate data, hence they model demand flows, rather than disaggregate user choices, as in random utility models (RUMs) commonly adopted for choice modeling within a an econometric framework (Ben-Akiva and Lerman, 1985; Cascetta; 2001; Train, 2002). Hensher and Ton (2000), more recently, have proposed an alternative approach based on MLFFNs calibrated on disaggregate stated preferences data (i.e. simulated choice contexts have been proposed to users). They have applied it and obtained results have been compared with RUMs.

Following an approach similar to the one of Hensher and Ton (2000) the authors have applied MLFFNs to simulate mode choice in an extra-urban context, by calibrating them against real data (Cantarella and de Luca, 2004). Results reported in all these papers show that proposed MLFFNs, with one intermediate (hidden) *PU* layer connecting inputs to the output PU layer can be a feasible and effective tool for travel demand analysis, their effectiveness being only slightly improved by increasing the number of intermediate layers. MLFFNs may in same cases outperform commonly adopted RUMs, but, the advantages of their application for travel demand forecasting demand may be argued since no clear interpretation of parameters is possible, differently of common practice RUMs.

This paper presents the application of MLFFNs with a layout different of the one used in literature, including explicit utility specification through a further intermediate (hidden) layer with a processing unit for each alternative. This way, both the utility function and the choice function are explicitly and separately specified, as in RUMs within an econometric framework. Moreover, utility parameters may be given an interpretation, useful for project appraisal.

After this section, the paper is organized as follows. Section 2 introduces the main notations and definitions, section 3 describes the proposed utility-based Multi-Layer FeedForward Networks for choice modeling, section 4 shows the models specified, the results of an application to a real case study, and the comparisons with a Hierarchical Logit model. Section 5, finally, summarizes the main conclusions and reports some research perspectives.

## 2. NOTATIONS AND DEFINITIONS

In this section Multi-Layer FeedForward Networks (MLFFNs) are briefly described mainly to introduce definitions useful in the following sections.

### 2.1 Description of MLFFNs

In broad sense, a MLFFN is a vector-valued function, $\mathbf{y} = \varphi(\mathbf{x})$, which may be considered a special type of Parallel Distributed Processing (PDP) model. As such, PDP model specifies a vector-valued function, $\mathbf{y} = \varphi(\mathbf{x})$, through the serial-parallel activation of processing units (*PUs*), each described by a real-valued function. At each *PU* outputs from upstream *PUs* (if any) are processed resulting in an output which is forwarded to downstream *PUs* (if any). From such a process, input values, **x**, are forwarded through intermediate (hidden) *PUs* to the output *PUs*, which provide the output values, **y**. *Output processing units* supply their outputs to the model-user; conventionally inputs are also represented by *input processing units* which just receive their inputs from the model-user and forward them to the downstream *PUs* without any transformation (without performing any process at all). All the other *PUs* are called *intermediate (hidden).*

At each *PU k* first a weighted combination of inputs $x_j$, received from each upstream PU $j$, is carried out: $z_k = \sum_j w_{jk}\, x_j + {}_{,}b_k$, where $w_{jk}$ is the weight given to connection $(j,k)$, $b_k$ is a constant (usually called bias), $z_k$ is the result of the combination. Then, the output value $y_k$ is computed from the result of the linear combination, $z_k$ , through a function (usually called activation function) and is forwarded to downstream *PUs*, (see Figures in section 3.2 for an example).

A Multi-layered FeedForward Network (MLFFN) is obtained when all *PUs* are grouped into layers, and the layers are sequentially ordered from the input one to the output one so that each *PU* is only connected with all those in the upstream layer (if any) and in the downstream layer (if any) but not with those in the same layer or in others (see Figure 1, in section 3.2). This architecture is the most used one for classification and (non-linear) regression analysis.

### 2.2 Application of MLFFNs

After input and output *PUs* have been given a meaning, the *specification* of a MLFFN requires the definition of the number of hidden layers, the number of *PUs* for each layer, the activation function for each PU (usually all *PUs* in the same layer have the same function), linear, logistic or hyperbolic tangent being the most used types. Usually, the parameters of the activation functions are chosen by the model-builder, so far the parameters to be calibrated against a sample of observations (usually called training data-set) include one weight for each connection and one bias for each *PU* except the input ones. Due to the high number of parameters over-fitting may well occur (and calibration techniques should be designed also taking into this issue).

The *calibration* of parameters can be performed by minimizing any distance between observed and predicted outputs; usually Euclidean distance is used, i.e. the (mean) sum of the square of differences (MSE). In this case a (local) minimum can be found through carefully designed gradient methods (such as the back-propagation algorithm that duly exploits the structure of the MLFFN by backwards updating weights and biases from the output layer to the input one). To improve convergence several copies (usually called number of epochs) of the calibration data-set are actually used for calibration; however it is well-known that an excessively high number of epochs may lead to over-training. Moreover, due to the existence of several local minima the obtained solution can be greatly affected by the starting values of weights and biases, thus the minimization algorithm is applied several times (usually called number of repetitions) each with a different initialization of parameters.Among all the solution obtained by this way one is chosen, for instance, the one with the least value of MSE computed over the calibration or the validation data set.

The *validation* of a calibrated MLFFN is very relevant due to the high number of parameters as well as its flexible structure. At this aim, as already said, some observations (hold-out or validation sample) are usually taken aside to check whether generalization vs. reproduction of observations is obtained. Cantarella and de Luca (2003) address all the above issues in details, with respect to the use of MLFFN for choice modeling.

# 3. MODEL SPECIFICATION

After a brief introduction to choice models for travel demand analysis, this section describes the proposed utility-based Multi-Layer FeedForward Networks for choice modeling, according to the definitions and notations previously introduced. Reference is mainly made to transportation mode choice, but the same approach can be followed to simulate choice of destination, of route, etc. (or choices in contexts different of transportation).

## 3.1 Choice models for travel demand analysis

This subsection presents a brief and formal description of choice models for travel demand analysis. In a broad sense a choice model tries to capture how attributes, $\boldsymbol{x}$, such as level-of-service features (measured for the current state or assumed for a design scenario) and user socio-economic characteristics affect choice fractions, $\boldsymbol{p}$:

$$\boldsymbol{p} = p(\boldsymbol{x}) \qquad \text{with} \quad \boldsymbol{p} \geq \boldsymbol{0} \quad \boldsymbol{1}^T\boldsymbol{p} = 1$$

*Utility-based* choice models express the relation between attributes and choice fractions by explicitly introducing a (systematic) utility value, $\boldsymbol{v}$, for each choice alternative as a function of attributes:

$\boldsymbol{v} = v(\boldsymbol{x})$ utility function
$\boldsymbol{p} = p(\boldsymbol{v})$ choice function

The meaning of the systematic utility value depends on the framework within which the choice model is embedded, for instance the mean value of perceived utility considered as a random variable within random utility theory. Utility-based choice models allow to give an interpretation to some parameters of the models, useful for project appraisal.
For each choice alternative, $m$ (for instance transport mode: car, bus...), the systematic utility is often specified as a linear combination of attributes: $v_m = \sum_j \beta_j x_{mj}$, where $\beta_j$ are duly defined parameters to be calibrated against a set of observations, commonly assumed generic regard to choice alternative. Of course the choice function, $\boldsymbol{p} = p(\boldsymbol{v})$, may well contain other parameters to be calibrated. *Behavioral* choice models, such as RUMs, which are derived from explicitly assumptions about user choice behavior, are distinguished from *non-behavioral* choice models, such as MLFFNs.

## 3.2 Utility-based MLFFNs for choice modeling

This paper deals with the specification of MLFFNs for choice modeling. The proposed lay-out is made-up by two intermediate layers besides the input and output ones (figure 1), expressing the combination of two functions, according to the utility-based approach introduced above:

$\boldsymbol{v} = v(\boldsymbol{x})$ utility function: from layer 0 to layer 1
$\boldsymbol{p} = p(\boldsymbol{v})$ choice function: from layer 1 to 3 through 2

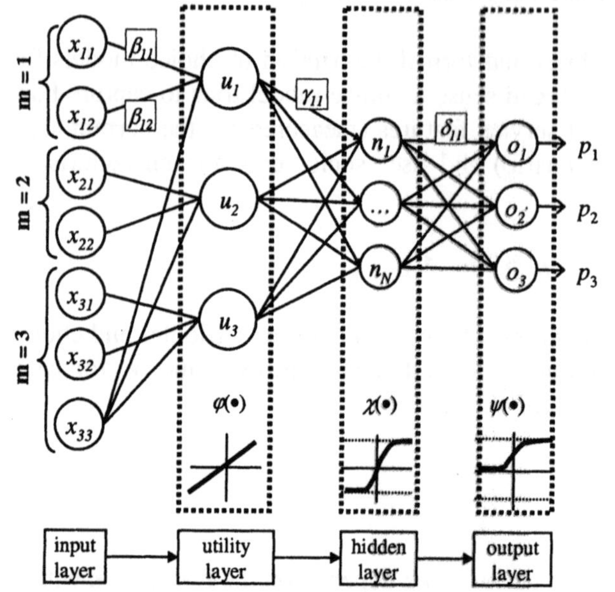


*Figure 1.* utility-based MLFFN architecture

[0] *Input layer.* This layer contains one *PU mj* for each attribute *j* (for instance: travel time, monetary cost, ...) and each mode *m*, it simply forwards the input value $x_{mj}$ (from the data-set given by the model user) to downstream *PUs*, without processing it.

[1] *Utility layer.* This layer contains one *PU m* for each mode *m*, which only receives input values from the upstream input *PUs mj* corresponding to the same mode *m* (figure 2). So far the input and utility layers are not fully connected.
Assuming the identity function as activation function, $\varphi(\bullet)$, the output, $v_m$, is given by a linear combination of attributes:

$$v_m = \sum_j \beta_{mj}\, x_{mj} + ASA_m$$

where
$\beta_{mj}$ is the weight associated to the connection between input PU *mj* (attribute *j* for mode *m*) and utility PU *m*;
$ASA_m$ is the bias (constant) associated to utility PU *m* (named after Alternative Specific Attribute from econometrics).

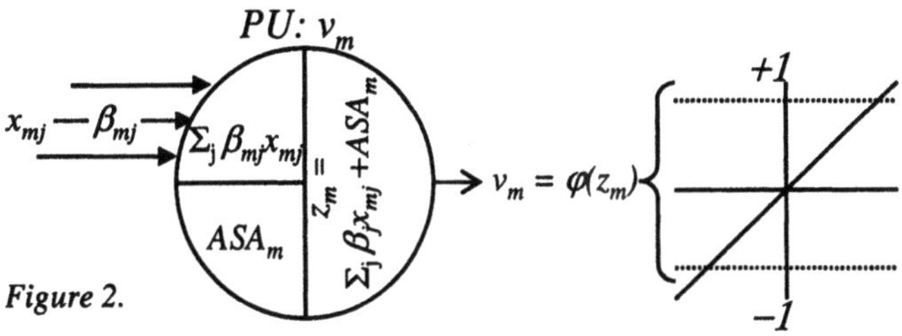


*Figure 2.*

The provided output $v_m$ is formally analogous to commonly adopted utility function, quoted in the previous sub-section. Once parameters such as weights and biases have been calibrated they may be given an interpretation.

[2] *Hidden layer*. The number of *PUs* in this layer is defined during the model-building stage. Each *PU n* in this layer is connected to (receives an input from) each *PU m* in the utility layer (figure 3). Assuming an activation function $\chi(\bullet)$, common to all *PUs* in this layer, the output, $y_n$, is given by:

$$y_n = \chi(\Sigma_m \gamma_{mn} v_m + c_n)$$

where

$\gamma_{mn}$ is the weight associated to the connection between utility *PU m* (for mode *m*) and hidden *PU n*;

$c_n$ is the bias (constant) associated to hidden *PU n*.

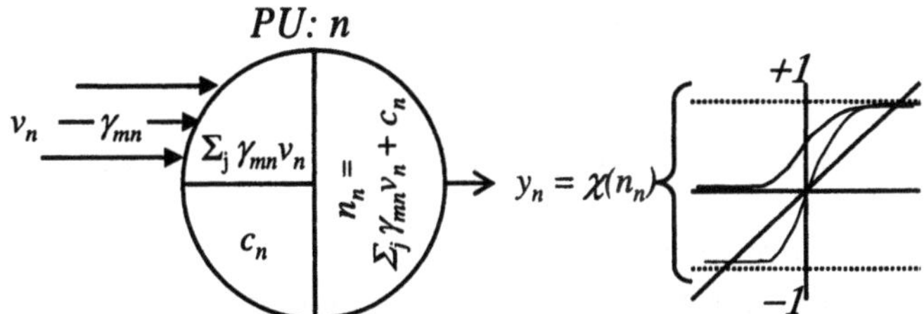


*Figure 3.*

It should be noted that the calibrated weights and biases in this layer should hardly be given a clear interpretation.

[3] *Output layer*. This layer contains one *PU m* for each mode *m*, which is connected to (receives an input from) each *PU n* in the hidden layer. Assuming an activation function $\psi(\bullet)$ with values in the range [0,1] and common to all *PUs* in this layer, the output, $p_m$, is given by:

$p_m = \psi_m(\Sigma_n \delta_{nm} y_n + b_m)$

where

$\delta_{nm}$ is the weight associated to the connection between hidden PU $n$ and output *PU m* (for mode $m$);

$b_m$ is the bias (constant) associated to hidden *PU n.*

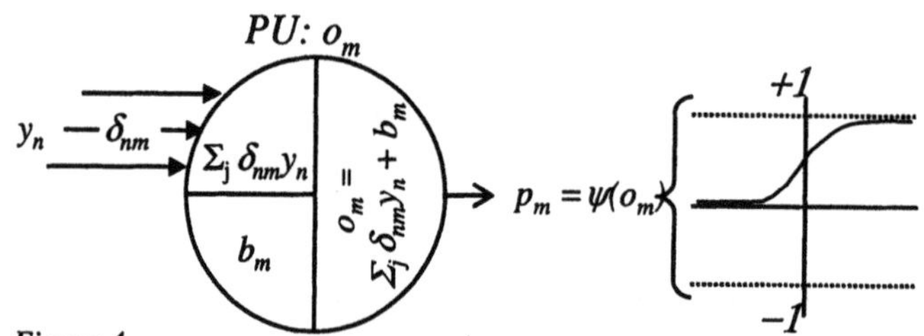


*Figure 4.*

It should be noted that the calibrated weights and biases in this layer should hardly be given a clear interpretation.

According to the proposed lay-out allows both the utility function and the choice function are explicitly and separately specified, as in random utility models (RUMs) within an econometric framework.

# 4. APPLICATIONS

This section presents a comparative analysis of some instances of the following modelling approaches:

[1] utility based (partially connected) multi-layer feedforward network models (MLFFN-UB),

[2] multi-layer feedforward (fully connected) networks(MLFFN-FC),

[3] a closed form random utility model (RUM): hierarchical logit (HL).

A common calibration data-set has been used, as well as the same validation data-set (hold out sample).

## 4.1 The case study and the calibration/validation data sets

The proposed models have been tested against real data (already used to analyze several choice models by Cantarella and de Luca 2002, 2004). The whole database contains 2,808 interviews referring to journeys of students

from outside the city of Salerno towards (the country-side location of) the University of Salerno, Italy. Although four transport modes may be available: car-as-driver, car-as-passenger, car-pool and bus, only users that have available all transport modes have been considered to make clearer the analysis by avoiding the effects of mode availability. In such a context car-as-passenger is no longer an available transport mode, the observations become 944, and mode market share are: car-as-driver (36.9%), bus (5.3%), car-pool (57.8%).
In order to avoid overfitting and overtraining in MLFFNs calibration, and to carry out more effective comparisons between RUMs and MLFFNs, the whole data set has been split into a calibration data set and a validation one, through random sampling. According to two criteria the calibration data set must large enough (*i*) to reproduce observed survey market shares (within a small error), (*ii*) to allow a stable estimation of RUM utility parameters. If either criterion is not satisfied, uncorrected model calibration and ineffective comparisons may result.
Analyzing the survey, as showed in Figure 5, the mode market shares become stable with about 700 observations (75%), which have been used to calibrate the models (almost the same size has been obtained by repeating 10 times such analysis with different observations). The remaining 244 observations have been used as validation data set. Clearly the mode market shares from the validation data set may be slightly different of those from the calibration data set.

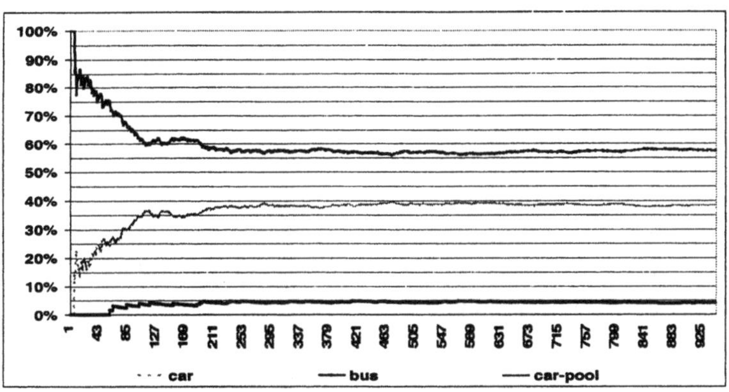


*Figure 5*. Mode share against number of observations

Origin (and destination) zoning was mainly city based. Several type of attributes have been used as showed in table 1. Levels of service (LoS) attributes were computed through a transportation network model.

*Table 1.* Attributes used in models specification

| | | | Type | Car | Bus | Car-pool |
|---|---|---|---|---|---|---|
| ***Level of service (LoS)*** | | | | | | |
| Time | Trip time | (h) | *Continuous* | • | • | • |
| Cost | Trip monetary cost | (€) | *Continuous* | • | • | • |
| ***Socio-economic (SE)*** | | | | | | |
| Gen | 1 if gender is female | - | *Binary* | • | - | - |
| ***Activity related and Land Use (AC+LU)*** | | | | | | |
| $ACT_{lenght}$ | Activity time length | (h) | *Continuous* | • | - | - |
| Freq | Weekly trip frequency | - | *Discrete* | - | - | • |
| ***Others*** | | | | | | |
| ASA | Alternative specific attribute | - | *Binary* | • | • | - |

## 4.2. Indices for model validation and comparison

The effectiveness of a model may be analyzed with respect to:

- interpretability: parameters may be given a clear interpretation,
- reproducibility: observed choices are well reproduced with reference both to calibration and validation data sets.

Moreover, the model should be able to generalize to other data sets, describing design scenarios (generalization). It is also relevant taking into account the efficiency of the model, namely computing resources needed to calibrate and to apply it. For RUMs statistical significance of parameters should also be tested. All the models developed have been compared by introducing descriptive indices calculated both for the validation data set.

*aggregate indices*

• $[P_{sim}\text{-}P_{real}]$ for each transport mode the differences between mode shares observed and simulated by the model have been evaluated.

• $MSE_{shares}$ is the mean square error between observed and simulated mode choice shares, this index takes a null value over the calibration data-set for any Logit model calibrated through maximum likelihood estimation (see for instance Train, 2002), thus it is useful only when referred to the validation data-set and/or to other choice models.

*disaggregate indices*

• $\%_{clearly\ right}$ is the percentage of users in the sample whose observed choices are given a probability greater than 0.90 (or any given threshold not less than 50%) by the model.

- $\%_{clearly\ wrong}$ is the percentage of users in the sample for whom the model gives a probability greater than 0.90 (or any given threshold not less than 50%) to a choice different from the observed one.
- $\%_{unclear} = 100 - (\%_{clearly\ right} + \%_{clearly\ wrong})$ the percentage of users for whom the model does not give a probability greater than 0.90 (or any given threshold not less than 50%) to any choice.
- Fitting Factor (FF) $= \Sigma_{\text{user}}\ p^{sim}{}_{user}\ /\ N_{users} \in [0,1]$, with FF = 1, meaning that the model perfectly simulates the choice actually made by each user, say with $p^{sim}{}_{user} = 1$. Let MAE be the mean absolute error, it clearly turns out $\text{MAE} = 2 \times (1 - \text{FF})$.
- %right is the percentage of users in the calibration data-set whose observed choices are given the maximum probability (whatever the value) by the model; clearly it is not less than $\%_{clearly\ right}$. This rather meaningless index is very often reported when describing RUM applications, thus will be reported in this paper too.

For RUMs only, consolidated indices and statistical tests have been also adopted: *t* student and pseudo $rho^2$.

All the results presented have been obtained through commercial software packages, HieLow (STRATEC) for RUMs and MATLAB® for MLFFNs.

## 4.3 Utility-based MLFFNs: analysis and comparisons

This section reports results of an application of two types of MLFFNs, non utility based vs. utility based (fully vs. partially connected), and HL (RUM) to the data set described above. For each choice model two different specifications are analysed, depending on hypotheses on input variable sets:

[1] *MLFFN/HL* with only Level of Service attributes (***LoS***);
[2] *MLFFN/HL* with Level of Service attributes, socio-economic and activity related attributes (***LoS + SE + AC***);

| | **MLFFN utility based** | **MLFFN non–utility based** | **RUM** |
|---|---|---|---|
| [1] ***LoS*** | *MLFFN-UB[1]* | *MLFFN [1]* | *HL[1]* |
| [2] ***LoS+SE+AC*** | *MLFFN-UB[2]* | *MLFFN [2]* | *HL[2]* |

Such specifications have been adopted to understand the impact of different combinations of variables on model goodness of fit and to let an easier interpretation of utility weight values obtained from MLFFN-UB calibration.

### 4.3.1 Utility based MLFFNs (MLFFN-UB)

To specify a MLFFN model the following main operational issues should be defined: calibration and validation data-sets, error function to minimize, parameters (weight and biases) initialization technique, input attributes, number of epochs and of starting conditions (initializations), selection of MLFFN architecture, computation of the parameters for the selected MLFFN architecture. All these issues have been addressed following the procedure (presented in details by Cantarella and de Luca, 2004) summarized in the figure 6, that easily allow to make operational a MLFFN model, once inputs and outputs have been defined.
As regard split of survey data-set into calibration and validation data-sets (Step 0), a 75% threshold proved to be effective, thus the remaining 25% of the survey data set has been used for validation, as already said. Concerning calibration algorithm issues (Step 2), 120 repetitions (initializations) and less than 1000 epochs allow to cover the solution space, and to obtain good reproduction capabilities as well as to minimize the effects of over-training, whichever the architecture is. Finally, the most efficient MLFFN architectures have been selected (step 3) by considering one hidden layer and increasing number of processing units (PUs = 15, 30, 45, 60), since some results suggest that two hidden layers do not improve effectiveness; several activation functions have been tested in the hidden layers and in the output one. As regard utility based MLFFN-UB the additional utility layer has been characterized by linear activation function (see Figure 7 in next section). Finally, several calibrations have been carried out by varying the epoch number (50, and from 100 to 1000 by 100). Once a MLFFN architecture has been selected, the set of parameters with the best value of error function computed on the validation data set has been chosen (Step 4).

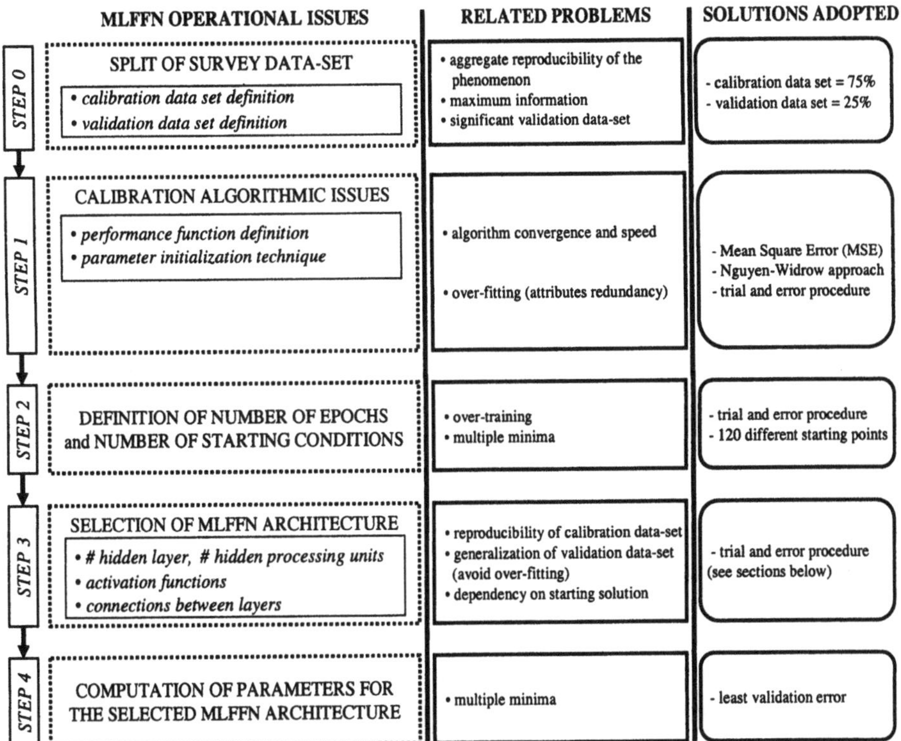


*Figure 6.* Operational issues, related problems and solutions adopted

As already highlighted two different models are proposed, with only ***LoS*** attributes (MLFFN-UB[1] with 6 input processing units) and with ***LoS, SE*** and ***AC*** attributes (MLFFN-UB[2] with 9 input processing units). The hyperbolic tangent activation function in the hidden layers and sigmoid in the output one have lead to the best results. Figure 7 proposes the two architectures specified and calibrated. Main characteristics of specified utility based MLFFN models are showed in table 2.

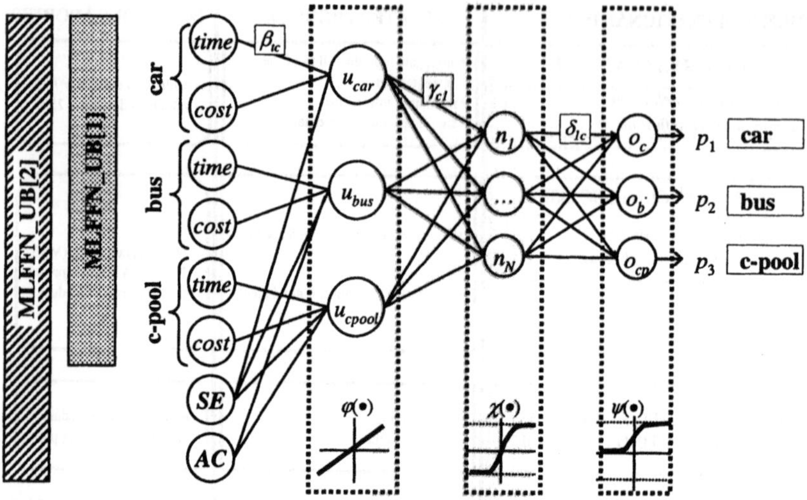


*Figure 7.* Utility based MLFFN architecture ($p_k \in [0,1]$)

As regard MLFFN-UB[1], 60 processing units in the hidden layers and 1000 epochs have been necessary to get satisfactory reproducibility and, at the same time, to avoid over-training and over-fitting. By introducing more attributes, MLFFN-UB[2], the number of processing units in the hidden layer decreases while the same number of epoch are necessary to calibrate model parameters.

The main experimental evidences are reported in table 3 and summarized in figure 8.

*Table 2.* Main characteristics and training outputs for MLFFN-UB models

| Survey = 944 | MLFFN-UB[1] | MLFFN-UB[2] |
|---|---|---|
| Attributes | ***LoS*** | ***LoS+SE+AC*** |
| Input PUs | 6 | 6+1+2 |
| Output PUs | 3 | 3 |
| Hidden layers | 1 | 1 |
| Hidden PUs | 60 | 30 |
| Calibration set (75%) | 700 | 700 |
| Validation set (25%) | 244 | 244 |
| # Epoch | 1000 | 1000 |

Both architectures show good capabilities in reproducing market shares prediction. Such a result is confirmed by analyzing disaggregate indices, see $\%_{clearly\ right}$ and $\%_{unclear}$, where unclear predictions number is over 50% of all hold-out sample.

*Table 3.* Main comparison indices for MLFFN-UB models

| | **MLFFN-UB[1]** | **MLFFN-UB[2]** |
|---|---|---|
| Attributes | ***LoS*** | ***LoS+SE+AC*** |
| $\%_{clearly\ right}$ | 34% | 41% |
| $\%_{unclear}$ | 61% | 51% |
| $\%_{clearly\ wrong}$ | 5% | 8% |
| $[P_{sim}-P_{real}]^{car}$ | -3.3% | -2.8% |
| $[P_{sim}-P_{real}]^{c\text{-}pool}$ | -0.8% | +0.1 |
| $[P_{sim}-P_{real}]^{bus}$ | +4.0% | +2.7% |
| MS_mse | 2.77E-03 | 1.40E-3 |
| FF | 70% | 69% |
| %right | 78% | 76% |

By introducing more attributes, ***SE*** and ***AC*** attributes, the most effective architecture shows a slightly better fit allowing to achieve more than 40% of clearly right predictions, a better value of FF and very satisfactory market shares. As expected, ***SE-AC*** attributes introduce more parameters that allow segmentation of users and, as a consequence, a better reproduction of single user choices (see $\%_{clearly\ right}$). On the other hand it should be noted an increase of clearly wrong predictions due to the segmentation is introduced.

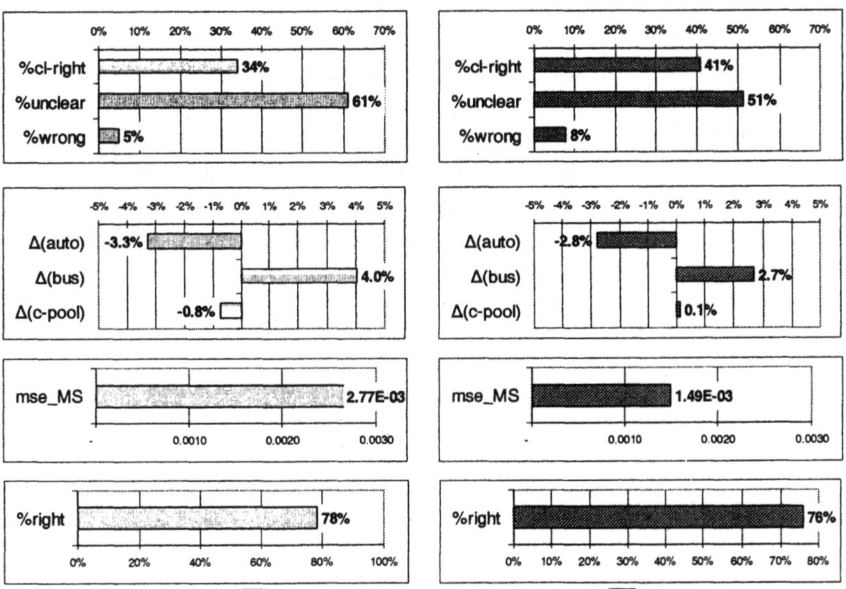


*Figure 8.* MLFFN-UB[1] vs MLFFN-UB[2]

Both models show similar value of indices, but the MLFFN-UB[2] outperforms MLFFN-UB[1], allowing to minimize market shares error and to obtain more clearly right predictions, without a significant increase of clearly wrong predictions. Such results are not surprising and, one more time, stress how choice models are influenced by the adoption of ***SE*** attributes. It is worth noting that the value of %right for MLFFN-UB[1] is better than the value for MLFFN-UB[2], although only slightly; this result confirms that %right is poor index.

As regard interpretation, an analysis of weight values is proposed for both models, MLFFN-UB[1] and for MLFFN-UB[2]. As described in section 3.2, the connections weights between input layer and utility layer may be interpreted as coefficient of systematic utility. These values, computed after calibration, represent the parameter that we wish to analyze. It should be noted, finally, that we compute specific coefficients for each attribute, and not generic as RUMs. It should be remembered that calibrated weights from input to utility layer include a scale factor, due to the identity activation function. In other words, if all these weights are multiplied by a common factor $\alpha$, the MLFFN still provides the same results if all the weights of the connections from the utility layer are divided by the factor $\alpha$.

As regard MLFFN-UB[1], in table 4, for each transport mode the ratio between weight associated to travel time and weight associated to travel cost is reported. All the ratios assume reasonable values and positive sign, meaning that all weights are negative according to what we expected. In particular, car has the least ratio, which is doubled ratio for bus and increases up to 6.0 for car-pool. Such results highlight that travel time is much more relevant only for car-pool, and this is consistent with a transport mode where users divide travel costs (in the case study 2.5 students on average share the same car) and travel time depend on destination of each passenger and, therefore, can be much longer than car. Analogous remark may be extended to bus, where travel costs are lower than car but greater than car-pool.

*Table 4*. MLFFN-UB[1] utility functions parameters

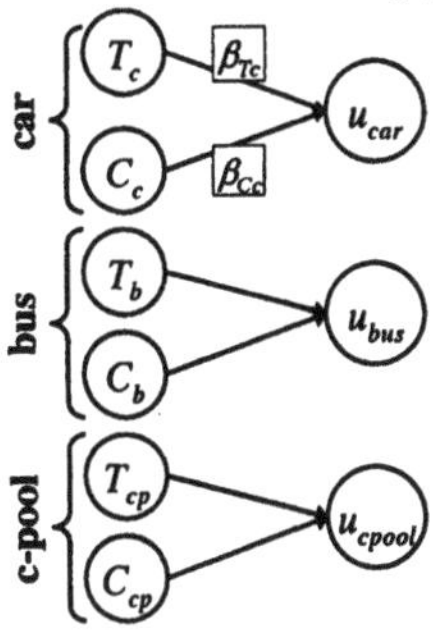


| **MLFFN-UB[1]** | | **Transport mode (x)** | |
|---|---|---|---|
| **ratio** | **car** | **bus** | **car-pool** |
| $\beta_{Tx}/\beta_{Cx}$ **(euro/h)** | 0.17 | 0.34 | 6.0 |

Analyzing MLFFN-UB[2], time/cost ratios sensibly change for bus and car-pool (see table 5). As regard car, the ratio decreases but not significantly.

*Table 5*. MLFFN-UB[2] utility functions parameters

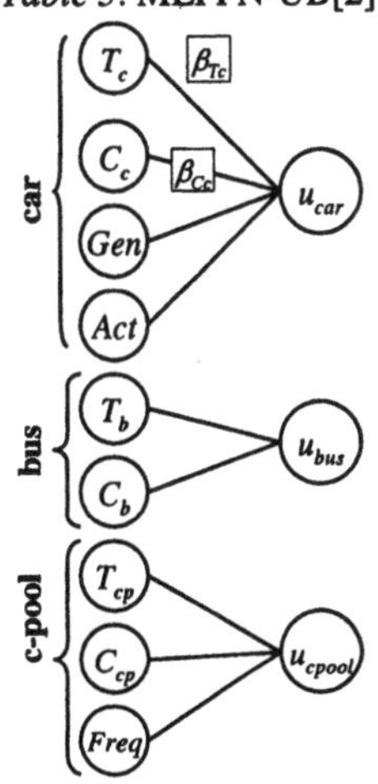


| **MLFFN-UB[2]** | | **Transport mode (x)** | |
|---|---|---|---|
| **ratio** | **car** | **bus** | **car-pool** |
| $\beta_{Tx}/\beta_{Cx}$ **(euro/h)** | 0.11 | 7.29 | 4.5 |
| $\beta_{gender}/\beta_{Cx}$ | 0.15 | - | - |
| $\beta_{act}/\beta_{Cx}$ | - | - | 0.36 |
| $\beta_{freq}/\beta_{Cx}$ | 0.15 | - | - |

Such results are a consequence of ***SE-AC*** attributes that allows to interpret better the choices introducing segmentation between users. The main consequence regards bus time/cost ratio that assumes a higher and more reasonable value. It allows, in fact, to simulate in a more realistic way the travel time "disutility" that is perceived by transit users.

### 4.3.2 Comparisons with MLFFNs non utility-based

This section presents results for two non utility-based MLFFNs, with only ***LoS*** attributes (MLFFN[1] with 6 input processing units) and with ***LoS*** and ***SE*** attributes (MLFFN[2] with 9 input processing units). The hyperbolic tangent activation function in the hidden layer and sigmoid in the output one have lead to the best results. In figure 9 and table 6 are showed the architectures and the main characteristics of the most effective MLFFNs. For both models 30 processing units in the hidden layers and 1000 or 500 epochs have been sufficient to avoid over-training and over-fitting

Following the procedure summarized in figure 3 the main experimental evidences are reported below (table 7 and figure 10).

Both architectures show good capabilities in reproducing market shares prediction, considerations similar to those in the previous sub-section hold in this case too.

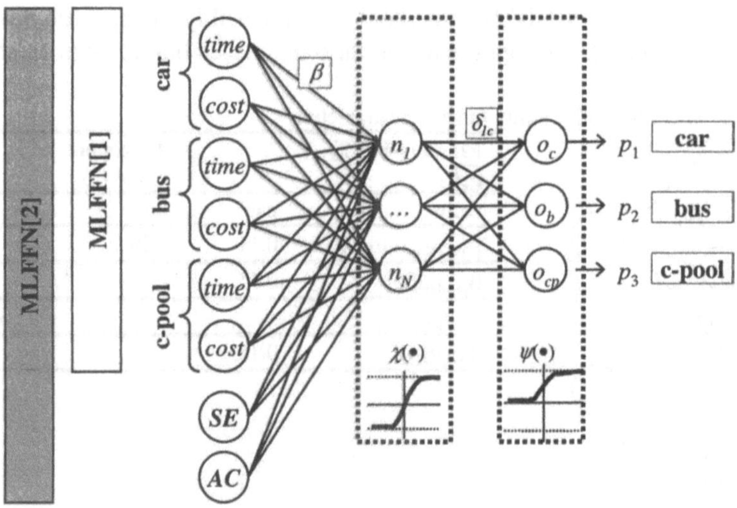


*Figure 9.* MLFFN architecture ($p_k \in [0,1]$)

*Table 6.* Main characteristics and training outputs for MLFFN models

| Survey =944 | MLFFN[1] □ | MLFFN[2] ■ |
|---|---|---|
| Attributes | *LoS* | *LoS+SE+AC* |
| Input PUs | 6 | 6+1+2 |
| Output PUs | 3 | 3 |
| Hidden layers | 1 | 1 |
| Hidden PUs | 30 | 30 |
| Calibration set (75%) | 700 | 700 |
| Validation set (25%) | 244 | 244 |
| Epoch | 1,000 | 500 |

*Table 7.* Main comparison indices for MLFFN_UPC models

| | MLFFN[1] □ | MLFFN[2] ■ |
|---|---|---|
| $\%_{clearly\ right}$ | 38.9% | 41.4% |
| $\%_{unclear}$ | 54.5% | 50.4% |
| $\%_{clearly\ wrong}$ | 6.6% | 8.2% |
| $[P_{sim}-P_{real}]^{car}$ | -2.4% | -2.3% |
| $[P_{sim}-P_{real}]^{c\text{-}pool}$ | +5.3% | -0.9% |
| $[P_{sim}-P_{real}]^{bus}$ | -2.9% | +3.2% |
| MS_mse | 4.2E-03 | 1.60E-03 |
| FF | 68% | 69% |
| %right | 75% | 76% |

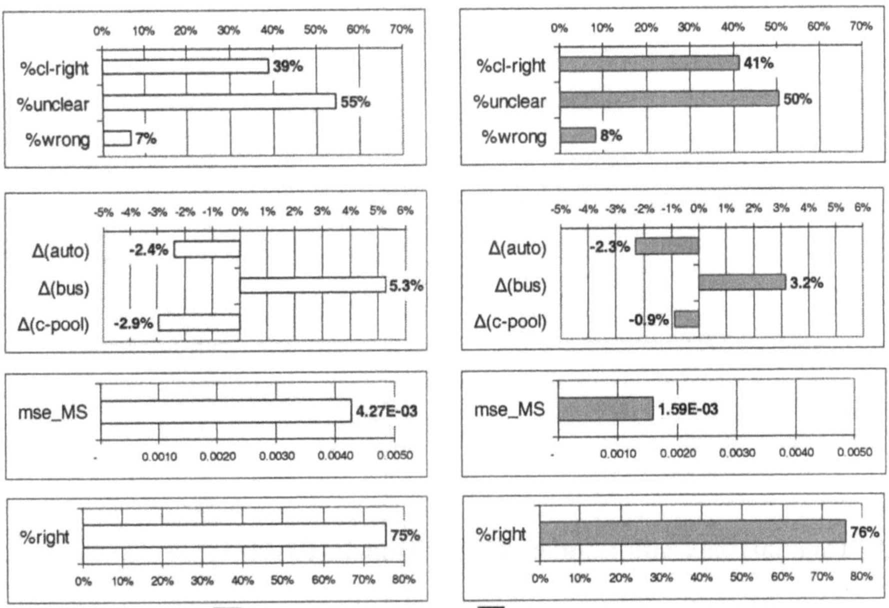


*Figure 10.* MLFFN[1] □ vs MLFFN[2] ■

More interesting is the comparison between the best MLFNN non-utility based and the best MLFFN-UB (see figure 11). The two models are characterized by a different architecture that leads to a considerable different number of parameters. In MLFFN[2] the weights are 270 (9×30), while in MLFFN-UB[2] the weights are only 99 (9+3×30).

Although MLFFN[2] is characterised by much more parameters, the utility based approach guarantees the same capabilities in reproducing disaggregate choices. The two models show very similar performances as regard clearly right prediction, and %right, and MLFFN-UB[2] reproduces market shares slightly better.
In conclusion, the utility based MLFFN model does not suffer from the smaller number of parameters. The utility layer, presumably, combined to SE attributes allow a better segmentation of user choices.

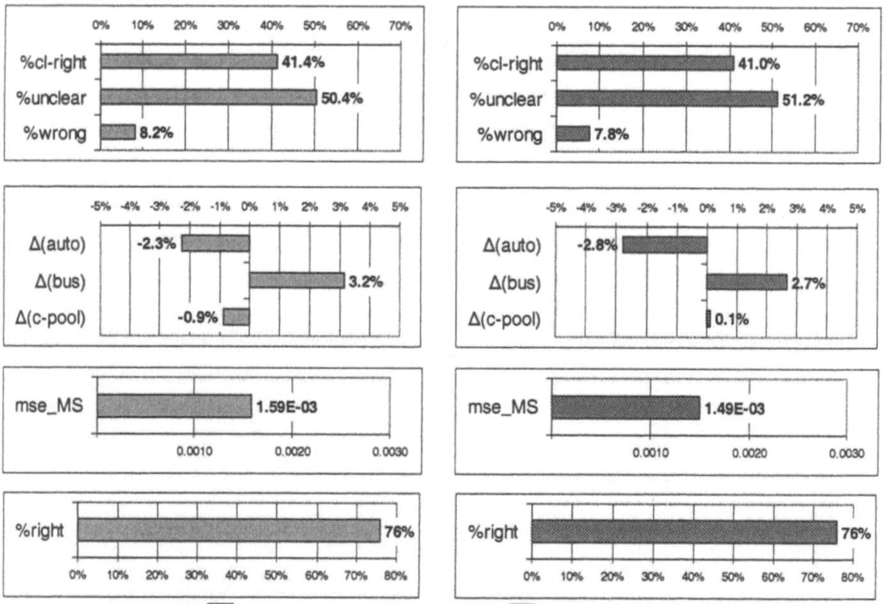


*Figure 11.* MLFFN[2] *vs* MLFFN-UB[2]

### 4.3.3 Comparisons with Random Utility Models (RUM)

The random utility model specified is a Hierarchical Logit model (HL), which overcome the assumption of non-correlation of perceived utility of the Multinomial Logit model, still retaining a closed analytical expression. The expression of the overall choice probability of the generic alternative $p_k$ is obtained as the product of probability $p_{k/g}$ of choosing elementary alternative $k$ within the predefined group (or nest) $g$ containing $m_g$ alternatives (expressed by a MultiNomial Logit model with parameter $\theta$), multiplied by the probability $p_g$ of choosing group $g$ (expressed by a MultiNomial Logit model with parameter $\theta_o$). The choice probability for alternative $k$ is:

$$p_k = p_{k/g} \cdot p_g = \frac{exp(v_k / \theta)}{\sum_{m_g \in g} exp(v_{m_g} / \theta)} \cdot \frac{exp(\delta Y_g)}{\sum_G exp(\delta Y_g)}$$

where:

$Y_g = ln \sum_{m_g \in g} exp(v_{m_g} / \theta)$ is the so called inclusive utility also known as logsum variable;

$\delta = \theta / \theta_o \in [0,1]$ is the ratio of parameters $\theta$ and $\theta_o$ associated to the two choice levels. Parameters to calibrate are those with utility function, say $\beta_j$, plus parameter $\delta \in [0,1]$.

Other different RUMs may be proposed (such as Probit or Mixed-Logit), they have not been considered since they do not provide significant improvements of model effectiveness, but their calibration is much more computer demanding (requiring simulation) rather than closed-form ones.

In our case study, the utility structure and the choice mechanism corresponding to a single-level HL models are represented by the choice tree shown in table 8, where, car-pool and bus perceived utility are correlated, as confirmed by results. The main characteristics of the model specified are briefly described below.

*Table 8.* Main characteristics of HL models

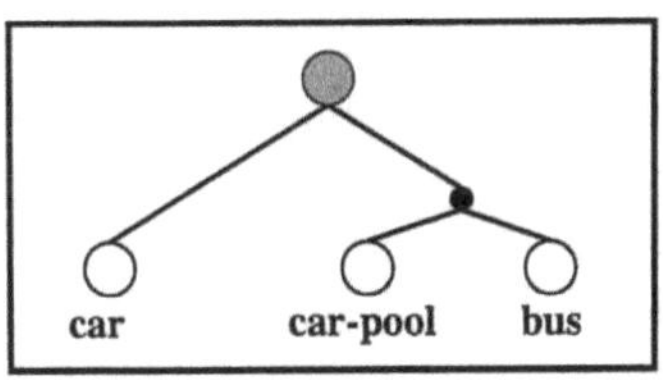


| Survey =944 | HL[1] | HL[2] |
|---|---|---|
| Attributes | *LoS+ASA* | *Los+SE+AC+ASA* |
| Input attributes | 8 | 11 |
| Output | 3 | 3 |
| Calibration set (75%) | 700 | 700 |
| Validation set (25%) | 244 | 244 |
| *Log-likelihood* ($\beta = 0$) | -769 | -769 |
| *Log-likelihood* ($\beta = \beta^*$) | -485 | -451 |
| $\rho^2$ | 0.36 | 0.41 |
| $\delta$ | - | 0.8 |
| $\beta$ *- trip monetary cost* | *-1.14* | *-1.18* |

All coefficients are statistically significant and assume realistic values (table 9). In this case the role of ***SE*** and ***AC*** attributes is relevant.

*Table 9.* Calibration results for HL models

| | Attribute | *X* | Mode | HL[1] ratio $\beta_X/\beta_{cost}$ | HL[2] ratio $\beta_X/\beta_{cost}$ |
|---|---|---|---|---|---|
| *LoS* | Time | Trip time | *car, car-pool, bus* | 0.37 | 0.42 |
| *SE* | Gen | 1 if gender is male | *car* | - | 1.09 |
| *AC-LU* | $ACT_{lenght}$ | Activity time length | *car* | - | 1.38 |
| *AC-LU* | Freq | Weekly trip frequency | *car-pool* | - | 0.22 |
| *ASA* | ASA | Alternative specific attribute | *car* | 1.73 | 2.70 |
| *ASA* | ASA | Alternative specific attribute | *bus* | 1.78 | 0.37 |

The comparisons between the MLFFNs and RUMs have been carried out following the criteria introduced and already used in the previous sub-sections: reproducibility and interpretability.

As regard reproducibility, the main evidences are reported in table 10 and figure 12. Both MLFFNs (MLFFN[2] and MLFFN-UB[2]) clearly outperforms HL model as regards %clearly right, with similar %clearly wrong predictions. As regards market shares prediction, the MLFFNs show slight better capabilities. This result is noteworthy since one of the main features of RUMs is the capability to reproduce aggregate market shares. Analogous considerations may be carried out by analyzing %right and FF indices. The results and analysis proposed confirm the goodness of MLFFN approach.

*Table 10.* Main comparison indices for HL model

| | HL[2] | MLFFN[2] | MLFFN-UB[2] |
|---|---|---|---|
| $\%_{clearly\ right}$ | 13.1% | 41.4% | 41.0% |
| $\%_{unclear}$ | 81.6% | 50.4% | 51.2% |
| $\%_{clearly\ wrong}$ | 5.3% | 8.2% | 7.8% |
| $[P_{sim}-P_{real}]^{car}$ | -2.1% | -2.3% | -2.8% |
| $[P_{sim}-P_{real}]^{c\text{-}pool}$ | 1.4% | -0.9% | +0.1 |
| $[P_{sim}-P_{real}]^{bus}$ | 1.1% | +3.2% | +2.7% |
| MS_mse | 2.13E-3 | 1.60E-3 | 1.40E-3 |
| FF | 61% | 69% | 69% |
| %right | 73% | 76% | 76% |

As regard interpretation, an analysis of weight values is proposed for both models, MLFFN-UB[1] and MLFFN-UB[2], each one compared with the correspondent HL. It is not worthless to notice that the two approaches, MLFFN vs HL, are characterized by a significantly different number of parameters.

With reference to MLFFNs, in table 11 reports the ratio between weight associated to travel time and weight associated to travel cost for each transport mode. It should be remembered that ***LoS*** attributes are generic for HL, thus the correspondent ratios do not depend on the mode; an attempt has been made to calibrate HL models with mode specific ***LoS*** coefficients, but obtained results are rather poor: values of coefficients do not even show the expected signs.

To allow a comparison, for MLFFN-UBs, it has been computed a weighted mean ratio by averaging the time/cost ratios, each multiplied by the correspondent market share ($P_k$):

$$\textit{Mean}\ [\beta_T / \beta_C] = \Sigma_k\ [\beta_{Tk} / \beta_{Ck}] \cdot P_k \qquad \forall k \in I_M$$

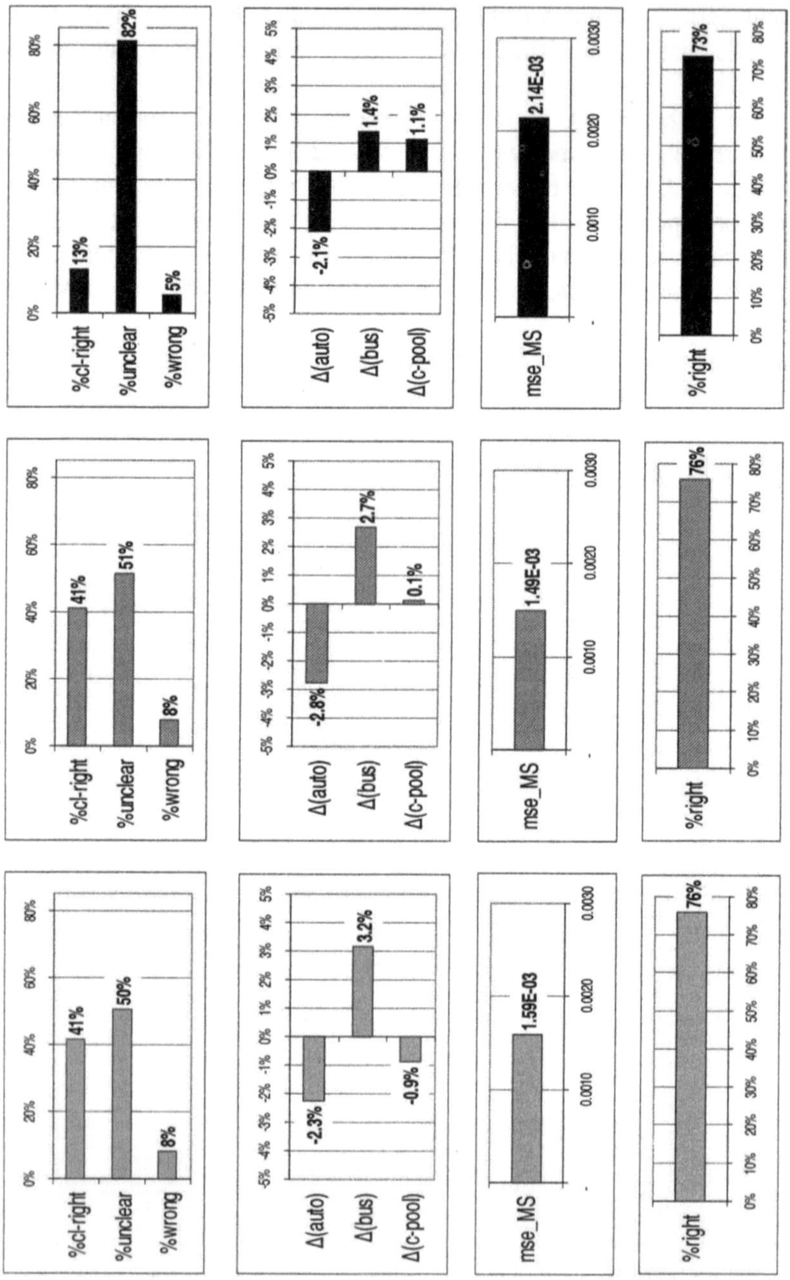


*Figure 12*. MLFFN[2] vs MLFFN-UB[2] vs HL[2]

*Table 11.* Comparisons between parameters ratios for HL and MLFFN models

| | | HL[1] | MLFFN-UB[1] | | | HL[2] | MLFFN-UB[2] | | |
|---|---|---|---|---|---|---|---|---|---|
| | **Attribute** | **ratio** $\beta_X/\beta_{cost}$ | $\beta_X/\beta_{cost}$ | | | $\beta_X/\beta_{cost}$ | $\beta_X/\beta_{cost}$ | | |
| | | generic | car | bus | c-pool | generic | car | bus | c-pool |
| ***LoS*** | Time | 0.37 | 0.17 | 0.34 | 6.0 | 0.42 | 0.11 | 7.29 | 4.5 |
| ***SE*** | Gen | - | - | - | - | 1.09 | 0.15 | - | - |
| ***AC-LU*** | $ACT_{lenght}$ | - | - | - | - | 1.38 | - | - | 0.36 |
| | Freq | - | - | - | - | 0.22 | 0.15 | - | - |

Analysing models with ***LoS*** attributes only, [1], it can be noted that the value of time of HL[1], 0.37 €/h, is almost equal to the value of time for bus in MLFFN-UB[1], but is twice the value for car and much lesser than the value for car-pool. This difference still remains significant when it is compared with *Mean* $[\beta_T/\beta_C]$ = 1.14 €/h (≈ 3×0.37). This result is consistent with considerations already made in sub-section 4.3.1.

Analysing models with ***LoS*** and ***SE*** attributes, [2], the value of time of HL[2], 0.42 €/h, is greatly different of values in MLFFN-UB[2], four times greater than the value for car, and at least ten times lesser than the value for bus and car-pool. This difference is less significant when it is compared with *Mean* $[\beta_T/\beta_C]$ = 0.98 €/h (≈ 2.3×0.42). Not surprisingly the values for ***SE*** attributes are quite different, in particular, those for gender and activity length are much lesser.

All these considerations suggest that, as already pointed out, the role of ***SE*** is relevant; in addition it seems that MLFFN-UBs support an explanation of observed choices based more on ***LoS*** attributes than ***SE***. This issue is surely worth of further analysis based on more experimental evidences. The case of MLFFN-UBs with generic ***LoS*** attributes will be addressed in a future paper.

# 5. CONCLUSIONS AND RESEARCH PERSPECTIVES

Recently attempts have been made to apply MLFFNs to travel demand analysis. The first papers on this topic followed an aggregate approach trying to simulate directly demand flows, thus resulting into not very effective models. More recently, Hensher and Ton (2000) have suggested a disaggregate approach trying to simulate user choices as in consolidated modeling approaches to travel demand analysis, based on random utility models (RUMs). In a previous paper (Cantarella and de Luca, 2004) the authors have described in details how a MLFNN can be used for choice modeling to support travel demand analysis, with rather effective results when compared with RUMs. Still, MLFFNs, adopted in literature, should be considered black-box models that do not allow any kind of parameter interpretation, thus making rather arguable their application to support demand forecasting for project appraisal.

This paper has presented utility-based MLFFN models (MLFFN-UBs) different of those existing in literature, since they explicitly include utility specification through an intermediate layer with a processing unit for each alternative. This way, both the utility function and the choice function are explicitly and separately specified, as in RUMs within an econometric framework. Moreover, parameters of the utility function may be given an interpretation useful for project appraisal. Calibrated utility parameters for the analyzed real case study show signs as expected and reasonable values.

The proposed MLFFN-UBs architecture has fewer parameters than other MLFFNs, still the utility-based approach guarantees the same capabilities in reproducing disaggregate choices and shows better performances in market shares simulation. It is worth noting that the more the parameters are the more likely the over-fitting may raise up. With regard to the analysed case study, proposed MLFFN-UBs outperform RUMs.

So far reported results support the use of MLFFN-UBs both from a theoretical and applicative point of view: MLFFN-UBs are able to generalize satisfactorily mode choices, simulating aggregate mode shares as well as single user mode choice, and may outperform a (closed form) RUM with same attributes.

Object of a future paper will be a formal analysis of conditions (if any) assuring that a duly defined MLFFN-UB may include as a special case a Logit model (or other RUMs) and an in-depth comparison with fully connected MLFFNs with respect to common mathematical features. Some issues remain open, such as the analysis of mathematical features, continuity and monotonicity, of choice fractions from MLFFN-UBs against utility values, and how user surplus, useful for project evaluation, may be estimated within a MLFFN-UB approach. Calibration method seem worth of further research work, regarding constraints on parameter values, for including generic utility parameters consistently with an econometric frameworks, as well as different calibration function, allowing a statistical analysis of parameters estimation.

*An early and shorter version of this paper has been presented at the Fourth International Symposium on Uncertainty Modeling and Analysis (ISUMA 2003), Maryland, USA.*

# ACKNOWLEDGEMENTS

*Research partially supported by the Ministry of Education, University and Research of Italy (EU) under grant n. 2003089420_004/2003, and by the University of Salerno, Italy (EU) fund year 2003 and 2004.*

# REFERENCES


Ben Akiva, M., Lerman, S., 1985, Discrete choice analysis theory and application to travel demand. MIT press, Cambridge, MA.

Cantarella G.E., de Luca S. 2002. Artificial neural networks vs random utility models for simulating transportation mode choice behaviour, Proceedings of the 13th Mini-EURO Conference, Bari, 2002.

Cantarella G.E., de Luca S. 2004. Multilayer feedforward networks for Transportation mode choice analysis: an analysis and a comparison with random utility models, technical report, Department of Civil Engineer, University of Salerno (submitted to Trans. Research C).

Cascetta, E., 2001, Transportation systems engineering: theory and methods. Kluwer Academic Publishers, Dordrecht, The Netherlands.

Hensher, D., Ton, T.T., 2000. A comparison of the predictive potential of Artificial Neural Networks and Nested Logit models for commuter mode choice. Transportation Research 36E, 155-172.

Mozolin, M., Thill, J.C. and Lynn Usery, E., 2000. Trip distribution forecasting with multiplayer perceptron neural networks; a critical evaluation. Transportation Research 34B, 53-73.

Nijkamp, P., Reggiani, A., Tritapepe, T., 1996. Modelling inter-urban transport flows in Italy: a comparison between neural network analysis and Logit analysis. Transportation Research 4C, 323-338.

Train, K., 2002, Discrete Choice Methods with Simulation. Cambridge University Press.

Schintler, L.A., Olurotimi, O., 1998. Neural networks as adaptive Logit models. Neural networks in Transport systems, Himanen, V., Nijkamp, P., Reggiani, A., (eds.), Ashgate, Brookfield.

Reggiani, A., Tritapepe, O., 1998. Neural networks and Logit models applied to commuters' mobility in the Metropolitan area of Milan. Neural networks in Transport systems, Himanen, V., Nijkamp, P., Reggiani, A., (eds.), Ashgate, Brookfield.

Shmueli, D., Salomon, I., Shefer, D., 1996. Neural network analysis of travel behavior: evaluating tools for prediction. Transportation Research 4C, 151-166.

Shmueli, D., 1998. Applications of neural networks in transportation planning. Progress in Planning 50, 141-204.

# Chapter 17

# HETEROGENEITY IN COMMUTER DEPARTURE TIME DECISION: A PROSPECT THEORETIC APPROACH

**Metin Senbil and Ryuichi Kitamura**

## 1. INTRODUCTION

Commuter departure time choice analysis has gained importance in the last twenty years. Underlying this development is the desire to better understand the behavioral mechanisms behind peak-period road congestion. If these behavioral mechanisms are better understood, congestion relief measures may be better coupled with commuters' decision processes, thus will be more effective in achieving their objectives.

The commuter departure time studies up to now can be divided into two mutually exclusive groups with respect to the behavioral paradigms they adapt (see Senbil and Kitamura, 2003). The first group of studies has been based on the notion of random utility maximization, and implicitly assumes the network conditions are known to commuters (e.g., de Palma et al. 1983, Hendrickson and Planck 1984, Mahmassani and Herman, 1984). The second group of studies relaxes this assumption considerably and acknowledges the roles of heuristics in human decision, and adopts the concept of bounded rationality proposed by Simon (1955) (e.g., Mahmassani and Chang, 1987; Mahmassani and Jou, 1998; Mahmassani and Liu, 1999; Jou, 2001; Jou and Kitamura, 2002).

This second group of researches has generally emphasized the framing of the departure time choice problem where the work start time is a key element. An indifference band has been proposed and applied by Mahmassani and co-researchers to repeated departure time choices in experimental and real settings. Jou and Kitamura (2002) further develop a decision frame by incorporating certain elements of prospect theory (Kahneman and Tversky, 1979) such as zero asset positions for reference points and differentiation of gains and losses. Jou and Kitamura (2002) also propose that the upper bound of the utility of a commute trip materializes when its arrival time coincides with the preferred arrival time. Incorporating this maximum utility as a constraint into the empirical formulation of utility functions, however, is yet to be achieved.

This line of research has been extended by Senbil and Kitamura (2004) to be more compatible with prospect theory. This parallels Avineri and Prashker (2003) who employ cumulative prospect theory and incorporate learning mechanisms. This study conducts an experiment on route choices with random travel times. For this reason, they employ two scenarios: the first scenario is a choice between two routes with different Normally distributed travel times, the second scenario is a change of the first scenario in a way that one of the routes yields two different mean travel times with different dispersion parameters probabilistic (0.5 vs 0.5). The mean travel times of both of the routes are same both of the scenarios. Participants of the experiment are repetitively subjected to one of the scenarios randomly and a selection between two routes has been done by them without no prior information on the means travel times just their accumulated knowledge of the network. This study attempts to improve the cumulative prospect theory, a later development of the original prospect theory, by injecting learning mechanism into the decision process. Avineri and Prashker (2003) report that increasing travel time variability of a route decreases a travelers sensitivity to the route, that is to say, participants show tendency to choose the route with high variance of travel times; thus increasing travel time variability of a route increases its attractiveness.

The study by Senbil and Kitamura (2004) has estimated the parameters of two value functions associated with two decision frames, one of which was originally proposed by Jou and Kitamura (2002). They conclude that prospect theory contribute to a better understanding of commuters' departure time decisions by offering the framing of decisions into gains and losses. The study, however, leaves outside its scope the weight function, another important element of prospect theory. Thus this study has the aim of extending Senbil and Kitamura (2004) on how commuters value outcomes and weight these outcomes in their decision making. In this regard, both observed and unobserved heterogeneity is introduced into the analytical framework to account for possible differences in the value function, which evaluates arrival time at work with respect to reference points. In other words, different commuters are assumed to have different valuations of being late or early, and parts of the differences may be accounted for by their measured attributes (observed heterogeneity), but the rest may be attributable only to unobserved factors (unobserved heterogeneity). For example, some employees might be risk averse and depart fairly early to rule out the possibility of arriving late, while some might be risk seekers who leave their homes at times that might result in late arrivals. These differences in behavior must be accounted for if one wishes to predict behavioral changes in response to a potential policy packet which may induce departure time changes by commuters.

As in Senbil and Kitamura (2004), two alternative decision frames are adopted in this study to represent different ways of editing the departure time choice problem. These frames are different in their definitions of gain and loss regions, and represent two possible ways commuters may view consequences of arrivals at work at different clock times. Weight function is considered within a broader model which we name as contingency adjustment model. In this regard, prior and posterior probabilities are introduced. Prior probability is the expectation of gains as acceptable gains. As this expectation changes to a new one with the realized arrival time. Rigorous econometric methods are applied to a data set that contains three-day records of commuting for each respondent of a survey, conducted in Otsu City, Japan, in 2002. The study reveals properties of an important dimension of departure time choice as a decision under uncertainty, and thereby contributes to the state-of-art in travel decision analysis.

## 2. PROSPECT THEORY

Prospect theory is concerned with risky choices "such as whether or not to take an umbrella and whether or not to go to war" (Kahneman and Tversky, 2000, p.2), which "are made without advance knowledge of their consequences" (Kahneman and Tversky, 2000, p.2). There are also prospect theory studies concerning decisions made under uncertainty such as Tversky and Fox (1995). The basic premise of prospect theory is that the postulates of utility theory developed by Von Neumann and Morgenstern (1947) are incomplete and consequently inapplicable (see Allais, 1953, and Kahneman and Tversky, 1979) in many situations.

For example, it has been found by experiments that the choice depends on the position in which the decision maker is placed and whether he is to gain or to lose, relative to that position, by making a decision. A typical example is given as follows:

Two offers are made to an individual:
*A*: Receive $1,000 with a probability of 0.85
*B*: Receive $800 for sure

Although the expected utility of *A* is larger than that of *B* ($\$1,000 \times 0.85 = \$850 > \$800$), the individual tends to choose *B*, because he tends to avert risk when he is gaining some asset (certainty effect). If these options are turned reverse, taken on the negative scale with the same probabilities so that the individual is to lose with each alternative, then the individual tends to choose alternative *A*. The individual becomes a risk seeker when he is to lose some asset, or, is in a loss region. He then prefers a probabilistic loss to a certain loss, even when the former has a larger expected loss.

In order to address this, and many other flaws of expected utility theory as given in detail by Kahneman and Tversky (1979), they have developed prospect theory. Editing, evalution and choice are subsequent stages of prospect theory. In the editing phase, prospects are organized by coding, combination, segregation and cancellation. The editing phase produces a subset of initial prospects. During coding, prospects are separated into gain and losses using the reference point, a neutral position that "[corresponds to current asset position, in which case gains and losses coincide with the actual amounts that are received or paid" '(Kahneman and Tversky, 1979, p.276). The evaluation phase is characterized with two scaling functions, i.e., the value and weight functions. The value function scales a deviation from an asset position (gain or loss) to a subjective value. The weight function rescales the objective probability to a subjective probability. Finally in the choice phase, individual chooses the prospect which has the highest value.

The properties of the value function (Figure 1) can be summarized as:

1. The domain of value function is defined as the deviation from a reference point, which is the zero asset position.
2. The value function is concave for gains and convex for losses.
3. The value function is steeper for losses than for gains.

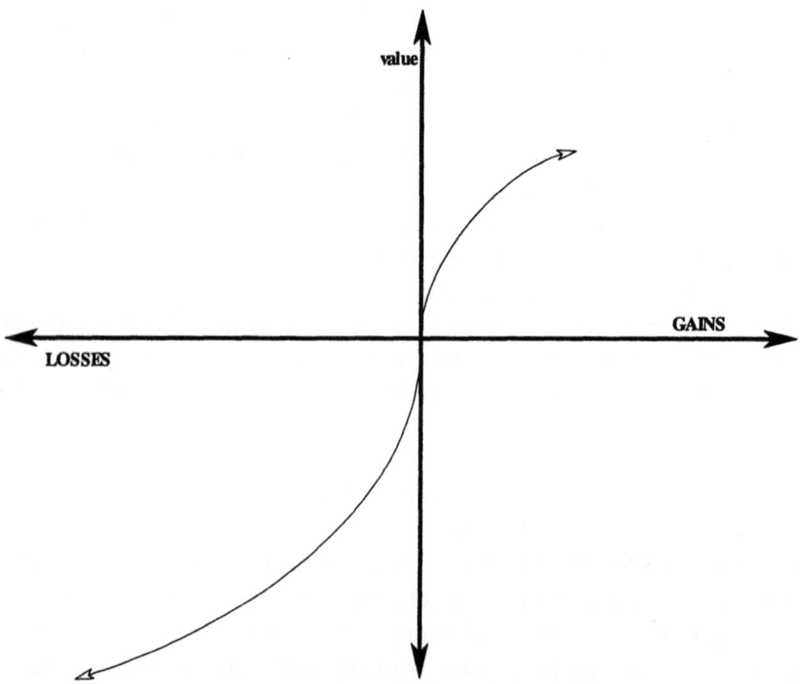


Figure 1. Typical value function

Kahneman and Tversky (1992) propose the following value function:

$$V = \begin{cases} x^{\alpha} & x \geq 0 \\ -\lambda(-x)^{\beta} & x < 0 \end{cases} \qquad (1)$$

where $x$ refers to the outcome expressed as a deviation from the reference point, $\lambda$ is a scale parameter for losses, and $\alpha$ and $\beta$ are parameters for the gain region and loss region, respectively.

The other major element of prospect theory, the weight function "measures the impact of events on the desirability of the prospects, and not merely the perceived likelihood of these events" (Kahmenan and Tversky, 1979, p.?). The range of weight function, $\pi$, is normalized to the same range of probability. The properties of the weight function are:

1. Weight function is an increasing function in its domain.
2. If p is low $\Rightarrow \pi(p) > p$.
3. If p is high $\Rightarrow \pi(p) < p$.
4. Subcertainty: $0 < p < 1 \Rightarrow \pi(p) + \pi(1-p) < 1$.
5. Subproportionality: $0 < p,q,r < 1 \Rightarrow \dfrac{\pi(pr)}{\pi(p)} < \dfrac{\pi(pqr)}{\pi(pq)}$.
6. Subadditivity *a*: $\pi(p+\Delta) - \pi(p) \leq \pi(1) - \pi(1-\Delta)$
7. Subadditivity *b*: $\pi(\Delta) - \pi(0) \geq \pi(p+\Delta) - \pi(p)$

The second and third properties of the weight function refer to overweighting for small probabilities and underweighting for large probabilities. Overweighting is evident in the case of lottery tickets, and represents risk seeking behavior in case of gains and risk aversive behavior in case of losses. The fourth property, subcertainty, implies that the sum of complementing probabilities, when scaled by the weight function, falls short of 1.0. Subproportionality represents decreasing relative sensitivity; in other words, "for a fixed ratio of probabilities, the ratio of the corresponding decision weights are closer to unity when the probabilities are low than when they are high" (Kahneman and Tversky, 1979, p.282). The subadditivity feature of the weight function implies that a difference in probability values affects the weight more at the extreme ends of the probability scale (i.e., 0 or 1) than in medium probability ranges.

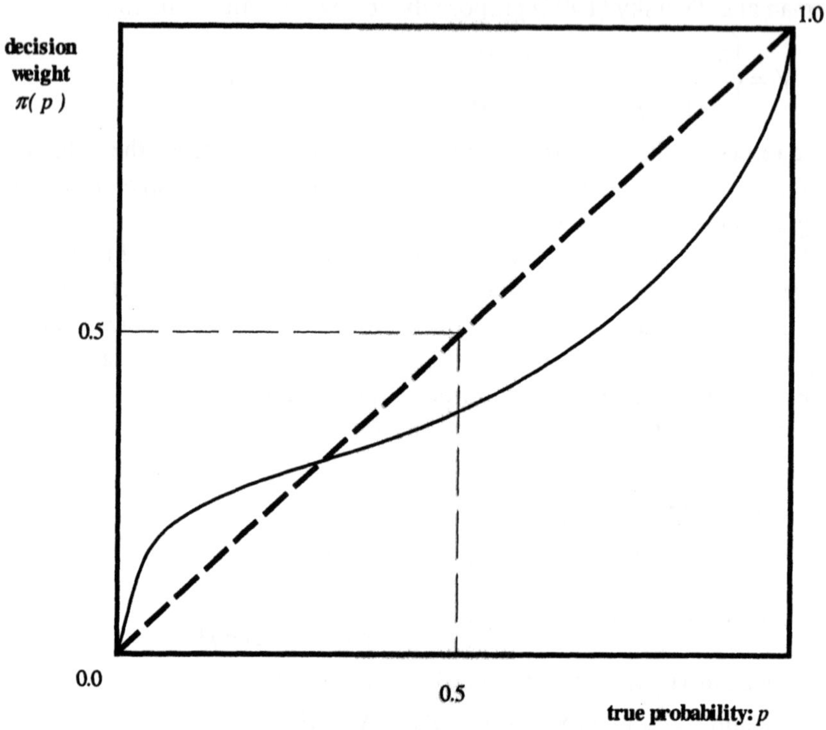


Figure 2. Typical weight function

A functional form of the weight function is given as

$$\pi(p)=\frac{p^{\gamma}}{\left(p^{\gamma}+(1-p)^{\gamma}\right)^{\frac{1}{\gamma}}} \tag{2}$$

by Kahneman and Tversky (1992) with a single parameter γ. Another two-parameter specification is proposed by Prelec (1998) as

$$\pi(p)=\exp\left(-\lambda\left[-\ln(p)\right]^{\theta}\right) \tag{3}$$

Different specifications of weight function including those given by Eq. 2 and 3 are reviewed in Bleichrodt and Pinto (2000) who propose a parameter free weight function estimation.

Kahmeman and Tversky estimate the parameters of value and weight functions by using data obtained from controlled experiments in which subjects are presented with a variety of packets of risky prospects by the rules-of-thumb proposed by Tversky and Kahneman (1992). In their numerous studies, Kahneman and Tversky emphasize the importance of experiments in the analysis of decision making by individuals as they offer information that cannot be captured in any of the conventional data collection methodologies, e.g., panel surveys, time series observation, and cross section surveys. By offering subjects risky prospects vs. certain prospects simultaneously and having them choose one from risky prospect and one from certain prospects between which they are indifferent, Kahneman and Tversky computes the certainty equivalents of risky prospects and consequently estimate the parameter values of both value and weight functions by using indifference relationships. By using risky prospects, $a$ and $b$ having certainty equivalents $c$ and $d$ respectively, we can give some of these algebraic relationships as follows:

1. $$\left.\begin{array}{l} V(a)\pi(p) \approx c \\ V(b)\pi(p) \approx d \end{array}\right\} \frac{V(a)}{V(b)} = \frac{c}{d},$$

2. $$\left.\begin{array}{l} V(a)\pi(p_1) \approx c \\ V(a)\pi(p_2) \approx d \end{array}\right\} \frac{\pi(p_1)}{\pi(p_2)} = \frac{c}{d}.$$

The parameters of the value function are estimated by regressing the risky prospects on their corresponding certainty equivalents. The parameters of the weight function, on the other hand, are determined by regressing the probability of a prospect on the ratio of certainty equivalent to the prospect.

# 3. DECISION FRAMING

A commuter's choice of departure time is contingent on the arrival at the workplace relative to reference time points that constitute the framework for his decision (Jou and Kitamura, 2002). Namely, a departure time is assumed to be chosen using reference time points around expected arrival times. Although there are many studies in the transportation literature that recognize the uncertainty and risks involved in commuting (for example, Bonsall, 2001, Avineri and Prascker, 2003), there has been no rigorous examination of how the commuter views the possibilities of being late or early for work. Also, there have not been many studies that recognize the effect of framing on decisions (Fujii and Kitamura, 2001, 2004, are exceptions that emphasize decision framing).

Reference points are established in time through experience and as well as being imposed as constraints (e.g., work starting time). A learned commuter is expected to know the variability as well as the average of commute travel times associated with a particular departure time, because he must have experienced various events that may occur, such as congestion, accidents, road maintenance work, etc. If relevant information becomes available prior to departure by, e.g., television or radio, a learned commuter will be able to predict with improved accuracy how long his usual commute will take. If the usual departure time and route is not satisfactory in his decision frame, the commuter may change his departure time; if this shift does not seem satisfactory, he may switch his route to take another route. Decision process does not end with this. Once departed from home for work, the remaining portion of the journey might also be subject to sporadic decisions on switching to another route in the case, lane changes as well as speed changes with respect to the relative arrival expectation relative to the reference points.

It is proposed in this study that coupling constraints (Hägerstrand, 1971) for work activity constitute two reference points along the time axis, i.e., the earliest acceptable arrival time (EAT), before which the commuter feels loss, and the latest tolerable arrival time (LAT), after which he also incurs loss. It is further proposed that the commuter holds a preferred arrival time (PAT). Arriving at work at PAT achieves the highest utility for the commuter. It is reasonable to expect that PAT is before LAT, providing the commuter with some time to adjust or prepare himself for the work environment. It is in general assumed that the commuter is satisfied when he arrives between EAT and LAT. The gain region is assumed to lie between EAT and LAT. Outside the gain region is the loss region; the commuter is not satisfied with an arrival that falls in this region.

## 3.1. Symmetric Decision Frame

Symmetric decision frame (Figure 3) is experienced when gain nad loss regions are symmetric about PAT. In this decision frame, PAT is not devised as a reference point; rather it represents the commuter's preferences that arriving at this time point achieves the highest value. The effect of PAT on commuter behavior might be more prominent when the commuter is still motivated to achieve an arrival at PAT even after he is guaranteed to arrive within the gain region. PAT might be regarded as a pseudo-reference point (or a secondary reference point) such that when a commuter secures a gain arrival, he might be motivated to achieve an arrival at PAT, thus might feel petty loss in the gain region. In this regard, PAT is introduced as a *pseudo* reference point because it is not a threshold between gain and loss regions, thus is not treated as a reference point in the study, but still is expected to act as a singular time point in departure time decision as noted above. Arrivals are classified into two groups with respect to PAT; those before PAT are early arrivals, and those after PAT late arrivals.

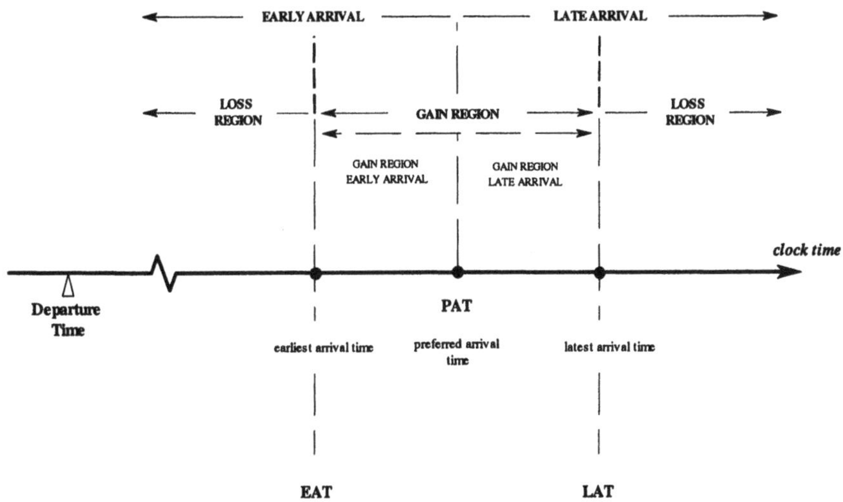


Figure 3. Symmetric decision frame

## 3.2. Asymmetric Decision Frame

In the second approach (Figure 4), gains are assumed to exist between PAT and LAT as before. The amount of gain increases as one moves from LAT to PAT. Losses are assumed to materialize when the commuter arrives before EAT or after LAT. The treatment of the region between EAT and PAT is different from the first approach.

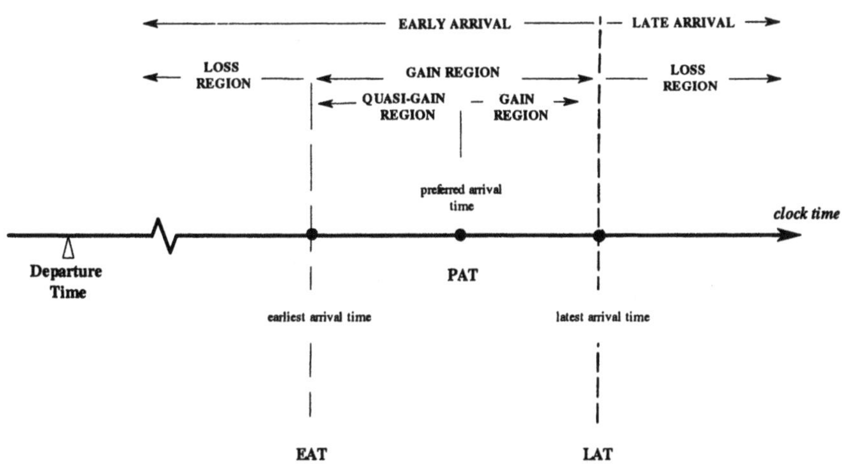


Figure 4. Asymmetric decision frame

The second approach is motivated by the consideration that the gain may be viewed differently by the commuter between early and late arrivals. In the symmetric decision frame discussed above, gain is assumed to be monotonically increasing toward PAT. In the problem editing of the second approach, the nature of gain is differentiated before and after PAT. When the arrival is after PAT, the mount of gain increases according to a concave value function as one moves away from reference point LAT back toward PAT.

When the commuter arrives between EAT and PAT, on the other hand, he may focus on the fact that he could have arrived at PAT by leaving home later, which is feasible under usual circumstances for the morning commute. In this sense, the commuter may focus on the decline in gain caused by arriving earlier than PAT. Namely, he would evaluate an early arrival with respect to the difference between the gain of arriving at PAT and the gain at the earlier arrival time, and this difference is likely viewed as a loss. If this is the case, then, the value function should behave as if in a loss region, and therefore should be convex. This region is thus named the "*quasi*-gain region." The commuter is assumed to feel absolute loss if his arrival is earlier than EAT.

In sum, the rationale for establishing the quasi-gain region is as follows. When the commuter evaluates an arrival between EAT and PAT, he feels gains increasingly as he approaches to PAT, which has the maximum value. Thus, the value function is concave between EAT and PAT on the assumption that the gain of an arrival is evaluated with respect to the decline in gain from that at PAT. The region between EAT and PAT thus represents perceived losses in the second approach.

## 4. THE DATA

The data are compiled from a survey that was conducted in Otsu City in Shiga Prefecture, Japan, in May 2002. The questionnaire was mailed to one thousand randomly selected resident drivers, and 260 of them completed questionnaires and sent them back (response rate = 0.26).

The survey consists of three parts which respectively addressed:

i. General commute information:
   a. Average, longest and shortest commute durations,
   b. Work start time,
   c. Latest tolerable time for arrival at the work place,
   d. Tolerance by managers on tardiness.

ii. Respondents' socio-economic and demographic characteristics:
   a. Sex, age, marital status, annual income and employment type,
   b. Existence of children below 15 years of age.

iii. Commute characteristics on three consecutive survey days:
   a. Whether the commute was a usual one or not,
   b. Whether the route used was same as the previous one or not,
   c. Departure time from residence and arrival time at the work place,

d. Whether the respondent would depart at the same time in the next commute or not,
e. Whether the respondent would use the same route in the next commute or not.

After screening for item non-responses on commuting information, the number of respondents in the sample decreased from the initial 260 to 226. Summary statistics of the data set are given in Table 3.

When we have a close look at the individual attributes, we notice two points to report. The first one of the ratio of car commuters within the sample: the sample is dominated with the male car commuters: 74% of 226 commuters while the same ratio of male car commuters is higher than the similar ratio (69 %) of 1788 car commuters in Otsu city in a larger household travel survey, the Kei-Han-Shin household travel survey for 2000 and the second one is the average age of the drivers on the sample. The second point is the average age of the commuters: which is approximately 50: thus we can expect that most of the commuters are experienced in driving and most of them have developed comprehensive driving as well as commute habits. Thus the information that they supply for their commutes in our survey are result of at least 5 years of experience.

Reported LAT either coincided with WST or was before that time for 64% of commuters. Approximately half of these commuters-LAT≤WST (or 31% of all commuters) had to arrive at the work place before WST, i.e., LAT was before WST. . Thus, taking WST as a reference point might create serious problems in applying prospect theory to the case of departure time choice. Besides, the location of PAT is assumed to be between EAT and LAT in proposed decision frames in this study. But there are seven cases that have PAT later than LAT: PAT>LAT and this number increases to 28 when we include the cases where PAT coincides with LAT. To estimate the value functions based on the hypothesized decision frames, we have to eliminate these cases (28 in total). Because we assume monotonically increasing function(s) in between EAT and LAT and meet at PAT which is the highest value point for an arrival.

The mode of commute trip durations on three consecutive days in our survey is consistent with the larger household travel survey data at 30 min. Most of the respondents in our survey reported their official work start times as 8:30 AM, and departure times from home mostly one hour before work start times (the mode is at 7:30 AM).

| | | N | Minimum | Maximum | Mean | Std. Deviation |
|---|---|---|---|---|---|---|
| Individual characteristics | Sex (1 indicates male respondent) | 225 | 0 | 1 | .74 | .44 |
| | Age | 225 | 24 | 80 | 49.88 | 11.63 |
| | Income [a] | 175 | 50 | 2700 | 633.24 | 386.49 |
| | Marital status (1 indicates married person, 0 indicates else marital statuses) | 220 | 0 | 1 | .87 | .33 |
| | Children below 15 (1 indicates existence of children below 15 years of age, 0 indicates else situations) | 215 | 0 | 1 | .28 | .45 |
| General Commute Information | Longest commute duration (mode= 60) [b] | 225 | 10 | 300 | 67.22 | 43.70 |
| | Shortest commute duration (mode= 20) [b] | 225 | 4 | 180 | 31.67 | 20.30 |
| | Average commute duration (mode= 30) [b] | 225 | 6 | 210 | 39.36 | 23.95 |
| | Preferred Arrival Time (mode= 8:30) [c] | 225 | 3:50 | 16:40 | 8:44 | 1:22 |
| | Work Start Time (mode= 8:30) [c] | 225 | 4:00 | 17:00 | 8:57 | 1:20 |
| | Latest Arrival Time (mode= 9:00) [c] | 225 | 5:30 | 18:00 | 9:14 | 1:34 |
| | Managers are tolerable to tardiness is indicated by 1. | 224 | 0 | 1 | .60 | .49 |
| | Flex-time time policy ( 1 indicates Flex-time policy is being practiced at the workplace, 0 indicates else situation). | 223 | 0 | 1 | .25 | .43 |
| Commute trips on three consecutive days | Departure Time (mode= 7:30) [d] | 226 | 2:00 | 16:10 | 7:57 | 1:24 |
| | Arrival Time (mode= 8:20) [d] | 226 | 3:00 | 16:23 | 8:36 | 1:21 |
| | Commute duration (mode= 30) [e] | 226 | 10 | 150 | 37.98 | 21.05 |
| | Usual commute (1 indicates a usual commute, 0 indicates else) | 226 | 0 | 1 | .96 | .19 |
| | Same route (1 indicates that same route has been used as the previous time, 0 indicates else) | 226 | 0 | 1 | .92 | .28 |
| | Same Departure time next day? (1 indicates that the same departure time will be used next time, 0 indicates else) | 226 | 0 | 1 | .77 | .42 |
| | Route change next day? (1 indicates a different route will be used next time, 0 indicates else) | 226 | 0 | 1 | .15 | .35 |

a. in 10000 Yen units.

b. in minutes.

c. clock time (24 hours)

d. clock time (24 hours)

e. in minutes.

Table 1. Profiles of the Data Set

An important point that should be highlighted is in the last section of Table 1. Almost all of the respondents (96%) replied that their commutes were usual ones. At the same time, only approximately three quarters (77%) of the respondents reported they would not change departure times on the next occasions. The result suggests that, even when commuters are making "usual commute" to usual work places for usual work starting times, about one quarter of commuters are changing their departure times. Also interesting is the result that 92% of the respondents indicated that they commuted on the same route as the previous commute, while 15% indicated that they would take a different route on the next occasion. The two responses are not mutually consistent and indicates that the respondents tended to overstate their intentions to switch commute routes in the survey.

# 5. CONTINGENCY ADJUSTMENT MODEL

Although it is unrealistic to assume that a commuter is aware of the mathematical probability distribution function of arrival times of his commute trips, it is reasonable to expect that he has a fairly accurate estimate of arrival time given a departure time. Through experience, the commuter has probably acquired an assessment of the variation in trip durations, and he bases his predictions of travel times on this accumulated knowledge. At the same time, it is anticipated that his departure time choice is also influenced by day-to-day variations in commute trip duration. In this section, we propose a procedure, which shall be called the "contingency adjustment model," to estimate the weight function based on the relationship between the long-term distribution and short-term realizations of commute trip durations. Before presenting the model, however, discussions are due on the method adopted in this study to estimate the perceived distribution of trip durations that a commuter has.

The assumptions postulated in this study is that the commuter has accurate perceptions of the shortest, average and longest commute trip durations, given the departure time. In other words, we assume that the commuter who has made a considerable number of commute trips has well-founded perceptions of the mean trip duration and its dispersion expressed in terms of the shortest and longest trip durations. Furthermore, since the duration of a trip is influenced by numerous random elements, it is reasonable by the central limit theorem to assume that commute trip durations have a normal distribution (Figure 5). The approach taken in this study, then, is to estimate the mean and variance parameters of the normal distribution based on the shortest, average and longest commute trip durations as reported by a commuter in a survey.

Communicating the concept of travel time variability in a survey is not a trivial task (Cook, et al., 1999). Commuters may perceive the variability in commute trip duration in terms of quantities other than the shortest and longest trip durations. For example, a commuter may have the perception, "It normally takes 30 minutes, but takes more than 45 minutes about once a week" (see, e.g., Abdel-Aty et al., 1995) The approach taken in this study is based on the consideration that survey respondents will easily be able to indicate shortest and longest durations of their commutes. Although one would ideally ask for their estimates of the variance in travel times, the concept of variance is far too alien a notion for typical respondents to offer an estimate. On the other hand, questions addressing the frequencies of delays as in the above quote, are unlikely to be relevant to all respondents without customization. The approach adopted in this study is thus a compromise under limitations in survey administration.

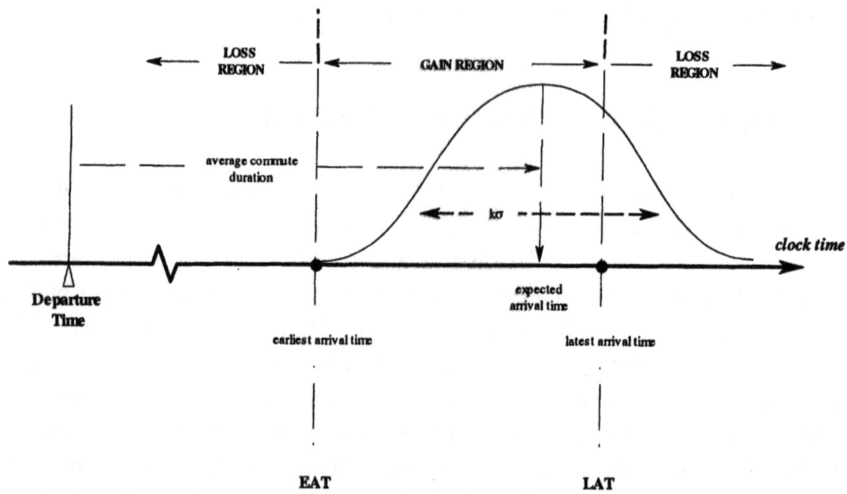


Figure 5. Arrival time distribution conditioned on departure time

The normal curve in Figure 5 shows the distribution of arrival times given the departure time, average trip duration, and the variance of trip durations. In order to construct a measure of the perceived variance of trip durations for each respondent, we introduce the assumptions that the perceived distribution of trip durations also has a normal distribution, and further that its standard deviation, σ, is proportional to the difference between the longest and shortest trip durations as indicated by the respondent. Suppose the reported shortest duration corresponds to the 2.5 percentile value and the longest duration to the 97.5 percentile value, and therefore the range defined by the two covers the central 95% of the distribution. If these assumptions are agreeable, then we would use the relation that the difference between the shortest and longest trip durations equals to 3.92 times the standard deviation. With this relation, the standard deviation of the perceived distribution of trip durations can be estimated for each respondent.

Let the perceived average, longest and shortest trip durations be *at*, *lt*, and *st*. Also let the departure and arrival times denoted by *DT* and *AT* for each commute. Then, the perceived distribution of arrival times is

$$\mathrm{N}\left(DT + at, [(lt - st)/3.92]^2\right) \tag{4}$$

With this distribution, the probability, $\tilde{p}_0$, that the commuter will arrive between EAT and LAT can be evaluated and used as a measure of his risk aversiveness. The complement of this probability, $\tilde{q}_0 = 1 - \tilde{p}_0$, serves as a measure of his risk proneness. Namely, letting $f(t|DT)$ be the normal probability density function of perceived arrival times, given departure time *DT*, we have

$$\begin{cases} \tilde{p}_0 = \int_{EAT}^{LAT} f(t\,|\,DT)dt \\ \tilde{q}_0 = 1 - \tilde{p}_0 \end{cases} \tag{5}$$

where *t* represents clock time.

Unfortunately, the earliest arrival time is not available in the data set. EAT is therefore estimated by adding the reported shortest commute duration to the departure time. With this estimation, it is possible to obtain an EAT that is placed after a LAT when the commuter chooses a departure time that results in an arrival later than the LAT at the earliest. This may be interpreted as a case where the commuter is oblivious of the risk of being late. In this situation, $\tilde{p}_0$ and $\tilde{q}_0$ cannot be defined by Eq. (5).

The discussions so far of this section aid in the formulation of the weight function. If the perceived distribution of trip durations can be represented as discussed above, then $\tilde{p}_0$ defined as in Eq. (5) serves as a measure of the subjective probability of gain which the commuter will perceive prior to the departure at *DT*. Thus this probability shall be called the "expected gain probability." Now, suppose the commuter arrives at work at time *AT*. Recall that *AT* could not have been determined beforehand. Given *AT*, the commuter may have a posterior assessment of the probability of gain in which the arrival time, *AT*, is, after the fact, viewed more as a fixed constant than a random variate. The commuter may conceive a posterior distribution of arrival times which is centered at the realization arrival time, *AT*. He may then reassess the probability of gain associated with the departure time using this posterior distribution, i.e.,

$$\begin{cases} p' = \int_{EAT}^{LAT} g(t \mid AT)dt \\ q' = 1 - p' \end{cases} \tag{6}$$

where $g(t|AT)$ is the posterior distribution of arrival times given $AT$. We shall call $p'$ the "realized gain probability."

In probing the relationship between the two, we shall introduce another proposition that the relationship between the expected gain probability, $\hat{p}$ as a prior probability, and the realized gain probability, $p'$, as a posterior probability, corresponds to the relationship between the objective (or "true") probability, $p$, and the decision weight, $\pi(p)$. If this proposition is accepted, then the weight function discussed in Section 2 can be obtained by regressing $p'$ on $\hat{p}$

The rationale behind the above proposition is as follows. As noted earlier, typical commuters repeat trips to the same work place about the same time of the day over and over. It would then be logical to assume that they have accurate assessments of the distribution of commute trip durations, at least for those departure times they often choose. It then follows that the probability of gain as they perceive is also accurate. Thus $\hat{p}$ can be expected to closely approximate $p$. Note that $\hat{p}$ can be regarded to represent long-term beliefs that a commuter holds about the distribution of his commute trip durations.

Now, it is also expected that a commuter's departure time decision is influenced by the outcomes experienced in the immediate past. For example, it is conceivable that a commuter leaves earlier than before following a commute which took longer than usual. Researchers have often assumed that the trip duration predicted by a commuter can be expressed as a weighted average of trip durations experienced in the past, with heavier weights assigned to trip durations experienced more recently (e.g., Horowitz, 1984; Friesz et al., 1984; Smith, 1984; Cascatta, 1989; Cascatta and Canterella, 1991 and van Berkum and van der Mede 1998). In fact if the commuter were to behave solely on the basis of his long-term beliefs, then there would be no shifts in departure time or switching of commute routes. Empirical observations, including the ones presented in Table 1 of this study, lend no support to such a view. The realized gain probability, $p'$, may be interpreted as one of the inputs on which the commuter's short-term adjustments are made. In other words, the prior probability based on long-term beliefs is adjusted, or *weighted*, using the short-term experience as represented by $p'$.

In sum, the commuter has long-term beliefs about commute trip durations, or, arrival times, that he has formed through experience. At the same time, his departure time decision is adjusted under day-to-day variations in travel time. This short-term adjustment is motivated by the discrepancy between the long-term, prior gain probability, $p_g^0$ and the short-term, posterior probability, $p_g^b$. Then the relationship between the two, obtained from the collection of ($p_g^0$,$p_g^b$) as data, can be utilized to describe short-term decisions based on long-term data. We shall call this "contingency adjustment model" of departure time choice. It is proposed that the relationship between $p_g^0$ and $p_g^b$ be adopted as an estimate of the weight function in prospect theory.

Scatter plots of ($p_g^0$,$p_g^b$) and ($q_g^0$,$q_g^b$) are produced using the data set described earlier and given in Figures 6 and 7, respectively. Figure 7 exhibits that plots are more dispersed in moderate to high probability ranges. When the realized gain probability is linear-regressed on the expected gain probability, a slope coefficient of 0.82 is obtained ($t$ = 27.14; constant = 0.09, $t$ = 3.96; $R^2$ = 0.74, adjusted $R^2$ = 0.54). We also applied nonlinear regression to expected and realized probability pairs. The regression equation is specified as in Eq. (2). The $\gamma$ coefficient is estimated at 0.84 and is highly significant ($t$ = 33.6), and indicates subproportionality and subcertainty.

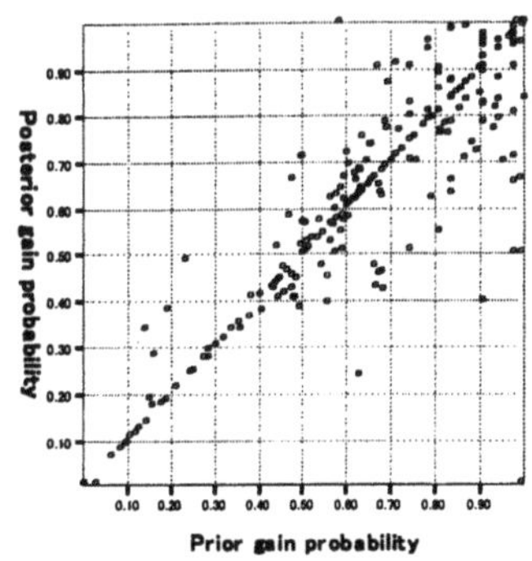


Figure 6. Prior and posterior gain probabilities

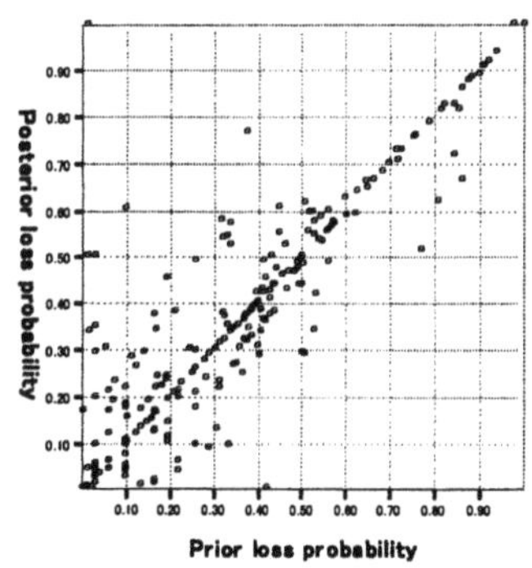


Figure 7. Prior and posterior loss probabilities

Now, turning to the loss probability, [illegible] a linear regression analysis yields the constant term as 0.09 (t = 7.60) and the slope coefficient as 0.82 (t = 27.14), with similar goodness-of-fit measures as the model for gains. A striking difference between the models for gains and losses emerges at this point that gains are not as underweighted as losses for moderate probabilities. This is also evident from results of nonlinear regression analysis. The analysis indicates that $\gamma$ coefficient (= 0.77, $t$ = 26.3) is smaller than that for gains (= 0.84) and significant. The two $\gamma$ coefficients are significantly different from each other at a (one-tailed) 95% confidence level. The regression analyses of [illegible] and [illegible] have thus yielded estimates of the weight function whose parameters are consistent with the properties prescribed by Kahmenan and Tversky (1979), suggesting the practical usefulness of the contingent adjustment approach proposed here.

## 6. HETEROGENEITY IN THE WEIGHT FUNCTION

Heterogeneity in the weight function is now examined by introducing both observed and unobserved heterogeneity into the weight function. For the simplicity of analysis, a linear weight function is adopted. This, however, limits the scope of analysis to the investigation of over- and under-weighting properties; examination of the other properties, such as subproportionality and subadditivity, while incorporating observed and unobserved heterogeneity, remains as a future task.

The model takes on the form, [illegible]= [illegible]$\alpha + \theta' X + \nu$). Note that over-weighting refers to the case where [illegible] [illegible]$> 1$ and under-weighting to [illegible] [illegible]$< 1$. Unobserved heterogeneity is represented by the individual-specific random effect, $\upsilon$, which is assumed to have a normal distribution. We specify the model as (Greene, 2002):

$$
\begin{aligned}
&y_{it} = \alpha + \theta' X_{it} + \nu_{it} \\
&\nu_{it} = \varepsilon_{it} + \upsilon_i \\
&\mathrm{E}(\nu_{it}) = 0 \\
&\mathrm{V}(\nu_{it}) = \sigma^2 = \sigma_\varepsilon^2 + \sigma_\upsilon^2 \\
&\mathrm{Cov}(\nu_{it}, \nu_{is}) = \sigma_\upsilon^2 \\
&\mathrm{Cov}(\nu_{it}, \nu_{js}) = 0 \quad \forall t, s, \quad i \neq j
\end{aligned}
\tag{7}
$$

where the subscripts $i$ and $j$ refer to the individual, subscripts $t$ and $s$ refer to time, $y_{it}$ = [illegible]/ [illegible] $X_{it}$ is the vector of explanatory variables, $\theta$ is a vector of coefficients, $\alpha$ is a constant,.$\varepsilon_{it}$ is a purely random normal error term, $\upsilon_i$ is an individual specific error component as noted earlier, and $\sigma_\varepsilon^2$ and $\sigma_\upsilon^2$ are their respective variances. assumed to be the generic error term and error term that is caused by individual unobserved heterogeneity respectively.

In addition to the linear regression model, we also propose another model based on discrete choice by which we separate the sample into two with respect to over-weighting ( $\tilde{p}'/\tilde{p} > 1 \rightarrow 1$) and under-weighting ( $\tilde{p}'/\tilde{p} < 1 \rightarrow 0$). In this model, we exclude cases with equal posterior to prior probability and We estimate the parameters by using random effects probit model:

$$\tilde{p}'/\tilde{p} = \alpha + \boldsymbol{\theta}'\mathbf{X}_{\mathrm{it}} + \nu_{it}$$

$$\nu_{it} = \varepsilon_{it} + \upsilon_i$$

$$\tilde{p}'/\tilde{p} > 1 \rightarrow y_{it} = 1$$

$$\tilde{p}'/\tilde{p} < 1 \rightarrow y_{it} = 0 \tag{8}$$

$$\Pr\left(y_{it} \middle| \alpha + \boldsymbol{\theta}'\mathbf{X}_{\mathrm{it}} + \upsilon_i\right) = \Phi\left(\alpha + \boldsymbol{\theta}'\mathbf{X}_{\mathrm{it}} + \upsilon_i\right)$$

$$\sigma^2 = \sigma_\upsilon^2 + \sigma_\varepsilon^2$$

$$\rho = \frac{\sigma_\upsilon^2}{\sigma_\upsilon^2 + \sigma_\varepsilon^2}$$

The random effects probit model produces correlation, $\rho$, between individual choices as a result of unobserved heterogeneity, $\upsilon$. The unobserved heterogeneity is assumed to be distributed as normal and integrated out of the likelihood function:

$$\lambda\left(y_{i1}, y_{i2}, \cdots y_{iT} \middle| \alpha, \boldsymbol{\theta}\right) =$$

$$\int_{-\infty}^{+\infty} \left[\prod_t^T f\left(y_{i1}, y_{i2}, \cdots y_{iT} \middle| \alpha, \boldsymbol{\theta}, u_i\right)\right] \phi(u_i) du_i \tag{9}$$

The results of model estimations, obtained using LIMDEP 8.0, are presented in Table 2. The linear regression model accounts for 77% of the total variation (adjusted $R^2$ = 0.63) and is highly significant ($F[228, 389]$ = 5.61, $p < 0.005$). Note that cases where $\tilde{p}$ is not properly defined because the commuter departed too late to arrive at or before LAT, are excluded in this estimation.

On the other hand, random effects probit model is significantly improved over the simpler specifications with fewer variables. The cases that are employed for linear regression, 621 cases for 215 commuters, decreases approximately 50% in the random effects probit model estimation, 315 cases for 155 commuters. In both of the models, we have parameters of unobserved heterogeneity that are significantly different from zero. Both of the models reveal different signs for arrivals in gain and loss regions: linear regression model reveals over-weighting in the gain region while probit regression reveals under-weighting. The same contradictory results is obtained for the constant term-positive for linear regression and negative by probit regression.

None of the individual attributes have significant coefficients but existence of children under 15 years of age. Besides, some of the trip attributes—commute trip duration variability, ratio of commute trip duration and use of the same route as previous commute—have significant coefficients. Specially, use of the same route contributes to over-weighting-a result suggested by both of the models. Probit regression suggests that tolerated individuals increase their weigthts, which is also supported by the linear regression although the estimated coefficient value is insignificant in the linear regression.

Commute trip duration weighted by average commute trip duration has a positive coefficient estimate. This variable is a ratio that computed by dividing commute trip duration to the average commute trip durations. If the ratio is one, then trip is completed as expected and then corresponding ratio of posterior to prior probability is increased by 0.15. In terms of choice making, this variable can cause difficulty about how a commuter would know that he will end up with the average trip duration, although there are instances that a trip duration can be guessed beforehand with the help of the weather forecasts, traffic broadcasts and the ITS applications. These information supplies might help commuters guess their commute durations in terms of known average trip durations, thus might increase the expectation of arriving at a certain point.

| Variable | | Random Effects Linear Regression | | Random Effects Probit Regression | |
|---|---|---|---|---|---|
| | | coefficient | t-value | coefficient | t-value |
| Sex | 1 represents male, 0 represents female. | 0.02 | 0.69 | 0.27 | 0.60 |
| Age group | Ten year intervals from 20 years of age are given integer values from 1 to 6. | 0.01 | 1.06 | 0.05 | 0.34 |
| Income Group | Income is indexed in increasing integer values from 1 (lowest quartile in Japan) to 4 (highest quartile in Japan). | -0.01 | -0.80 | 0.11 | 0.71 |
| Child | 1 represents the existence of children below 15 years of age, 0 represents else cases. | 0.04 | 1.91 | 0.38 | 0.93 |
| Departure time | Time passed from midnight to the departure time in minutes ( X1000) | 0.12 | 0.80 | 0.01 | 0.30 |
| Commute trip duration / average known duration | Ratio of commute trip duration to known average commute duration.( X100) | 0.35 | 2.21 | 1.23 | 1.89 |
| Usual commute | 1 represents a usual commute as everday, 0 represents else cases. ( X100) | 0.37 | 0.10 | -28.34 | -0.42 |
| Same route as previous commute | 1 represnetsThe same route has been taken as the previous time, 0 repsesents else cases. | 0.06 | 2.35 | 1.01 | 2.01 |
| Flexitime policy | 1 represents flexitime policy is applied in the work place, 0 represents else cases. | 0.02 | 0.59 | 0.21 | 0.40 |
| Tolerance | 1 represents tardiness is tolerable, 0 represents else cases. | 0.03 | 1.23 | 0.54 | 1.97 |
| Commute trip duration variability | Time difference between longest and shortest known commute durations. ( X100) | 0.14 | 2.16 | 0.01 | 0.72 |
| LAT is before WST | 1 represents LAT is before WST, 0 represents else cases. | -0.03 | -0.95 | -0.44 | -0.98 |
| PAT is after WST | 1 represents PAT is after WST, 0 represents else cases. | -0.01 | -0.26 | -0.78 | -1.06 |
| Arrival at PAT | 1 represents arrival at PAT, 0 represents else cases. | -0.03 | -1.71 | 0.35 | 1.04 |
| Arrival in gain/loss | 1 represents arrival in gain region, 0 represents arrival in loss region. | 0.68 | 18.23 | -2.63 | -3.90 |
| Constant | | 0.15 | 2.01 | -1.70 | -1.86 |
| $\rho$ | Correlation due to individual specific errors<br>$H_o$: $\sigma_u$=0 (rejected) $H_1$: $\sigma_u \neq 0$ (accepted) with respect to Lagrange Multiplier Test ($\sim\chi^2_1$=74.13, p<0.005) | 0.41 | | 0.72 | 7.60 |
| $R^2$ | | 0.77 | | | |
| Adjusted $R^2$ | | 0.63 | | | |

| | Random Effects Linear Regression | Random Effects Probit Regression |
|---|---|---|
| **Random effects linear regression** | | |
| Ordinary Least Squares: F[ 15,602] (prob) = 25.43 (p=0.005) | | |
| Generalized Least Squares with Random Effects: F[228, 389] = 5.61 (p=0.005) | | |
| **Random effects probit regression: LogLikelihood with only constant term** | | |
| L1: Loglikelihood function with constant term | | -218.13 |
| L2: Loglikelihood function without random effects | | -180.13 |
| L3: Loglikelilhood function with random effects | | -164.98 |
| # of individuals | 215 | 155 |
| # of observations | 621 | 315 |

Table 2. Linear weight function with heterogeneity

Note that the effect of the use of the same route as previous commute has turned out be positive and significant for both of the models. eans if the driver uses the same route as previous time (however we do not know the route taken previous time is the generally used route) is positive. This outcome suggests that commuters have propensity to increase the weights given to posterior probability when they continue to use the same route.

## 7. VALUING THE ARRIVALS

The other element of prospect theory is the value function. Value function maps outcomes of a choice to different values in gain and loss regions. During an evaluation, choice is made by weighting the values of risky outcomes with the weight function. We have already estimated the weight function by nonlinear regression which yielded $\gamma$ parameter 0.84 and 0.77 for gain and loss regions respectively (without any observed and unobserved heterogeneity). In this section, we estimate the value function, by using individual satisfaction on every commute and by controlling for individual heterogeneity as well. As stressed above when introducing contingency updating model, weight function implicitly is assumed to be a daily updating of individual evaluation of probability. Thus we separate the estimation of the two functions as Kahneman and Tversky 1992).

We formulate the value function, $V$ for the both of the decision frames as Eq. 6 with $t$ defined as the time difference from a reference point, $\mathbf{X}$ as observed variables, $\boldsymbol{\beta}$ vector of parameters, A as the time interval between EAT and LAT, B as the time interval between PAT and LAT.

$$V = \begin{cases} 0 & AT = EAT \vee AT = LAT \\ \boldsymbol{\beta}\mathbf{X} + \alpha \ln\left(\dfrac{t}{A}100\right) & EAT < AT < PAT; A = PAT - EAT \\ \boldsymbol{\beta}\mathbf{X} + \alpha \ln\left(\dfrac{t}{B}100\right) & PAT < AT < LAT; B = LAT - PAT; DF1 \\ \boldsymbol{\beta}\mathbf{X} + \alpha\left(\dfrac{t}{A}2.14\right)^2 & EAT < AT < PAT; A = PAT - EAT; DF2 \\ \boldsymbol{\beta}\mathbf{X} + \alpha \ln\left(\dfrac{t}{A+B}100\right) & AT < EAT \vee AT > LAT \end{cases} \tag{10}$$

With respect to this formulation, the value function is assumed to be equal to zero at reference points which are EAT and LAT. Time deviations, $t$, are computed from EAT and LAT for both the symmetric decision frame (DF1) and asymmetric decision frame (DF2). The time deviation in the gain region is normalized by dividing with the time interval , $A$ for gains in early side and *B* for gains on the late side arrivals; both of the values are increased to take values between zero and 100 for DF1. For DF2, this value in the early-gain region is assumed to be equal to 2.14 as its square and logarithm of 100 are equal. Thus, the value function given by Eq. 6 achieves the maximum value at PAT. As noted in Section 3, the values are assumed to increase decreasingly from reference points in DF1- this is guaranteed by logarithmic transformation. The same assumption is maintained for DF2 but for the early-gain arrivals, where the value increases increasingly-this is guaranteed by power function.

The stated intention about the next departure time reported by the respondent at the end of each survey day is used in the analysis. The intention is binary, notably, whether commuter intends to change or not to change the departure time the following day. In order to account for gains, we assumed value/or utility prevails with "not to change departure time" in the gain region, that is to say, they are given value of one. To account for losses, we assumed value/or utility prevails with "change departure time" in the loss region, thus they are given value of one.

The value functions of both decision frames are estimated by using the binary Probit model that controls for random effects specification for unobserved heterogeneity ($V$ refers to the observed value given in Eq. 9 and $U$ refers to the unobserved value, the other terms and the likelihood function are same as Eq. 8 and 9):

$$
\begin{aligned}
&U_{it} = V_{it} + \nu_{it} \\
&\nu_{it} = \varepsilon_{it} + \upsilon_i \\
&U_{it} > 0 \Rightarrow y_{it} = 1 \\
&U_{it} \leq 0 \Rightarrow y_{it} = 0 \qquad\qquad (11) \\
&\Pr\left(y_{it} \middle| V_{it} + \upsilon_i\right) = \Phi\left(V_{it} + \upsilon_i\right) \\
&\sigma^2 = \sigma_\upsilon^2 + \sigma_\varepsilon^2 \\
&\rho = \frac{\sigma_\upsilon^2}{\sigma_\upsilon^2 + \sigma_\varepsilon^2}
\end{aligned}
$$

The estimation is done by using LIMDEP 8.0 Econometrics Software Package (Greene, 2002). The estimated parameter values are given in Table 3. The estimation results of value function coefficients are more or less similar for both of the decision frames, even the Log likelihood functions are approximately equal. Although we conclude that both of the decision frames with different value function specifications are similar at least in coefficient values, the value function becomes different in the quasi-gain region of DF2 as the time difference is transformed by power function (with power assumed to be equal to two).

Similar to the results of the weight function estimation, few of personal attributes display observed heterogeneity significantly. Males as well as high income earners decrease values, however the coefficient values are insignificant. On the other hand, increasing age make commuters more sensitive to valuing: values increase as one gets older, the same can also be said for commuters having children younger than 15 years of age. These coefficients of both age group and existence of children below 15 years of age are closely associated with life stage. For example, those who are older might have well established decision frames (we stressed this point in the data section too), so they might be more sensitive to values than those whose decision frames might be subject to change. Also having children younger than 15 might place household to act in accordance as children have tight schedules such as care, formal schooling etc.

| Variable | | Symmetric Decision Frame coefficient | Symmetric Decision Frame t-value | Asymmetric Decision Frame coefficient | Asymmetric Decision Frame t-value |
|---|---|---|---|---|---|
| Sex | 1 represents male, 0 represents female. | -0.44 | -1.09 | -0.49 | -1.19 |
| Occupation | 1 represents paid workers enrolled in government and private sectors, 0 represents else. | 0.35 | 1.13 | 0.36 | 1.14 |
| Age group | Ten year intervals from 20 years of age are given integer values from 1 to 6. | 0.35 | 2.28 | 0.33 | 2.10 |
| Income Group | Income is indexed in onvreasing integer values from 1 (lowest quartile in Japan) to 4 (highest quartile in Japan). | -0.08 | -0.61 | -0.08 | -0.59 |
| Child | 1 represents the existence of children below 15 years of age, 0 represents else. | 0.87 | 2.34 | 0.83 | 2.17 |
| Dummy for gain region | 1 represents arrivals in gain region, 0 repsents else. | -0.87 | -1.13 | -1.12 | -1.38 |
| Dummy for loss region | 1 represents arrivals in loss region, 0 repsents else. | -2.75 | -3.18 | -2.84 | -3.23 |
| Quasi-Gain | 1 represents arrival in early gain region, 0 represents else. | | | 2.14 | 3.74 |
| Arrival | 1 represents early arrival, 0 represents late arrival and 2 represents arrival at PAT. | -0.03 | -0.19 | -0.07 | -0.50 |
| Flexitime policy | 1 represents flexitime policy is applied in the work place, 0 represents else. | -0.53 | -1.51 | -0.48 | -1.33 |
| Tolerance | 1 represents tardiness is tolerable, 0 represents else. | 0.94 | 3.33 | 0.96 | 3.22 |
| Commute trip duration variability | Time difference between longest and shortest known commute durations. (X100) | -0.49 | -0.35 | -0.45 | -0.32 |
| LAT is before WST | 1 represents LAT is before WST, 0 represents else. | 0.25 | 0.71 | 0.24 | 0.65 |
| PAT is after WST | 1 represents PAT is after WST, 0 represents else. | 0.40 | 0.49 | 0.32 | 0.38 |
| Commute trip duration / average known duration | Ratio of commute trip duration to known average commute duration. | -1.24 | -2.28 | -1.28 | -2.25 |
| Time Deviation from Reference points | Time deviation from reference point transformed either by natural logarithm (DF1 and DF2) or by power function (quasi-gain region in DF2). | 0.54 | 3.94 | 0.59 | 4.01 |
| ρ | Correlation due to individual specific errors | 0.66 | 8.88 | 0.67 | 8.86 |
| L1: Loglikelihood function with constant term: | | -306.95 | | -306.95 | |
| L2: Loglikelihood function without random effects: | | -289.30 | | -290.72 | |
| L3: Loglikelilhood function with random effects: | | -255.10 | | -255.27 | |
| # of Individuals | | 198 | | 198 | |
| # of observations | | 594 | | 594 | |

Table 3. Value function with heterogeneity

The coupling constraints such as flexitime policy and tolerance on tardiness cause different effects. Flexitime policy (although insignificant) decreases (absolute) values, but tolerance on tardiness increases absolute values significantly. Tolerance although is a good thing on the side of commuters might be another way of controlling the workers. Both flexitime and tolerance differs from one another significantly. The first one is something that no parties- the paid worker and the manager loses, because flexiworker makes up the time that is vested to him in the morning. But the tolerance is something that bases on goodwill but the manager gradually (and possibly increasingly) loses while commuter enjoys fixed gains. So the two constraints although might seem similar to each other differs totally from one another. Commuter who is tolerated might choose to liquidize the tolerance when needed by gaining that tolerance by his compliance with the coupling constraints.

The timely location of WST, such as after LAT or before PAT show increasing effect on values. But coefficients to these variables are insignificant too. On the other hand, as expected, the ratio of commute trip duration to known average commute trip has a significant coefficient that decreases absolute values. Even this ratio is one, value decreases by 1.24 (DF1) and 1.28 (DF2). This reflects close association of commute duration with disutility. But note that the coefficient of this ratio turned out to be positive for weight function properties estimated in the previous section. The commute trip duration variability which is observed by the time difference between known shortest and longest trip times has a positive effect on the absolute value. But this variable is also insignificant.

## 8. CONCLUSIONS

This study follows the line of research of Senbil and Kitamura, 2004 and Kitamura and Jou, 2003. In this study, we employ two decision frames in order to access the value and weight functions, two pivotal elements of the Prospect theory. Weight function has been associated with a new model coupled as contingency adjustment model. This new model finds its theoretical background on individual updating of perceived likelihood of any arrival time conditioned on a certain departure time. In this regard, the weights are assumed to be realized with respect to a comparison between expected (at the departure time) and realized arrival times. Commuter holds an expected arrival time at a departure time which is established by his commute history, but the same time, everyday is taken as another episode of risks and uncertainties: commuter reevaluates his chances and takes actions, such as listening the radio broadcast carefully when it is raining, or lane changing to make a null probability a possible gain arrival etc. Although our nonlinear regression estimation of the weight function complies with the basic premise of the Prospect theory, we need know other structural elements, the effects of observed and unobserved heterogeneity, affecting behavioral responses to the expected probability. For this reason, we employ linear and probit regressions that control for heterogeneity but only refer to over- and under- weighting of probabilities. The weight function yielded results that are significant for trip attributes but not for most of the commuter attributes.

The value function is devised by using two decision frames, the first one is symmetric about the preferred arrival time for gains and losses, the second one is not symmetric. The estimation of the value function by the binary Probit model yielded approximately similar results for both of the decision frames. In both of the decision frames, it is significantly true that commuters are responsive to the time deviations from reference points in their decision frames. Thus, we can say the arrival points in the gain region are not equal in values and the choice of the departure time is strongly conditioned on the possible arrival times.

# REFERENCES


Abdel-Aty, M.A., Kitamura, R., Jovanis, P.P. Investigating effect of travel time variability on route choice using repeated-measurement stated preference data. Transportation Research Record 1995; 1493: 39-45.

Avineri, E., Prashker, J.N. Sensitivity to uncertainty: the need for a paradigm shift. Paper presented at the 82nd Annual Meeting of the Transportation Research Board, Washington, D.C., 2003.

Ben-Akiva, M., Lerman, S. Discrete Choice Analysis: Theory and Application to Travel Demand, Cambridge, Mass.: The MIT Press , 1985.

Bleichrodth, H., Pinto, J. L. A parameter-free elicitation of the probability weighting in medical decision analysis. Management Science 2000; 46:1485-1496.

Bonsall, P. Predicting "Traveller's Reponse to Uncertainty." In *Travel Behavior Research: The Leading Edge*, David Hensher, Elsevier Science, Oxford, U.K., 2001.

Cascetta, E. Stochastic process approach to the analysis of temporal dynamics in transportation networks, Transportation Research B 1989; 23:1-17.

Cascetta, E., Canterella, G. E. A day to day and within-day dynamic stochastic assignment model, Transportation Research A 1991; 25:277-291.

Cook, A.J., Jones, P., Bates, J.J., Polak, J., Haigh , M. Improved methods of representing travel time reliability in SP experiments. Proceedings 27th European Transport Forum, PTRC, London; 1999.

de Palma, A., Ben Akiva, M., Lefevre, C., Litinas, N. Stochastic equilibrium model of peak period traffic congestion. Transportation Science 1983; 17: 430-453.

Greene, W. H. LIMDEP Version 8.0, Econometric Software, Inc., N.Y., 2002.

Fujii, S., Kitamura, R. Decision frame for departure time choice under uncertainty. Infrastructure Planning Review 2001 ; 18 : 491-495. (In Japanese)

Friesz, T.L., Bernstein, D., Mehta, N.J., Tobin, R.L., Ganjalizadeh, S. Day-to-day dynamic network disequilibria and idealized traveller information systems. Operations Research 1994; 42:1120-1136.

Fujii, S., Kitamura, R. Drivers' mental representation of travel time and departure time choice in uncertain traffic network conditions. Networks and Spatial Economics 2004, (forthcoming).

Hägerstrand, T. What about people in regional science? Papers of The Regional Science Association 1970; 24:7-21.

Hendrickson, C. and E. Plank (1984) The flexibility of departure times for work trips. Transportation Research A 1984; 18:887-902.

Horowitz, J.L. (1984) The stability of stochastic equilibrium in a two-link transportation network, Transportation Research B ,18:13-28.

Jou, R. Modeling the impact of pre-trip information on commuter departure time and route choice, Transportation Research B 2000; 35:887-902.

Jou, R., Kitamura R. Commuter Departure Time Choice: A Reference Point Approach, Mimeograph, 2002.

Luce, R. D. Individual Choice Behavior, New York: Wiley, 1959.

Kahneman, D., Tversky, Prospect theory: an analysis of decision under risk, Econometrica 1979; 47:263-291.

Kahneman, D. Tversky, A. "Choices, Values and Frames" In Choice, Values and Frames, Kahneman, D. Tversky, A., ed. Cambridge University Press, Cambridge, U.K., 2000.

Prelec, D. "Compound Invariant Weighting Functions in Prospect Theory" In Choice, Values and Frames, Kahneman, D. Tversky, A., ed. Cambridge University Press, Cambridge, U.K., 2000.

Mahmassani, H., Herman, R. Dynamic user equilibrium departure time and route choice on idealized traffic arterials. Transportation Science 1984; 18: 362-453.

Mahmassani, H. S. and G-L. Chang Experiments with the departure time choice dynamics of urban commuters. Transportation Reseach B 1986; 20:297-320.

Mahmassani, H. S., Chang, G-L. On boundedly rational user equilibrium in transportation systems. Transportation Science 1987; 21:89-99.

Mahmassani, H. S., Liu, Y. Dynamics of commuting decision behavior under advanced traveller information systems. Transportation Research C 1999; 7: 91-107.

Rosenbloom, S. Peak-period traffic congestion: a state-of-the-art analysis and evaluation of effective solution. Transportation 1978; 7:167-191.

Senbil, M. Kitamura, R. Reference points in commuter departure time choice: a prospect theoretic test of alternative decision frames. Journal of Intelligent Transportation Systems 2004. 8:19-31.

Small, K. A. The scheduling of consumer activities: work trips. American Economic Review 1982; 72: 467-479.

Smith, M. J. (1984) The stability of dynamic model of traffic assignment: An application of method of lyapunov, Transportation Research 1984; 18: 245-252.

Tversky, A. and D. Kahneman Advances in prospect theory: cumulative representation of uncertainty. Journal of Risk and Uncertainty 1992; 195-230.

Tversky, A., Fox, C. Weighting risk and uncertainty. Journal of Risk and Uncertainty 1995; 5: 297-323.

van Berkum, E.C. and P.H.J. van der Mede The impact of dynamic traffic information: Modelling approach and empirical results. In: *Travel Behaviour Research: Updating the State of Play*, J. de D. Ortúzar, D. Hensher and S. Jara-Diaz, eds. Oxford, UK: Pergamon, Elsevier. 1998

Von Neumann, J., Morgenstern, O. Theory of Games and Economic Behavior, New York: Wiley, 1944.

# Chapter 18

# Importance of Fuzzy Sets Definitions for Fuzzy Signal Controllers

**Maria Alice P. Jacques, Daliana B. L. M. Santos, Matti Pursula, and Iisakki Kosonen**

## 1. INTRODUCTION

The operation of fuzzy signal controllers in comparison with traditional pretimed or vehicle-actuated control modes, has provided better traffic operations according to the usually adopted performance measures as is the case with delay and number of stops. Previous studies have demonstrated however, that fuzzy signal controllers' performance is highly affected by the choice of their decision-making logic and defuzzification interface.

Other fuzzy controller aspects recognised as important in controller operation are the fuzzy sets adopted, their respective membership functions, as well as the rule base defined. Nevertheless, specific assessment of the impact of fuzzy set definitions on traffic signal performance has not been sufficiently explored in the literature.

This research aims therefore to study the implications of small modifications to some fuzzy sets' parameters on fuzzy signal controllers. For this purpose a basic fuzzy signal controller developed at the Helsinki University of Technology is used. Fuzzy sets related to the controller input variables (in this case queue of the halted traffic and number of arrivals at the approach receiving green indication) have their parameters slightly modified, specifically in terms of the values for which the membership function is equal to zero and one. That is, the shape of the membership function related to each fuzzy set was not modified, only their limiting values. Six different modifications were evaluated, along with the basic situation, forming a set of seven studied situations.

The evaluation of the impact of the above-mentioned modifications was carried out in two stages. Initially, the surface of control responses was developed for all the seven situations through MATLAB software, along with the corresponding matrix of control responses. The matrices were then evaluated to verify whether they differ from each other. A descriptive analysis of the general shape of the surface of control responses was also performed. In the second stage, the effective impact of the different situations on traffic performance was assessed through simulation studies generated from HUTSIM software, developed at the Helsinki University of Technology. Simulation runs were performed for an isolated junction of two one-way streets, operating under three different traffic volume levels.

# 2. FUZZY SETS AND FUZZY SIGNAL CONTROLLERS

This section provides general concepts linked to the subject studied. These are related to fuzzy set definition, terminology and fuzzy controllers.

## 2.1. Fuzzy set definition and terminology

According to Zadeh, "a fuzzy set is a class of objects with a continuum of grades of membership. Such a set is characterised by a membership function which assigns to each object a grade of membership between zero and one". In other words, a fuzzy set can be defined as a set of elements belonging to a given universe of discourse, which is represented by ordered pairs of elements and their respective grade of membership to the set considered.

The following relation can present this definition:

$$A = \{(u, \mu_A(u)) \,/\, u \in U\}$$

In the above relation, the fuzzy set $A$ is defined in a universe of discourse $U$, where $\mu_A(u)$ is the degree of membership of the element $u \in U$ in $A$, defined by the membership function $\mu_A$ . Therefore, the set of all points $u \in U$ is the support of the fuzzy set $A$. For the case in which the support of a fuzzy set is a single point in the universe of discourse, the fuzzy set is referred to as fuzzy singleton.

Fuzzy sets are used to characterise each term belonging to the term set related to linguistic variables. The meaning of a fuzzy linguistic term is defined by the membership function assigned according to the intended use of this term.

## 2.2. Fuzzy controllers

Fuzzy logic controller (FLC) is a controller based on fuzzy logic whose algorithm converts the "linguistic control strategy based on expert knowledge into an automatic control strategy". The principal components of this type of controller are: fuzzification interface, knowledge base, decision-making logic, and defuzzification interface. Each of these components is fairly detailed in the literature. Thus, only the aspects relevant to the ongoing controllers' analysis and related to the fuzzy sets' role in fuzzification interface will be presented in this section.

In the fuzzification interface of an FLC, the system's attributes (linguistic variables) are defined. These are linked to the state of the process (input variables) and to the control action (output variables). The linguistic values of these variables (set of terms) are labels of fuzzy sets. The partition of the universe of discourse related to a linguistic variable is not unique and, although optimal partition can be achieved by a heuristic method, the basic principle can be the use of real life linguistic terms. The membership function of each fuzzy set is another important aspect related to the fuzzification interface. Its shape is quite free and must be defined by the controller´s developer. For instance, the Fuzzy Logic Toolbox of MATLAB (FLT-M) software includes 11 built-in membership function types and allows the user to create his/her own membership function in the case that the built-in types do not suit his/her needs.

When the shape of the fuzzy sets is determined, several other parameters have to be adjusted. In terms of triangle or trapezoidal-shaped functions, for example, these parameters are basically the points of the universe of discourse for which the membership function reaches its minimum and maximum values, zero and one.

Membership functions are usually constructed upon expert judgement. All the same, even when different experts agree upon the universes of discourse and their corresponding partition to be associated with the different linguistic variables related to a particular control problem, they can differ regarding the shape of the membership functions and/or the parameters of these functions.

## 2.3. Fuzzy signal controllers

For traffic control purposes, the literature presents the development of different fuzzy controllers, referred to as fuzzy signal controller. The first controller of this type was developed by Pappis and Mamdani. This controller, as well as those to follow, is based on the fuzzy extension principle. Different fuzzy signal controllers' performance, according to the usually adopted measures of traffic performance, has been accessed by different authors with positive results.

In an earlier paper the evaluation of different fuzzy traffic signal controllers for isolated intersections showed that there are differences not only among their linguistic variables but also among the partition of variables' universes of discourse (fuzzy sets labels). Tables 1 and 2 present, respectively, the linguistic variables and related labels of fuzzy sets of the traffic signal controllers evaluated, where: Model 1 is the fuzzy logic controller developed by Pappis and Mamdani; Model 2 and Model 3 are fuzzy traffic signal controllers for isolated intersections developed by Kim and Favilla, Machion, and Gomide, respectively; Model 4 refers to the controller developed by

Trabia, Kaseko and Ande; and Model 5 is the fuzzy signal controller built by Niittymäki and Pursula.

In addition, Table 2 shows that the same label refers to different range of values of the corresponding universe of discourse in distinct models analysed. Triangle-shaped (trapezoidal) functions are commonly used in these controllers.

In the present paper, small differences on the membership functions' parameters are investigated as they can impact upon a fuzzy signal controller's performance.

*Table 1.* Linguistic variables and fuzzy sets at the controllers studied

| Model | Linguistic variables |
|---|---|
| 1 | A- No. of arrivals at the arm with the right of way, estimated for each of the 10 seconds ahead. |
| | Q- No. of vehicles at the queue corresponding to the halted traffic, estimated for each of the 10 seconds ahead. |
| | T- Time elapsed from the end of the current green period. |
| | E- Extension to be given to the current green period. |
| 2 | S.LEFT- No. of left-turning vehicles present within a 61 meter-distance behind the stop-line at the phase having the right of way. |
| | S.THRU- Identical, for through vehicles at the phase having the right of way (at the same or compatible phase as the variable S.LEFT) |
| | C.LEFT- Identical, for the left-turning vehicles at the halted traffic phase. |
| | C.THRU- Identical, for the through vehicles at the phase having the halted traffic (at the same or compatible phase as the variable C.LEFT. |
| | E- Extension to the current green time. |
| 3 | A- No. of arrivals at the approach having the right of way. |
| | Q- No. of vehicles queued at the approaches with the halted traffic. |
| | E- Extension to the current green time. |
| | RQ- Residual queue at the end of the green phase. |
| | QV- Queue variation during the green phase. |
| | ULV- Upper-limit variation for the membership functions of E. |
| 4 | Omax- The maximum approach flow within the previous time interval $\Delta_t$, expressed in vehicles/s/lane. |
| | Qmax- The maximum queue length within $\Delta_t$, in vehicles/s/lane. |
| | TRgreen- The green traffic intensity within $\Delta_t$, in vehicles/s/lane. |
| | TRred- The red traffic intensity within $\Delta_t$, in vehicles/s/lane. |
| | E- Extension, which is not properly a linguistic variable. |
| 5 | A- No. of vehicles within the approach zone (100 m) during the green. |
| | Q- No. of vehicles with speed less than 5 km/h within the approach zone on the red signals. |
| | EXT- Extension to the current green period |

Source: Adapted from Analysing Different Fuzzy Traffic Signal Controllers for Isolated Intersections, Jacques, et. al.

*Table 2.* Fuzzy sets and respective membership functions at the controllers studied

| Model | Linguistic variables | Fuzzy sets (*) | M. F. Shape |
|---|---|---|---|
| | *A* | none (≤ 4); a few (≤ 5); few (0-6); medium (0-7); many (0-8); too many (1-9). | Triangular |
| 1 | *Q* | very small (4-12); small (8-16); small plus (12-20); medium (16-24); long (20-28); very long (24-32). | Triangular |
| | *T*<br>E | very short (≤ 3); short (1-5); medium (3-7); long (5-9); very long (≥ 7) | Triangular |
| | S.LEFT | small (≤ 4); medium (2-7); large (5-9); very large (≥ 8) | Trapezoidal |
| | S.THRU | small (≤ 5); medium (3-8); medium plus (6-11); large (10-22); very large (≥ 20) | Trapezoidal |
| 2 | C.LEFT | small (≤ 4); medium (2-7); large (5-9); very large (≥ 8) | Trapezoidal |
| | C.THRU | small (≤ 5); medium (3-14); large (10-22); very large (≥ 20) | Trapezoidal |
| | E | short = 3s; short plus = 5s; medium = 7s; long = 10s | Singleton |
| | A | almost none (≤ 3); few (0-6); many (3-9); too many (≥ 6) | Triangular |
| 3 | Q | very small (≤ 3); small (0-6); medium (3-9); long (≥6) | Triangular |
| | E | very short (≤ 4); short (1-7); medium (4-10); long (≥7) | Triangular |
| | RQ | small (≤ 4); medium (2-8); large (≥6) | Trapezoidal |
| | QV | small (≤ 4); medium (2-9); large (≥7) | Trapezoidal |
| | ULV | decrease (≤ 0); keep (-2 - +2); increase (≥0) | Triangular |
| | Omax | zero (≤0.17); small (0 – 0.25); medium (0.08-0.33); big (≥ 0.25). | Trapezoidal |
| | Qmax | zero (≤0.74); small (0.26 – 1.74); medium 1.26-3.26); big (≥ 2.74). | Trapezoidal |
| 4 | TRgreen | zero (≤0.6 7); small (0.23 – 1.73); medium (1.27-3.27); big (≥ 2.73). | Trapezoidal |
| | TRred | zero (≤0.74); small (0.26 – 1.74); medium(1.26-3.26); big(≥ 2.74). | Trapezoidal |
| | E | E = -1 ( not to extend the green); E = +1 (extend the green) | |
| 5 | A | zero (≤ 2); a few (1-9); medium (4-12); many (≥ 8) | Triangular |
| | Q | a few (0-10); medium (5-15); too long (≥ 8) | Triangular |
| | EXT | zero = 0; short = 5s; medium = 10s; long = 15s | Singleton |

* These are the values of the universes of discourse for which the membership function value for a given fuzzy set is zero. At these points the values of the membership functions start growing or stop decreasing in relation to the function's maximum value.

Source: Adapted from Analysing Different Fuzzy Traffic Signal Controllers for Isolated Intersections, Jacques, et. al.

# 3. METHODOLOGY FOR THE STUDY

The following research steps were developed: definition of the basic intersection to be simulated with HUTSIM software; definition of the basic fuzzy signal controller and generation through MATLAB software of its corresponding control table and surface; definition of the modifications to be performed in the input variables' membership functions' parameters and generation of the corresponding control tables and surfaces; simulation with HUTSIM of different traffic volume levels for the previously generated control tables; and analysis of the simulation results. Each of these steps is presented in the following sections.

## 3.1. Basic intersection for the study

The intersection to be considered in this study is an isolated junction of two one-way streets, operating under different traffic volume levels, as shown in Table 3. These volumes include only passenger cars. The intersection has two lanes per approach and its legs are 600 meters long.

*Table 3.* Intersection operation to be considered

| Volume Level | Volume (veh/h) | |
|---|---|---|
| | Minor Street Approach | Major Street Approach |
| Low | 200 | 600 |
| Medium | 400 | 1200 |
| High | 600 | 1600 |

## 3.2. Basic signal controller

The basic signal controller was developed at Helsinki University of Technology, and has been used in previous research work. It has the following linguistic input variables: ***queue*** at the approach with the halted traffic, with the terms small, medium, long, and any; and ***arrivals*** at the approach receiving green signal indication, with the values of zero, small, medium, long, and any. The controller's output variable is the ***extension*** to be given to the green time, having the following possible values: zero, short, medium, and long. All the values for the aforementioned linguistic values are labels of fuzzy sets whose membership functions are triangular or trapezoidal-shaped, as Table 4 indicates. The basic case is referred to as Case 01 in Table 4.

Triangular functions are implemented in MATLAB software according to three parameters: the first is the value for which the membership function value is set equal to zero; the second has the membership function equal to one; and the last has membership function value equal to zero. For the trapezoidal-shaped functions, four parameters are required. The first and last values have membership function value equal to zero; for the second and third values, the membership function value is equal to one. The rule base is of the type: IF Queue AND Arrival THEN Extension. Thirteen rules form the rule base, linked by the aggregation connective ALSO. The fuzzy operators used to implement the connectives are: minimum for the connective AND and maximum for the connective ALSO. The fuzzy implication function was the minimum, and the defuzzification method adopted was the centre of gravity. The shape of the controller response for this basic signal controller is shown in Figure 2.

The adopted knowledge base is shown in Table 5. It was defined in order to capture the objectives of the fuzzy signal control and to provide continuous change in the control action. Rules number 11, 12 and 13 were introduced for the sake of the continuity of the control's surface under de centre of gravity defuzzification method.

*Table 4.* Fuzzy sets considered in each case studied

| Variable | Case | Fuzzy sets | | | | |
|---|---|---|---|---|---|---|
| | | Small | Medium | Long | Any | |
| Queue | 01 | [0 05 10] | [05 10 15] | [10 15 20 20] | [0 0 20 20] | |
| | 02 | [0 04 08] | [04 08 12] | [08 12 20 20] | [0 0 20 20] | |
| | 03 | [0 05 10] | [05 10 15] | [10 15 20 20] | [0 0 20 20] | |
| | 04 | [0 04 08] | [04 08 12] | [08 12 20 20] | [0 0 20 20] | |
| | 05 | [0 06 12] | [06 12 18] | [12 18 20 20] | [0 0 20 20] | |
| | 06 | [0 05 10] | [05 10 15] | [10 15 20 20] | [0 0 20 20] | |
| | 07 | [0 06 12] | [06 12 18] | [12 18 20 20] | [0 0 20 20] | |
| | | Zero | Few | Medium | Long | Any |
| Arrivals | 01 | [0 0 05] | [0 05 10] | [05 10 15] | [10 15 20 20] | [0 0 20 20] |
| | 02 | [0 0 05] | [0 05 10] | [05 10 15] | [10 15 20 20] | [0 0 20 20] |
| | 03 | [0 0 04] | [0 04 08] | [04 08 12] | [08 12 20 20] | [0 0 20 20] |
| | 04 | [0 0 04] | [0 04 08] | [04 08 12] | [08 12 20 20] | [0 0 20 20] |
| | 05 | [0 0 05] | [0 05 10] | [05 10 15] | [10 15 20 20] | [0 0 20 20] |
| | 06 | [0 0 06] | [0 06 12] | [06 12 18] | [12 18 20 20] | [0 0 20 20] |
| | 07 | [0 0 06] | [0 06 12] | [06 12 18] | [12 18 20 20] | [0 0 20 20] |
| | | Zero | Short | Medium | Long | |
| Extension | All | [0 0 05] | [0 05 10] | [05 10 15] | [10 15 20 20] | |

*Table 5.* Knowledge base adopted

| Rule Number | *If* "Queue" is ... | *and* "Arrivals" is ... | *then* "Extension" is ... |
|---|---|---|---|
| 1 | Any | Zero | Zero |
| 2 | Small | Few | Short |
| 3 | Medium | Few | Zero |
| 4 | Long | Few | Zero |
| 5 | Small | Medium | Medium |
| 6 | Medium | Medium | Short |
| 7 | Long | Medium | Short |
| 8 | Small | Many | Long |
| 9 | Medium | Many | Medium |
| 10 | Long | Many | Medium |
| 11 | - | Few | Short |
| 12 | - | Medium | Medium |
| 13 | - | Many | Long |

The computer simulation will be performed with HUTSIM, and the operation of the fuzzy controller will consider the following procedures:
a) minimum green time equal to 5 seconds;
b) maximum number of extensions to be given in sequence for a signal group equal to 5;
c) if one extension is calculated as being less than or equal to 2 seconds, the extension is given but at its end no other extension is possible.

## 3.3. Modifications to the membership functions' parameters for the input variables

Based on the basic fuzzy signal controller previously described, the following situations will be investigated:
Case 02 – Reduction of the queue limits by 20%
Case 03 – Reduction of the arrival limits by 20%
Case 04 – Reduction of the queue and arrivals limits by 20%
Case 05 – Increasing of the queue limits by 20%
Case 06 – Increasing of the arrival limits by 20%
Case 07 – Increasing of the queue and arrival limits by 20 %

The above-mentioned modifications lead to the fuzzy sets shown in Table 4. It is important to point out that in all cases studied, the limits of the universes of discourses were kept unchanged as well as the fuzzy sets related to the variable extension. This was done in order to simplify the analysis to be performed and took into account that the modifications on the input variables´s fuzzy sets should be, in principle, less effective in changing the controller response than changes on the output variable. That is, if the changes in the input variable prove to affect the controller response it is possible to assume that the modifications on the output variable will also be effective.

The shape of the membership functions was also kept unchanged for all cases studied.

# 4. ANALYSIS OF CONTROLLER RESPONSE

A visual analysis of the control tables, generated from the control surfaces, shows that there are some differences on the extension value for different pairs of queue and arrival. As can be seen from the comparison of the surfaces shown in Figures 1 and 2, these differences are specially noted for the situations related to the arrival intervals from zero to five and greater than 15, for all possible queue values. The importance of the differences observed in the fuzzy signal controller response and traffic performance can be evaluated based on the simulation study described in the next section.

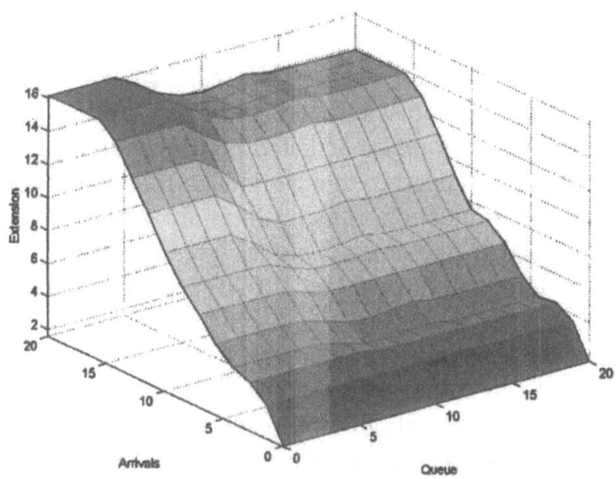


*Figure 1.* Control surface related to Case 01

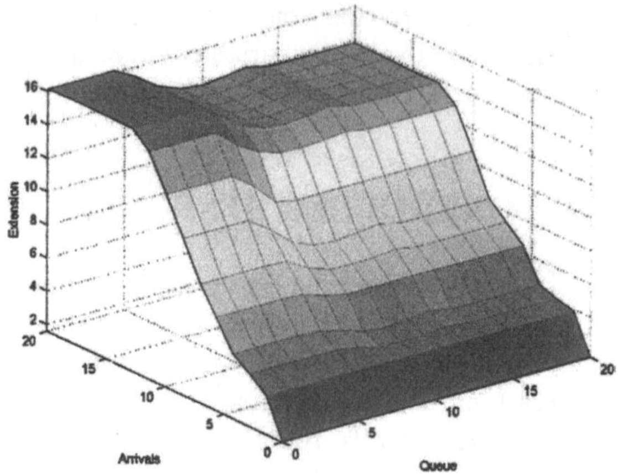


*Figure 2.* Control surface related to Case 03

# 5. SIMULATION STUDY WITH HUTSIM SOFTWARE

Control tables generated for each of the case studied were introduced through HUTSIM software. This software provides, among other features, microscopic traffic simulation under actuated control strategies, including the fuzzy signal controller considered in this work. The following conditions were considered during the simulation runs: minimum green time of 5 (five) seconds, maximum number of extensions to the green indication equal to five, and after extensions less than or equal to 2 (two) seconds the current green is ended.

Ten independent one-hour simulation runs for each of the cases considered were performed and refer to the three different traffic volume levels shown in Table 03. The simulation results for the controller response, averaged over the simulation hours, are shown in Table 06, while those regarding the traffic performance are presented in Table 07.

The coefficient of variation (CV), expressed in percentage, was calculated in order to verify the variability of the controller response and traffic performance along the ten independent one-hour simulation runs for each case studied. Figures 3 to 5 show the CV for the results related to the major street, whilst Figures 6 to 8 show the results for the minor street.

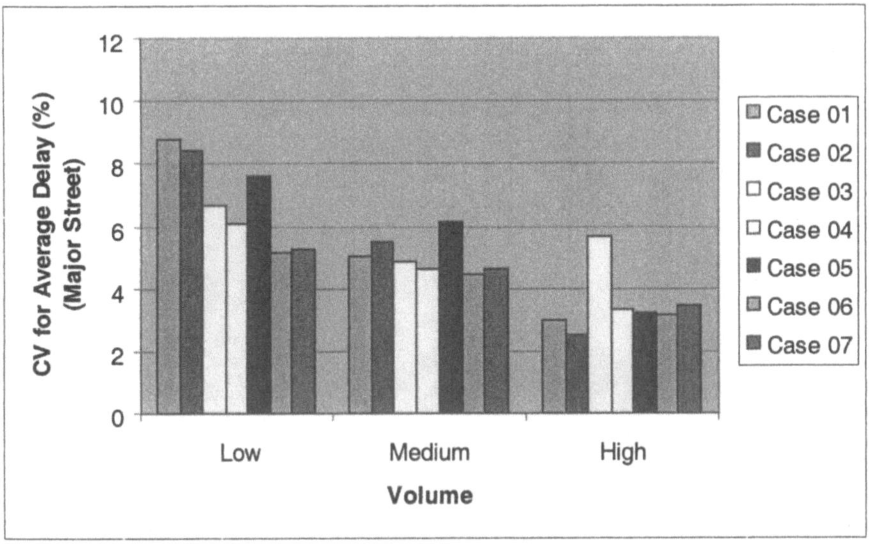


*Figure 3.* Coefficient of variation for average delay at Major Street

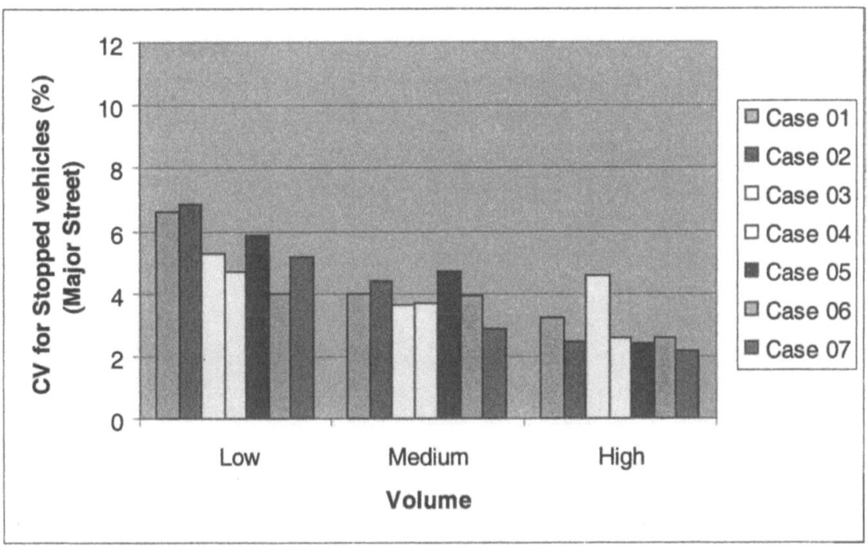


*Figure 4.* Coefficient of variation for stopped vehicles at Major Street

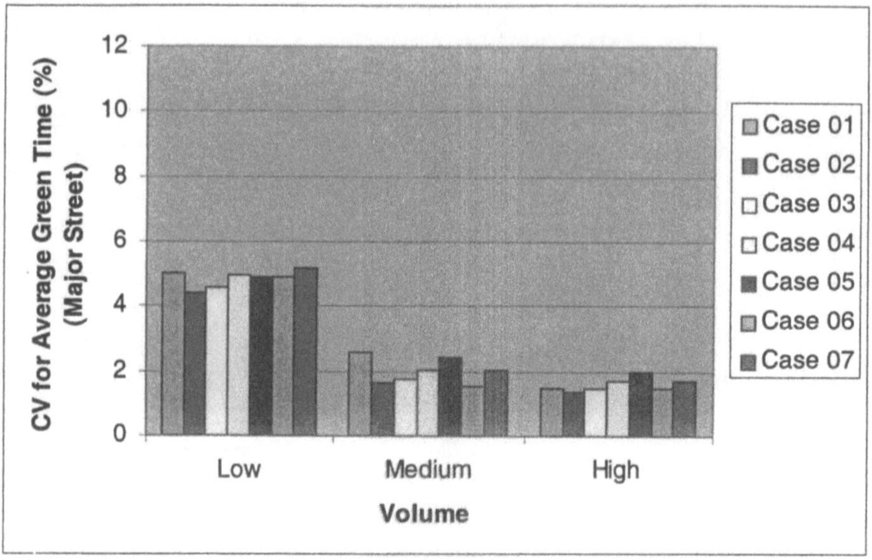


*Figure 5.* Coefficient of variation for average green time at Major Street

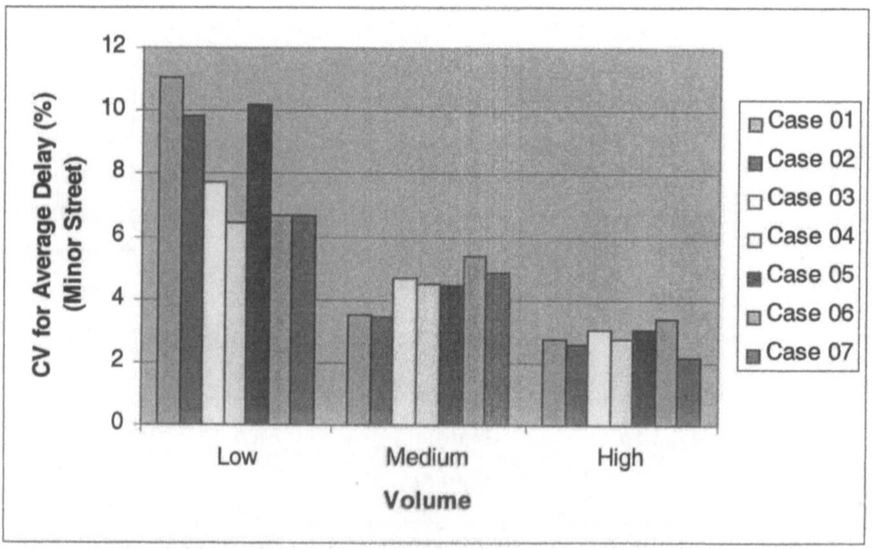


*Figure 6.* Coefficient of variation for average delay at Minor Street

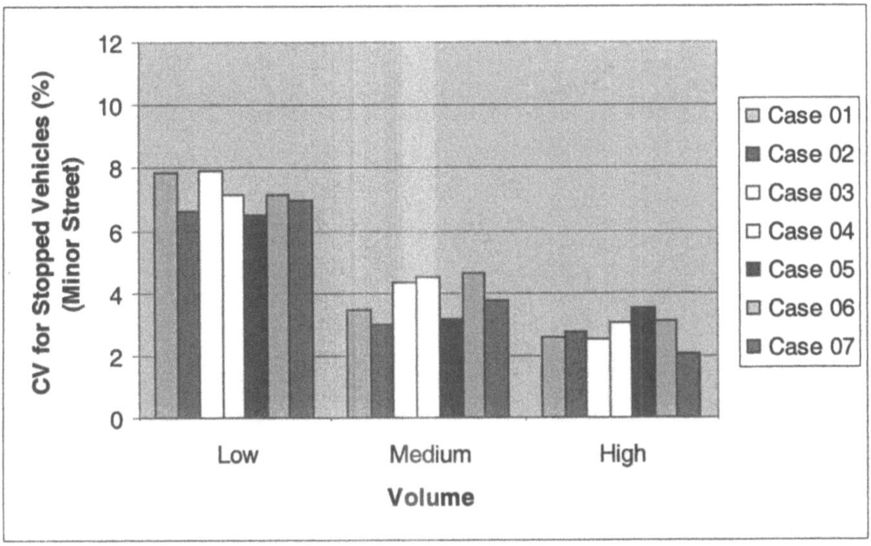


*Figure 7.* Coefficient of variation for stopped vehicles at Major Street

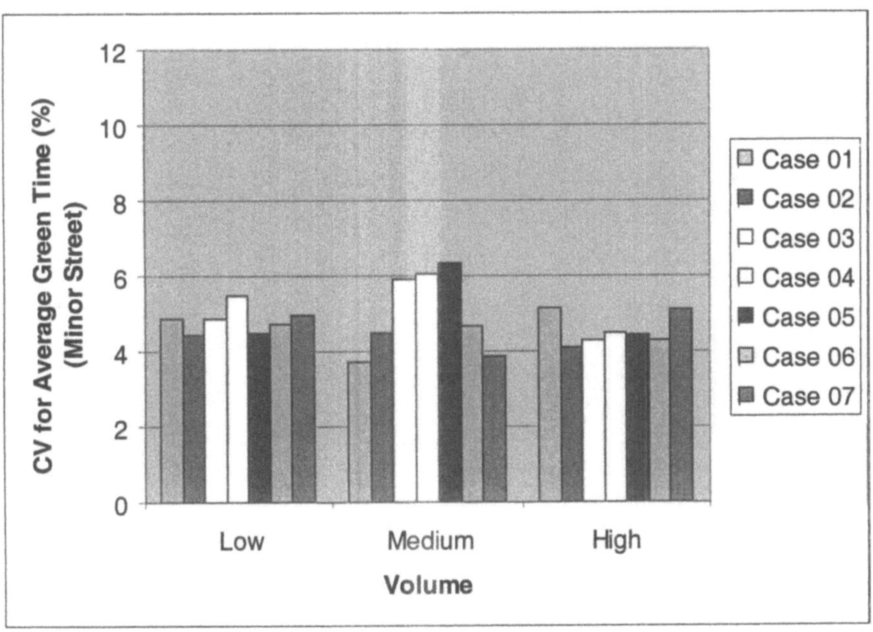


*Figure 8.* Coefficient of variation for average green time at Major Street

*Table 6.* Simulation results related to the controller's operation in the cases studied

| Case | Major street: average green time (s) | | | Minor street: average green time (s) | | | Intersection: average cycle (s) | | |
|---|---|---|---|---|---|---|---|---|---|
| | Volume level | | | Volume level | | | Volume level | | |
| | Low | Medium | High | Low | Medium | High | Low | Medium | High |
| 01 | 19.01 | 31.15 | 41.02 | 10.42 | 16.78 | 22.93 | 41.44 | 59.88 | 75.97 |
| 02 | 18.98 | 30.97 | 40.77 | 10.50 | 16.33 | 22.88 | 41.48 | 59.25 | 75.61 |
| 03 | 19.55 | 34.83 | 45.71 | 10.62 | 17.84 | 24.97 | 42.18 | 64.59 | 82.68 |
| 04 | 19.48 | 34.94 | 45.08 | 10.59 | 17.58 | 24.60 | 42.09 | 64.41 | 81.70 |
| 05 | 18.93 | 31.44 | 41.39 | 10.46 | 16.94 | 22.94 | 41.39 | 60.28 | 76.35 |
| 06 | 18.15 | 28.99 | 37.97 | 10.38 | 15.94 | 21.63 | 40.55 | 56.87 | 71.60 |
| 07 | 18.09 | 29.00 | 38.20 | 10.35 | 15.88 | 22.03 | 40.46 | 56.81 | 72.19 |
| F-value | 3.965 | 141.977 | 199.988 | 0.408 | 7.793 | 13.720 | 4.053 | 52.498 | 81.635 |

*Table 7.* Simulation results related to traffic performance

| Case | Major street: average delay (s/veh) | | | Minor street: average delay (s/veh) | | | Intersection: average delay (s/veh | | |
|---|---|---|---|---|---|---|---|---|---|
| | Volume level | | | Volume level | | | Volume level | | |
| | Low | Medium | High | Low | Medium | High | Low | Medium | High |
| 01 | 6.93 | 11.66 | 16.22 | 10.63 | 15.88 | 21.22 | 7.82 | 12.70 | 17.57 |
| 02 | 7.02 | 11.56 | 16.28 | 10.60 | 16.14 | 21.45 | 7.88 | 12.68 | 17.69 |
| 03 | 6.96 | 11.33 | 16.37 | 10.75 | 17.68 | 22.90 | 7.87 | 12.92 | 18.13 |
| 04 | 6.89 | 11.37 | 16.40 | 10.69 | 17.69 | 23.06 | 7.81 | 12.92 | 18.22 |
| 05 | 6.98 | 11.70 | 16.08 | 10.57 | 16.35 | 21.91 | 7.84 | 12.85 | 17.65 |
| 06 | 7.16 | 11.58 | 15.81 | 10.11 | 15.61 | 20.46 | 7.88 | 12.60 | 17.09 |
| 07 | 7.21 | 11.54 | 15.79 | 10.15 | 15.60 | 20.54 | 7.93 | 12.56 | 17.08 |
| F-value | 0.600 | 0.558 | 1.879 | 0.828 | 15.289 | 28.516 | 0.255 | 0.916 | 8.923 |
| Case | Major street: stopped vehicles (%) | | | Minor street: stopped vehicles (%) | | | Intersection: stopped vehicles (%) | | |
| | Volume level | | | Volume level | | | Volume level | | |
| | Low | Medium | High | Low | Medium | High | Low | Medium | High |
| 01 | 37.41 | 47.31 | 56.16 | 53.73 | 60.40 | 65.58 | 41.38 | 50.59 | 58.72 |
| 02 | 37.83 | 46.81 | 55.94 | 53.42 | 61.37 | 66.03 | 41.60 | 50.45 | 58.68 |
| 03 | 37.65 | 45.06 | 54.28 | 53.05 | 62.53 | 66.20 | 41.34 | 49.42 | 57.49 |
| 04 | 37.29 | 44.98 | 54.06 | 52.84 | 62.69 | 67.14 | 41.02 | 49.39 | 57.58 |
| 05 | 37.60 | 47.54 | 55.36 | 53.30 | 61.21 | 66.30 | 41.41 | 50.92 | 58.33 |
| 06 | 38.73 | 48.20 | 55.48 | 52.19 | 60.82 | 66.16 | 42.00 | 51.36 | 58.34 |
| 07 | 39.22 | 48.17 | 55.57 | 52.03 | 61.14 | 66.01 | 42.34 | 51.42 | 58.41 |
| F-value | 1.189 | 5.421 | 2.416 | 0.275 | 1.274 | 0.640 | 1.262 | 2.769 | 1.682 |

# 6. ANALYSIS OF THE SIMULATION RESULTS

The simulation results were submitted to an Analysis of Variance, whose F-values are shown in Tables 6 and 7. These values were calculated for each variable studied at each of the volume levels considered. The values presented in shadowed cells are those statistically significant at the 5% risk level, as $F_{critical}$ is equal to 2.25.

The evaluation of the impacts produced on the controller response shows that this response is effectively affected by the modifications made at the fuzzy sets. This is especially true for the medium and high traffic volume level as shown in Table 6. At medium volume the average cycle length varies from 56.81s to 64.59s, which means that the number of cycles per hour is modified from about 63 in Case 7 to 56 in Case 3. For the high volume level the number of cycles per hour varies from around 50 to 44. Based on Figures 5 and 8 it is possible to verify that the variability of the average green time along the 10 one-hour simulation runs is different for each case and volume level studied. At Major Street, the average green time is more stable for case 2 for all volume levels. However, at the Minor street the average green time is more stable for case 2 when the volume is low and high, and for case 1 when the volume is medium.

Based on the results, it is also possible to verify that while the modifications made to the input variables' fuzzy sets impacted significantly upon most of the variables associated with the controller response, they only impact on traffic performance basically when the volume level is medium or high. For these volume levels, the percentage of stopped vehicles at the minor street and the delay at the major street are significantly affected by the modifications made to the input fuzzy sets. The results of the CV shown in Figures 3, 4, 6 and 7 can explain these results as the CV for the performance measures related to the low traffic volume is much higher than its counterpart for the medium and high volumes. At the Major Street, the more stable results for the average delay were found for case 6 for medium volume and for case 2 for high volume. In the case of Minor Street, case 2 produces the least CV for medium volume and case 7 for the high volume. Regarding the percentage of stopped vehicles, the more stable results are provided by case 7 at the Major Street for the medium and high traffic volumes, and at the Minor Street for high volume. For the medium traffic volume at the Minor Street, the least CV was found for case 2.

Other analysis to be performed is related to the generic impact of the modifications at the fuzzy sets limits on the controller response and traffic performance. Remembering that case 4 corresponds to the smallest extreme values for the fuzzy sets, case 1 is the basic case and case 7 is related to the biggest fuzzy sets' extreme values, some findings can be extracted from Tables 6 and 7. As the fuzzy set limits increase:

- the green times for both streets decrease;
- the average delay for the low traffic volume increases at the Major street and decreases at the Minor Street;

- while there is no trend observed for average delay for the medium volume at the Major Street, this performance measure decreases at the Minor Street;
- the average delay decreases at both streets for high volume levels;
- the percentage of stopped vehicles at the Major Street increases for the low and medium traffic volumes. No trend in the percentage of stopped vehicles was found for the high volume at the Major Street and for all volume levels at the Minor Street.

# 7. CONCLUSIONS

The study conducted showed that the definitions of the fuzzy sets associated to the input variables of a given fuzzy signal controller must be taken very carefully as they affect both the actual controller response and the traffic performance.

Although not studied here, it is possible to infer from the results that if the controller and traffic operations are sensitive to small variations in the input variables' fuzzy sets, they should also be very much influenced by modifications to the output variable fuzzy sets.

For the situation analysed in this work, it was found that increasing the fuzzy sets limits causes general improvement for the traffic performance at the Minor Street. These, as well as other numerical results are of course dependent on the characteristics of the intersection and control strategy in question.

Thus, in order to allow for the traffic control objectives at a given intersection to be well captured by the fuzzy signal controller, special care must be taken in defining the fuzzy sets linked to all control linguistic variables. A very straightforward method for capturing these objectives from expert knowledge must therefore be applied. In light of this, a study geared towards defining a procedure to build fuzzy sets and rules from traffic control experts' knowledge of a given city is being conducted by this research team.

# 8. ACKNOWLEDGEMENTS

The authors wish to thank Peter Pecovnik, from the Laboratory of Transportation Engineering of Helsinki University of Technology, for his support with the simulation runs using HUTSIM. This project has been financially supported by CNPq and CAPES, the Brazilian agencies for R&D.

# REFERENCES


Jacques M. A. P., Pursula M., Niittymäki J., Kosonen I. Analysis of Different Defuzzification Methods Applied to Fuzzy Signal Controllers. Proceedings of the XII Panamerican Traffic & Transport Engineering Conference; 2002 November 18-21; Quito. CD-Rom.

Jacques M. A. P., Pursula M., Niittymäki J., Kosonen I. The Impact of Different Approximate Reasoning Methods on Fuzzy Signal Controllers. Proceedings of the 13th Mini-EURO Conference; 2002 June 10-13; pp. 184-192.

Niittymäki, J. *Fuzzy traffic signals – control principles and applications.* Espoo: Helsinki University of Technology, Laboratory of Transportation Engineering, Publication 103, 2002.

Zadeh L. A. Fuzzy Sets. Information and Control 1965; 8: 338-353.

Lee C. C. Fuzzy Logic in Control Systems: Fuzzy Logic Controller – Part I. IEEE Transactions on Systems, Man, and Cybernetics 1990; Vol. 20, 2: 404-418.

Pappis C. P., Mamdani H. A Fuzzy Logic Controller for a Traffic Junction. *IEEE* Transactions on Systems, Man and Cybernetics 1997; Vol. SMC-7, 10: 707-717.

Jacques M. A. P., Niittymäki J, Pursula M. Analysing Different Fuzzy Traffic Signal Controllers for Isolated Intersections. Preprints of TRB 81st Annual Meeting; 2002 January 13-17; Washington, D.C, CD-Rom.

Niittymäki J., Nevala R. Fuzzy Adaptive Traffic Signal Control – Principles and Results. Proceedings of Joint 9th IFSA World Congress and 20th NAFIPS International Conference; 2001 July; Vancouver.

Kim S. *Application of Petri Networks and Fuzzy Logic to Advanced Traffic Management Systems.* New York: Polytechnic University, Ph.D Thesis, 1994.

Favilla Jr., J., Machion A., Gomide F. Fuzzy Traffic Control: Adaptive Strategies. Proceedings of the 2nd IEEE International Conference on Fuzzy Systems; 1993 March; San Francisco: 1: 506-511.

Trabia, M.B., Kaseko M. S., Ande M. A Two-stage Fuzzy Logic Controller for Traffic Signals. Transportation Research Part C 1999; 7: 353-367.

Niittymäki, J., Pursula M. Signal Control Using Fuzzy Logic. Fuzzy Sets and Systems 2000, 116: 11-22.

Kosonen I. *HUTSIM – Urban Traffic Simulation and Control Model: Principles and Applications*, Espoo: Helsinki University of Technology, Transportation Engineering, Publication 100, 1999.

# Chapter 19

# RELIABILITY EVALUATION OF REALISTIC STRUCTURES USING FEM

**Jungwon Huh, Achintya Haldar, and Seung Y. Lee**

## 1. INTRODUCTION

The finite element method (FEM)-based approach is commonly used to realistically study the behavior of complicated real structures. It is a very powerful tool commonly used in many different engineering disciplines to analyze structural systems. With this approach, it is straightforward to consider complicated geometric arrangements, various sources of nonlinearity, boundary or support, and connection conditions, different materials, and load path to failure.

To study the behavior of real steel frame structures, material and geometric nonlinearities need to be considered in the formulation. Furthermore, the structural members are connected to each other by various types and forms of connections. For steel structures, these connections are usually modeled as fully restrained (FR). However, extensive experimental studies indicate that they are partially restrained (PR) or semi-rigid. Thus, consideration of realistic rigidity of connections is warranted to study the behavior of steel frames. The connection rigidity adds another major source of nonlinearity in the formulation. Another major weakness of ordinary steel frames is their inability to transfer horizontal loads, e.g., high winds, strong earthquakes, and ocean waves, etc., effectively because of their relative flexibility. Reinforced concrete (RC) shear walls can be used to increase the lateral stiffness of flexible steel frames. This dual system is very common in seismically active regions. The behavior of the dual systems can be studied effectively by representing them with finite elements.

The deterministic study of the nonlinear behavior of real structural systems using finite elements is an efficient and logical approach. However, most of the parameters required for the deterministic evaluation are random or uncertain in nature. The uncertainty in modeling the flexible connection behavior cannot be overlooked. It is expected that a more realistic modeling of connection conditions would considerably increase the amount of uncertainty in the evaluation of the structural behavior (Haldar and Mahadevan, 2000a; Huh and Haldar, 2002). The implications of proper consideration of the connection conditions and the uncertainty in modeling them are comprehensively addressed in this study.

The presence of reinforced concrete shear walls in a flexible steel frame adds several other sources of uncertainty. Furthermore, the evaluation of the propagation of uncertainties from the finite element level to the structural level is also expected to be very challenging. Integrating the finite element method and the first-order reliability method (FORM) (Haldar and Mahadevan, 2000a), a reliability method commonly used in the profession, the authors developed a new method. They called it a stochastic finite element method (SFEM)-based algorithm (Haldar and Mahadevan, 2000b). The method is briefly discussed in the following sections to evaluate the reliability of a simple steel frame with FR connections, a steel frame with PR connections, and a steel frame with RC shear walls systems under the static loading condition.

## 2. DETERMINISTIC FEM FORMULATION

For the problem under consideration, the iterative strategy is required to incorporate both the nonlinear behavior of the structure and the uncertainty in modeling all the parameters in the formulation. Since the FEM is an integral part of the algorithm, the efficiency of the deterministic FEM being used is very important in the iterative strategy. In this study, the assumed stress-based FEM is used for the basic deterministic FEM representation of the nonlinear structure. In this approach, an explicit form of the tangent stiffness is formulated, satisfying joint equilibrium and displacement compatibility. The method is efficient, economical, and accurate, particularly for frame structures, since fewer elements are needed to model the frame and the numerical integration to obtain the tangent stiffness can be completely eliminated. Details of the deterministic FEM can not be provided here due to lack of space; however, they are widely available in the literature (Haldar and Nee, 1989; Kondoh and Atluri, 1987; Haldar and Gao, 1997).

Since the problem under consideration is essentially nonlinear even when the load is small, the following linear iterative strategy is used:

$$\mathbf{K}^{(n)}\,\Delta\mathbf{D}^{(n)} = \mathbf{F}^{(n)} - \mathbf{R}^{(n-1)} \tag{1}$$

where $\mathbf{K}^{(n)}$, $\Delta\mathbf{D}^{(n)}$, and $\mathbf{F}^{(n)}$ are the tangent stiffness matrix, the displacement increment vector, and the external load vector at the $n^{th}$ iteration, respectively; and $\mathbf{R}^{(n-1)}$ is the internal force vector at the $(n-1)^{th}$ iteration.

Using the assumed stress-based FEM, the tangent stiffness matrix and the internal force vector can be calculated as:

$$\mathbf{K} = \mathbf{A}_{\sigma do}^{T}\mathbf{A}_{\sigma\sigma}^{-1}\mathbf{A}_{\sigma do} + \mathbf{A}_{ddo} \tag{2}$$

and

$$\mathbf{R} = -\mathbf{A}_{\sigma do}^{T}\mathbf{A}_{\sigma\sigma}^{-1}\mathbf{R}_{\sigma} + \mathbf{R}_{do} = -\mathbf{A}_{\sigma do}^{T}\mathbf{R}_{A\sigma} + \mathbf{R}_{do} \tag{3}$$

where $\mathbf{A}_{\sigma\sigma}$ is the elastic property matrix, $\mathbf{A}_{\sigma do}$ is the transformation matrix, $\mathbf{A}_{ddo}$ is the geometric stiffness matrix, and $\mathbf{R}_{do}$ is the homogeneous part of the internal nodal force vector. They cannot be described further here but can be found elsewhere (Haldar and Nee, 1989; Kondoh and Atluri, 1987; Haldar and Gao, 1997).

## 2.1. Incorporation of PR Connections into the FEM

As mentioned earlier, the presence of partial connection rigidity needs to be incorporated into the deterministic analysis of structures to capture their realistic behavior. In general, the relationship between the moment $M$, transmitted by the connection, and the relative rotation angle $\theta$ is used to represent the flexible behavior of a connection. Among the many alternatives (Richard model, piecewise linear model, polynomial model, exponential model, B-Spline model, etc.), the Richard four-parameter moment-rotation model is chosen here to represent the flexible behavior of a connection. It is expressed as (Richard and Abbott, 1975; Huh and Haldar, 2002):

$$M = \frac{(k-k_p)\theta}{\left(1+\left|\dfrac{(k-k_p)\theta}{M_0}\right|^{N}\right)^{1/N}} + k_p\theta \tag{4}$$

where $M$ is the connection moment, $\theta$ is the relative rotation between the connecting elements, $k$ is the initial stiffness, $k_p$ is the plastic stiffness, $M_0$ is the reference moment, and $N$ is the curve shape parameter. These parameters are identified in Figure 1.

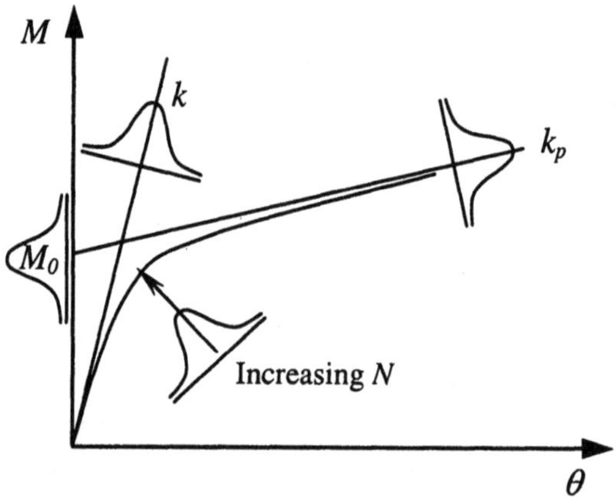


Figure 1. M-θ curve using the Richard Model

In the finite element representation, all the members are represented by the appropriate beam-column elements. To represent a PR connection, an ordinary beam-column element can be used for the numerical analysis, however, its stiffness needs to be updated at each iteration since the stiffness representing the partial rigidity depends on the rotation $\theta$. This can be accomplished by updating the Young's modulus as (Richard and Abbott, 1975; Huh and Haldar, 2002):

$$E_C(\theta) = \frac{l_C}{I_C} K_C(\theta) = \frac{l_C}{I_C} \frac{\partial M(\theta)}{\partial \theta} \tag{5}$$

where $l_C$, $I_C$, and $K_C(\theta)$ are the length, the moment of inertia, and the tangent stiffness of the connection element, respectively. $K_C(\theta)$ is calculated using Equation (4) and can be shown to be:

$$K_C(\theta) = \frac{dM}{d\theta} = \frac{(k - k_p)}{\left(1 + \left|\frac{(k - k_p)\theta}{M_0}\right|^N\right)^{\frac{N+1}{N}}} + k_p \tag{6}$$

The basic FEM formulation of the structure remains unchanged even in the presence of PR connections. Although it appears to be complicated, the evaluation of the behavior of steel frames in the presence of PR connections is simple using the algorithm discussed here.

## 2.2. Incorporation of RC Shear Walls into the FEM

The incorporation RC shear wall in a steel frame is a little more challenging. The basic steel frame is represented by two-dimensional (2D) beam-column elements and the shear walls are represented by four-node plane stress elements. The static governing equation for the combined system can be represented in the incremental form as:

$$\mathbf{K}_T^{(n)}\,\Delta\mathbf{D}^{(n)} = \mathbf{F}^{(n)} - \left[\mathbf{R}^{(n-1)} + \mathbf{K}_{sh}^{(n-1)}\,\mathbf{D}^{(n-1)}\right] \qquad (7)$$

where $\mathbf{K}_T^{(n)} = \mathbf{K}^{(n)} + \mathbf{K}_{sh}^{(n)}$, $\mathbf{K}_{sh}^{(n)}$ is the global tangent stiffness matrix of the shear walls at the $n^{\text{th}}$ iteration, $\mathbf{K}_{sh}^{(n-1)}\,\mathbf{D}^{(n-1)}$ is the internal force vector of the shear walls at the $(n\text{-}1)^{\text{th}}$ iteration, and $\mathbf{K}^{(n)}$, $\Delta\mathbf{D}^{(n)}$, $\mathbf{F}^{(n)}$, and $\mathbf{R}^{(n-1)}$ are defined earlier. Using the assumption the stress-based finite element method, Equations (2) and (3) can be used to define the tangent stiffness matrix and the internal force vector of the frame required to solve Equation (7).

A four-node plane stress element is used to incorporate the presence of RC shear walls in the steel frame. An explicit expression for the stiffness matrix of the plate elements is necessary for efficient reliability analysis. To achieve this, the shape of the shear wall is restricted to be rectangular. Two displacement (horizontal and vertical) dynamic degrees of freedom (DDOFs) are used at each node point. Based on an extensive literature review and discussions with experts on finite element methods, it was concluded that the rotation at a node point could be overlooked (Lee and Haldar, 2003). To incorporate the shear wall stiffness into the frame structure, the components of the shear wall stiffness are added to the corresponding frame stiffness components in Equation (7). The explicit form of a stiffness matrix of a 4-node plane stress element can be obtained as (Lee, 2000):

$$\mathbf{K}_{sh} = \frac{t}{4\gamma}\mathbf{A}^T\mathbf{E}\mathbf{A} + \frac{t}{12\gamma}\mathbf{B}^T\mathbf{E}\mathbf{B} + \frac{t\gamma}{12}\mathbf{C}^T\mathbf{E}\mathbf{C} \qquad (8)$$

where $2a$ and $2b$ are the long and short dimensions of the rectangular shear wall, respectively, $t$ is the thickness of the wall, $\gamma$ is the ratio of $b$ and $a$; i.e., $\gamma = b/a$. The matrixes **A**, **B**, **C**, and **E** in Equation (8) can be represented as:

$$\mathbf{A} = \begin{bmatrix} -\gamma & 0 & \gamma & 0 & \gamma & 0 & -\gamma & 0 \\ 0 & -1 & 0 & -1 & 0 & 1 & 0 & 1 \\ -1 & -\gamma & -1 & \gamma & 1 & \gamma & 1 & -\gamma \end{bmatrix} \tag{9}$$

$$\mathbf{B} = \begin{bmatrix} 0 & 0 & 0 & 0 & 0 & 0 & 0 & 0 \\ 0 & 1 & 0 & -1 & 0 & 1 & 0 & -1 \\ 1 & 0 & -1 & 0 & 1 & 0 & -1 & 0 \end{bmatrix} \tag{10}$$

$$\mathbf{C} = \begin{bmatrix} 1 & 0 & -1 & 0 & 1 & 0 & -1 & 0 \\ 0 & 0 & 0 & 0 & 0 & 0 & 0 & 0 \\ 0 & 1 & 0 & -1 & 0 & 1 & 0 & -1 \end{bmatrix} \tag{11}$$

and

$$\mathbf{E} = \frac{E_c}{1-\nu^2} \begin{bmatrix} 1 & \nu & 0 \\ \nu & 1 & 0 \\ 0 & 0 & \frac{1-\nu}{2} \end{bmatrix} \tag{12}$$

where $E_c$ is the modulus of elasticity and $\nu$ is the Poisson's ratio of RC shear walls.

Different types of shear walls are used in practice. The reinforced concrete (RC) shear wall is appeared to be the most widely used and is considered in this study. Thus, two additional parameters, namely, the modulus of elasticity and the Poisson ratio of concrete, are necessary in the deterministic formulation as in Equation (12). The tensile strength of concrete is very small compared to its compressive strength. Cracking may develop at a very early stage of loading. The behavior of a RC shear wall before and after cracking can be significantly different and needs to be considered in any realistic evaluation of the behavior of shear walls. The subject of cracking in RC panels has been extensively researched and reported in the literature (Gupta and Akbar, 1983; Liauw and Kwan, 1985; Vecchio, 1989; Lefas et. al., 1990; Inoue et. al., 1997). It was observed that the degradation of the stiffness of the shear walls occurs after cracking and can be considered effectively by reducing the modulus of elasticity of the shear walls. Based on the experimental research reported by Lefas et al. (1990), the degradation of the stiffness after cracking can vary from 40% to 70% of the original stiffness depending on the amount of reinforcement and the intensity of axial loads. In this study, the behavior of a shear wall after cracking is considered by introducing the degradation of the shear wall stiffness based on the observations made by Lafas et al. (1990). The shear wall is assumed to develop cracks when the tensile stress in concrete exceeds the prescribed value. The rupture strength of concrete, $f_r$, according to the American Concrete Institute (ACI, 1999) is assumed to be $f_r = 7.5 \times \sqrt{f_c'}$, where $f_c'$ is the compressive strength of concrete.

Once the explicit form of the stiffness matrix of shear walls is obtained using Equation (8), the information can be incorporated in Equation (7) to study the static behavior of the combined system. The finite element representation of the RC shear walls is kept simple in order to minimize the number of basic random variables present in the SFEM formulation. More sophisticated methods can be attempted in future studies, if desired. Reliability evaluation procedures are emphasized in this paper.

The governing equation of the combined system consisting of steel frame and RC shear walls, i.e., Equation (7), is solved using the modified Newton-Raphson method with the arc-length procedure. The deterministic formulation of the problem discussed above is expected to be very accurate and efficient. The formulation now needs to be incorporated in the reliability analysis in the context of the SFEM.

# 3. STOCHASTIC FINITE ELEMENT FORMULATION

Haldar and Mahadevan (2000b) discussed the basic SFEM-based algorithm used in this study. The SFEM to be used here is based on the first-order reliability method (FORM). In the context of FORM, a limit state function or performance function is required. Without losing any generality, the limit state function $g$ can be expressed in terms of the set of basic random variables $\mathbf{x}$ (e.g., loads, material properties and structural geometry), the set of displacements $\mathbf{u}$ and the set of load effects $\mathbf{s}$ (except the displacements, such as internal forces). The displacement $\mathbf{u} = \mathbf{QD}$, where $\mathbf{D}$ is the global displacement vector and $\mathbf{Q}$ is a transformation matrix. The limit state function can be expressed as $g(\mathbf{x}, \mathbf{u}, \mathbf{s}) = 0$. For reliability computation, it is convenient to transform $\mathbf{x}$ into the standard normal space $\mathbf{y} = \mathbf{y}(\mathbf{x})$ such that the elements of $\mathbf{y}$ are statistically independent and have a standard normal distribution. An iteration algorithm is used to locate the design point (the most likely failure point) on the limit state function using first-order approximation. During each iteration, the structural response and the response gradient vectors are calculated using the finite element models discussed in the previous section. The following iteration scheme can be used for finding the coordinates of the design point:

$$\mathbf{y}_{i+1} = \left[ \mathbf{y}_i^T \alpha_i + \frac{g(\mathbf{y}_i)}{|\nabla g(\mathbf{y}_i)|} \right] \alpha_i \tag{13}$$

where

$$\nabla g(\mathbf{y}) = \left[ \frac{\partial g(\mathbf{y})}{\partial y_1}, \Lambda, \frac{\partial g(\mathbf{y})}{\partial y_n} \right]^T \text{ and } \alpha_i = -\frac{\nabla g(\mathbf{y}_i)}{|\nabla g(\mathbf{y}_i)|} \tag{14}$$

To implement the algorithm and assuming the limit state equation has a general form of $g(\mathbf{x},\mathbf{u},\mathbf{s}) = 0$, the gradient $\nabla g(\mathbf{y})$ of the limit state function in the standard normal space can be derived as (Haldar and Mahadevan, 2000b):

$$\nabla g(\mathbf{y}) = \left[ \frac{\partial g}{\partial \mathbf{s}} \mathbf{J}_{s,x} + \left( \mathbf{Q} \frac{\partial g}{\partial \mathbf{u}} + \frac{\partial g}{\partial \mathbf{s}} \mathbf{J}_{s,D} \right) \mathbf{J}_{D,x} + \frac{\partial g}{\partial \mathbf{x}} \right] \mathbf{J}_{y,x}^{-1} \tag{15}$$

where $\mathbf{J}_{i,j}$'s are the Jacobians of transformation (e.g., $\mathbf{J}_{s,x}=\partial \mathbf{s}/\partial \mathbf{x}$). Once the coordinates of the design point $\mathbf{y}^*$ are evaluated with a preselected convergence criterion, the reliability index $\beta$ can be evaluated as:

$$\beta = \sqrt{(\mathbf{y}^*)^T (\mathbf{y}^*)} \tag{16}$$

The evaluation of Equation (15) will depend on the problem under consideration and the performance functions used. It is necessary to determine the three partial differentials ($\partial g/\partial \mathbf{s}$, $\partial g/\partial \mathbf{u}$, and $\partial g/\partial \mathbf{x}$) and four Jacobians ($\mathbf{J}_{y,x}$, $\mathbf{J}_{s,x}$, $\mathbf{J}_{s,D}$, and $\mathbf{J}_{D,x}$). They are evaluated in the following sections in the context of the reliability analysis of a simple steel frame, a steel frame with PR connections, and a steel frame with RC shear walls.

## 3.1. Limit State Functions

Reliability is always estimated with respect to a performance function or a limit state. Both strength and serviceability limit states are necessary to study the underlying risk of realistic frame structures. The following limit states are used for this study.

### 3.1.1. Strength Limit State Functions

According to the Load and Resistance Factor Design (LRFD) design guidelines (AISC, 2001) published by the American Institute of Steel Construction (AISC), the strength limit state functions for members in a two-dimensional steel frame can be defined as:

$$g(\mathbf{x},\mathbf{u},\mathbf{s}) = 1.0 - \frac{P_u}{P_n} - \frac{8}{9}\left(\frac{M_{ux}}{M_{nx}}\right); \quad if\ \frac{P_u}{\phi P_n} \geq 0.2 \tag{17}$$

and

$$g(\mathbf{x},\mathbf{u},\mathbf{s}) = 1.0 - \left(\frac{P_u}{2P_n} + \frac{M_{ux}}{M_{nx}}\right); \quad if\ \frac{P_u}{\phi P_n} < 0.2 \tag{18}$$

where $\phi$ is the resistance factor, $P_u$ is the required tensile and compresssive strength, $P_n$ is the nominal tensile and compressive strength, $M_{ux}$ is the required flexural strength, and $M_{nx}$ is the nonminal flexural strength. $P_u$ and $M_{ux}$ in Equations (17) and (18) are unfactored load effects. Nominal axial load and bending moment capacity of a steel member can be calculated using the procedures suggested in the AISC's LRFD design guidelines (AISC, 2001).

For the steel frame with RC shear walls system considered in this study, it can be noted that some of the members in the frame are connected to the shear walls. The shear walls are expected to prevent local and lateral torsional buckling of steel members, thus improving their strength. Therefore, to consider the strength failure probability of the weakeest steel members in the dual system, this study considers the failure of steel members where shear walls are not present.

### 3.1.2. Serviceability Limit State Function

The vertical deflection at the midspan of a beam and the lateral displacement at the top of the frame are considered to be the two serviceability performance functions in this study. For the serviceability criterion, the limit state function is represented as:

$$g(\mathbf{x},\mathbf{u},\mathbf{s}) = 1.0 - \frac{\delta}{\delta_{limit}} \tag{19}$$

where $\delta$ is the calculated displacement component and $\delta_{limit}$ is the prescribed maximum or the allowable value of the displacement component. The allowable or the prescribed maximum value of the displacement component is generally suggested in design guidelines, as discussed in Section 4.

## 3.2. Evaluation of Partial Derivatives and Jacobians

### 3.2.1. Evaluation of Partial Derivatives

To implement the algorithm, the three partial derivatives ($\partial g/\partial \mathbf{s}$, $\partial g/\partial \mathbf{u}$, and $\partial g/\partial \mathbf{x}$) in Equation (15) need to be evaluated using both the strength and serviceability limit state functions.

The strength limit state function is considered first. Since the strength limit state functions represented by Equations (17) and (18) do not explicitly contain any displacement component, $\partial g/\partial \mathbf{u}$ is zero. The basic random variables in the limit state functions should be defined for the calculation of $\partial g/\partial \mathbf{x}$. The Young's modulus $E$, sectional area $A$, yield stress $F_y$, plastic modulus $Z_x$, and the moment of inertia of a cross-section $I$ along with the external force $F$ are considered to be basic random variables in this study. Thus, $\partial g/\partial \mathbf{x}$ can be expressed as:

$$\frac{\partial g}{\partial \mathbf{x}} = \left[ \frac{\partial g}{\partial E} \quad \frac{\partial g}{\partial A} \quad \frac{\partial g}{\partial I} \quad \frac{\partial g}{\partial Z_x} \quad \frac{\partial g}{\partial F_y} \right] \tag{20}$$

And $\partial g/\partial \mathbf{s}$ can also be derived by taking the partial derivatives with respect to $P_u$ and $M_{ux}$ as:

$$\frac{\partial g}{\partial \mathbf{s}} = \left[ \frac{\partial g}{\partial P_u} \quad \frac{\partial g}{\partial M_{ux}} \right] \tag{21}$$

For the serviceability limit state represented by Equation (19), the three partial differentials can be expressed as follows:

$$\frac{\partial g}{\partial \mathbf{x}} = \frac{\partial g}{\partial \mathbf{s}} = \mathbf{0} \tag{22}$$

and

$$\frac{\partial g}{\partial \mathbf{u}} = \left[\frac{\partial g}{\partial \delta} \quad \mathbf{0}\right] \tag{23}$$

where $\partial g/\partial \delta = -1/\delta_{limit}$. The actual $\delta_{limit}$ depends on the structure under consideration.

### 3.2.2. Evaluation of Jacobians

As discussed previously, it is also necessary to determine four Jacobians of transformation to evaluate $\nabla g(\mathbf{y})$. Because of the triangular nature of the transformation, $\mathbf{J}_{y,x}$ and its inverse are easy to compute. Since $s$ is not an explicit function of the basic random variables $\mathbf{x}$, $\mathbf{J}_{s,x} = \mathbf{0}$. $\mathbf{J}_{s,D}$ and $\mathbf{J}_{D,x}$, however, are not easy to compute since $\mathbf{s}$, $\mathbf{D}$, and $\mathbf{x}$ are implicit functions of each other. The adjoint variable method (Arora and Haug, 1979; Haldar and Huh, 1998) is used to compute the product of the second term in Equation (15) directly rather than evaluating its constituent parts. An adjoint vector $\lambda$ can be introduced such that:

$$\lambda^T \mathbf{K}^{(n)} = \mathbf{Q}\frac{\partial g}{\partial \mathbf{u}} + \frac{\partial g}{\partial \mathbf{s}}\mathbf{J}_{s,D} \tag{24}$$

After some mathematical manipulation, it can be shown that:

$$\left(\mathbf{Q}\frac{\partial g}{\partial \mathbf{u}} + \frac{\partial g}{\partial \mathbf{s}}\mathbf{J}_{s,D}\right)\mathbf{J}_{D,x} = \lambda^T \mathbf{K}^{(n)} \mathbf{J}_{D,x} = \lambda^T \left(\frac{\partial \mathbf{F}}{\partial \mathbf{x}} + \frac{\partial \mathbf{R}^{(n-1)}}{\partial \mathbf{x}}\right) \tag{25}$$

For the strength limit state given represented by Equations (17) and (18), $\partial g/\partial \mathbf{u}$ is zero and $\partial g/\partial \mathbf{s}$ was derived in Equation (21). On the other hand, for the serviceability limit state, $\partial g/\partial \mathbf{s}$ is zero and $\partial g/\partial \mathbf{u}$ was derived in Equation (23).

At this stage, $\mathbf{J}_{s,D}$ needs to be estimated. Generally, when the strength limit state function is considered, the internal force vector $\boldsymbol{\sigma}$ is the only contribution of the load effects $\mathbf{s}$ and can be expressed as $\mathbf{s} = \mathbf{A}\boldsymbol{\sigma}$, where $\mathbf{A}$ is a transformation matrix with constant elements. Thus, $\mathbf{J}_{s,D}$ can be obtained as:

$$\mathbf{J}_{s,D} = \frac{\partial \mathbf{s}}{\partial \mathbf{D}} = \mathbf{A}\frac{\partial \boldsymbol{\sigma}}{\partial \mathbf{D}} = \mathbf{A}\left[\frac{\partial \boldsymbol{\sigma}}{\partial \mathbf{d}} \quad \mathbf{0}\right] \tag{26}$$

where **d** is the nodal displacement vector in the global coordinate for the element. In Equation (25), $\partial \mathbf{F}/\partial \mathbf{x}$ is easily obtained since the explicit dependence of **F** on the basic random variables is known, assuming the external load is not affected by the structural response.

Using Equation (3) and the fact that $\mathbf{R}_{do}$ and $\mathbf{A}^{T}{}_{\sigma do}$ are not functions of basic random variables, the derivative of **R** with respect to **x**, namely $\partial \mathbf{R}/\partial \mathbf{x}$ in Equation (25), can be expressed as:

$$\left.\frac{\partial \mathbf{R}}{\partial \mathbf{x}}\right|_{D,\sigma} = -\mathbf{A}^{T}_{\sigma do} \frac{\partial \mathbf{R}_{A\sigma}}{\partial \mathbf{x}} \tag{27}$$

where $\partial \mathbf{R}_{A\sigma}/\partial \mathbf{x}$ can be expressed for a beam-column element as:

$$\frac{\partial \mathbf{R}_{A\sigma}}{\partial \mathbf{x}} = \begin{bmatrix} \frac{\partial \mathbf{R}_{A\sigma}}{\partial E} & \frac{\partial \mathbf{R}_{A\sigma}}{\partial A} & \frac{\partial \mathbf{R}_{A\sigma}}{\partial I} & \mathbf{0} \end{bmatrix} \tag{28}$$

The $\lambda$ estimated for the serviceability and strength limit state can then be associated with the FEM algorithm by substituting $\lambda$ into the last part of Equation (25). Therefore the first part of Equation (25) is now available in a simple explicit form for the evaluation of the gradient of the limit state function in Equation (15).

## 3.3. Consideration of PR Connections in SFEM

The task now is to incorporate the uncertainties in the PR connection conditions into the unified SFEM formulation previously discussed. The four parameters of the Richard model, namely, $k$, $k_p$, $M_0$, and $N$, are considered to be basic random variables for connection elements. The following expression is considered for connection elements instead of Equation (28):

$$\frac{\partial \mathbf{R}_{A\sigma}}{\partial \mathbf{x}} = \begin{bmatrix} \frac{\partial \mathbf{R}_{A\sigma}}{\partial k} & \frac{\partial \mathbf{R}_{A\sigma}}{\partial k_p} & \frac{\partial \mathbf{R}_{A\sigma}}{\partial M_0} & \frac{\partial \mathbf{R}_{A\sigma}}{\partial N} & \mathbf{0} \end{bmatrix} \tag{29}$$

The components of Equation (29) can be shown to be a function of $K_C(\theta)$ expressed by Equation (6), i.e.:

$$\frac{\partial \mathbf{R}_{A\sigma}}{\partial \eta_i} = 2\frac{\partial K_C(\theta)}{\partial \eta_i}\left[0 \quad \left(2\,{}^1\theta^* + {}^2\theta^*\right) \quad -\left({}^1\theta^* + 2\,{}^2\theta^*\right)\right]^T \tag{30}$$

where ${}^i\theta^*$ is the relative rotation of node $i$ ($i$ = 1 or 2) and can be obtained by subtracting the rigid rotation from the total rotation, $\eta_i = k$, $k_p$, $M_0$, and $N$, and

$$\frac{\partial K_C(\theta)}{\partial k} = \frac{1 - Na^N}{\left[1 + a^N\right]^{\frac{(2N+1)}{N}}} \tag{31}$$

$$\frac{\partial K_C(\theta)}{\partial k_p} = \frac{-(N+2)a^N - 1}{\left[1 + a^N\right]^{\frac{(2N+1)}{N}}} + 1 \tag{32}$$

$$\frac{\partial K_C(\theta)}{\partial M_0} = \frac{(N+1)a^N}{\left[1 + a^N\right]^{\frac{(2N+1)}{N}}} \frac{(k - k_p)}{M_0} \tag{33}$$

$$\frac{\partial K_C(\theta)}{\partial N} = \left[ -\frac{a^N (N+1)\log a}{N\left(1 + a^N\right)^{\frac{(2N+1)}{N}}} + \frac{-N + (N+1)\log(1 + a^N)}{N^2\left(1 + a^N\right)^{\frac{(N+1)}{N}}} \right]\left(k - k_p\right) \tag{34}$$

where

$$a = \left|\frac{(k - k_p)\theta}{M_0}\right| \tag{35}$$

The derivations of Equations (31) through (35) are given in more detail in Haldar and Mahadevan (2000b). As previously stated, a beam-column element is introduced for each connection. For such an element, Equation (29) needs to be evaluated. Once Equation (29) is evaluated, the rest of the steps are the same as those for an ordinary beam-column element. All the quantities required for the computation of $\nabla g(\mathbf{y})$ in Equation (15) for a two-dimensional frame with PR connections are now available in a simple explicit form. Therefore, the reliability index and the corresponding failure probability can be calculated using the FORM analysis presented in Equations (13) through (16).

## 3.4. Consideration of RC Shear Walls in SFEM

As mentioned in section 3.1.1, only steel members where RC shear walls are not present are investigated for the strength limit state in this study. The steel members are expected to be weaker in strength in this case. Thus, although the parameters in Equations (17) and (18) are expected to be influenced by the presence of shear walls, the partial derivatives with respective to the random variables related to shear walls, namely, $E_c$ and $\nu$, need not be evaluated. For the serviceability limit state in Equation (19), since the partial derivatives with respective to $E_c$ and $\nu$ are also zero, they also need not be evaluated.

To consider the presence of shear walls, on the other hand, the global tangent stiffness **K** in Equations (24) and (25) should be substituted by $\mathbf{K}_T = \mathbf{K} + \mathbf{K}_{sh}$ where **K** and $\mathbf{K}_{sh}$ were defined in Equations (2) and (8). This requires the calculation of the derivatives of internal forces, $\partial\mathbf{R}/\partial\mathbf{x}$ in Equation (25), with respect to $E_c$ and $\nu$ in the evaluation of Jacobians. They can be derived as:

$$\frac{\partial \mathbf{K}_{sh}}{\partial \mathbf{x}} = \left[ \frac{\partial \mathbf{K}_{sh}}{\partial E_c} \quad \frac{\partial \mathbf{K}_{sh}}{\partial \nu} \right] \tag{36}$$

in which

$$\frac{\partial \mathbf{K}_{sh}}{\partial E_c} = t \left\{ \frac{1}{4\gamma} \mathbf{A}^T \mathbf{E}' \mathbf{A} + \frac{1}{12\gamma} \mathbf{B}^T \mathbf{E}' \mathbf{B} + \frac{\gamma}{12} \mathbf{C}^T \mathbf{E}' \mathbf{C} \right\} \tag{37}$$

and

$$\frac{\partial \mathbf{K}_{sh}}{\partial \nu} = E_c t \left\{ \frac{1}{4\gamma} \mathbf{A}^T \mathbf{E}'_{,\nu} \mathbf{A} + \frac{1}{12\gamma} \mathbf{B}^T \mathbf{E}'_{,\nu} \mathbf{B} + \frac{\gamma}{12} \mathbf{C}^T \mathbf{E}'_{,\nu} \mathbf{C} \right\} \tag{38}$$

where the matrices **A**, **B**, **C**, and **E** are already difined in Equations (9) through (12), and other two matrices can be expressed as:

$$\mathbf{E}' = \frac{1}{1-\nu^2}\begin{bmatrix} 1 & \nu & 0 \\ \nu & 1 & 0 \\ 0 & 0 & \dfrac{1-\nu}{2} \end{bmatrix} \tag{39}$$

and

$$\mathbf{E}'_{,\nu} = \begin{bmatrix} \dfrac{2\nu}{(1-\nu^2)^2} & \dfrac{1+\nu^2}{(1-\nu^2)^2} & 0 \\ \dfrac{1+\nu^2}{(1-\nu^2)^2} & \dfrac{2\nu}{(1-\nu^2)^2} & 0 \\ 0 & 0 & -\dfrac{1}{2(1+\nu)^2} \end{bmatrix} \tag{40}$$

Finally, the gradient of a limit state function is now available in an explicit form for a frame and shear wall structural system. The reliability index and the corresponding failure probability can thus be calculated using the FORM analysis presented in Equations (13) through (16).

## 4. NUMERICAL EXAMPLES

To elaborate the proposed SFEM and to investigate the effect of PR connection conditions and the presence of RC shear walls on the overall reliability of steel frames, the following two examples are considered. In the first example, a two-story one-bay steel frame is considered. The connections in the frame are first considered to be FR type, a common assumption in analyzing such a frame. Connections are then considered to be PR type. The corresponding reliabilities of the frame in the presence of FR and PR types are evaluated using the method discussed here and compared. In the second example, the reliabilities of a two-story two-bay steel frame without and with the presence of shear walls are evaluated and compared. In both examples, all loads are applied statically.

### 4.1. Example 1: Effect of PR Connections on Reliability Analysis

To investigate the effect of PR connection conditions on the overall structural reliability, a two-story one-bay steel frame, shown in Figure 2, with FR connections is considered first. All the beams are made of W24×55 and all the columns are made of W14×74. Grade A36 steel is used for this illustrative example. The frame is subjected to dead, live, and horizontal loads. The uncertainties associated with all the random variables are given in Table 1.

To consider the effects of different rigidities in the PR connections located at *b*, *c*, *g*, and *h* (refer to Figure 2); three $M$-$\theta$ curves shown in Figure 3 are considered. The probabilistic descriptions of the four parameters of the Richard model representing the three curves are listed in Table 2.

The following four cases of connection conditions are considered for discussion purposes:

- Case 1: a steel frame without PR connection, i.e., connection in the frame are assumed to be rigid or FR type, representing the standard practice in the profession
- Case 2: a steel frame with PR connections represented by $M$-$\theta$ Curve 1. It represents the realistic behavior of a FR type connection.
- Case 3: a steel frame with PR connections with intermediate rigidity represented by $M$-$\theta$ Curve 2
- Case 4: a steel frame with PR connections with low rigidity represented by $M$-$\theta$ Curve 3

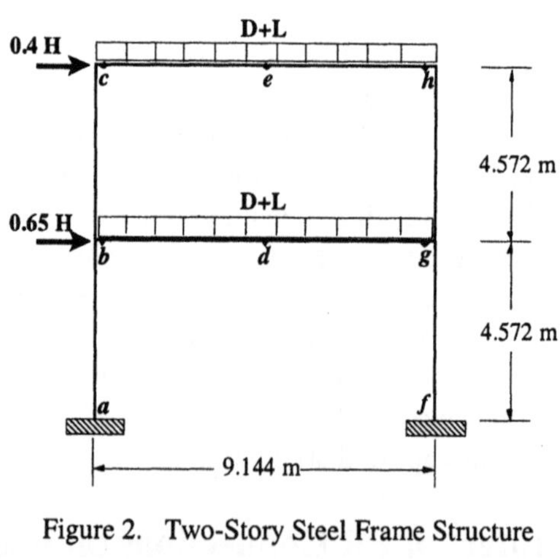


Figure 2. Two-Story Steel Frame Structure

*Table 1. Basic random variables in the steel frame*

| Random Variables | Nominal Value | Mean/Nominal | C.O.V | Distribution |
|---|---|---|---|---|
| E (kN/m$^2$) | $2.000\times10^8$ | 1.00 | 0.06 | LN |
| $A^b$ (m$^2$) | $1.045\times10^{-2}$ | 1.00 | 0.05 | LN |
| $I^b{}_x$ (m$^4$) | $5.619\times10^{-4}$ | 1.00 | 0.05 | LN |
| $Z^b{}_x$ (m$^3$) | $2.196\times10^{-3}$ | 1.00 | 0.05 | LN |
| $A^c$ (m$^2$) | $1.406\times10^{-2}$ | 1.00 | 0.05 | LN |
| $I^c{}_x$ (m$^4$) | $3.313\times10^{-4}$ | 1.00 | 0.05 | LN |
| $Z^b{}_x$ (m$^3$) | $2.065\times10^{-3}$ | 1.00 | 0.05 | LN |
| $F_y$ (kN/m$^2$) | $2.606\times10^5$ | 1.05 | 0.10 | LN |
| D (kN/m) | 32.10 | 1.05 | 0.10 | LN |
| L (kN/m) | 23.34 | 1.00 | 0.25 | Type I |
| H (kN) | 44.50 | 0.78 | 0.37 | Type I |

Note: b = Beam, c = Column, LN = Log-normal

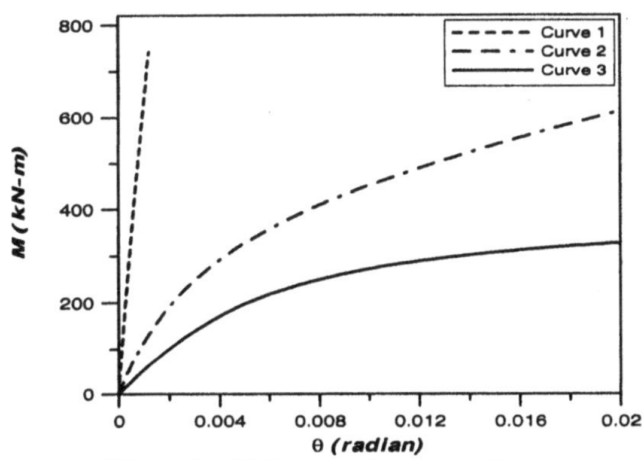


Figure 3. M-θ curves for connections

*Table 2. Statistical Description of the Four Parameters in the Richard Model*

| Random variables | Mean Value | | | COV | Distribution |
|---|---|---|---|---|---|
| | Curve 1 | Curve 2 | Curve 3 | | |
| $k$ (kN·m/rad) | $1.13\times10^6$ | $1.47\times10^5$ | $5.65\times10^4$ | 0.15 | Normal |
| $k_p$ (kN·m/rad) | $1.13\times10^5$ | $1.13\times10^4$ | $1.13\times10^3$ | 0.15 | Normal |
| $M_0$ (kN·m) | 508.64 | 452.12 | 339.09 | 0.15 | Normal |
| $N$ | 0.50 | 1.00 | 1.5 | 0.05 | Normal |

Both the strength and serviceability limit states discussed earlier (Equation (17) or (18), and Equation (19), respectively) are considered in this example. For the serviceability performance function, the permissible lateral displacement at the top of the frame is considered not to exceed $h/400$, i.e., 22.86 mm, and the allowable vertical deflection at the beam's mid-span is considered to be $L/360$ under an unfactored live load, i.e., 25.4 mm. Considering all the random variables given in Tables 1 and 2, the corresponding reliability indexes and the probabilities of failure for both limit states of the four cases are evaluated using the proposed SFEM. The results are summarized in Table 3 in terms of reliability indices. In Table 3, $\beta_1$, $\beta_2$, and $\beta_3$ represent the reliability indices corresponding to $M$-$\theta$ curves 1, 2, and 3, respectively.

Based on the results of this example, it can be stated that the proposed algorithm can be used to estimate the probability of failure of a steel frame structure in the presence of FR and PR connections under static loading conditions. Several important observations can be made from the results shown in Table 3. Assuming a reliability index of 3.0 is acceptable, the reliability indices for Case 1 indicate that the frame is safe for both the strength and serviceability limit states, representing the normal practice in the profession. However, for Cases 2, 3, and 4, when connections are assumed to be PR types of different rigidities, the reliability indices change for both the strength and serviceability limit states. This is expected since redistribution of moments takes place in the frame due to the presence of PR connections. The reliability indices for Cases 1 and 2 for both the strength and serviceability limit states are very similar. The results are expected since Case 2 is a realistic representation of the idealized FR connection.

*Table 3. Reliability indexes for frame without and with PR connections*

| Limit State | | Strength Limit State | | Serviceability Limit State | |
|---|---|---|---|---|---|
| Location | | *Beam at $e$ or $h$ | Column at $h$ | Drift at $c$ | Deflection at $e$ |
| Load | | D+L+H | D+L+H | D+L+H | L |
| Rigid Connection | Case 1 | $\beta = 3.608$ | $\beta = 3.123$ | $\beta = 3.057$ | $\beta = 4.155$ |
| PR Connection | Case 2 | $\beta_1 = 3.581$ | $\beta_1 = 3.131$ | $\beta_1 = 2.910$ | $\beta_1 = 4.038$ |
| | Case 3 | $\beta_2 = 3.065$ | $\beta_2 = 3.773$ | $\beta_2 = 2.200$ | $\beta_2 = 3.599$ |
| | Case 4 | $\beta_3 = 2.525$ | $\beta_3 = 5.071$ | $\beta_3 = 1.511$ | $\beta_3 = 3.021$ |

* Location for the beam : $e$ for Case 1 only & $h$ for others

In this particular example, the bending moment of the beam decreased at the column ends and increased at the mid-span, i.e., the location of the design moment for the beam shifted from *h* to *e*. The reliability indices for the strength limit state decreased for the beam and increased for the column as the rigidity of the connections decreased, making the beam more prone to failure than the columns. Thus, for the frame under consideration, the lower rigidity in the connections has a beneficial effect on the column and a detrimental effect on the beam. It is important to note that the reliability indices for the lateral deflection limit state decrease significantly as the rigidity of the connections is reduced. Assuming the reliability index lower than 3.0 is not acceptable, the frame with all three connection rigidities will not satisfy the serviceability requirement. The observation indicates that the serviceability limit state is expected to be more critical than the strength limit state for frames with PR connections.

## 4.2. Example 2: Effect of RC Shear Walls on Reliability Analysis

The reliability of a steel frame without and with the presence of RC shear walls are evaluated in this example using the SFEM algorithm discussed earlier.

### 4.2.1. Reliability evaluation of a steel frame without shear walls

A two-story two-bay frame without the shear walls, shown in Figure 4, is considered first. Grade A36 steel is used, and all columns are made of a W14×61 section and all beams are made of a W18×86 section. The statistical characteristics of the cross-sectional and material properties required for the reliability analysis are given in Table 4. The frame is subjected to dead, live and horizontal loads and the statistical properties of these loads are also given in Table 4.

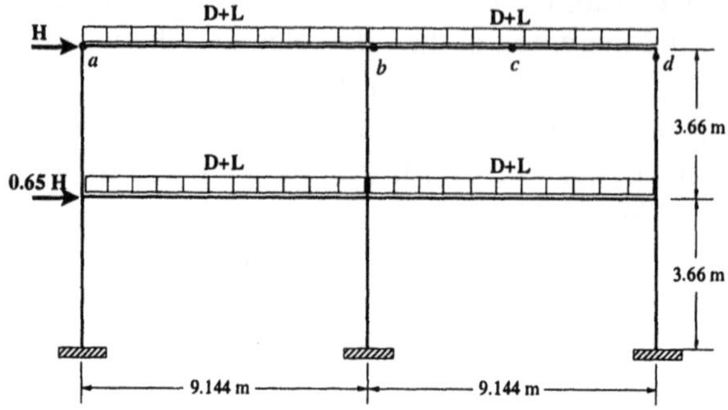


Figure 4. A steel frame without shear walls

*Table 4. Basic random variables in the steel frame*

| Random Variables | Nominal Value | Mean/Nominal | C.O.V | Distribution |
|---|---|---|---|---|
| E (kN/m$^2$) | 1.999×10$^8$ | 1.00 | 0.06 | LN |
| $A^b$ (m$^2$) | 1.632×10$^{-2}$ | 1.00 | 0.05 | LN |
| $I^b_x$ (m$^4$) | 6.368×10$^{-4}$ | 1.00 | 0.05 | LN |
| $Z^b_x$ (m$^3$) | 3.048×10$^{-3}$ | 1.00 | 0.05 | LN |
| $A^c$ (m$^2$) | 1.155×10$^{-2}$ | 1.00 | 0.05 | LN |
| $I^c_x$ (m$^4$) | 2.664×10$^{-4}$ | 1.00 | 0.05 | LN |
| $Z^b_x$ (m$^3$) | 1.671×10$^{-3}$ | 1.00 | 0.05 | LN |
| $F_y$ (kN/m$^2$) | 2.482×10$^5$ | 1.05 | 0.10 | LN |
| D (kN/m) | 35.02 | 1.05 | 0.10 | LN |
| L (kN/m) | 24.81 | 1.00 | 0.25 | Type I |
| H (kN) | 38.25 | 0.78 | 0.37 | Type I |

Note: b = Beam, c = Column, LN = Log-normal

For the strength limit state, the reliability of the most critical beam at node *b* and the most critical column at node *d* are evaluated using the proposed algorithm using the performance functions represented by Equations (17) and (18). For the serviceability limit state, the horizontal drift of the top floor at node *a* and the vertical deflection of the beam at the midspan at node *c* are checked. In Equation (19), the prescribed horizontal drift at the top floor is considered to not exceed *h*/400, where *h* is the height of the frame. Thus, $\delta_{\text{limit}}^{drift}$ is equal to 18.3 mm in this example. Similarly, the prescribed vertical deflection in the midspan of the beam is considered to be *l*/360 under the unfactored live load, where *l* is the span length of the beam. In this case, $\delta_{\text{limit}}^{deflection}$ is considered to be 25.4 mm.

Considering all the random variables given in Tables 4, the corresponding reliability indexes and the probabilities of failure at different node points are evaluated using the SFEM. The results are summarized in Table 5. The reliability indices for both the beam and column for the strength limit state are found to be less than 3.0, i.e., they are weak in strength.

*Table 5. Reliability indexes for frame without shear walls*

| Limit State | Locations | Load Combination | $\beta$ |
|---|---|---|---|
| Strength | Beam, Node *b* | D+L+H | 2.792 |
| | Column, Node *d* | D+L+H | 2.807 |
| Serviceability | Drift, Node *a* | D+L+H | 4.522 |
| | Deflection, Node *c* | L | 5.434 |

### 4.2.2. Reliability evaluation of a steel frame with shear walls

The frame shown in Figure 4 is then reinforced with shear walls as shown in Figure 5. The compressive strength of concrete for shear wall, $f_c'$, is considered to be $2.068\times10^4$ kN/m$^2$.

The statistical properties of two additional variables related to the shear walls, $E_c$ and $\nu$, are given in Table 6. The building is assumed to contain 5 similar frames connected by rigid diaphragms at the floor levels. Only the center frame of the building is assumed to have shear walls. Although the physical thickness of the shear wall is 12.7 cm, considering the presence of 5 similar frames and the rigid behavior of diaphragms, the effective thickness per frame is assumed to be 2.54 cm in this study. The combined system is also subjected to the three static loads given in Table 5. After the tensile stress of each shear wall exceeds the prescribed tensile stress of concrete, the degradation of the shear wall stiffness is assumed to be reduced to 40% of the original stiffness.

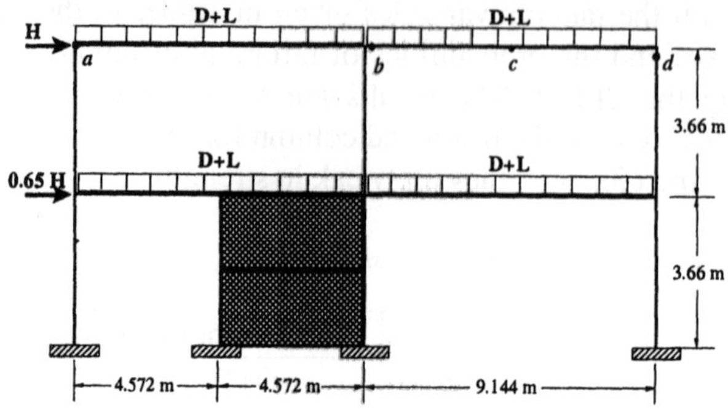


Figure 5. Steel frame with RC shear walls

*Table 6. Basic random variables in RC shear walls*

| Random Variables | Nominal Value | Mean/Nominal | C.O.V | Distribution |
|---|---|---|---|---|
| $E_C$ (kN/m$^2$) | $2.137\times10^7$ | 1.00 | 0.18 | LN |
| $\nu$ | 0.17 | 1.00 | 0.10 | LN |

The probability of failure of the combined system is calculated using the proposed algorithm. For the strength limit state, the probability of failure of a column at node *d* in Figure 5 is estimated. For the serviceability limit state, the horizontal deflection at the top of the combined system (point *a* in Figure 5) is evaluated. The results are summarized in Table 7.

*Table 7. Reliability indexes for frame with shear walls*

| Limit State | Locations | Load Combination | β |
|---|---|---|---|
| Strength | Column, Node *d* | D+L+H | 3.051 |
| Serviceability | Drift, Node *a* | D+L+H | 7.708 |

The results in this example clearly indicate that the proposed algorithm can be used to estimate the probability of failure of a combined system consisting of flexible steel frame and RC shear walls under static loading conditions. For the reliability analysis of the frame without shear walls, the reliability indexes for the beams and columns are similar for the strength limit state, satisfying the intent of the AISC's LRFD code. The reliability of the column did not change significantly due to the presence of shear walls. However, the horizontal drift at the top of the frame reduced significantly and the probability of failure of the combined system in serviceability became almost zero. This is expected. For the combined system, the controlling limit state has changed from serviceability to strength. This simple example clearly demonstrates the beneficial effect of shear walls in carrying horizontal loads. It also demonstrates that the proposed algorithm can be used to estimate the reliability of a complicated structural system under static loading conditions, broadening the application potential of reliability methods.

# 5. CONCLUSIONS

A very efficient nonlinear finite element-based reliability analysis algorithm is presented to evaluate the reliability of complicated real structural systems. The authors called it a stochastic finite-element-based approach. It integrates the stress-based finite element method and the first-order reliability method. All major sources of nonlinearity and uncertainty can be incorporated in the algorithm. Both strength and serviceability limit states can be used for the reliability evaluation. The algorithm has been found to be accurate and efficient in evaluating reliabilities. The connections in steel frames are routinely considered to be FR type. However, they are essentially PR type with different rigidities. The behavior of steel frames change significantly if the connection behavior is considered realistically. The four-parameter Richard model is used to express the partial rigidity of connection conditions in this study. Steel frames with PR connection conditions and steel frames reinforced with RC shear walls are emphasized in this study. In steel frames with PR connections, the serviceability limit state can become the controlling limit state. The consideration of uncertainties in modeling the PR connections is also important in the reliability evaluation of steel frames. A steel frame can be weak in resisting lateral loads. RC shear walls can be used to improve their lateral rigidities. The shear walls are represented by four-node plane stress elements in this study. The SFEM is extended to evaluate the reliability of such a combined system. The procedure is clarified with the help of two examples. The example demonstrates how the SFEM-based reliability method can be used to evaluate risk of real structural systems capturing their realistic mechanical behavior. The procedure will be useful in developing the performance-based design guidelines under consideration by the profession.

## 6. ACKNOWLEDGEMENTS

This paper is based on work partly supported by the National Science Foundation under Grant CMS-9526809. Financial support received from the American Institute of Steel Construction (AISC), Chicago, is appreciated. Any opinions, findings, conclusions, or recommendations expressed in this publication are those of the authors and do not necessarily reflect the views of the sponsors.

# REFERENCES

American Concrete Institute, *Building Code Requirements for Structural Concrete (318-99) and Commentary (318R-99).* Farming Hills, MI., 1999.

American Institute of Steel Construction (AISC), *Manual of Steel Construction: Load and Resistance Factor Design, 3rd Edition,* Chicago, Il., 2001.

Arora J.S. and Haug E.J., "Methods of design sensitivity analysis in structural optimization", *AIAA Journal*, 17(9), 1979, pp. 970-974.

Gupta A.K. and Akbar H., "Cracking in reinforced concrete analysis", *Journal of Structural Engineering, ASCE,* 107(ST1), 1983, pp. 1735-1746.

Haldar A. and Gao L., "Reliability evaluation of structures using nonlinear SFEM", *Uncertainty Modeling in Finite Element, Fatigue, and Stability of Systems,* A. Haldar, A. Guran, and B.M. Ayyub, eds., World Scientific Publishing Co., River Edge, N.J., 1997, pp. 23-50.

Haldar A. and Huh J., "Modeling of Material Uncertainties using the Nonlinear Stochastic Finite Element Method", *PROBAMAT-21st Century: Probabilities and Materials, NATO ASI Series,* Kluwer Academic Publishers, Netherlands, 1998, pp. 299-310.

Haldar A. and Mahadevan S., *Probability, Reliability and Statistical Methods in Engineering Design,* Wiley, New York, 2000a.

Haldar A. and Mahadevan S., *Reliability assessment using stochastic finite element analysis*, Wiley, New York, 2000b.

Haldar A. and Nee K.M., "Elasto-plastic large deformation analysis of PR steel frames for LRFD", *Computers and Structures*, 34(5), 1989, pp. 811-823.

Huh J. and Haldar A., "Seismic reliability on non-linear frames with PR connections using systematic RSM", *Probabilistic Engineering Mechanics*, 17(2), 2002, pp. 177-190.

Inoue N., Yang K., and Shibata A., "Dynamic nonlinear analysis of reinforced concrete shear wall by finite element method with explicit analytical procedure", *Earthquake Engineering and Structural Dynamics*, 26, 1997, pp. 967-986.

Kondoh K. and Atluri S.N., "Large-deformation, elasto-plastic analysis of frames under non-conservative loading, using explicitly derived tangent stiffnesses based on assumed stresses", *Computational Mechanics*, 2(1), 1987, pp. 1-25.

Lee S.Y. and Haldar A., "Reliability of Frame and Shear Wall Structural Systems. I: Static Loading", *Journal of Structural Engineering, ASCE,* 129 (2), 2003, pp. 224-232.

Lee S.Y., *Static and dynamic reliability analysis of frame and shear wall structural systems*, PhD dissertation, Dept. of Civil Engineering and Engineering Mechanics, University of Arizona, Tucson, AZ, 2000.

Lefas D., Kotsovos D., and Ambraseys N., "Behavior of reinforced concrete structural walls: strength, deformation characteristics, and failure mechanism", *ACI Structural Journal,* 87(1), 1990, pp. 23-31.

Liauw T.C. and Kwan K.H., "Static and cyclic behaviors of multistory infilled frames with different interface conditions", *J. Sound and Vibration,* 99(2), 1985, pp. 275-283.

Richard R.M. and Abbott B.J., "Versatile elastic-plastic stress-strain formula", *Journal of Engineering Mechanics, ASCE,* 101(EM4), 1975, pp. 511-515.

Vecchio F.J., "Nonlinear finite element analysis of reinforced concrete membranes", *ACI Structural Journal,* 1989, pp. 26-35.

# Chapter 20

# SIMULATION IN RISK-BASED CODIFIED ENGINEERING DESIGN

**Achintya Haldar**

## 1. INTRODUCTION

In spite of improvements in our basic understanding, analytical capabilities, and computational power, the presence of uncertainty in engineering designs cannot be avoided. Engineering designs consist of many interconnected activities, and some of them require the prediction of events that may occur during the lifetime of the systems being designed. Design activities include the selection of design loads, possible combinations of loads during the lifetime of the structure, evaluating load effects for a load or a load combination, and selecting the critical load effects for which the structure needs to be designed or proportioned satisfying prescribed performance criteria. In general, proportioning the elements is considered to be engineering design, and it appears to be the easiest task in the whole design process. Every phase besides the last phase of proportioning the elements cannot be predicted with certainty, and this is one of the main reasons why structures fail from time to time. The uncertainty in predicting future loads, load effects, and resistance has been recognized by the profession. It is a major challenge to satisfy design requirements in the presence of uncertainty. Risk-based design guidelines and codes are being developed and promoted worldwide in all major engineering disciplines. To help develop such guidelines, several reliability evaluation procedures with various degrees of sophistication were proposed (Haldar and Mahadevan, 2000a, b). Selecting design loads and load combinations is the weakest link in the design process. The American Institute of Steel Construction (AISC) provided major leadership in this area by introducing the Load and Resistance Factor Design (LRFD) codes as early as 1986. The AISC also demonstrated the dynamic and progressive characteristics of the concept by updating it twice since its inception. The third edition of the LRFD code was published in 2001. Similar design guidelines for concrete (ACI, 2002a), masonry (ACI, 2002b), and wood (ASCE, 1995: AWC, 1996) are now available, reflecting the risk-based design concept. The load combinations and load factors used in these guidelines are identical and in the future are expected to be the only option worldwide. The uncertainty in the material behavior and other resistance-related parameters of structural elements needs to be incorporated in designs considering specific applications and satisfying an underlying risk (ASCE, 2002).

It is generally believed that all major sources of uncertainty are incorporated in the reliability-based design guidelines. This took multi-disciplinary efforts spanning more than two decades. The analysis or evaluation of load effect is still an open area and hopefully will be addressed in the near future. The advancement in computational power has yet to be fully utilized. The available reliability-based design codes are very similar to the earlier deterministic codes. The nature or amount of uncertainty in the design variables and advanced reliability concepts used in developing these codes generally remain unknown to designers. Thus, it may be difficult for an experienced design engineer to consider the presence of levels of uncertainty different than those used to develop the reliability-based design guidelines. Furthermore, the propagation of uncertainties from the variable level to the structural level is expected to be different for various design applications (Haldar and Mahadevan 2000b). The area is still being developed, and a typical engineer in a design office is not expected to be familiar with recent developments. In most cases, the design guidelines were developed to consider the behavior of elements of complex structural systems satisfying many explicit performance criteria. Some of the performance criteria are applicable at the local element level, e.g., the overall stress condition in a member, and some others are applicable at the overall structural level, e.g., the lateral deflection at a prescribed height of a structure. To satisfy all major performance criteria and for the purpose of uniformity, the system reliability may need to be evaluated using the information on element level reliabilities. In most cases, the evaluation of system reliability is very complicated and difficult. Reliability is always evaluated for a specific performance function and a reference or allowable value is required to formulate a performance function. The selection of the reference value for a given performance function is controversial and may not have been developed or accepted by the profession.

For the design of nuclear power plants in mid-seventies, before the development of the LRFD concept, multiple analyses of structures to determine the most critical load effects were practiced. Three analyses were generally conducted by using: (i) the most probable values of the variables present in the problem, (ii) the least likely values, and (iii) the most conservative values. This way the implication of the presence of uncertainty in the problem can be indirectly studied. However, the information may not be of practical use since the underlying reliability remains unknown. It will be very desirable if the multiple analyses concept can be integrated with the LRFD concept.

Schuëller (2002), in his Euro-SiBRAM'2000 keynote address, commented that essentially the uncertainty management was the main topic in the risk-based codified design. It may be more rational to manage uncertainty by solving the problems several times instead of solving it only once. The challenge is to convince the deterministic colleagues the progressive nature of the multiple analyses, particularly considering the advancement in the computational power.

Uncertainty associated with most of the design variables is now available in the literature. With the help of experts from many disciplines, several reliability methods are available to estimate the underlying reliability considering the uncertainty in the design variables (Haldar and Mahadevan, 2000a, b). Thus, underlying risk in any engineering design can be evaluated without using any prescribed load and resistance factors. This allows an engineer to design more or less conservative way than the current codified approach, considering the problem under consideration and the willingness of the owner to accept the corresponding risk. Simulation, particularly Monte Carlo simulation method can be used for the reliability evaluation. The simulation approach is relatively simple and does not require sophisticated statistical and probabilistic knowledge required for other reliability evaluation techniques. Since simulation is multiple deterministic analyses satisfying the underlying uncertainty in the design variables, it can also be used for the design purpose. Thus, simulation is a viable alternative to the codified approach due to the significant advancement in computer technology and computational power. However, the question remains whether the simulation-based design concept is mature enough to be considered as an alternative to the currently available codified approach.

Even if one assumes that the simulation approach is mature, there are many issues that need to be addressed before it can be accepted as an alternate design method. Some of the issues may not be simulation-related; they may be related to the weaknesses in the current deterministic approaches. Since the deterministic analysis is an essential element in simulation, the limitations in the deterministic procedures are also expected to be present in the simulation-based approaches. Also, if history is an indication, a strong resistance to the implementation of simulation-based engineering design can be expected from current designers. The engineering profession is reluctant to adopt a new approach in most cases. Lack of familiarity and resources in terms of money and time to learn a new approach are the main reasons for the resistance. Many issues that are avoided in the codified approaches need broader discussion in the profession, such as lack of data, the correlation characteristics of input information, lack of information on reference value to be used in the performance function, legal issues, software, and the efficiency and accuracy of the deterministic algorithm to be used for simulation.

The engineering community, in general, accepts the presence of uncertainties in engineering designs and acknowledges the necessity of explicitly incorporating them in the design whenever possible. The current design codes try to address the uncertainty-related issues conservatively, but practicing engineers may not be aware of this. They must know that the design value for the load-related design variables is selected to be above the mean value and the design value for the resistance-related design variables is selected to be below the mean value. The uncertainty characteristics, in terms of mean, standard deviation, and the underlying distribution for all the design variables, are in general different. It is essential that they know the basic concept and the assumptions behind the current codified approaches. They must be encouraged to use the simulation approach if the basic assumptions of the codified approach are not satisfied.

Simulation is very powerful but it is an approximate technique. The accuracy in simulation increases with the increase in the number of simulation cycles used. However, it is not possible to predict the minimum number of simulation cycles required for a specific problem to satisfy the accuracy requirement before conducting the experiment. It depends on the underlying probability of failure which is unknown at the beginning of the experiment. However, using sufficient number of cycles, the simulation technique will estimate the system reliability considering the corresponding reference values, nonlinear behavior, static as well as dynamic response to the loading, correlation characteristics of random variables, etc. However, even for a given number of simulation cycles, the outcomes of the simulation for a specific problem could be different depending on the characteristics of computer-generated random numbers. There is no uniqueness in the simulation outcomes. The outcomes are computer-specific. Thus, the error in estimating probability of failure should be an integral part of simulation.

Another fundamental drawback is the time or cost or efficiency of simulation. Huh and Haldar (2001, 2002) reported that simulating 100,000 cycles in a supercomputer (SGI Origin 2000) to estimate the reliability of a one-bay two-story steel frame subjected to only 5 second of an earthquake loading may take more than 23 hours. The efficiency of simulation can be improved by using variance reduction techniques (VRTs), which can be grouped in several ways (Haldar and Mahadevan, 2000a). One approach is to consider whether the variance reduction method alters the experiment by altering the input scheme, by altering the model, or by special analysis of the output. The VRTs can also be grouped according to description or purpose (i.e., sampling method, correlation methods, and special methods). Haldar and Mahadevan (2000a) noted that VRTs increase the computational difficulty for each simulation, and a considerable amount of expertise may be necessary to implement them. Even experts may not know some of these approaches, and practicing engineers may not be able to use them. The most desirable feature of simulation, its basic simplicity, is thus lost.

Based on the above discussions, it is clear that the simulation approach provides a very reasonable alternative to the commonly used codified approach. However, there are still some issues need to be addressed before it can be adopted. Issues related to the efficiency and accuracy of the deterministic algorithm to be used in simulations, the appropriate way to quantify the randomness, information to be used to define the statistical characteristics of design variables, defining appropriate performance functions and the selection of reference values, evaluating correlation characteristics of random variables present in complex systems, simulation of random variables versus random field, simulation of multi-variate random variables, system reliability, the effect of load combinations, time dependent reliability, available software to implement the simulation-based concept, etc., need further evaluation. Documentation of case studies will also help in this endeavor. A world body of distinguished scholars on reliability-based design addressed these issues in the International Colloquium Euro-SiBRAM'2002 (Marek, et al., 2002) to help to formulate the future direction in the simulation-based design. The opinions, comments, observations and recommendation made by the attending international scholars and practitioners are presented here.

## 2. SIMULATION CONCEPT

Lewis and Orav (1989) wrote, "Simulation is essentially a controlled statistical sampling technique that, with a model, is used to obtain approximate answer for questions about complex, multi-factor probabilistic problems." They added, "It is this interaction of experience, applied mathematics, statistics, and computing science that makes simulation such a stimulating subject, but at the same time a subject that is difficult to teach and write about." Theoretical simulation is usually performed numerically with the help of computers, allowing a more elaborate representation of a complicated engineering system than can be achieved by physical experiments, and it is often cheaper than physical models. It allows a designer to know the uncertainty characteristics being considered in a particular design, to use judgment to quantify randomness beyond what is considered in a typical codified design, to evaluate the nature of implicit or explicit performance functions, and to have control of the deterministic algorithm used to study the realistic structural behavior at the system level.

The method commonly used for this purpose is called the Monte Carlo simulation technique. In the simplest form of the basic simulation, each random variable in a problem is sampled several times to represent the underlying probabilistic characteristics. Solving the problem deterministically for each realization is known as a simulation cycle, trial, or run. Using many simulation cycles will give the probabilistic characteristics of the problem, particularly when the number of cycles tends to infinity. Using computer simulation to study the presence of uncertainty in the problem is an inexpensive experiment compared to laboratory testing. It also helps evaluate different design alternatives in the presence of uncertainty, with the goal of identifying the optimal solution. The use of simulation in engineering design was strongly advocated by Marek et al. (2001, 2003), Schuëller in his keynote address, and others. Elishakoff (2001) wrote an interesting essay on Monte Carlo simulation.

The gathering of distinguished scholars on risk-based design in Euro-SiBRAM'2002 was a significant development and helped to promote the simulation-based engineering design as an alternative to the classical codified approach. However, there are still many challenges that need to be addressed first. In all fairness, similar challenges also exist in the current codified approach, and the discussions in the colloquium helped to identify them. Thus, these discussions are also expected to help improve the codified approach.

## 3. STEPS IN SIMULATION

Reliability evaluation using Monte Carlo simulation technique requires the execution of a series of sequential steps. The success of implementing the Monte Carlo simulation in design will depend on how accurately each step is addressed. Haldar and Mahadevan (2000a) identified the following six essential steps: (1) defining the problem in terms of all the random variables; (2) quantifying the probabilistic characteristics of all the random variables in terms of their probability density functions and the corresponding parameters; (3) generating values of these random variables; (4) evaluating the problem deterministically for each set of realizations of all the random variables, or simply numerical experimentation of the problem; (5) extracting probabilistic information from $N$ such realizations; and (6) determining the accuracy and efficiency of the simulation. It is not necessary to discuss these steps in detail in this chapter.

## 4. BACKGROUND INFORMATION ON CURRENT CODIFIED APPROACH

One of the earliest codified approaches was published in 1916 on Report on Recommended Practice and Standard Specifications for Concrete and Reinforced Concrete was issued by several professional organizations including American Concrete Institute, American Institute of Architects, American Railway Engineering Association, American Society of Civil Engineers, and American Society for Testing and Materials. However, the risk-based engineering design concept has been under development for only over the last four decades. It required a multi-disciplinary research effort. The introduction of the load and resistance factor design (LRFD) approach for the design of steel structures by the American Institute of Steel Construction (AISC) was an important development in civil engineering (AISC, 1986). The preliminary attempts were very simplistic in nature. In the developing the first edition of the LRFD code, the load effects and the resistance were assumed to be lognormally distributed and the first-order second moment (FOSM) method (Haldar and Mahadevan, 2000a) was used to estimate the reliability and the corresponding load and resistance factors. In the later editions (AISC, 1994, 2001), the advanced first-order second-moment (AFOSM or FORM) approach was used to accommodate more complex design situations and included probability distributions other than the lognormal distribution. Simulation technique can also be used to extract similar information on risk and reliability. For the reliability evaluation methods applying simulation technique, the representation of individual variables need not be limited to parametric distributions. Bounded histograms, piecewise-uniform distributions and other non-parametric distributions can be considered as documented by Marek et al. (1995 and 2001) and Mack (2002). Further attention should be given to non-parametric distributions considering the scientific acceptability and limitations of such representations and the development of required databases.

## 4.1. Deficiencies in the Current Risk-based Codified Approach

Some of the major advantages of the current risk-based codified approach have been well publicized and accepted in the profession. However, it is based on several major assumptions. Some of these assumptions are identified next.

The LRFD method was based on reliability analysis of isolated simple structural elements and was calibrated to achieve levels of reliability similar to conventional allowable stress-based design guidelines used at that time. The use of isolated simple structures to derive the safety factors is related to the basic design philosophy common to all codified design procedures. There are several advantages to isolated member approach including (Bjorhovde et al., 1978): (1) in deterministic design methods that use factors of safety, it is not practical to prepare detailed requirements for each structural configuration; (2) the characteristics of the individual members and connections themselves are independent of the framework; and (3) most research has been devoted to the study of such elements, and theoretical and experimental verification of their performance is readily available. Nevertheless, the performance of a member is directly dependent on its location in a structural configuration and on its relationship or connection with other members in the framework (Mahadevan and Haldar, 1991). An important objective of reliability-based design methods is to reduce the scatter of nonuniform risk levels produced under various load combinations, but Mahadevan and Haldar observed that in many cases it fails to do so. The codified approach also fails to consider the statistical correlations among the design variables.

Popper (1982) wrote, "The fundamental idea underlying scientific determinism is that the structure of the world is such that every future event can in principle be rationally calculated in advance, if only we know the laws of the nature and the present state of the world." The nonlinear state of the structure needs to be considered appropriately in estimating the probability of failure. But since the code does not address the minimum analytical requirement for deterministic evaluation, this area has been overlooked. Haldar and Mahadevan (2000b) advocated the use of nonlinear stochastic finite element approach for this purpose.

In the current codified approach, the reference or permissible or allowable value is required for the reliability evaluation, but in many situations the reference values are unknown. In defining the serviceability requirement for steel structures, the latest LRFD code (AISC, 2001) states "Deformation in structural members and structural systems due to service loads shall not impair the serviceability of the structures." The reference value for the fatigue-related problem has yet to be established. The information on the critical crack size or the damage accumulation function has yet to be developed (Zhao et al., 1994). Time-dependent reliability has been generally overlooked.

In any case, the advanced reliability concepts used in developing the LRFD codes generally remain unknown to designers. Furthermore, for a particular design application, it may be difficult for an experienced design engineer to consider the presence of levels of uncertainty different than those used in developing the reliability-based design guidelines. As mentioned earlier, in most cases, these guidelines were developed to consider the behavior of elements of complex structural systems satisfying explicit performance criteria. However, the strength-related performance functions are generally applicable at the element level and serviceability-related performance functions are related to the system level. As mentioned earlier, the evaluation of system reliability from the element-level reliabilities is not simple. Thus, satisfying performance criteria for both strength and serviceability performance functions may not provide similar risks as observed by many in the past (Ellingwood et al., 1980, Gao et al., 1996).

## 5. IMPLEMENTATION OF SIMULATION-BASED ENGINEERING DESIGN

The international scholars active on simulation (Schuëller and Spanos 2001, Marek et al., 2002) agree that simulation is an attractive alternative to codified approach particularly for large real structural systems. However, several issues need to be addressed before advocating for its immediate implementation. Some of the questions are as follows: (1) is the simulation-based design concept mature enough to be considered as an alternative to the currently available codified approach? (2) At present, should designers have the option to use either the simulation-based approach or the codified approach? (3) What is the future of simulation-based design considering the advancement in computer and information technology? They are discussed in the following sections.

## 5.1. Is the simulation-based design concept mature enough to be considered as an alternative to the currently available codified approach?

As mentioned earlier, numerous researchers have already advocated and identified the potential for simulation-based engineering designs (Schuëller and Spanos 2001, Sundararajan 1995, Marek et. al 1995). All the efforts can be grouped into two major categories: (a) improvements of the simulation methods beyond the basic Monte Carlo method to obtain a higher order of efficiency and accuracy, and (b) attempts to introduce it as a structural reliability assessment concept to the designers (Marek et al., 1995, 2001). At present, designers serve mainly as an interpreter of codes. In most codes, they are given option to use alternative methods with the responsibility of defending them when necessary. Designers rarely use this option for the fear of legal ramifications. The use of simulation in design is expected to showcase his/her creativity and leadership role in the profession. The general consensus of the colloquium participants was that Monte Carlo simulation is a versatile method that is mature enough to be used for reliability analysis of large complicated real engineered structures.

## 5.2. At present, should engineers have the option to use either the simulation-based approach or the codified approach?

To address this question, it is essential to identify what has been done so far in the international communities. Theoretically, designers should have the option to use either approach. In the Czech Republic, CSN 7314 01-1998 (Appendix A) is one of the pilot codes allowing the Monte Carlo simulation as a design tool.

It was pointed out during the colloquium that in Canada, a 14 km long bridge with span length of 250 m was recently built. The code would not cover this design, and simulation was used. It was suggested that simulation could be used to design large projects. The reliability-based design is very common for offshore structures.

In Europe, highway and railway companies use simulation for assessment purposes. This is a good development. Once professionals are familiar with using simulation to evaluate existing structures, it will be simpler for them to use it for design.

In the U.S., the general feeling is that we are safe if we design according to the design code. Any deviation from the codified requirements is discouraged to avoid the liability and insurance issues. This is not entirely true. Designers should use all available means to satisfy performance requirements, according to a judge. The automotive industry satisfied the code requirements. However, they should have used simulation to address all the issues related to the problem. We need to interact with designers, stating that they need to use all available information to make their design safe. This argument can be used to promote simulation.

Some of the developments in the use of simulation in engineering design are very encouraging. Simulation could be used in design in some countries, but it is also necessary to look at its legal ramification. Unlike in Europe, in the U.S.A. a code is not a government document. It is developed by the profession and its acceptance is voted by the users and developers. It was pointed out in the colloquium that in some countries, code guidelines must be followed to the letters, and other countries permit alternative methods if they are better. We need to change the mentality and laws to implement simulation in design. In Europe, two tendencies currently exist: Anglo-Saxon – more or less free to do anything, and middle-European – fixed or obligatory requirements. Current Euro-code is obligatory. It is a product of about 20 years of work from many different countries, and the developers of codified approach may not advocate simulation because of all the time invested in the current system. Also, the current code does not address the serviceability and durability issues in a probabilistic manner. In some cases, such as corrosion, good theoretical models exist which can be used with simulation benefiting the profession.

In the U.S., the Accreditation Board of Engineering and Technology (ABET) now requires that all civil engineering undergraduate students demonstrate knowledge of the application of probability and statistics to engineering problems, indicating its importance in civil engineering education. Thus, the LRFD concept and simulation can be introduced to them at the same time.

## 5.3. What is the future of simulation-based design considering the advancement in computer and information technology?

Haldar (2002), Schuëller (2002) and others commented that Monte Carlo simulation is a versatile method that is mature enough to be used for reliability analysis of large engineered structures. However, the reliability community has no influence on hardware development, but has to keep up with software development in terms of parallel processing or computer farming. We need to reduce the dimensions of the problem, for example by using Karhunen-Loeve expansion. There is room for improvement in the variance reduction techniques. We need to provide these options to designers, and we need to implement new data management schemes integrated with the Internet in the new generation of computer codes. Schuëller stated that this is a way to bring the concept to practicing engineers. They do not need to know the theory behind it very well. Another option could be to bring probability-based code to finite element method-based algorithms like ADINA, ANSIS, and NASTRAN. The programmers are working on these, but they are not fully developed yet. We need to develop more reliability-based software.

# 6. GENERAL COMMENTS

Unfamiliarity with the simulation-based design concept may be main reason behind why it is not used regularly in every day designs. Commercial aircrafts are now being designed using the simulation-based approach. Many participants of the colloquium admitted that they were essentially deterministic persons but now had been re-educating themselves in reliability-based design. They favored simulation-based performance assessment. They pointed out that during the Northridge earthquake of 1994, cracks developed in steel structures, and simulation could be used to study their behavior. Simulation could be the best approach for nonlinear problems where superposition may not work. Simulation can improve the understanding of uncertainty in a problem. Simulation can be used for sensitivity analysis. Simulation is an assessment method rather than a design method, and will fit very well with the performance-based design being gradually implemented in many countries. Simulation is only as good as the data, and reliable data on variables must be available in design offices from a legal point of view.

## 6.1. Improvement in Efficiency in Simulation

The Monte Carlo Simulation technique becomes attractive when the presence of randomness comes from many different sources and the theoretical solution becomes very tedious and impractical. On the other hand, it is very inefficient if the underlying probability to be estimated is relatively small, say less than $10^{-3}$ (Schuëller, 2002). Thus, for large complicated problems with low probability of failure, the efficiency of the basic Monte Carlo Simulation technique needs to be improved. In Section 1, some of the commonly used variance reduction techniques to increase efficiency are discussed. In recent years, controlled Monte Carlo Simulation technique has been proposed (Schuëller, 2002). In contrast to the importance sampling VRT, this methodology is self adapting and does not require detailed a-priori information.

Schuëller (2002) also advocated using parallel processing to increasing the efficiency. According to him "As MCS is based on the generation of independent samples to be used for subsequent computations, the procedure is ideally suited for parallel processing. This implies that more than one processor is used for the same task of an analysis." He commented that the parallelization should be possible from user level. He continued "One possible way to allow this user-supported parallelization is to offer the user a module group, which handles all tasks concerning parallelization, among which the following are found to be most important. In order to reduce the required programming especially the communication tasks need to be robust. Naturally, - not directly apparent to the user – the commands utilize available parallel processing library packages and systems, e.g. the Parallel Virtual Machine (PVM) package allows for a straight-forward implementation, i.e. on real parallel processing computers as well as on a number of workstations connected by network."

## 6.2. Education

The presence of uncertainty in engineering designs must be acknowledged by the profession and the concept needs be integrated in design courses. Reliability assessment methods using simulation can contribute to the transition from deterministic to probabilistic way of thinking of students as well as designers. A pilot international project titled TERECO (TEaching REliability COncepts), sponsored by Leonardo da Vinci Agency in Europe (Marek et al. 2001), was very successful. Education is important to bring the concept to students and practicing engineers, and it needs to be addressed in terms of undergraduate and graduate education. Graphical representation is an important tool for undergraduate education, but students need to be kept motivated beyond this so that they can apply the concept to real problems.

## 6.3. Software Development

To implement the risked-based design concept, either theoretical or simulation-based, in the engineering design, computer programs must be available to engineers. Special application-based computer programs need to be developed for this purpose. As an alternative, the commercially available deterministic computer programs should be modified to give an option to use risk-based design concept.

Schuëller (2002) discussed the subject in detail in his keynote address. According to him, "In order to keep the software on Computational Stochastic

Structural Analysis flexible and generally applicable, it has to be based on MCS. Furthermore, it should be based on modular programming. In general a program is developed by different groups and for different capabilities. Therefore, a natural modular structure is automatically given. However, in many cases this structure is not sufficiently observed by programmers or, even worse, modules are available, but different data handling procedures within the modules do not allow an efficient transfer of data." He encouraged developing modular programming transparent to all users. He continued "A second task is, to allow the user to access these modules at any time of the current analysis. Finally, the third task is to organize the easy transfer of data from one module to the other, i.e., input data as matrices, vectors as well as parameters for the calculations within the distinct modules. Efficiency is also a major issue. Fourteen major sources of uncertainty have been identified for performance-based seismic design. Parallel computing and smart simulation are necessary for this purpose. Simulation-based assessment can be used for this type of complicated problem.

# 7. CONCLUSIONS

The general consensus of the participants of Euro-SiBRAM'2002 was that Monte Carlo simulation is a versatile method that is mature enough to be used for reliability analysis of large engineered structures. The simulation-based approaches are being increasingly used as an alternative to codified approaches in some countries. However, it may suffer same deficiencies as the current risk-based codified approach. It is essential that the limitations of the current codified approaches be known to the practicing engineers so that they can use simulation whenever necessary to address the situations. Education and documentation of case studies are expected to accelerate the implementation of simulation in engineering design. With the advancement in the information technology, efficiency in simulation can be improved significantly. Another alternative will be to add simulation-based code to currently available finite element algorithms. There is some advancement in this area also. Many deterministic professionals are in the process of re-educating themselves in the simulation area. Legal aspects of using simulation in engineering design also need to be addressed worldwide.

Many other valuable ideas, suggestions and opinions were presented and discussed at ten sessions of the colloquium. Interested readers are requested to refer to the colloquium proceedings (Marek et al. 2002).

## ACKNOWLEDGMENTS

The project was support by the Grant Agency of the Czech Republic (Projects No. 103/01/1410 and 105/01/0783). Any opinions, findings, conclusions, or recommendations expressed in this publication are those of the authors and do not necessarily reflect the views of the sponsor.

To develop the paper, comments made by many participants of the colloquium, are summarized here and they were not individually identified. More detailed comments made by them can be found in the proceedings of the colloquium (Marek et al., 2002).

# REFERENCES


ACI, Building Code Requirements for Structural Concrete (318-02), Farmington Hills, MI., 2002a.

ACI, Building Code Requirements for Masonry Structures and Specification for Masonry Structures – 2002, ACI 530-02/ASCE 5-02/TMS 402-02, reported by the Masonry Standards Joint Committee, 2002b.

AISC, Manual of Steel Construction Load and Resistance Factor Design, 1st Edition, 2nd Edition, and 3rd Edition, 1986, 1994, 2001.

ASCE, Standard for Load and Resistance Factor Design (LRFD) for Engineered Wood Construction, ASCE 16-95, 1995.

ASCE, A panel session on Past, Present and Future of Reliability-based Structural Engineering Worldwide, Civil Engineering Conference and Exposition, Washington, D.C., 2002.

American Wood Council, The Load and Resistance Factor Design (LRFD) Manual for Engineered Wood Construction, Washington, D.C., 1996.

Bjorhovde, R., Galambos, T.V., and Ravindra, M.K., LRFD Criteria for Steel Beam-Columns, Journal of the Structural Engineering, ASCE, 104(9), 1371-1388, 1978.

Czech Institute of Standards, CSN 73 1401 - 1998 Design of Steel Structures, Prague, Czech Republic.

Elishakoff, I., Essay on the Role of the Monte Carlo Method in Stochastic Mechanics, Monte Carlo Simulation, Schueller, G.I., and Spanos, P., (Editors), Balkema, Rotterdam, 619-627, 2001.

Ellingwood, B., Galambos, T.V., MacGregor, J.G., and Cornell, C.A., Development of a Probability Based Load Criterion for American Standard A58: Building Code Requirements for Minimum Design Loads in Buildings and Other Structure, Special Publication 577, National Bureau of Standards, Washington, D.C., 1980.

Gao, L., Haldar, A., and Shome, N., Strength and Serviceability Requirements in Seismic Design Using Nonlinear SFEM, International Union of Theoretical and Applied Mechanics (IUTAM), Advances in Nonlinear Stochastic Mechanics, A. Naess and S. Krenk (Eds), Kluwer Academic Publishers, 1996, pp. 179-188.

Haldar, A., Basic Simulation Concepts Applicable in Codified Design, Euro-SiBRAM'2002 Colloquium Proceedings, Marek P., Haldar A., Guštar M. and Tikalsky P., Editors, ITAM Academy of Sciences of Czech Republic, Czech Republic, 2002.

Haldar, A., and Mahadevan, S., *Probability, Reliability, And Statistical Methods In Engineering Design*, John Wiley & Sons, New York, NY., 2000a.

Haldar, A., and Mahadevan, S., *Reliability Assessment Using Stochastic Finite Element Analysis*, John Wiley & Sons, New York, NY., 2000b.

Huh, J., and Haldar, A., Stochastic Finite Element-Based Seismic Risk Evaluation for Nonlinear Structures, Journal of the Structural Engineering, ASCE, 2001, 127(3), 323-329.

Huh, J., and Haldar, A., Seismic Reliability of Nonlinear Frames with PR connections using Systematic RSM, Probabilistic Engineering Mechanics, 2002, 17(2), 177-190.

Lewis, P.A.W., and Orav, E.J., Simulation Methodology for Statisticians, Operations Analysts, and Engineers, Vol. 1, Wadsworth & Brooks/Cole Advanced Books & Software, Pacific Grove, California, 1989.

Mack, Y.P., Statistical Issues on Simulation Techniques in Structural Engineering, Euro-SiBRAM'2002 Colloquium Proceedings, Marek P., Haldar A., Guštar M. and Tikalsky P., Editors, ITAM Academy of Sciences of Czech Republic, Czech Republic, 2002.

Mahadevan, S., and Haldar, A., "Stochastic FEM-Based Validation of LRFD", *Journal of Structural Engineering, ASCE,* 117(5), 1393-1412, 1991.

Marek, P., Brozzetti, J., and Gustar, M., *Probabilistic Assessment of Structures using Monte Carlo Simulation,* Academy of Sciences of the Czech Republic, Praha, Czech Republic, 2001.

Marek, P., Brozzetti, J., Gustar, M., and Tikalsky., *Probabilistic Assessment of Structures using Monte Carlo Simulation, Background, Exercises and Software*, 2nd Edition, Academy of Sciences of the Czech Republic, Praha, Czech Republic, 2003.

Marek P., Guštar M., and Anagnos, T., *Simulation-Based Reliability Assessment for Structural Engineers,*CRC Press, Boca Raton, Florida, 1995.

Marek P., Haldar A., Guštar M. and Tikalsky P., Editors, Euro-SiBRAM'2002 Colloquium Proceedings, ITAM Academy of Sciences of Czech Republic, Prosecka 76, 19000 Prague 9., Czech Republic, 2002.

Popper, K.R., *The Open Universe*, Hutchinson, London, 1982.

Schuëller, G.I., Past, Present & Future of Simulation-based Structural Analysis, Euro-SiBRAM'2002 Colloquium Proceedings, Marek P., Haldar A., Guštar M. and Tikalsky P., Editors, ITAM Academy of Sciences of Czech Republic, Czech Republic, 2002.

Schuëller, G.I., and Spanos, P.D., Editors, *Monte Carlo Simulation*, A.A. Balkema Publishers, 2001.

Sundararajan, C. (Raj), Editor, *Probabilistic Structural Mechanics Handbook: Theory and Industrial Applications,* Chapman & Hall, New York, 1995.

Zhao, Z., Haldar, A., and Breen, F.L., "Fatigue Reliability Evaluation of Steel Bridges," Journal of the Structural Division, ASCE, 120(5), 1608-1623, 1994.

# Chapter 21

# SYSTEM IDENTIFICATION AT LOCAL LEVEL UNDER UNCERTAINTY

**Hasan Katkhuda and Achintya Haldar**

## 1. INTRODUCTION

Health assessment of structures in use is a major challenge to the profession. Visual inspections are routinely used for this purpose. They cannot be effective if the defects are not visible to the naked eye or hidden behind some obstacles such as false ceilings, fire proofing material or other obstructions (Katkhuda, et. al 2003). Furthermore, defects which do not alter the behavior of structures are expected to be present in large structural systems. Thus, locating them using visual inspections is a waste of resources and may forced to make some unnecessary maintenance decisions. An objective health assessment technique is urgently required to locate structural behavior altering defects at the local level. If such a method can be developed, the whole structure need not be inspected, the effect of defects on structural behavior can be easily evaluated, and the improvement in the structural behavior just after a repair action can also be studied very effectively. It is obvious that there is a need for a simple, inexpensive nondestructive evaluation (NDE) procedure that can be used routinely for in-service health assessment of existing structures at the local element level without disrupting their normal operation.

To locate defects at the local level, a structure needs to be represented by elements and well developed finite element method can be used for this purpose. The solution of the inverse problem will identify the properties of all the finite elements in the formulation. The inverse problem can be set up as a static problem or as a dynamic problem. In dynamic formulation, since the inertia and damping forces must be considered, they provide more constraints. The dynamic formulation is used in developing the proposed method. The discussion clearly leads to developing a finite element-based system identification technique using dynamic response information.

System identification (SI) is a multidisciplinary research area and the existing literature is very extensive (Doebling, et. al 1996, and Housner, et. al 1997). Available SI techniques can be broadly divided in to two categories: frequency domain and time domain. Frequency domain approaches are very common. Instead of using an enormous amount of response data in time domain, the structural properties can be described in terms of frequencies and mode shapes and changes in them can be used to detect defect. Since frequencies and mode shapes indicate the overall structural properties, the information can be used to decide whether the structure is defective or not, but the defects can not be located at the element level. Furthermore, the frequencies may not change significantly even in the presence of a major defect, e.g., the loss of a member of the structure, particularly considering the noises in the response data. To meet the objective of this study, the time domain SI techniques will be most appropriate.

Several time domain approaches have been proposed (Hoshiya and Maruyama 1987, Wang and Haldar 1994, and Koh et. al 2000). A typical SI technique has three components: input excitation, the system to be identified, and the output response information. By knowing input excitation and output response information, the third component, i.e., the system can be identified. Outside the control laboratory environment, the measurement of exciting force could be very costly and may contain so much noise that the solution of inverse problem concept cannot be applied. The desirability of a SI technique will be significantly improved, if a system can be identified using only noise-laden response information. The concept is expected to be very challenging since two of the three components in a SI approach are unknown.

Wang and Haldar (1994) and Wang (1995) proposed a linear time domain finite element-based SI approach and identified structures using response information only. They called it Iterative Least Square with Unknown Input (ILS-UI). In developing the dynamic equation of motion, they assumed viscous type damping and identified shear-type buildings, the simplest mathematical representation of structures.

## 2. ILS-UI METHOD

The governing equation of motion of a structure using viscous damping can be written in matrix form as (Wang and Haldar 1994):

$$\mathbf{M}\,\ddot{\mathbf{x}}(t) + \mathbf{C}\,\dot{\mathbf{x}}(t) + \mathbf{K}\,\mathbf{x}(t) = \mathbf{f}(t) \tag{1}$$

where $\mathbf{M}$ is the mass matrix; $\mathbf{C}$ is the viscous damping matrix; $\mathbf{K}$ is the stiffness matrix; $\ddot{\mathbf{x}}(t)$, $\dot{\mathbf{x}}(t)$, and $\mathbf{x}(t)$ are vectors containing the dynamic responses in terms of acceleration, velocity and displacement respectively; and $\mathbf{f}(t)$ is the excitation force vector. The acceleration of the structure is measured and then integrated successively to obtain the velocity and displacement time histories.

Assuming **M** is known mass matrix, Equation (1) can be rewritten as:

$$[\mathbf{C}:\mathbf{K}]\begin{bmatrix}\dot{\mathbf{x}}(\mathbf{t})\\ \mathbf{x}(\mathbf{t})\end{bmatrix}=\mathbf{f}(t)-\mathbf{M}\,\ddot{\mathbf{u}}(t) \quad (2)$$

For *N*- Dynamic Degrees of Freedoms (DDOFs) system, and suppose the responses of the structure are measured for a duration of ($h.\ \Delta t$), where $h$ is the total number of sample points, and $\Delta t$ is the constant time increment; Equation (2) can be rearranged in matrix form as:

$$\mathbf{A}(t)_{(N.h)\times L}\,\mathbf{P}_{L\times 1}=\mathbf{F}(t)_{(N.h)\times 1} \quad (3)$$

where **A**(t) is an $(N\,.\,h)\times L$ matrix composed of the system responses vectors of velocity and displacement; $L$ is the total number of unknowns; **P** is a $L\times 1$ vector composed of the unknown system parameters- damping and stiffness- that need to be known at the element level; and **F**(t) is $(N\,.\,h)\times 1$ vector composed of input excitation and inertia forces at any time $t$.

For mathematical convenience, Equation (3) can also be expressed as:

$$\sum_{s=1}^{L}\mathbf{A}_{rs}\,\mathbf{P}_{s}=\mathbf{F}_{r} \qquad r=1,2,\ldots,h\times N \quad (4)$$

The total error, *Er,* in the estimation of the system parameters can be expressed as:

$$Er=\sum_{r=1}^{h\times N}\left(\mathbf{F}_{r}-\sum_{s=1}^{L}\mathbf{A}_{rs}\,\mathbf{P}_{s}\right)^{2} \quad (5)$$

To minimize the total error, Equation (5) can be differentiated with respect to each one of the $\mathbf{P}_q$ parameters as:

$$\frac{\partial Er}{\partial \mathbf{P}_{q}}=\sum_{r=1}^{h\times N}\left(\mathbf{F}_{r}-\sum_{s=1}^{L}\mathbf{A}_{rs}\,\mathbf{P}_{s}\right)\mathbf{A}_{rq}=0 \qquad q=1,2,\ldots,L \quad (6)$$

Equation (6) gives $L$ simultaneous equations. The solution of Equation (6) will give all $L$ unknown parameters to be estimated. The unknown parameter vector, **P** can be evaluated as:

$$\mathbf{P}_{L\times 1} = \left(\mathbf{A}^{\mathrm{T}}{}_{L\times(N.h)}\,\mathbf{A}_{(N.h)\times L}\right)^{-1}\mathbf{A}^{\mathrm{T}}{}_{L\times(N.h)}\,\mathbf{F}^{\mathrm{T}}{}_{(N.h)\times 1} \tag{7}$$

It is relatively simple to solve for the system parameters vector **P** provided that the force vector **F**(t) and **A**(t) are known. However, as mentioned earlier, the input excitation is not known; thus, the force vector **F**(t) becomes a partially unknown vector.

To address this issue, Wang and Haldar (1994) proposed an iterative procedure. Since the input excitation is not available at any time, they assumed the input excitation force **f**(t) to be zero for $p$ time points to start the iterative process. This $p$ time points should be kept to a minimum without compromising the convergence or the accuracy of the method. They observed that $p$ can be only 2 time points if the structure is excited at any floor; and only 4 time points if the structure is excited at the base representing earthquake motion.

As mentioned earlier, they applied the concept to identify shear-type buildings and extensively verified the method using computer generated output response information. The representation of a multi story building as a shear-type structure is the simplest mathematical representation and consists of many assumptions. This type of building deflects under shear forces only. The total mass of the structure is lumped at the floor levels; the girders/floors are assumed infinitely rigid compared to columns; all the columns in a floor are represented by one column, and the deformation of the structure is considered to be independent of the axial force present in the columns. Thus, in this representation, an $N$-story building is represented by $N$ dynamic degree of freedoms, i.e., each floor level has only one horizontal displacement. The mass matrix of a shear-type building is diagonal since the mass is lumped at each story. It can be represented as:

$$\mathbf{M} = \text{diag.}\,(m_1, m_2, \ldots, m_N) \tag{8}$$

The corresponding damping and stiffness matrices can be shown to be:

$$\mathbf{C}=\begin{bmatrix} c_1+c_2 & -c_2 & 0 & 0 & \cdots\cdots & 0 & 0 \\ -c_2 & c_2+c_3 & -c_3 & 0 & \cdots\cdots & 0 & 0 \\ \cdots\cdots & \cdots\cdots & \cdots\cdots & \cdots\cdots & \cdots\cdots & \cdots\cdots & \cdots\cdots \\ 0 & 0 & 0 & 0 & \cdots\cdots & c_{N-1}+c_N & -c_N \\ 0 & 0 & 0 & 0 & \cdots\cdots & -c_N & c_N \end{bmatrix} \tag{9}$$

and

$$\mathbf{K}=\begin{bmatrix} k_1+k_2 & -k_2 & 0 & 0 & \cdots\cdots & 0 & 0 \\ -k_2 & k_2+k_3 & -k_3 & 0 & \cdots\cdots & 0 & 0 \\ \cdots\cdots & \cdots\cdots & \cdots\cdots & \cdots\cdots & \cdots\cdots & \cdots\cdots & \cdots\cdots \\ 0 & 0 & 0 & 0 & \cdots\cdots & k_{N-1}+k_N & -k_N \\ 0 & 0 & 0 & 0 & \cdots\cdots & -k_N & k_N \end{bmatrix} \tag{10}$$

where $m_i$, $c_i$, and $k_i$ ( i =1, 2, ... , $N$ ) is the mass, damping, and stiffness respectively, at the $i^{th}$ DDOF of the building.

**A**(t) matrix in Equation (3) for shear-type buildings will take the following form:

$$\mathbf{A}(t)=\left[\begin{array}{cccc:cccc} \dot{x}_1 & \dot{x}_1-\dot{x}_2 & \cdots & 0 & x_1 & x_1-x_2 & \cdots & 0 \\ 0 & \dot{x}_2-\dot{x}_1 & \cdots & 0 & 0 & x_2-x_1 & \cdots & 0 \\ \cdots & \cdots & \cdots & \cdots & \cdots & \cdots & \cdots & \cdots \\ 0 & 0 & \cdots & \dot{x}_{N-1}-\dot{x}_N & 0 & 0 & \cdots & x_{N-1}-x_N \\ 0 & 0 & \cdots & \dot{x}_N-\dot{x}_{N-1} & 0 & 0 & \cdots & x_N-x_{N-1} \end{array}\right] \tag{11}$$

The unknown system parameters vector **P** for shear-type building will be:

$$\mathbf{P}=\left[c_1\ c_2\ \ldots\ c_N\ \vdots\ k_1\ k_2\ \ldots\ k_N\right]^{T} \tag{12}$$

And the **F**(t) vector can be expressed as:

$$\mathbf{F}(t_i) = \begin{bmatrix} f_1(t_i) - m_1\ddot{x}_1(t_i) \\ f_2(t_i) - m_2\ddot{x}_2(t_i) \\ \cdots \\ f_N(t_i) - m_N\ddot{x}_N(t_i) \end{bmatrix} \tag{13}$$

Using ILS-UI approach, using only response information, $N$ number of stiffness and $N$ number of damping parameters can be identified. Although, the ILS-UI approach is a significant improvement over the other methods available at that time, it cannot be used to detect defects at the element level, i.e., in columns and beams. Thus, the finite element representation of a structure needs to be improved to make it more realistic. A structure needs to be represented by elements, potential locations of defects. Only then the elements can be classified as defect-free or defective.

For the ease of illustration, two dimensional frames are considered next. In such a frame, all the elements are represented by uniform two dimensional beam elements. For a two dimensional beam element, there are three Dynamic Degrees of Freedom (DDOFs) at each node. Two are translational DDOFs; one is along the length of the element (x axis) and the other is perpendicular to the x axis, i.e., along the y axis, and the third DDOF represents the rotation of the node.

The mass matrix in Equation (8) for a shear-type building needs to be modified to represent three DDOFs at each node, giving a total of six DDOFs for an element. It can be shown to be (Cook et. al. 1989):

$$\mathbf{M}^i = \text{diag}\left(\frac{\overline{m}_i}{2}\begin{bmatrix} 1 & 1 & \frac{L_i^2}{39} & 1 & 1 & \frac{L_i^2}{39} \end{bmatrix}\right) \tag{14}$$

Where $L_i$ is the length of the $i^{th}$ element and $\overline{m}_i$ is the corresponding mass per unit length.

For viscous damping, Equation (9) remains the same. However, the stiffness matrix in Equation (10) needs to be changed. The stiffness matrix $\mathbf{K}^i$ for the $i^{th}$ beam element of uniform cross section (constant flexural stiffness or constant $EI$) is given by:

$$\mathbf{K}^i = \frac{E_i I_i}{L_i}\begin{bmatrix} A_i/I_i & 0 & 0 & -A_i/I_i & 0 & 0 \\ 0 & 12/L_i^2 & 6/L_i & 0 & -12/L_i^2 & 6/L_i \\ 0 & 6/L_i & 4 & 0 & -6/L_i & 2 \\ -A_i/I_i & 0 & 0 & A_i/I_i & 0 & 0 \\ 0 & -12/L_i^2 & -6/L_i & 0 & 12/L_i^2 & -6/L_i \\ 0 & 6/L_i & 2 & 0 & -6/L_i & 4 \end{bmatrix} \tag{15}$$

where $E_i$, $I_i$, and $A_i$ are the Young's modulus, moment of inertia, and area of the cross-section of the $i^{th}$ element of the beam element, respectively.

The global stiffness matrix **K** for the frame can be assembled from the stiffness matrices of all the elements using the direct stiffness method as:

$$\mathbf{K}_{N\times N} = \sum_{i=1}^{ne} \mathbf{K}^{i}{}_{N\times N} \tag{16}$$

The unknown system parameters vector **P** for a frame can be expressed as:

$$\mathbf{P} = \left[ c_1 \;\; c_2 \;\; \ldots \;\; c_{ne} \;\vdots\; k_1 \;\; k_2 \;\; \ldots \;\; k_{ne} \right]^{\mathrm{T}} \tag{17}$$

where *ne* is the total number of elements required to represent the whole plane frame, and $k_i = E_i I_i / L_i$ is the unknown stiffness parameter for the $i^{th}$ beam element that needs to be identified.

If the damping is considered to be viscous type, Equation (17) indicates that for a total of *ne* elements, $2 \times ne$ numbers of parameters need to be identified. At this early stage of the development of the method, only the degradation of stiffness of elements is tracked to detect defects. The information on damping is not used. Thus, the identification of damping values of the elements may not be important and the efficiency of the algorithm can be significantly improved if they are not identified, particularly for a large structural system.

The efficiency of the ILS-UI method can be significantly increased by considering Rayleigh-type damping, i.e., the damping is proportional to mass and stiffness, in the dynamic formulation (Ling 2000, and Ling and Haldar 2004). The proposed algorithm is finite element based and the mass and stiffness matrices of a structural system will be readily available. The incorporation of the Rayleigh-type damping in the dynamic formulation is not expected to introduce any additional problem. For the Rayleigh-type damping, the damping matrix **C** in Equation (9) can be represented as:

$$\mathbf{C} = \alpha \mathbf{M} + \beta \mathbf{K} \tag{18}$$

where factor $\alpha$ is the mass-proportional damping coefficient and $\beta$ is the stiffness-proportional damping coefficient. They can be evaluated using the standard procedure from the information on the first two undamped frequencies of the structure as suggested by Clough and Penzien (1993).

Incorporating Equation (18) in to Equation (1) will result:

$$\mathbf{K}\,\mathbf{x}(t) + (\alpha\mathbf{M} + \beta\mathbf{K})\dot{\mathbf{x}}(t) = \mathbf{f}(t) - \mathbf{M}\,\ddot{\mathbf{x}}(t) \tag{19}$$

As mentioned earlier, the mass matrix is assumed to be known. Lumped and consistent mass matrices are commonly used. For more realistic representation of the distribution of masses in a structure, the consistent mass matrix is generally preferred and is used in this study. Thus, the lumped mass matrix used for shear-type buildings needs to be modified. The consistent mass matrix for the i$^{th}$ element can be represented as (Cook et. al., 1989):

$$\mathbf{M}^i = \frac{\overline{m}_i L_i}{420}\begin{bmatrix} 140 & & & & & \\ 0 & 156 & & & \text{Sym.} & \\ 0 & 22L & 4L^2 & & & \\ 70 & 0 & 0 & 140 & & \\ 0 & 54 & 13L & 0 & 156 & \\ 0 & -13L & -3L^2 & 0 & -22L & 4L^2 \end{bmatrix} \tag{20}$$

where $\mathbf{M}^i$ is the consistent mass matrix for the i$^{th}$ beam element of uniform cross section, $L_i$ is the element length and $\overline{m}_i$ is the mass per unit length.

The global mass matrix **M** for a typical frame can be assembled from the information on mass matrices of all the elements as:

$$\mathbf{M}_{N\times N} = \sum_{i=1}^{ne} \mathbf{M}^i_{N\times N} \tag{21}$$

Equations (15) and (16) representing element and structural stiffness matrices, respectively, will remain the same. However, the **A**(t) matrix for the Rayleigh-type damping can be shown to be:

$$\mathbf{A}(t)_{(N.h)\times L} = [\,\mathbf{R}^1\mathbf{x}(t)\;\mathbf{R}^2\mathbf{x}(t)\ldots\mathbf{R}^{ne}\,\mathbf{x}(t)\mathbf{R}^1\dot{\mathbf{x}}(t)\;\mathbf{R}^2\dot{\mathbf{x}}(t)\ldots\mathbf{R}^{ne}\dot{\mathbf{x}}(t)\;\mathbf{M}\dot{\mathbf{x}}(t)] \tag{22}$$

where $\mathbf{R}^i$ is the 6×6 matrix containing all the terms in the square bracket in Equation (15) for the i$^{th}$ element; $N$ is the total number of DDOFs, $L$ is the total number of unknown parameters, and $h$ is the total number of equally-spaced sample points in the dynamic response information.

The unknown system parameters vector **P** for the Rayleigh-type damping can be shown to be:

$$\mathbf{P} = [k_1, k_2, \ldots, k_{ne}, \beta\,k_1, \beta\,k_2, \ldots, \beta\,k_{ne}, \alpha]^T \tag{23}$$

The improved efficiency in using the Rayleigh-type damping can be observed by comparing Equations (17) and (23). In Equation (17), the total number of parameters to be identified is $2 \times ne$, where $ne$ is the total number of elements in the structural system. However, using Equation (23), only $(ne + 2)$ numbers of parameters need to be identified. For a structural system containing a large number of structural elements, the use of Rayleigh-type damping is expected to significantly improve the efficiency of the algorithm. In all the subsequent discussions, only Rayleigh-type damping will be considered.

As mentioned earlier, the algorithm is iterative in nature. A structure needs to be represented by finite elements. Initially, the basic finite element representation should kept simple without compromising the dynamic characteristics of the structure. It will be shown later with the help of examples that the finite element model can be refined, if necessary, to locate defect spots more accurately. It needs to be emphasized that basic objective of the finite element representation is not to accurately evaluate the structural responses but to provide a platform to compare structural responses as the structure degrade or deteriorate with time. The degradation will be captured by tracking the changes in the identified structural parameters. Structural responses in term of acceleration time histories must be available at node points in the finite element representation. The velocity and displacement time histories can be obtained by integrating the acceleration measurements successively. As mentioned earlier, the consistent mass matrix is assumed to be known. At this stage, all the information required to develop the **A**(t) matrix is available. The **A**(t) matrix for a two dimensional frame structure is shown in Equation (22). The unknown system parameters in Equation (23) can be evaluated by solving the $L$ simultaneous equations as shown in Equation (7). However, the iteration process cannot be initiated since it is assumed that the input excitation force **f**(t) is unknown. Since the input excitation is not available, the iteration can be started by assuming it is zero at time $t_i$, $i = 1, 2, \ldots, p$, where $p \leq h$. Wang and Haldar (1994) showed that $p$ could be only two points if the structure is excited at any DDOF, and only four points if the structure is excited at the base representing seismic motion. Later the authors (Katkhuda, et. al. 2004a) observed that the algorithm produces better and more accurate results if the excitation information is assumed to be zero at all time points $h$ instead to $p$ time points to start the iteration. With this assumption, the **F** vector in Equation (7) can be obtained and a first estimation of the unknown system parameters **P** can be evaluated. Using Equation (19) and the estimated system parameters **P**, the information on the input excitation force **f**(t) can be generated at all time points $h$. Using the information on the generated input excitation and Equation (7), the estimation on the system parameters can be updated. The algorithm will iterate until the system parameters are evaluated with a pre-determined accuracy. The convergence criterion is set with respect to the evaluated input excitation. The procedure will continue until there is a convergence in the input excitation with a predetermined tolerance ($\varepsilon$). The tolerance ($\varepsilon$) is set to be $10^{-8}$ in this study. The convergence requires $\left|\mathbf{f}^{i+1} - \mathbf{f}^{i}\right| \leq 10^{-8}$ applied for all time points *h*.

It is interesting to note that the algorithm not only identifies unknown stiffness parameters of all the elements and the two Rayleigh damping coefficients, it also identifies the time history of the unknown excitation force.

Obviously, the algorithm needs to be verified at this stage. Initially, the responses of structures were theoretically obtained using commercially available computer programs. To simulate realistic field conditions, artificially generated white noise was added to the theoretical response information. Using noise-free and noise contaminated response information, various structures were identified. Recently, defect-free and defective fixed ended and simply supported beams with uniform cross section were tested in the laboratory. The algorithm correctly identified the beams (Vo and Haldar, 2004). Several examples are given in the following section to better illustrate the algorithm and its capabilities.

## 3. NUMERICAL EXAMPLES

### Example 1

A defect-free two dimensional steel frame excited by a blast load is identified in this example (Katkhuda, et. al., 2003). A three story plane steel frame, shown in Figure 1, is considered. The frame consists of 9 members; six columns and three beams. The height of the columns in each floor is 3.66 m and the bay width is 9.14 m. W18x71 of grade A36 steel section is used for all the beams and columns. Assuming the bases are fixed; the structure is represented by 18 DDOFs; 3 DDOFs at each node. The masses of the three beams $m_1$, $m_2$, and $m_3$ are assumed to be 97.92 kg-sec$^2$/m, and the masses of the columns $m_4$ to $m_9$ are assumed to be 39.19 kg-sec$^2$/m. The beam stiffnesses $k_1$, $k_2$, and $k_3$ are considered to be 10651 kN-m, and the column stiffnesses $k_4$ to $k_9$ are considered to be 26611 kN-m.

The first two natural frequencies of the frame $f_1$ and $f_2$ were found to be 6.62 and 23.27 Hz, respectively. Assuming 3% damping for the first two undamped frequencies, the Rayleigh damping coefficients $\alpha$ and $\beta$ are estimated to be 1.9427 and 0.0003194, respectively.

The structure is assumed to be excited by a blast force applied horizontally at the top floor at node 1, as shown in Figure 1. The blast force is assumed to be a triangular pulse of magnitude 22 kN acting for a duration of 0.05 sec. The theoretical responses of the frame in terms of displacements, velocities and accelerations are calculated at each DDOF using a commercially available computer program ANSYS (2001). Once the theoretical responses are evaluated, the information about the input uncertain blast force is completely ignored. Considering responses from 0.02 to 0.123 sec, recorded at 0.001 sec time intervals providing 104 time points the stiffnesses of all 9 members are identified. The results are summarized in Table 1. The maximum error in the stiffness identification is observed to be only 0.17%. The error is extremely small compared to other available methods presently available in the literature, even when input excitation information was used in the identification process (Toki, et al. 1989). Considering the practical aspect of the problem, the noise in the response information cannot be avoided. To simulate the presence of noise in the response information, numerically generated white noises with intensities of 5% of the root mean square values of the responses at all dynamic degrees of freedom are added to the computer generated theoretical response information.

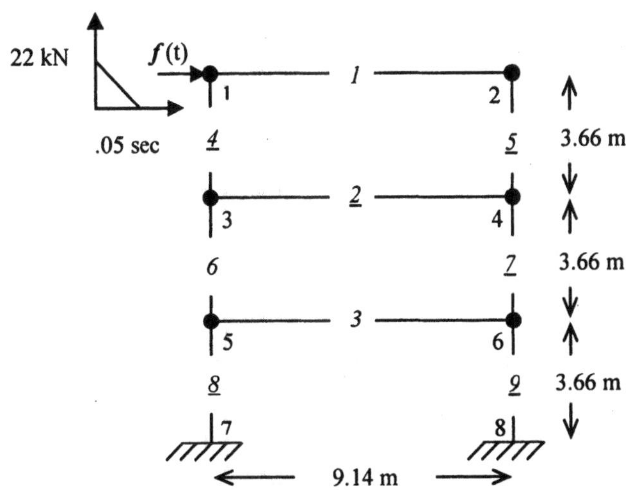


Figure 1. Three stories frame and the blast load used for example 1

The frame is again identified using the noise-contaminated response information. The results are shown in Table 1. As expected, the error in the stiffness identification went up, but the maximum error is observed to be only 1.46%. The results in Table 1 indicate that the algorithm is very accurate and robust, and is capable of identifying systems even in the presence of noise in the response information.

*Table 1. Stiffness (EI/L) identification for the frame excited by one load for example 1.*

| Member | Initial Theoretical Value (kN-m) | Identified Noise- Free (kN-m) | Error % | Identified Noise-Included (kN-m) | Error % |
|---|---|---|---|---|---|
| $k_1$ | 10651 | 10633 | 0.17 | 10495 | 1.46 |
| $k_2$ | 10651 | 10639 | 0.11 | 10501 | 1.41 |
| $k_3$ | 10651 | 10640 | 0.10 | 10510 | 1.32 |
| $k_4$ | 26611 | 26582 | 0.11 | 26493 | 0.44 |
| $k_5$ | 26611 | 26589 | 0.08 | 26489 | 0.46 |
| $k_6$ | 26611 | 26598 | 0.05 | 26475 | 0.51 |
| $k_7$ | 26611 | 26598 | 0.05 | 26472 | 0.52 |
| $k_8$ | 26611 | 26599 | 0.05 | 26480 | 0.49 |
| $k_9$ | 26611 | 26599 | 0.05 | 26477 | 0.50 |

## Example 2

A real structure can be excited by multiple forces acting simultaneously. Although the information on the input excitation forces is not required for the algorithm, the question remains if the algorithm can identify a structure when it is excited by more than one force. The two dimensional steel frame discussed in Example 1 is considered again, however, it is now excited by multiple forces applied at the superstructure (Katkhuda, et al. 2004b).

The same three stories frame used in example 1 is excited by two harmonic forces: $f_1(t) = 44.4 \sin(40\pi t)$ kN applied horizontally at the top floor at node 1, and $f_2(t) = 44.4 \sin(65\pi t)$ applied horizontally at the second floor at node 3, as shown in Figure 2.

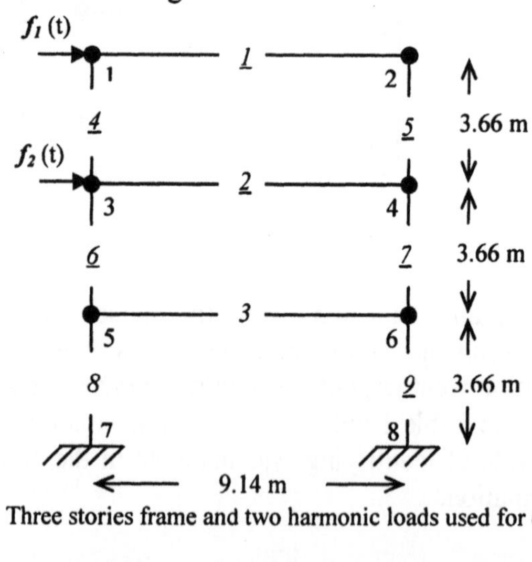


Figure 2. Three stories frame and two harmonic loads used for example 2

The theoretical responses in terms of displacements, velocities and accelerations are calculated at each DDOF of the structure using ANSYS. As in Example 1, once the theoretical responses are evaluated, the information on the two input harmonic forces is completely ignored. Using only responses from 0.02 to 0.87 sec recorded at 0.01 sec time interval providing a total of 86 time points, the structure is identified. The results are shown in Table 2. The maximum error in the stiffness identification is found to be only 0.94%. To simulate the presence of noise in the response information, numerically generated white noises with intensities of 2% of the root mean square values of the responses at all dynamic degrees of freedom are added to the computer generated theoretical response information. The frame is identified with noise contaminated response information. In this case, the maximum error is found to be 0.97%. It can be concluded that the method is capable of identifying the structure when it is excited by more than one load simultaneously.

*Table 2. Stiffness (EI/L) identification for the frame excited by two loads for example 2*

| Member | Initial Theoretical Value (kN-m) | Identified Noise- Free (kN-m) | Error % | Identified Noise- Included (kN-m) | Error % |
|---|---|---|---|---|---|
| $k_1$ | 10651 | 10550 | 0.94 | 10548 | 0.97 |
| $k_2$ | 10651 | 10556 | 0.89 | 10559 | 0.86 |
| $k_3$ | 10651 | 10558 | 0.87 | 10555 | 0.90 |
| $k_4$ | 26611 | 26382 | 0.86 | 26372 | 0.90 |
| $k_5$ | 26611 | 26382 | 0.86 | 26377 | 0.88 |
| $k_6$ | 26611 | 26387 | 0.84 | 26395 | 0.81 |
| $k_7$ | 26611 | 26387 | 0.84 | 26398 | 0.80 |
| $k_8$ | 26611 | 26393 | 0.82 | 26382 | 0.86 |
| $k_9$ | 26611 | 26393 | 0.82 | 26387 | 0.84 |

**Example 3**

The damage state evaluation of structures just after major natural events like strong earthquakes or high winds is a major concern. The method presented here can also be used for this purpose. The stiffness evaluation of a large structure excited by an earthquake load is illustrated in this example.

A four story two bay two dimensional steel frame, as shown in Figure 3, is considered. The frame consists of 20 members; 12 columns and 8 beams. The height of the columns in each floor is 3.66 m and each bay width is 9.14 m. W18x71 of grade A36 steel section is used for all the members. Assuming the bases are fixed; the structure is represented by 36 DDOFs; 3 DDOFs at each node. The masses of all the beams and columns are assumed to be 97.92 kg-sec$^2$/m and 39.19 kg-sec$^2$/m, respectively. The beam and column stiffnesses are estimated to be 10650.13 kN-m and 26625.34 kN-m, respectively. The first two frequencies, $f_1$ and $f_2$, of the structure were found to be 4.71 and 15.63 Hz, respectively. For an equivalent modal damping of 3% of the critical for the first two modes, the Rayleigh damping coefficients α and β are found to be 1.365105 and 0.000469435, respectively. The structure is excited by the El Centro earthquake of 1940 applied at its base. The time history of the earthquake is shown in Figure 4.

The theoretical responses of the frame are calculated in terms of displacements, velocities and accelerations at all nodes using ANSYS. The responses are recorded at 0.01 sec time interval. Using responses from 1.52 to 2.37 sec providing 86 time points, the structure is identified. After the theoretical responses are evaluated, the information on the input force is completely ignored. Using the response information only, the elements of the frame are identified. The theoretical and identified stiffnesses of the 20 elements frame are shown in Table 3. The maximum error in the stiffness identification is found to be only 0.024%.

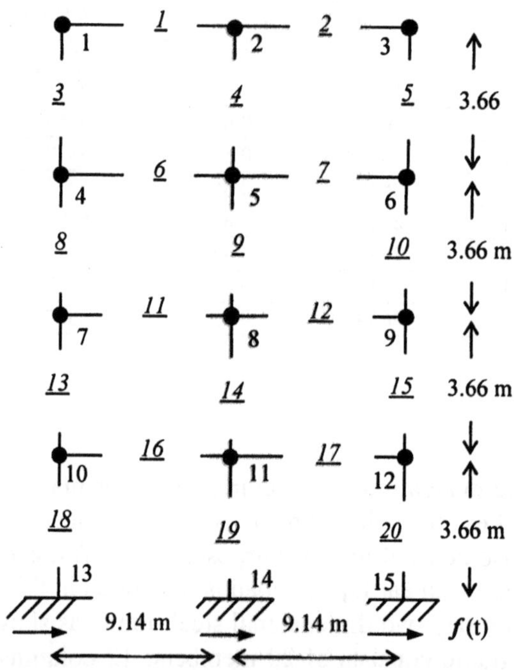


Figure 3. Four stories frame excited with El Centro Earthquake load for example 3

To consider the presence of noise, a numerically generated white noise with intensity of 5% of the root mean square values of the responses observed at all the dynamic degrees of freedom are added to the theoretical responses. As expected, the maximum error in the stiffness identification went up to 1.99%. It is still relatively small. This example clearly demonstrates the method can identify large structural systems excited by earthquake loadings, even in the presence of noise.

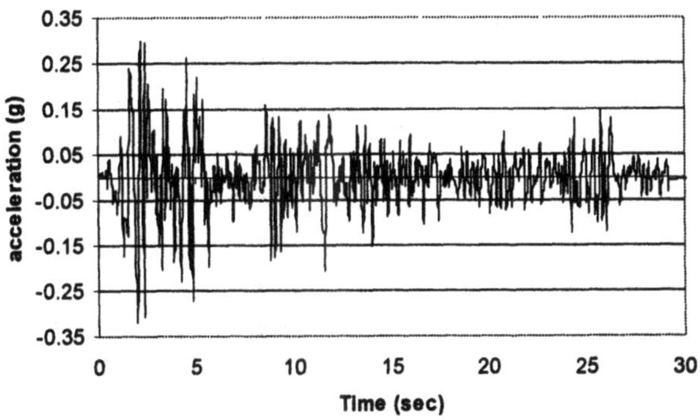


Figure 4. El Centro Earthquake time history

*Table 3. Stiffness (EI/L) identification for the frame excited by earthquake load for example 3.*

| Member | Initial Theoretical Value (kN-m) | Identified Noise-Free (kN-m) | Error % | Identified Noise-Included (kN-m) | Error % |
|---|---|---|---|---|---|
| $K_1$ | 10650.13 | 10647.61 | 0.024 | 10438.49 | 1.987 |
| $K_2$ | 10650.13 | 10647.61 | 0.024 | 10438.49 | 1.987 |
| $K_3$ | 26625.34 | 26619.27 | 0.023 | 26110.59 | 1.933 |
| $K_4$ | 26625.34 | 26618.7 | 0.025 | 26104.29 | 1.957 |
| $K_5$ | 26625.34 | 26619.27 | 0.023 | 26110.59 | 1.933 |
| $K_6$ | 10650.13 | 10647.86 | 0.021 | 10440.11 | 1.972 |
| $K_7$ | 10650.13 | 10647.86 | 0.021 | 10440.11 | 1.972 |
| $K_8$ | 26625.34 | 26619.71 | 0.021 | 26097.56 | 1.982 |
| $K_9$ | 26625.34 | 26619.85 | 0.021 | 26097.29 | 1.983 |
| $k_{10}$ | 26625.34 | 26619.71 | 0.021 | 26097.56 | 1.982 |
| $k_{11}$ | 10650.13 | 10647.86 | 0.021 | 10441.18 | 1.962 |
| $k_{12}$ | 10650.13 | 10647.86 | 0.021 | 10441.18 | 1.962 |
| $k_{13}$ | 26625.34 | 26619.6 | 0.022 | 26102.15 | 1.965 |
| $k_{14}$ | 26625.34 | 26619.62 | 0.021 | 26104.93 | 1.955 |
| $k_{15}$ | 26625.34 | 26619.6 | 0.022 | 26102.15 | 1.965 |
| $k_{16}$ | 10650.13 | 10647.86 | 0.021 | 10440.37 | 1.970 |
| $k_{17}$ | 10650.13 | 10647.86 | 0.021 | 10440.37 | 1.970 |
| $k_{18}$ | 26625.34 | 26619.59 | 0.022 | 26098.42 | 1.979 |
| $k_{19}$ | 26625.34 | 26619.66 | 0.021 | 26095.62 | 1.990 |
| $k_{20}$ | 26625.34 | 26619.59 | 0.022 | 26098.42 | 1.979 |

**Example 4**

In the three examples discussed so far, it was shown that the method is capable of identifying small and large structures excited by one, two or earthquake loadings, even in the presence of noise in the response information. However, all the structures are considered to be defect free. It is important at this stage to demonstrate how the method can identify defective elements in a structure.

The same two bays four-story steel frame considered in example 3 and shown in Figure 3 is used here. Defects in the frame are introduced in the following way. The stiffness of one beam, element 2, is reduced by 5% of its initial value to simulate defects in it. At the same time, the stiffness of one column, element 15, is reduced by 2% to simulate defects in it. As before, the theoretical responses of the defective frame at all node points excited by the El Centro earthquake are calculated using ANSYS, the responses are recorded at 0.01 sec time intervals. Using responses from 1.52 to 2.37 sec, providing a total of 86 time points, the frame is identified.

The identified stiffnesses of all the elements are given in Table 4. For the noise-free case, the results indicate that $k_2$ and $k_{15}$ decreased by 5.067% and 2.043%, respectively, more than the other elements indicating the defects are in elements 2 and 15. A numerically generated white noise with intensity of 5% of the root mean square values of the responses observed at all the dynamic degrees of freedom are added to the theoretical responses to consider the presence of noise. The results of the noise-contaminated responses are also shown in Table 4. For this case, the stiffnesses of $k_2$ and $k_{15}$ decreased by 5.779%, and 2.672%, respectively, indicating the presence of defects in them. This example demonstrates that the method can identify defective elements in a frame excited by earthquake loading, even in the presence of noise in the response information.

**Example 5**

To increase the application potential of the algorithm, it will be very desirable if it can identify the location of the defect more accurately within a defective element. The following example illustrates how the method can be used to locate defective spot in a defective element.

A five-story plane steel frame, as shown in Figure 5, is considered here. The frame consists of 15 members; 10 columns and 5 beams. In this illustrative example, the beam in the third floor is assumed to contain a defect. The nature and the location of the defect will be discussed in detail later. To locate the defect spot in the defective element, the beam is represented by six equal length finite elements as shown in Figure 5. For this finite element representation, the frame now consists of 20 elements. Of course, the defective beam can be represented by any number of elements depending upon the accuracy required for the detection.

*Table 4. Stiffness (EI/L) identification for the defective state frame for example 4.*

| Member | Initial Theoretical Value (kN-m) | Identified Noise-Free (kN-m) | Effect % | Identified Noise-Included (kN-m) | Effect % |
|---|---|---|---|---|---|
| $K_1$ | 10650.13 | 10641.60 | 0.080 | 10562.07 | 0.827 |
| $K_2$ | 10650.13 | **10110.46** | **5.067** | **10034.63** | **5.779** |
| $K_3$ | 26625.34 | 26604.84 | 0.077 | 26411.72 | 0.802 |
| $K_4$ | 26625.34 | 26605.32 | 0.075 | 26409.18 | 0.812 |
| $K_5$ | 26625.34 | 26604.63 | 0.078 | 26411.00 | 0.805 |
| $K_6$ | 10650.13 | 10643.68 | 0.061 | 10575.60 | 0.700 |
| $K_7$ | 10650.13 | 10642.52 | 0.071 | 10574.39 | 0.711 |
| $K_8$ | 26625.34 | 26609.65 | 0.059 | 26439.88 | 0.697 |
| $K_9$ | 26625.34 | 26608.46 | 0.063 | 26446.07 | 0.673 |
| $k_{10}$ | 26625.34 | 26606.35 | 0.071 | 26436.37 | 0.710 |
| $k_{11}$ | 10650.13 | 10644.19 | 0.056 | 10576.17 | 0.694 |
| $k_{12}$ | 10650.13 | 10643.69 | 0.060 | 10575.61 | 0.700 |
| $k_{13}$ | 26625.34 | 26610.78 | 0.055 | 26440.25 | 0.695 |
| $k_{14}$ | 26625.34 | 26610.61 | 0.055 | 26435.29 | 0.714 |
| $k_{15}$ | 26625.34 | **26081.32** | **2.043** | **25914.04** | **2.672** |
| $k_{16}$ | 10650.13 | 10644.38 | 0.054 | 10576.66 | 0.690 |
| $k_{17}$ | 10650.13 | 10645.34 | 0.045 | 10577.58 | 0.681 |
| $k_{18}$ | 26625.34 | 26610.92 | 0.054 | 26450.24 | 0.658 |
| $k_{19}$ | 26625.34 | 26614.08 | 0.042 | 26457.37 | 0.631 |
| $k_{20}$ | 26625.34 | 26608.90 | 0.062 | 26448.12 | 0.666 |

In this example, W21x57 of grade A36 steel section is used for all the members. The height of the columns in each floor is 3.66 m and the bay width is 9.14 m. Assuming the bases are fixed; the structure is represented by 45 DDOFs; 3 DDOFs at each node. The masses of all the beams and columns are assumed to be 78.62kg-sec$^2$/m and 31.47 kg-sec$^2$/m, respectively, but the mass of each beam elements in the fourth floor, $m_7$ to $m_{12}$, will be reduced to 13.10 kg-sec$^2$/m. The beam and column stiffnesses are estimated to be 10650.25 kN-m and 26625.62 kN-m, respectively, but the stiffness of the beam elements, $k_7$ to $k_{12}$, will be 63901.50 kN-m.

The mass-proportional damping coefficient $\alpha$ and the stiffness-proportional damping coefficient $\beta$ for the Rayleigh damping are evaluated following the procedure discussed earlier. The first two frequencies, $f_1$ and $f_2$, of the structure were found to be 4.11 and 13.66 Hz, respectively. For an equivalent modal damping of 3% of the critical for the first two modes, α and β are found to be 1.192267 and 0.00053728, respectively.

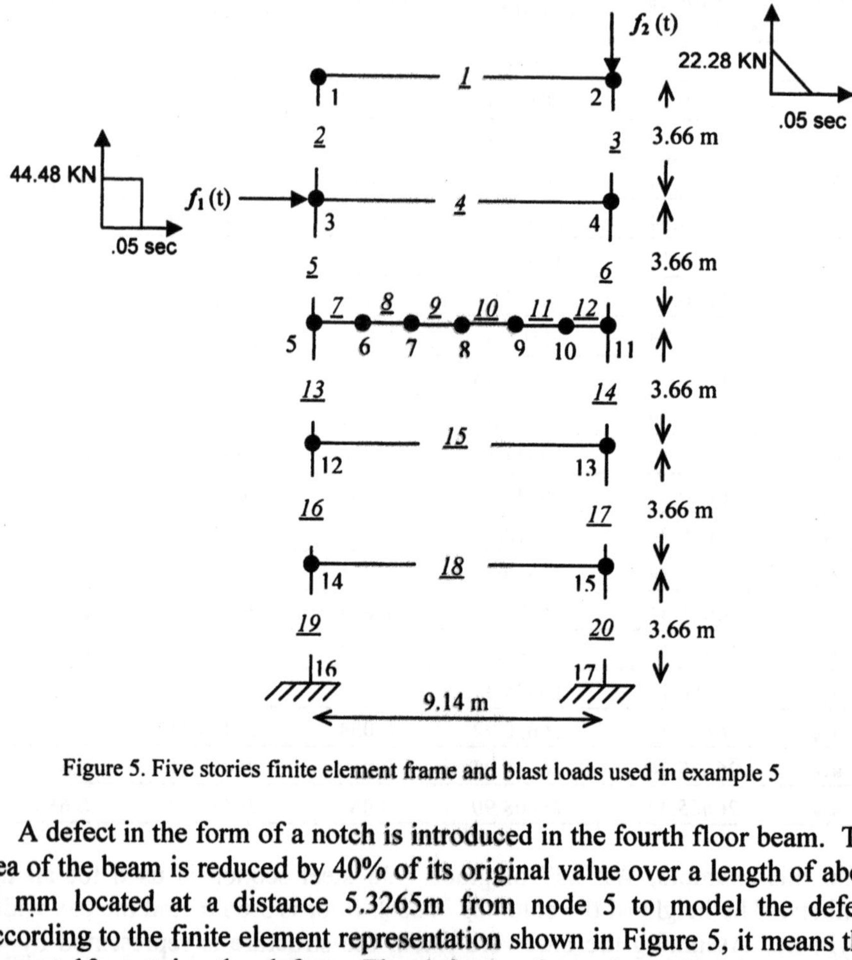


Figure 5. Five stories finite element frame and blast loads used in example 5

A defect in the form of a notch is introduced in the fourth floor beam. The area of the beam is reduced by 40% of its original value over a length of about 15 mm located at a distance 5.3265m from node 5 to model the defect. According to the finite element representation shown in Figure 5, it means that element 10 contains the defect. The defective frame is excited by two blast forces. A rectangular pulse with a magnitude of 44.48 KN and a duration of 0.05 sec is applied horizontally at node 3 at the forth floor. At the same time, a triangular pulse with a magnitude of 22.28 KN and a duration of 0.05 sec is applied vertically at node 2 at the fifth floor, as shown in Figure 5.

The theoretical responses of the frame, in terms of displacements, velocities and accelerations, are calculated at all nodes using ANSYS and recorded at 0.01 sec time intervals. The responses from 0.02 to 0.87 sec, providing 86 time points, are used for the identification purpose. The information on the two blast loads are ignored as stated earlier. All the elements of the frame are identified using the algorithm. The results are summarized in Table 5. The stiffness of element 10 reduced by about 1.756%, much more than the other elements indicating that the defect is in element 10. As before, to consider the presence of noise, a numerically generated white noise with intensity of 5% of the root mean square values of the responses at all DDOFs is added to the theoretical responses. For the noise-contaminated case, the maximum error in the stiffness identification is found to be 3.129%. This example demonstrates the capability of the method to identify defect spot in a defective element even when the structure is excited by multiple loadings.

*Table 5. Stiffness (EI/L) identification for spot defect-state for example 5*

| Member | Initial Theoretical Value (kN-m) | Identified Noise-Free (kN-m) | Effect % | Identified Noise-Included (kN-m) | Effect % |
|---|---|---|---|---|---|
| $k_1$ | 10650.25 | 10608.96 | 0.388 | 10468.63 | 1.705 |
| $k_2$ | 26625.62 | 26528.26 | 0.366 | 26313.51 | 1.172 |
| $k_3$ | 26625.62 | 26527.37 | 0.369 | 26323.05 | 1.136 |
| $k_4$ | 10650.25 | 10609.00 | 0.387 | 10550.42 | 0.937 |
| $k_5$ | 26625.62 | 26519.78 | 0.398 | 26279.60 | 1.300 |
| $k_6$ | 26625.62 | 26522.73 | 0.386 | 26303.32 | 1.210 |
| $k_7$ | 63901.50 | 63644.45 | 0.402 | 62986.54 | 1.432 |
| $k_8$ | 63901.50 | 63636.17 | 0.415 | 63031.25 | 1.362 |
| $k_9$ | 63901.50 | 63604.76 | 0.464 | 63135.51 | 1.199 |
| $K_{10}$ | 63901.50 | **62779.66** | **1.756** | **61902.15** | **3.129** |
| $K_{11}$ | 63901.50 | 63667.42 | 0.366 | 63031.92 | 1.361 |
| $K_{12}$ | 63901.50 | 63662.97 | 0.373 | 62953.37 | 1.484 |
| $K_{13}$ | 26625.62 | 26519.95 | 0.397 | 26294.01 | 1.245 |
| $K_{14}$ | 26625.62 | 26525.99 | 0.374 | 26184.37 | 1.657 |
| $K_{15}$ | 10650.25 | 10610.67 | 0.372 | 10550.90 | 0.933 |
| $K_{16}$ | 26625.62 | 26529.52 | 0.361 | 26117.08 | 1.910 |
| $K_{17}$ | 26625.62 | 26527.78 | 0.367 | 26164.26 | 1.733 |
| $K_{18}$ | 10650.25 | 10611.77 | 0.361 | 10544.48 | 0.993 |
| $K_{19}$ | 26625.62 | 26528.75 | 0.364 | 26293.53 | 1.247 |
| $K_{20}$ | 26625.62 | 26530.36 | 0.358 | 26293.37 | 1.248 |

# 4. SYSTEM IDENTIFICATION USING SUB-STRUCTURING APPROACH

In all the previous examples, it is assumed that the response information is available at all the dynamic degrees of freedom. However, for a large complicated structural system, the collection of dynamic responses at all DDOFs is not practical. To improve the practical application potential of the method, a system needs to be identified with limited response information. A sub-structuring approach can be used for this purpose where a part of a structure can be selected in such a way that the output responses are available at all DDOFs in that sub-structure to satisfy the requirements of the ILS-UI method discussed so far. In this way, not only the members of the sub-structure but also the input excitation information will be identified. The concept is now being developed and is very briefly discussed in this section.

To clarify the concept; consider the plane frame shown in Figure 6. Two situations can be envisioned. From the past experiences dealing with similar structures, defects, in any, are expected to be present in the roof beam or in the top floor column, as shown in the sub-structure shown in Figure 6. Thus, identifying a small part of the structure will provide the necessary information. In the second situation, suppose response information is available only at nodes 1, 2, and 3, as shown in Figure 6. It is necessary to identify Elements 1 and 2 without using excitation information. With the limited available response information, a sub-structuring approach is necessary. The basic ILS-UI method discussed earlier is still applicable, but it needs some modifications for the implementation purpose. The success of the approach depends on how the sub-structure is selected. In order to identify the stiffnesses of elements 1 and 2; a key node should be selected satisfying two requirements; (1) the key node should connect the elements to be identified, and (2) the point of application of the unknown input excitation force should be at the key node. The sub-structure shown in Figure 6 satisfies both requirements. In the case of earthquake loadings the sub-structure approach can be applied anywhere in the structure since all the nodes in the structure can be considered as a key node because the inertia forces resulted from the seismic load are applied to all the nodes as shown later in example 6.

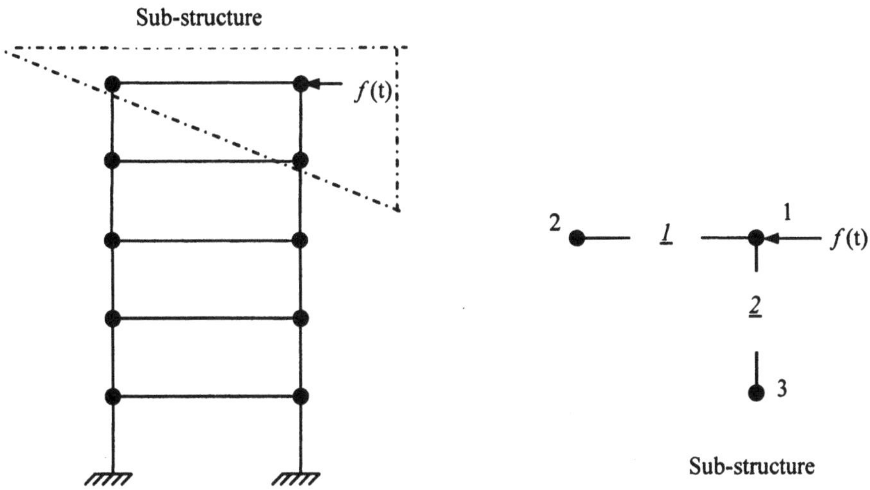


Figure 6. Sub-Structuring concept

In general the equation of motion for any sub-structure can be expressed as:

$$\mathbf{K}_{sub}\mathbf{x}_{sub}(\mathrm{t}) + \left(\alpha\mathbf{M}_{sub} + \beta\mathbf{K}_{sub}\right)\dot{\mathbf{x}}_{sub}(\mathrm{t}) = \mathbf{f}_{sub}(\mathrm{t}) - \mathbf{M}_{sub}\ddot{\mathbf{x}}_{sub}(\mathrm{t}) \tag{24}$$

Equation (24) can be rewritten in matrix form as:

$$\mathbf{A}(\mathrm{t})_{(Nkey.h)\times L}\,\mathbf{P}_{L\times 1} = \mathbf{F}(\mathrm{t})_{(Nkey.h)\times 1} \tag{25}$$

where matrix **A**(t) is $(Nkey\ .\ h) \times L$ ; *Nkey* is the number of DDOFs for the key node in the sub-structure, *h* is the total number of sample points, and *L* is the total number of unknown parameters in the sub-structure as mentioned earlier. Essentially Equation (25) is identical to Equation (3) except that the size of the unknown vector to be identified is reduced reflecting the availability of limited output response information. For the case shown in Figure 6, *Nkey* = 3 since the key node is node 1 and it has 3 DDOFs (two translation and one rotation as stated earlier). For, this particular example, Equation (25) can be expressed as:

$$\mathbf{A}(\mathrm{t})_{(3.h)\times L}\,\mathbf{P}_{L\times 1} = \mathbf{F}(\mathrm{t})_{(3.h)\times 1} \tag{26}$$

where **A**(t) matrix can be shown to be:

$$\mathbf{A}(t)_{(3.h)\times L} = [\,\mathbf{R}^1\mathbf{x}(t)\ \ \mathbf{R}^2\mathbf{x}(t)\ \ \mathbf{R}^1\dot{\mathbf{x}}(t)\ \ \mathbf{R}^2\dot{\mathbf{x}}(t)\ \ \mathbf{M}\dot{\mathbf{x}}(t)\,] \tag{27}$$

where $\mathbf{R}^1$ and $\mathbf{R}^2$ are the 6×6 matrixes containing all the terms in the square bracket in Equation (15) for elements 1 and 2, respectively; $\mathbf{x}(t)$ and $\dot{\mathbf{x}}(t)$ are vectors containing the displacement and velocity responses at nodes 1, 2, and 3 for the $h$ time points. The reason for including the responses at nodes 2 and 3 is that they are directly connected to the key node, and their responses are connected, in the context of the finite element model, with node 1. It can be noticed that the total number of DDOFs in this substructure *Nsub* is only 9.

The $\mathbf{x}(t)$ and $\dot{\mathbf{x}}(t)$ vectors can be expressed as:

$$\mathbf{x} = [x_1, y_1, \theta_1, x_2, y_2, \theta_2, x_3, y_3, \theta_3]^T \tag{28}$$

and

$$\dot{\mathbf{x}} = [\dot{x}_1, \dot{y}_1, \dot{\theta}_1, \dot{x}_2, \dot{y}_2, \dot{\theta}_2, \dot{x}_3, \dot{y}_3, \dot{\theta}_3]^T \tag{29}$$

Similarly, the unknown structural parameter vector to be identified is:

$$\mathbf{P} = [k_1, k_2, \beta\, k_1, \beta\, k_2, \alpha]^T \tag{30}$$

where $k_1$ and $k_2$ are the required unknown stiffness ($EI/L$) for elements 1, and 2, respectively, and $\alpha$ is the mass proportional damping coefficient and $\beta$ is the stiffness proportional damping coefficient need to be identified.

$\mathbf{F}(t)$ matrix can be expressed as:

$$\mathbf{F}(t)_{(3.h)\times 1} = \mathbf{f}(t)_{(3.h)\times 1} - \mathbf{M}\,\ddot{\mathbf{x}}(t)_{(3.h)\times 1} \tag{31}$$

where $\ddot{\mathbf{x}}(t)$ vector contains the acceleration responses at nodes 1, 2, and 3 for $h$ time points as follows:

$$\ddot{\mathbf{x}} = [\ddot{x}_1, \ddot{y}_1, \ddot{\theta}_1, \ddot{x}_2, \ddot{y}_2, \ddot{\theta}_2, \ddot{x}_3, \ddot{y}_3, \ddot{\theta}_3]^T \tag{32}$$

It can be noticed that the sub-structure satisfies all the requirements for the ILS-UI approach presented earlier. Thus, the two elements in the sub-structure can be identified using response information measured at only 3 nodes. The same iterative strategy discussed earlier can be used to solve Equation (26). To clarify all the steps necessary to implement the sub-structuring approach, a numerical example is given below.

## Example 6

The two bays four-story steel frame considered in examples 3 and 4, and shown in Figure 3, is considered. Suppose the available information indicates that elements 2 and 15 may contain some defects. Since the elements are widely separated, two sub-structures are necessary. To identify element 2, the required sub-structure is shown in Figure 7. In this representation, node 3 is the key node and the output response information must be available at nodes 2, 3, and 6. The defect-free frame is excited by the El Centro earthquake and the theoretical responses at nodes 2, 3, and 6 are calculated from 1.52 to 2.37 sec at 0.01 sec time intervals, providing a total of 86 time points.

Equation (24) for this sub-structure shown in Figure 7 can be expressed as:

$$\mathbf{A}(t)_{(3.86)\times 5}\,\mathbf{P}_{5\times 1} = \mathbf{F}(t)_{(3.86)\times 1} \tag{33}$$

where $\mathbf{A}(t)$ matrix can be shown to be:

$$\mathbf{A}(t)_{(3.86)\times 5} = [\,\mathbf{R}^2\mathbf{x}(t)\;\; \mathbf{R}^5\mathbf{x}(t)\;\; \mathbf{R}^2\dot{\mathbf{x}}(t)\;\; \mathbf{R}^5\dot{\mathbf{x}}(t)\;\; \mathbf{M}\dot{\mathbf{x}}(t)\,] \tag{34}$$

And the $\mathbf{x}(t)$ and $\dot{\mathbf{x}}(t)$ vectors can be expressed as

$$\mathbf{x} = [x_3, y_3, \theta_3, x_2, y_2, \theta_2, x_6, y_6, \theta_6]^T \tag{35}$$

and

$$\dot{\mathbf{x}} = [\dot{x}_3, \dot{y}_3, \dot{\theta}_3, \dot{x}_2, \dot{y}_2, \dot{\theta}_2, \dot{x}_6, \dot{y}_6, \dot{\theta}_6]^T \tag{36}$$

The unknown structural parameter vector to be identified is:

$$\mathbf{P} = [k_2, k_5, \beta k_2, \beta k_5, \alpha]^T \tag{37}$$

where $k_2$ and $k_5$ are the required unknown stiffness ($EI/L$) for elements 2, and 5, respectively, and $\alpha$ is the mass proportional damping coefficient and $\beta$ is the stiffness proportional damping coefficient need to be identified.

$\mathbf{F}(t)$ matrix can be expressed as:

$$\mathbf{F}(t)_{(3.86)\times 1} = \mathbf{f}(t)_{(3.86)\times 1} - \mathbf{M}\,\ddot{\mathbf{x}}(t)_{(3.86)\times 1} \tag{38}$$

where $\ddot{\mathbf{x}}(t)$ vector contains the acceleration responses at nodes 2, 3, and 6 for 86 time points as follows:

$$\ddot{\mathbf{x}} = \left[\ddot{x}_3, \ddot{y}_3, \ddot{\theta}_3, \ddot{x}_2, \ddot{y}_2, \ddot{\theta}_2, \ddot{x}_6, \ddot{y}_6, \ddot{\theta}_6\right]^T \tag{39}$$

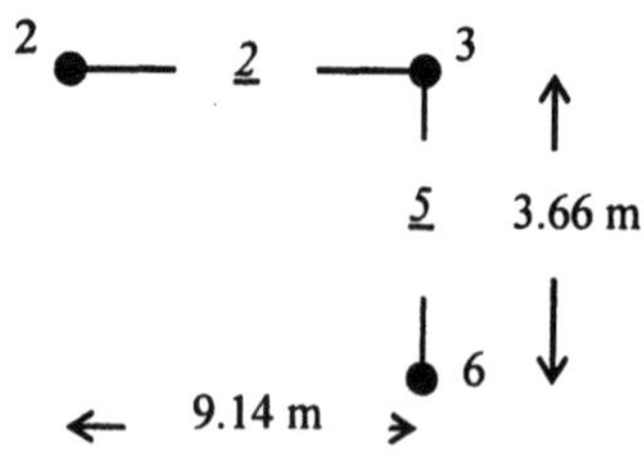


Figure 7. Sub-Structure 1 to identify the stiffnesses of element 2 for example 6

The stiffnesses of elements 2 and 5 are identified and are shown in Table 6. The algorithm identified both stiffnesses very accurately with a maximum error in identification of 0.03%. To simulate the presence of noise in the response information, numerically generated white noises with intensities of 5% of the root mean square values of the responses at all dynamic degrees of freedom of the sub-structure are added to the response information. The frame is identified with noise contaminated response information. In this case, the maximum error is found to be 1.897% as shown in Table 6.

To consider the presence of defects, the stiffness of element 2 is reduced by 5%. Again, the theoretical responses of the defective frame are evaluated at nodes 2, 3, and 6, and using the noise-free response information, stiffnesses of elements 2 and 5 are identified. The results are shown in Table 6. The stiffness $k_2$ reduces by 5.057 % indicating the defect is in element 2. As mentioned earlier, to consider the presence of noise in the response information, a numerically generated white noise with intensity of 5% of the root mean square values of the responses at all DDOFs in the sub-structure is added to the theoretical responses. The results for the noise-contaminated responses are shown in Table 6. As expected, the error in identification increased a little bit when noise is considered in the response information.

*Table 6. Stiffness (EI/L) identification for sub-structure model 1 to identify element 2.*

| **Members** | **Initial Theoretical Value** (kN-m) | **Identified Noise- Free** (kN-m) | **Effect %** | **Identified Noise- Included** (kN-m) | **Effect %** |
|---|---|---|---|---|---|
| | | **Defect-free state** | | | |
| $k_2$ | 10650.13 | 10646.93 | 0.030 | 10450.27 | 1.876 |
| $k_5$ | 26625.34 | 26619.03 | 0.024 | 26120.36 | 1.897 |
| | | **Defective State (5% Defect in member 2)** | | | |
| $k_2$ | 10650.13 | **10111.59** | **5.057** | **10013.72** | **5.976** |
| $k_5$ | 26625.34 | 26607.59 | 0.067 | 26425.30 | 0.751 |

As mentioned earlier; for the seismic excitation, all the nodes in the structure can be considered as key nodes for the sub-structure representation. Thus the sub-structure required to identify element 15 is shown in Figure 8. In this case, the key node is node 9. The stiffness of element 15 needs to be identified using response information available only at nodes 6, 8, 9, and 12. The defect-free frame is excited by the El Centro earthquake and the theoretical responses at nodes 6, 8, 9 and 12 are calculated from 1.52 to 2.37 sec at 0.01 sec time intervals, providing a total of 86 time points.

Equation (24) for this sub-structure shown in Figure 8 can be expressed as:

$$\mathbf{A}(t)_{(3.86)\times 7}\,\mathbf{P}_{7\times 1} = \mathbf{F}(t)_{(3.86)\times 1} \tag{40}$$

where **A**(t) matrix can be shown to be:

$$\mathbf{A}(t)_{(3.86)\times 7} = [\,\mathbf{R}^{12}\mathbf{x}(t)\ \ \mathbf{R}^{10}\mathbf{x}(t)\ \ \mathbf{R}^{15}\mathbf{x}(t)\ \ \mathbf{R}^{12}\dot{\mathbf{x}}(t)\ \ \mathbf{R}^{10}\dot{\mathbf{x}}(t)\ \ \mathbf{R}^{15}\dot{\mathbf{x}}(t)\ \ \mathbf{M}\dot{\mathbf{x}}(t)] \tag{41}$$

And the $\mathbf{x}(t)$ and $\dot{\mathbf{x}}(t)$ vectors can be expressed as:

$$\mathbf{x} = \left[x_9, y_9, \theta_9, x_8, y_8, \theta_8, x_6, y_6, \theta_6, x_{12}, y_{12}, \theta_{12}\right]^T \tag{42}$$

and

$$\dot{\mathbf{x}} = \left[\dot{x}_9, \dot{y}_9, \dot{\theta}_9, \dot{x}_8, \dot{y}_8, \dot{\theta}_8, \dot{x}_6, \dot{y}_6, \dot{\theta}_6, \dot{x}_{12}, \dot{y}_{12}, \dot{\theta}_{12}\right]^T \tag{43}$$

The unknown structural parameter vector to be identified for this sub-structure is:

$$\mathbf{P} = \left[ k_{12}, k_{10}, k_{15}, \beta\, k_{12}, \beta\, k_{10}, \beta\, k_{15}, \alpha \right]^{T} \tag{44}$$

where $k_{12}$, $k_{10}$ and $k_{15}$ are the required unknown stiffness ($EI/L$) for elements 12, 10 and 15, respectively, and $\alpha$ is the mass proportional damping coefficient and $\beta$ is the stiffness proportional damping coefficient need to be identified.

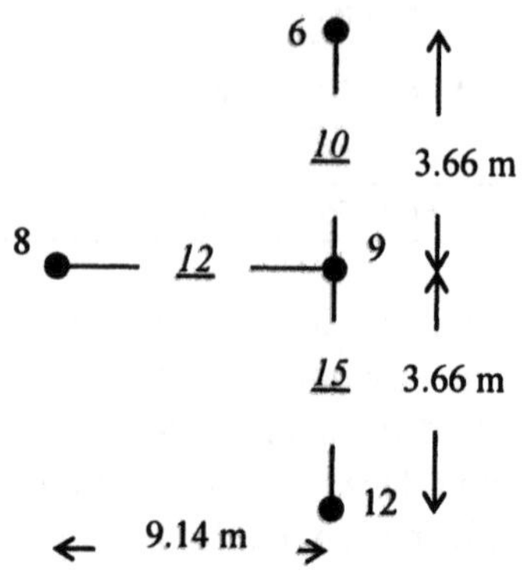


Figure 8. Sub-Structure 2 to identify the stiffnesses of element 15 for example6.

**F**(t) matrix can be expressed as:

$$\mathbf{F}(t)_{(3.86)\times 1} = \mathbf{f}(t)_{(3.86)\times 1} - \mathbf{M}\,\ddot{\mathbf{x}}(t)_{(3.86)\times 1} \tag{45}$$

where $\ddot{\mathbf{x}}(t)$ vector contains the acceleration responses at nodes 6, 8, 9 and 12 for 86 time points as follows:

$$\ddot{\mathbf{x}} = \left[ \ddot{x}_9, \ddot{y}_9, \ddot{\theta}_9, \ddot{x}_8, \ddot{y}_8, \ddot{\theta}_8, \ddot{x}_6, \ddot{y}_6, \ddot{\theta}_6, \ddot{x}_{12}, \ddot{y}_{12}, \ddot{\theta}_{12} \right]^{T} \tag{46}$$

The stiffnesses of elements 10, 12 and 15 are identified and are shown in Table 7. The algorithm identified the stiffnesses of the elements very accurately. To simulate the presence of noise in the response information, numerically generated white noises with intensities of 5% of the root mean square values of the responses at all DDOFs of the sub-structure are added to the response information. The frame is identified with noise contaminated response information. In this case, the maximum error is found to be 1.972% as shown in Table 7.

To consider the presence of defects, the stiffness of element 15 is reduced by 2%. Again, the theoretical responses of the defective frame are evaluated at nodes 6, 8, 9 and 12, and using the noise-free response information, stiffnesses of elements 10, 12 and 15 are identified. The results are shown in Table 7. The stiffness $k_{15}$ reduces 2.11 % indicating the defect is in element

15. As stated earlier, to consider the presence of noise in the response information, a numerically generated white noise with intensity of 5% of the root mean square values of the responses at all DDOFs in the sub-structure is added to the theoretical responses. The results for the noise-contaminated responses are shown in Table 7. As expected, the error in identification increased a little.

*Table 7. Stiffness (EI/L) identification for sub-structure model 2 to identify element 15*

| Members | Initial Theoretical Value (kN-m) | Identified Noise- Free (kN-m) | Effect % | Identified Noise-Included (kN-m) | Effect % |
|---|---|---|---|---|---|
| | | **Defect-free state** | | | |
| $k_{10}$ | 26625.34 | 26608.95 | 0.062 | 26438.17 | 0.703 |
| $k_{12}$ | 10650.13 | 10648.32 | 0.017 | 10440.75 | 1.966 |
| $k_{15}$ | 26625.34 | 26620.15 | 0.019 | 26100.25 | 1.972 |
| | | **Defective State (2% Defect in member 15)** | | | |
| $k_{10}$ | 26625.34 | 26616.28 | 0.034 | 26428.34 | 0.740 |
| $k_{12}$ | 10650.13 | 10643.69 | 0.061 | 10585.92 | 0.602 |
| $k_{15}$ | 26625.34 | **26061.68** | **2.11** | **25903.27** | **2.712** |

For the first sub-structure, only responses at 9 DDOFs are required to identify it. For the second sub-structure, only responses at 12 DDOFs are required to identify it. To identify the whole structure, responses at 36 DDOFs are required. The two examples indicate that the sub-structuring approach can be used to identify a part of a structure, even in the presence of noise, if the available response information is very limited. However, the sub-structure needs to be selected very carefully.

# 5. CONCLUSIONS

A system identification procedure is presented to identify the structural parameters at the local element level of framed structures. The structures are represented by finite elements. The procedure detects defects by tracking the changes in stiffness properties of each element. The most attractive feature of the procedure is that it does not require input excitation information for the identification purpose. It is capable of identifying defects even in the presence of noise. With the help of several numerical examples it is shown that the method can accurately identify structures excited by one or multiple dynamic loadings including seismic loading. Initially, the finite element representation can be kept very simple. However, if defective elements are identified, the finite element representation can be refined to locate the defect spot more accurately in the defective elements. Sub-structuring approach can be used if the available response information is limited. The procedure has the potential to be used as a nondestructive health assessment technique. It is expected to be simple and economical but reliable and accurate.

## ACKNOWLEGMENTS

This chapter is based on work partly supported by University of Arizona Foundation under a small grant program. Any opinions, findings, conclusions, or recommendations expressed in this paper are those of the authors and do not necessarily reflect the views of the sponsor.

# REFERENCES


ANSYS version 5.7.2001. The Engineering Solutions Company.

Cook R., Malkus D., and Plesha M., *Concepts and Applications of Finite Element Analysis*, third edition, John Wiley and Sons, 1989.

Clough R., and. Penzien J., *Dynamics of Structures*, second edition, McGraw-Hill, Inc., 1993.

Doebling S., Farrar C., Prime M., and Shevitz D., *Damage Identification and Health Monitoring of Structural and Mechanical Systems from Changes in Their Variation Characteristics: A Literature Review*, Report No. LA-13070-MS, Los Alamos National Laboratory, May 1996.

Housner G., Bergman L., Caughey T., Chassiakos A., Claus R., Masri S., Skelton R., Soong T., Spencer B., and Yao J., Structural Control: Past, Present and Future. Journal of Engineering Mechanics, ASCE, 1997, 123(9): 897-971.

Hoshiya M., and Maruyama O., Identification of Nonlinear Structural Systems. Proc. of ICAP 5, 1987, pp 182-189.

Katkhuda H., Martinez R., and Haldar A., Stiffness Identification under Uncertain Blast Loading. 4th International Symposium on Uncertainty Modeling and Analysis, University of Maryland, College Park, Maryland, 2003. pp. 22-27.

Katkhuda H., Martinez R., and Haldar A., A Novel Defect Identification and Structural Health Assessment Technique. Journal of structural Engineering, Apr.-Jun. 2004a, 31(1):1-8.

Katkhuda H., Martinez R., and Haldar A., Health Assessment at Local Level with Unknown Input Excitation. Journal of structural Engineering, ASCE, 2004b (accepted for publication).

Koh C., Hong B., and Liaw C. Parameter Identification of Large Structural Systems in Time Domain. Journal of Structural Engineering s, 2000, 126 (8): 957-963.

Ling X., *Linear and Nonlinear Time Domain System Identification at Element Level for Structural Systems with Unknown Excitation*, Report No. CEEM-001-101, Department of Civil Engineering and Engineering Mechanics, University Of Arizona, Tucson, Arizona, 2000.

Ling X., and Haldar A, Element Level System Identification with Unknown Input with Rayleigh Damping. Journal of Engineering Mechanics, ASCE, (to appear in August, 2004).

Toki K., Sato T., and Kiyono J. Identification of Structural Parameters and Input Ground Motion from Response Time Histories. Journal Structural. Engineering/ Earthquake Engineering, 1989 6(2): 413-421.

Vo P., and Haldar A., Health assessment of Beams - Theoretical and Experimental Investigation. Journal of Structural Engineering, Apr-Jun. 2004, 31(1):23-30.

Wang D., *An Element Level Time Domain System Identification Technique with Unknown Input Information*, Report No. CEEM-95-106, University of Arizona, Tucson, AZ, 1995.

Wang D., and Haldar A., An Element Level SI with Unknown Input Information. Journal of the Engineering Mechanics Division, ASCE, 1994,120 (1)1:59-176.

# Chapter 22

# UNCERTAINTY MODELING OF CHLORIDE CONTAMINATION AND CORROSION OF CONCRETE BRIDGES

**Zoubir Lounis**

## 1. INTRODUCTION

The maintenance of aging and deteriorating concrete bridge structures is recognized as one of the major challenges facing bridge owners and managers. Despite their better durability when compared to steel and timber bridge structures, reinforced and prestressed concrete structures are vulnerable to the damaging effects of corrosion induced primarily by chlorides (from deicing salts and seawater) and to a lesser extent by carbonation. It is estimated that one-third to one-half of the projected bridge rehabilitation costs in North America will be allocated for the rehabilitation of deteriorated bridge decks.

In most highway agencies, bridge maintenance management is based to a large extent on the results of bridge inspection combined with engineering experience and judgment for decision-making. Bridge maintenance management is a challenging task that involves the identification of optimal prioritization of bridge structures for maintenance and rehabilitation and the determination of the optimal rehabilitation strategy for each structure of a given bridge or a network of hundreds or thousands of bridges. To achieve this goal, there is a need to develop and integrate reliable and effective decision support models that include: (i) condition assessment models; (ii) deterioration prediction models; (iii) risk assessment models; and (iv) maintenance optimization models as illustrated in Figure 1.

Structural concrete (reinforced and prestressed) is the main constituent material of the majority of highway bridge structures in North America. Chloride-induced corrosion is identified as the main cause of deterioration of concrete bridge structures. The sources of chlorides are the seawater and deicing salts used during winter. The corrosion of the steel reinforcement leads to concrete fracture through cracking, delamination and spalling of the concrete cover, reduction of concrete and reinforcement cross sections, loss of bond between the reinforcement and concrete, reduction in strength (flexural, shear, etc.) and ductility. As a result, the safety and serviceability of concrete structures are reduced, and their useful service lives shortened.

The available information regarding the material properties, loading, deterioration processes, risk of failure, and design and maintenance costs are incomplete or uncertain. There are different sources of uncertainty with varying magnitudes that affect the predictions of the above models. These include: (i) physical uncertainty; (ii) model uncertainty; (iii) statistical uncertainty; and (iv) decision uncertainty. It is clear that the combination of these uncertainties lead to a considerable level of uncertainty in each model and in the overall bridge maintenance management system, in which decisions have to be made subject to uncertainty.

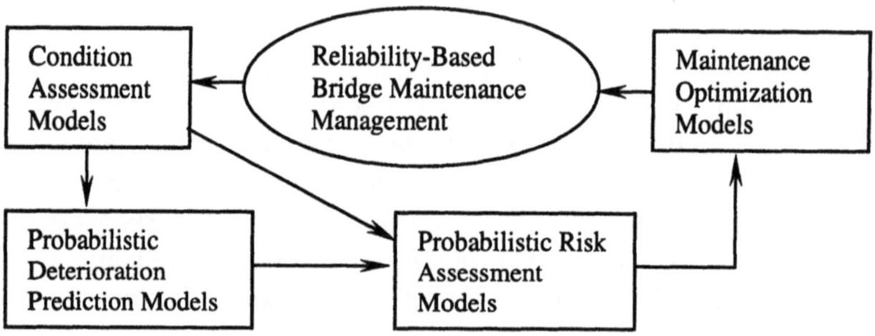


Figure 1. Decision Support Models for Bridge Maintenance Management

In this paper, the objective is the development of probabilistic deterioration models for the assessment of the level of chloride contamination of the concrete cover and the level of corrosion of the reinforcing steel of concrete structures under chloride attack from deicing salts. The proposed models take into account the uncertainty associated with the material properties, structure geometry and dimensions, applied environmental loads and corrosion resistance as well as the uncertainty associated with the analytical models. The prediction capability of the proposed probabilistic model is illustrated on an aging chloride-contaminated concrete bridge deck that was exposed to deicing salts for forty years for which field data were available.

# 2. DETERIORATION OF CONCRETE BRIDGES

## 2.1 Overview of Deterioration Models of Concrete Bridges

In the last two decades, highway agencies developed and implemented bridge management systems (BMS) for planning the inspection and maintenance of their bridges in order to ensure their reliability and minimize their life cycle costs. The effectiveness of a BMS, however, is highly dependent on the reliability of the deterioration models used. In the state-of-the-art BMS, stochastic deterioration models based on Markov chains were developed to predict the deterioration of different bridge components, including concrete deck slabs. The application of a stochastic deterioration model based on the discrete Markov chain for the prediction of cumulative damage in structures was first proposed by Bogdanoff (1978).

Despite their practicality and ease of updating, the Markov chain-based deterioration models have serious shortcomings, including: (i) they are qualitative prediction models, based on subjective condition ratings as opposed to quantitative models; (ii) do not consider all parameters that govern the component deterioration; (iii) assume constant rate of deterioration as the cumulative damage after a stress cycle is assumed to depend only on the length of the stress cycle and the initial condition of the structural element; and (iv) do not consider the entire historical performance of component (Lounis 2000).

These limitations may be acceptable for network-level analysis in which bridges are prioritized only for eligibility to maintenance funds. These models, however, have serious limitations for project-level analysis in which detailed and quantitative assessment of the levels of chloride contamination, corrosion, cracking, spalling, loss of bond and strength are critical for the assessment of the residual safety and identification of the appropriate and cost-effective rehabilitation strategy.

A more appropriate model for concrete bridge structures exposed to chlorides from deicing salts is the mechanistic model based on Fick's law of diffusion of chlorides and the concept of the "chloride threshold level" for corrosion initiation (Tuutti 1982; Kropp and Hilsdorf 1995). This model is able to predict the time and space variations of chloride contamination of concrete and the time to onset of corrosion of the reinforcing steel. Several other models, including empirical, analytical, numerical and statistical, have been developed to predict the levels of chloride contamination of concrete, times to onset of corrosion, cracking, spalling and failure (Weyers 1998; Lounis 2000).

The practicality and reliability of these predictive models, however, is limited because they are not able to effectively account for the considerable uncertainty in the governing variables, model, and condition assessment methods. These models can predict the chloride ingress, steel corrosion and concrete cracking, provided that the parameters are certain and the underlying assumptions are satisfied. Intuitively, given the considerable uncertainty in the governing parameters of the model, there is a considerable uncertainty in the structural response .

In the case of chloride contamination of concrete or corrosion of the reinforcing steel, this uncertainty is due to the heterogeneity of concrete, temporal and spatial variability of its properties, variability of the environment, concrete cover depth, chloride transport model and chloride threshold level, and measurement errors. Some of these shortcomings have been overcome through the use of probabilistic modeling of the chloride transport and corrosion initiation and solving the problem using reliability-based methods, such as Monte Carlo simulation, first-order or second-order reliability methods, or crossing theory (Enright and Frangopol 1998; Stewart and Rosowsky 1998; Lounis and Mirza 2001).

## 2.2 Imperative for Probabilistic Modeling of Deterioration

The prediction of the safety and serviceability of existing structures and the assessment of their maintenance needs is a very complex problem. This is due to the multitude of failure mechanisms and their interaction, which are very hard to quantify. For bridge structures, the main causes of failure may include aggressive environments, overstress due to heavy traffic load, accidental impacts, unsatisfactory design, protection, and construction, aging, and inadequate inspection and maintenance. Both the external effects and the material and structural parameters are time-dependent and random in nature. This requires the use of stochastic deterioration models to predict the structural response. Generally, a considerable level of uncertainty is associated with the predictive model and all its parameters. The sources of uncertainty can be identified as: physical uncertainty, statistical uncertainty, model uncertainty, measurement uncertainty and decision uncertainty.

The physical or inherent uncertainty is that identified with the inherent random nature of a basic variable such as: (i) variability of the structure geometry (e.g. concrete cover thickness, member depth, etc.); (ii) variability of the material properties (strength, diffusivity, etc.); (iii) variability of the micro-environment (e.g. surface chloride concentration on the deck); (iv) variability of the applied loads (e.g. traffic load and superimposed load); and (v) variability of the condition rating, etc.

The statistical uncertainty arises from modeling the parameters and /or performance indicators using simplified stochastic processes or random variables by using lower order of stochastic correlation of stochastic processes or assuming independence of random variables. This uncertainty arises also from the use of a limited sample size to estimate the statistical parameters that describe the probabilistic model of the governing parameters and performance indicators.

The model uncertainty results from the use of simplified physical models to describe the damage initiation or damage growth mechanisms, such as corrosion, cracking, spalling, collapse, etc. An example of such uncertainty arises in the modeling of the deterioration of concrete structure subjected to chloride attack from deicing salts, which is discussed further in this paper. This modeling uncertainty includes: (i) use of a simplified diffusion law to model the chloride transport mechanism; (ii) use of simplified chloride threshold level to define the corrosion resistance of concrete structures; and (iii) use of a simplified resistance degradation model in the propagation stage to assess the safety and serviceability of the structure.

The decision uncertainty is that associated with the definition of the acceptable level of damage or limit state or acceptable probability of failure for both serviceability and ultimate limit states. This is quite a complex problem due to its dependence on the risk of loss of life and injury, cost of repair and replacement, redundancy of the structure, and failure mode considered.

Probabilistic modeling of complex failure mechanisms of bridge structures has much to offer with regard to simplicity as compared with attempts of formulating purely deterministic models (Ditlevsen 1984; Melchers 1987; Mori and Ellingwood 1993; Frangopol et al 1997; Stewart and Rosowsky 1998; Lounis and Mirza 2001). Ditlevsen (1984) states: "*Probabilistic models are almost always superior to deterministic models of equal level of complexity in the sense that the former have considerable higher threshold of realism when dealing with phenomena taking place in uncertain environments*".

Therefore, given the considerable uncertainty that affect the material, structure, environment, loading, material performance and structural behavior, the imperative for probabilistic modeling of bridge deterioration cannot be ignored. In this paper probabilistic mechanistic models are proposed to predict the response of concrete bridge structures subjected to chloride attack.

# 3. UNCERTAINTY IN CHLORIDE TRANSPORT

## 3.1 Mechanism of Chloride Contamination of Concrete

Concrete is a porous material with pore spaces in the cement paste matrix and micro-cracks that provide a relatively easy access to aggressive agents, such as chlorides, oxygen, and water as shown in Fig.2. The penetration of these agents and their accumulation up to critical values at the steel level induce corrosion of the reinforcing steel, followed by concrete cracking, delamination, spalling and ultimately failure. The concrete cover provides both chemical and physical barriers to corrosion. The concrete pore water solution is naturally alkaline with a pH value about 13, which enables the formation and maintenance of a permanent protective passivating film on the steel surface (Bentur et al. 1995). The concrete cover with a depth ($d_c$) represents also a physical barrier against corrosion by providing a dense and impermeable concrete cover, which limits the penetration of aggressive agents as shown in Fig. 2.

The rate of penetration of chlorides into concrete is dependent primarily on the quality of concrete and more particularly on the water-cement ratio of the concrete mix and the presence of protective systems that delay or slow down the chloride ingress. The governing transport mechanisms of chlorides into structures are the ionic diffusion in saturated concrete and water absorption in partially saturated concrete. Chloride diffusion is a transfer of mass by random motion of free chloride ions in the pore solution resulting in a net flow from regions of higher concentration to regions of lower concentration (Kropp et al. 1995). The rate of chloride ingress is proportional to the concentration gradient and the diffusion coefficient of the concrete (Fick's first law of diffusion).

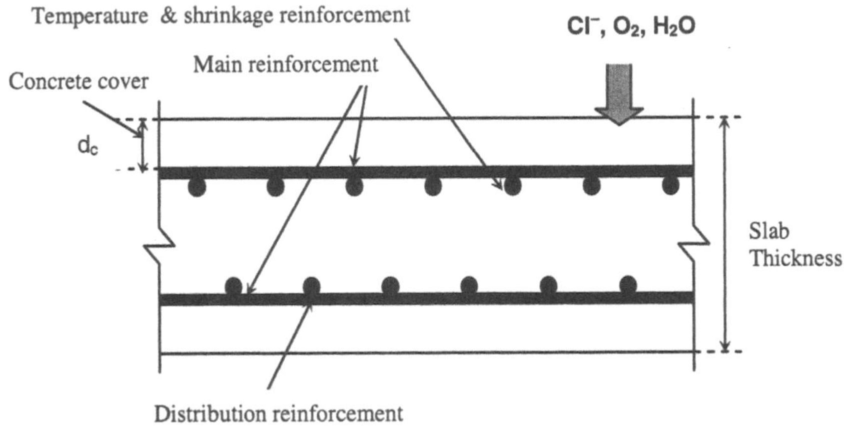


Figure 2. Typical reinforcement details of a bridge deck subjected to aggressive agents

However, in porous solids, such as concrete, moisture may flow via the diffusion of water vapor, as well as non-saturated or even saturated capillary flow may occur in finer pores (Kropp et al. 1995). Although chloride ingress into concrete is due to multiple transport mechanisms, Fick's law of diffusion may be applied to quantify the chloride ingress. A concentration gradient is considered as the common driving force. Given the fact that concrete is a heterogeneous and ageing material, temporal and spatial variability is associated with the diffusion coefficient. Since in the field, chloride ingress occurs under transient conditions, Fick's second law of diffusion can be used to describe the time variation of chloride concentration for one-dimensional flow, as follows:

$$\frac{\partial C}{\partial t} = \frac{\partial}{\partial x}[D\frac{\partial C}{\partial x}] \tag{1}$$

Under the assumption of a constant diffusion coefficient, and boundary condition specified as $C=C_s$ and the initial condition specified as $C=0$ for $x>0$, $t=0$, Crank's solution of Eq. (1) yields:

$$C(x,t) = C_S[1-\mathrm{erf}(\frac{x}{2\sqrt{Dt}})] \tag{2}$$

where $C_s$ is the chloride concentration at the surface; C(x,t) is the chloride concentration at depth x after time t; D is the diffusion coefficient; erf is the statistical error function; and t is the time of exposure.
The time variations of the chloride profiles are shown in Fig.3. Despite its simplicity and extensive use, this model has some shortcomings that are summarized in the next sections.

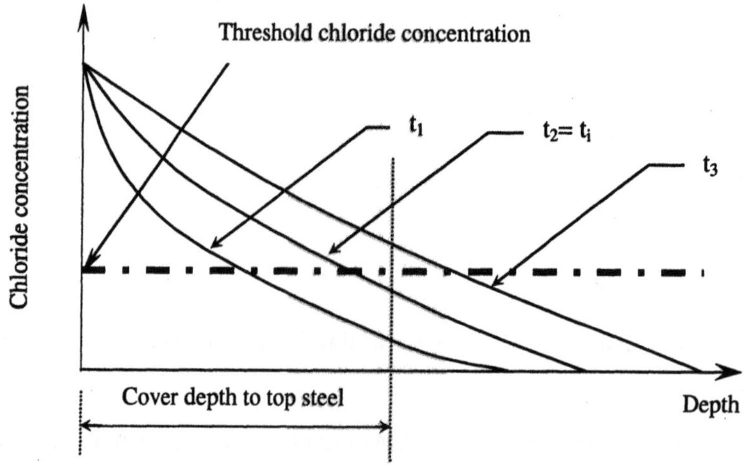


Figure 3. Chloride Concentration Profiles at Different Times

## 3.2 Uncertainty in Chloride Transport Model

The uncertainty in the chloride transport model results from the use of a simplified physical model or relationship between the basic variables to represent the actual phenomena, such as: (i) assumption of chloride transport mechanism governed by diffusion; (ii) use of simplified models of the diffusion coefficient and driving chloride concentration; and (iii) assumption of non-correlated variables. The introduction of the apparent values of governing parameters addressed to some extent the uncertainty associated with the diffusion model for chloride transport, however, the model does not consider the considerable spatial and temporal uncertainty associated with the values of the governing parameters.

## 3.3 Uncertainty in Governing Parameters

### 3.3.1 Diffusion Coefficient

The ingress of chlorides into concrete is a complex process that combines several transport mechanisms such as diffusion, capillary sorption (or convection), and permeation, which is influenced by several factors such as concrete mix, nonlinear chloride binding of the cement, temperature, curing, etc. The chloride diffusion coefficient is determined by fitting the solution of Fick's second law of diffusion to measured chloride profiles expressed in terms of total chloride concentrations (including both free and bound chlorides). Since only the chlorides dissolved in the pore solution (free chlorides) are responsible for the initiation of the corrosion process, this procedure yields only the value of the apparent diffusion coefficient "D" because chloride binding is not taken into account. The diffusion coefficient is not a constant but rather depends on time, temperature, and depth because of the heterogeneous nature and aging of concrete (Kropp et al. 1995; Neville 1995; Weyers 1998).

### 3.3.2 Surface Chloride Concentration

As concrete bridge structures are subjected to a continually changing chloride exposure, the surface chloride concentration is not constant but time-dependent. Using field data, it has been shown that the surface chloride concentration increases with age (square root law). This increase is relatively fast and reaches a quasi-constant concentration in about 5 years (Weyers 1998). Given the fact that the service life of concrete bridge structures (e.g. deck) varies between 20 to 50 years, it is therefore practical and reasonable to assume a constant surface chloride concentration. For bridge decks, the top surface is subjected to a continually changing chloride exposure. The chloride concentration at the concrete surface varies with the season, however at some shallow depth near the surface it can be assumed as a quasi-constant. In general, the values of the surface chloride concentration and "apparent" diffusion coefficient can be estimated from Eq. (2) by determining the best-fit curve through field data obtained from chloride profiles at different depths and exposure times.

# 4. UNCERTAINTY IN STEEL CORROSION MECHANISM

## 4.1 Mechanism of Steel Corrosion

The corrosion of conventional "carbon" or "black" steel embedded in concrete structures is considerably different from the corrosion of steel exposed to the atmosphere, as the reinforcing is protected by the concrete cover ("skin"), which provides a barrier, or protection that slows down the penetration of aggressive agents needed for the initiation and propagation of corrosion, namely chloride ions, water and oxygen. The corrosion of the reinforcing steel is assumed to start when the concentration of chlorides at the level of the reinforcement (chloride contamination over the concrete cover) has reached the so-called "chloride threshold level".

The corrosion of steel in concrete is an electrochemical process in which two separate, but coupled chemical reactions take place simultaneously at two different sites on the steel surface, namely the anode and cathode as shown in Fig. 4. The electrical potential differences are the driving force for the corrosion reaction. The potential difference may be generated by: (i) potential difference along the steel surface; (ii) difference in the concentration of ions in the pore solution along the steel concrete interface; and (iii) contact of dissimilar metals. Due to this potential difference, iron is oxidized at the anode according to:

$$Fe \rightarrow Fe^{2+} + 2\,e^{-} \tag{3a}$$

The released electrons move through the reinforcing steel towards the cathode, while ferrous ions are dissolved in the concrete pore solution according to:

$$O_2 + 2H_2O + 4\,e^{-} \rightarrow 4OH^{-} \tag{3b}$$

It is clear from Eq. (3b) that both moisture and oxygen should be present for the cathodic half-cell reaction to occur. The hydroxyl ions ($OH^-$) released at the cathode move towards the anode where they combine with dissolved ferrous ions to yield ferrous hydroxide $Fe(OH)_2$ (or rust). Given sufficient oxygen at the anodic sites, $Fe(OH)_2$ can be further oxidized into different corrosion products with higher volumes (up to six times the original volume). Such volume increase induces tensile stresses in the surrounding concrete, which in turn lead to the cracking, delamination, and spalling of the concrete cover. A simplified model of the service life of concrete structures exposed to chlorides illustrating the initiation and accumulation of the different damages is shown in Fig.5.

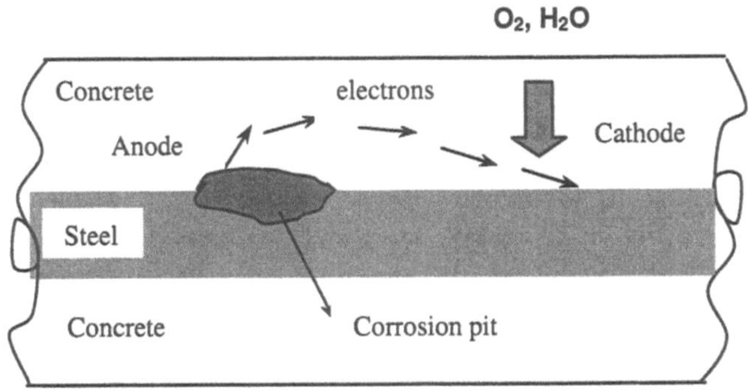


Figure 4. Schematic description of a corrosion cell in reinforced concrete

The exact role of chloride ions is not well understood yet (Rosenberg et al. 1989; Bentur et al. 1995). It is generally believed that the chloride ions become incorporated in the passive film that protects the steel, replaces some of the oxygen and increases both its conductivity and solubility (Rosenberg et al. 1989). As a result, the film loses its protective capacity, and ferrous chlorides form when chlorides react with iron. Chloride ions, therefore, act as catalysts of ion dissolution (Bentur et al. 1995; Rosenberg et al. 1989). The reactions consume $OH^-$ ions and then release the $Cl^-$ back into the solution. The process results in a concentration of $Cl^-$ and a reduction of pH at the point of corrosion initiation, which accounts for pitting corrosion. It is generally accepted that the concentration ratio of $Cl^-/OH^-$ in the pore solution determines the initiation of corrosion. Hausmann (1967) suggested 0.6 as the threshold value.

Tuutti (1982) proposed a model that describes the performance of concrete structures exposed to chlorides as a two-stage process (Fig. 5): (i) *Initiation stage*, which is defined as the time period from the initial exposure to chlorides until the onset of corrosion; and (ii) *Propagation stage*, which is the post-corrosion stage that corresponds to damage initiation (cracking, delamination, spalling, etc.) and damage accumulation until failure. The time to onset of corrosion ($t_i$) depends on the rate of ingress of chlorides into concrete, surface chloride concentration, depth of concrete cover, and the value of the threshold chloride level. Using Eq. (2) and assuming the same initial and boundary conditions, the time to onset of corrosion is determined as follows:

$$t_i = \frac{d_c^2}{4D[\mathrm{erf}^{-1}(1-\frac{C_{th}}{C_s})]^2} \quad (4)$$

where $C_s$: surface chloride concentration; $C_{th}$: threshold level of chloride concentration; D: chloride diffusion coefficient; and $d_c$: depth of concrete cover. From the above equation, kit is clear that the concrete cover depth is a key parameter in delaying the onset of corrosion and extending the durability of concrete structures.

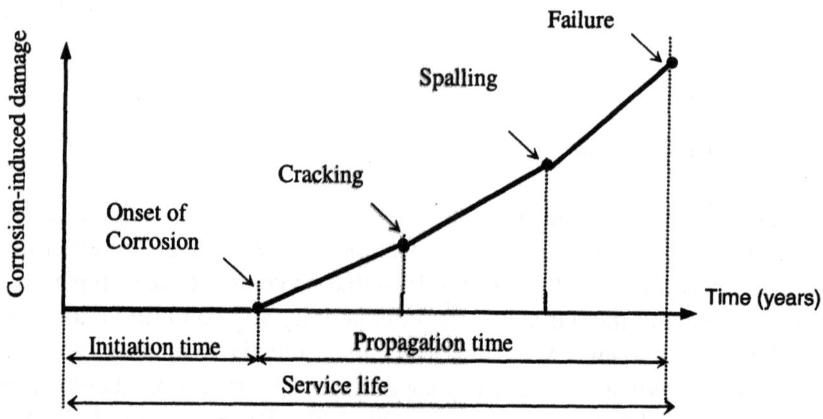


Figure 5. Service life model of reinforced concrete structures exposed to chlorides

## 4.2 Uncertainty in Corrosion Initiation Model

The uncertainty with the corrosion initiation model incorporates both the uncertainty associated with the chloride transport model as well as the uncertainty associated with the use of a simplified chloride threshold level to define the corrosion resistance of concrete structures subjected to chloride attack. As discussed earlier, there is great deal of uncertainty in the mechanisms of depassivation and breakdown of the protective film and onset of corrosion. The overall uncertainty includes: (i) assumption of chloride transport mechanism governed by diffusion; (ii) use of simplified models of the diffusion coefficient and driving chloride concentration; (iii) use of a single parameter (chloride threshold value) to define the resistance of concrete structures to chloride-induced corrosion; and (iv) assumption of non-correlated variables.

## 4.3 Uncertainty in Governing Parameters

### 4.3.1 Threshold Chloride Concentration

There is no consensus regarding the definition of a single value for the threshold chloride level. A considerable scatter of this threshold value is found in the literature (Rosenberg et al. 1989; Glass and Buenfeld 1995; Kropp et al. 1995; Pettersson 1996). The value of the threshold chloride concentration depends on several parameters, including:
concrete type (cements with high contents of tricalcium aluminate-$C_3A$ have a great capacity to bind chlorides, resulting in increased chloride threshold level);
source of chlorides, temperature and moisture content (higher temperature and moisture contents will decrease the threshold level);
type of reinforcing steel (conventional black, carbon steel, epoxy-coated steel, galvanized steel, or stainless steel);
concrete cover depth (thicker covers increase the threshold level by reducing the moisture and oxygen variations at the steel surface);
water-to-binder ratio (a lower ratio helps to stabilize the micro-environment at the steel level as the moisture permeability is decreased);
carbonation of concrete; and
presence of macro-cracks (reduces the threshold value).

Despite the fact that only the free chlorides induce steel corrosion, for practical purposes, the threshold content is given in terms of total chlorides (free and bound), as it is difficult to measure the free chlorides (Glass and Buenfeld 1995). In the literature, there is a considerable uncertainty associated with the value of the threshold concentration level "$C_{th}$" obtained from laboratory and field studies where $C_{th}$ was found to vary between 0.17% and 2.5% in terms of total chlorides by weight of cement for conventional black steel (Glass and Buenfeld 1995). Many highway agencies, however, use a total chloride threshold level of 0.2% by weight of cement (or 0.03% by weight of concrete or 0.7 kg/m$^3$).

### 4.3.2 Concrete Cover Depth

The mean value of the concrete cover depth and its coefficient of variation depend on the initial design, and quality control during construction. Wide variations from the mean are observed in the field as will be illustrated in the case study. It is generally measured using a covermeter.

# 5. DETERIORATION MODELING AND PREDICTION

## 5.1 Probabilistic Deterioration Modeling

As mentioned earlier, a considerable level of uncertainty is associated with the prediction of chloride contamination of concrete and reinforcing steel corrosion. In light of the above, it is clear that a deterministic prediction model can be quite inadequate owing to the considerable uncertainty in the prediction models, governing parameters and structural response. Therefore, the level of chloride contamination of concrete structures and corrosion of the reinforcing steel should be determined using probabilistic methods.

Using Eq. (2), the the level chloride contamination of the concrete cover at any given depth and time reaching a prescribed maximum value is formulated as the limit state function. Failure is defined as the event corresponding to the chloride concentration exceeding this maximum value, and the probability of failure is given by:

$$P_f = P[C_{max} - C(x,t) \leq 0 ] \quad (5)$$

The uncertainties associated with the surface chloride concentration and diffusion coefficient are considered by modeling them as random variables with probability density functions $f_{Cs}(c)$ and $f_D(D)$, respectively that are fitted to the data obtained from the field measurements of the chloride profiles. Similarly, using Eq. 2, the corrosion of the reinforcing at any given time is formulated as the limit state function. Failure is defined as the event corresponding to the chloride concentration at the steel level exceeding the chloride threshold value "$C_{th}$", and the probability of failure is given by:

$$P_f = P[C_{th} - C(d_c,t) \leq 0] \quad (6)$$

The uncertainties associated with the surface chloride concentration, diffusion coefficient, concrete cover depth, and threshold chloride level are considered by modeling them as random variables with probability density functions $f_{Cs}(c)$, $f_D(D)$, $f_{dc}(d_c)$, and $f_{Cth}(C_{th})$, respectively that are fitted to the data obtained from the field measurements of the chloride profiles, measurements of corrosion activity, and observed damage on the bridge structure.

## 5.2 Deterioration Prediction Using Monte Carlo Method

Monte Carlo methods are the most widely used techniques for uncertainty analysis, with a wide range of applications. These methods involve sampling at "random" from the distribution of inputs to simulate artificially a large number of experiments until a statistically significant distribution of the structure response is generated (Melchers 1987). The direct sampling Monte Carlo is the most widely used method, although not as efficient as those based on importance sampling. The probability of an event $g(\mathbf{x})\leq 0$ under consideration, typically termed "failure" (e.g. probability that the chloride concentration at the steel level exceeds a threshold level) may be expressed as (Melchers 1987):

$$P_f = \int\int \ldots \int I[g(\mathbf{x}) \leq 0] f_{\mathbf{x}}(\mathbf{x}) d\mathbf{x} \tag{7}$$

where I[ ]is an "indicator function" that equals 1 if [ ] is "true" and 0 if [ ] is "false". Eq. (4) represents the expected value of I[ ]. If $\mathbf{x}_j$ represents the j$^{th}$ vector of random observations from $f_{\mathbf{x}}$, then it follows directly from sample statistics that;

$$P_f \cong \sum_{j=1}^{N} I[g(\mathbf{x}_j) \leq 0)] / N \tag{8}$$

Eq. (5) represents an unbiased estimator of $P_f$ (Melchers 1987).

# 6. ILLUSTRATIVE EXAMPLE

## 6.1 Bridge Description

The proposed probabilistic modeling and the prediction capability of the Monte Carlo method are illustrated on the Dickson bridge in Montreal, Canada. This bridge was constructed in 1959, and had a total length of 366 m and width of 27 m. The superstructure of this bridge consisted of reinforced concrete T-girders in the end sections and a concrete deck on steel girders in the central section. This superstructure was severely deteriorated because of the inadequate quality control in construction and the aggressive environment resulting from the frequent use of de-icing salts in winter.

A detailed condition assessment that included detailed visual inspection, non-destructive and partial destructive evaluation was carried out in 1999 (i.e.

after 40 years) on the bridge superstructure prior to its demolition (Fazio 1999). Hundreds of data points were collected, which indicated a considerable variation in the parameters affecting the chloride contamination of the deck and corrosion of the top mat of reinforcing steel throughout the deck. The statistical distributions of the governing parameters were derived from the data collected (Lounis and Mirza 2001; Fazio 1999).

## 6.2 Field Survey

### 6.2.1 Concrete Cover Depth

The concrete cover depth was measured at 137 locations using a covermeter. The concrete cover depth ($d_c$) was found to be normally distributed ranging from 10 mm to 89 mm with an average of 36.6 mm and a standard deviation of 16.5 mm.

### 6.2.2 Diffusion Coefficient and Surface Chloride Concentration

As mentioned earlier, the "apparent" values of the diffusion coefficient and surface chloride concentration were obtained by regression analysis to best fit the solution given by Eq. (2) to the chloride profiles obtained from field data. The chloride content of powdered concrete samples was measured at 35 locations on the deck using the SHRP chloride analysis method known as the specific ion electrode technique (Fazio 1999). The "near surface" chloride concentration ($C_s$) was found to have a lognormal distribution with an average of 4.56 kg/m$^3$ and a standard deviation of 1.84 kg/m$^3$. The apparent diffusion coefficient (D) also had a lognormal distribution with an average of 0.51 cmm$^2$/year and a standard deviation of 0.16 cm$^2$/year.

### 6.2.3 Threshold Chloride Level

The chloride threshold level is determined by correlating chloride content measurements with electrochemical measurements of the steel embedded in concrete using changes in either half-cell potential or corrosion rate. The half-cell potential was measured at 137 locations using the conventional copper-copper sulfate half-cell and ASTM C876 criteria. The corrosion rate measurements were done with two different probes: 3-electrode linear polarization (3LP technique) and linear polarization device with controlled guard ring. The electrical resistivity was measured at 137 locations using the

Wenner four probe apparatus. The detection of delamination was investigated at 140 sites using a hammer.

Another method used was the direct measurement of the weight loss from which the corrosion initiation time and, consequently, the threshold level can be estimated by using the corrosion rate data. Other methods included visual inspection for rust stains, cracking and spalling, and the sounding technique using a hammer for delamination detection and correlating the results with the measured chloride contents. A combination of all these results yielded a lognormal distribution of the threshold chloride concentrations ($C_{th}$) with an average of 1.35 kg/m$^3$ and a standard deviation of 0.135 kg/m$^3$. The field data showed a considerable level of variability in all parameters that govern the chloride contamination of concrete and the corrosion of the reinforcing steel and are summarized in Table 1, in which μ and V represent the mean value and coefficient of variation, respectively (Lounis and Mirza 2001).

Table 1 - Summary of Data from Field Survey

| Variable | Distribution | μ | V |
|---|---|---|---|
| $d_c$ (cm) | Normal | 3.66 | 45% |
| $C_s$ (kg/m$^3$) | Lognormal | 4.56 | 40% |
| $D$ (cm$^2$/year) | Lognormal | 0.51 | 30% |
| $C_{th}$ (kg/m$^3$) | Lognormal | 1.35 | 10% |

## 6.3 Prediction of Chloride Contamination in Bridge Deck

In the assessment of the chloride contamination of the concrete cover of the bridge deck, diffusion is assumed to be the governing transport mechanism based on the so-called apparent diffusion coefficient D. The random variable vector is $\mathbf{x}=[C_s, D, d_c]^T$. Using the direct Monte Carlo simulation method, the chloride concentration at the steel level after 40 years is shown in Fig.6. This figure illustrates the skewed form of the distribution.

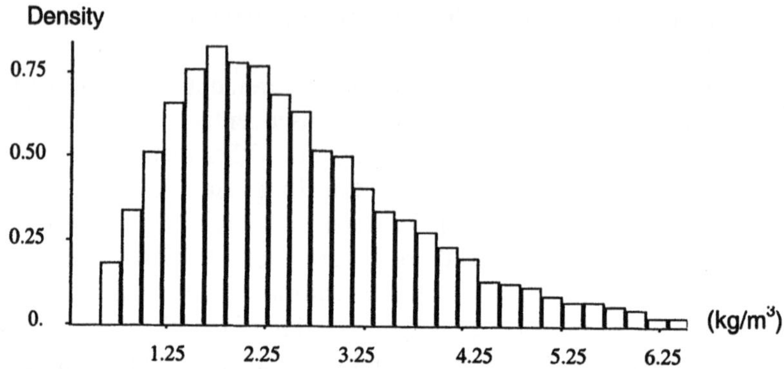


Figure 6. Distribution of Chloride Concentration at Reinforcement Level

It can be approximated by a gamma distribution with parameters 2 and 0.783, mean value of 2.57 kg/m$^3$ (0.71% by cement weight), standard deviation of 1.36 kg/m$^3$ (0.38% by cement weight) and a coefficient of variation of 0.53. The simulation results are very close to the field measurements that yielded a mean value of 0.73% by cement weight and a coefficient of variation of 0.72 (Lounis and Mirza 2001; Fazio 1999).

## 6.4 Prediction of Reinforcement Corrosion in Bridge Deck

The random variable vector is $\mathbf{x}=[C_s, C_{th}, D, dc]^T$. Using the direct Monte Carlo simulation method, the distribution of the time to onset of corrosion is generated and is shown in Fig.7. It has also a skewed distribution that was approximated by a lognormal model, with a mean of 10.23 years and a coefficient of variation of 100%.

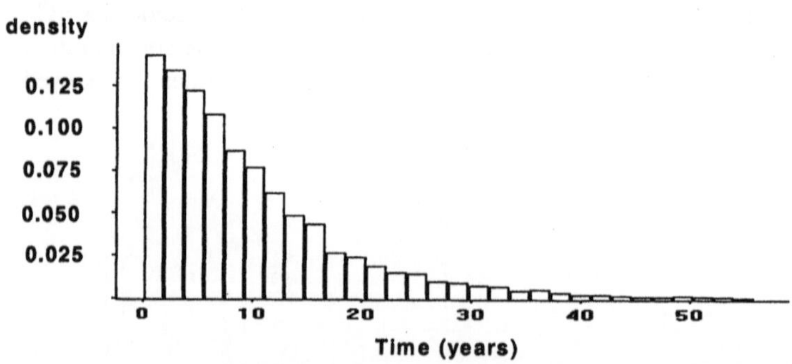


Figure 7. Distribution of Corrosion Initiation Time of Reinforcing Steel

If $t_i$ can be approximated by a lognormal distribution with mean $\mu_{ti}$ and coefficient of variation $V_{ti}$, it is possible to derive the following relationship for the time-dependent probability of reinforcement corrosion $P_f(t)$ as follows:

$$P_f(t) = \frac{1 - \mathrm{erf}(\beta_t / \sqrt{2})}{2} \tag{9a}$$

where

$$\beta_t = \beta(t) = \frac{\ln[(\mu_{ti}/t)\sqrt{1+V_{ti}^2}]}{\sqrt{\ln(1+V_{ti}^2}} \tag{9b}$$

# 7. CONCLUSIONS

This paper presented a reliability-based approach for the modeling of the deterioration and service life of concrete structures taking into account the uncertainties in the physical modeling, and variability of the material and structural parameters governing the chloride penetration into concrete and corrosion of the reinforcing steel. It illustrated the application of a probabilistic approach for the uncertainty modeling and prediction of chloride contamination of concrete and reinforcement corrosion in bridge structures that are subjected to the application of deicing salts during winter. The proposed probabilistic model provided very good predictions of the level of chloride contamination at different depths as well as the extent of corrosion of the reinforcing steel in the top mat of a deteriorated reinforced concrete bridge deck. The application of such an approach is required in the assessment of safety and serviceability of deteriorating concrete structures in order to ensure that the probability of failure is kept at an acceptable level. A reliability-based prediction of the service life of deteriorating concrete structures provides a rational decision support tool at both the initial design stage and during the operation and maintenance stage. The implementation of such an approach allows adequate control of the safety and serviceability of the structure throughout its service life and yields low life cycle cost.

# REFERENCES

Bentur, A. Diamond, S., and Berke, N.S. Steel corrosion in concrete. 1997, E & FN SPON.
Ditlevsen, O. "Probabilistic thinking: an imperative in engineering modeling." *Danmarks Tekniske Hojskole*, Lyngby, 1984, 1-23.

Bogdanoff, J.L. "A new cumulative damage model –Part 1." J. of Applied Mechanics, ASCE, Vol. 45, 1978, pp. 246-250.

Enright, M. P., and Frangopol, D. M. "Service-life prediction of deterioration concrete Bridges." J. Structural Engrg., ASCE, 124 (3), 1998, 309-317.

Fazio, R. *The Assessment and Prediction of Reinforcing Steel Corrosion on the Dickson Bridge*. M.Eng. Thesis, McGill University, 1999.

Frangopol, D.M., Lin, K.Y., and Estes, A.C. "Reliability of reinforced concrete girders under corrosion attack." *ASCE J. of Struct. Engrg.*, 123(3), 1997, 286-297.

Glass, K., and Buenfeld, N.R. "Chloride threshold levels for corrosion-induced deterioration of steel in concrete." In Chloride Penetration into Concrete, L.O. Nilsson and J.P. Ollivier (eds.), 1995, pp. 429-440.

Hausmann, D.A. "Steel corrosion in concrete: how does it occur?" *Material Protection*, Vol.6, pp.19-21.

Kropp, J., et al. "Transport Mechanisms and Definitions." In Performance Criteria for Concrete Durability, J. Kropp and H. Hisldorf, eds, E&FN SPON, London, 1995, pp. 4-14.

Lounis, Z . "Reliability-based Life Prediction of Ageing Concrete Bridge Decks." In Life Prediction and Aging Management of Concrete Structures, D. Naus, ed., RILEM Publications, Paris, 2000, pp. 229-238.

Lounis, Z, and Mirza, M.S. "Reliability-based Service Life Prediction of Deteriorating Concrete Structures." In Concrete under Severe Conditions, Banthia, N., et al., eds., Vancouver, 2001, pp. 965-972.

Melchers, R.E. Structural Reliability- Analysis and Prediction, Ellis Horwood Ltd., Chichester, England, 1987.

Mori, Y., and Ellingwood, B.. "Reliability-based service life assessment of aging concrete structures." *ASCE J. of Struct. Engrg.*, 119(5), 1993, 1600-1621.

Neville, A. "Chloride attack of reinforced concrete: an overview." Materials and Structures, Vol. 28, 1995, 63-70.

Pettersson, K, Chloride threshold values in reinforced concrete, in Durability of Concrete in Saline Environment, 1996, Alqvist & Wiksell Tryckei, Uppsala.

Rosenberg, A., Hanson, C. M., and Andrade, C. "Mechanisms of corrosion of steel in concrete." Material Science of Concrete I, J. Skalny, ed., American Ceramic Society, Westerville, Ohio, 1989, 285-313.

Stewart, M., and Rosowsky, D.V. "Structural safety and serviceability of concrete bridges subject to corrosion." J. Infrastructure Systems, ASCE, 4(4), 1998, 146-155.

Tuutti, K. (1982). "Corrosion of steel in concrete." Swedish Cement and Concrete Research Institute, Stockholm.

Weyers, R.E. "Service life model for concrete structures in chloride-laden environments." J. ACI Materials, 95(5), 1998, pp. 546-557.

# Chapter 23

# REDUNDANCY ANALYSIS OF STRUCTURAL SYSTEMS

**Tarek N. Kudsi**

## 1. INTRODUCTION

The reliability of engineering systems can be presented as a problem of supply versus demand. The reliability of engineering systems is achieved through the use of factors or margins of safety and adopts conservative assumptions in the process of design (Ang and Tang probability concepts, volume II). The existing supply and the required demand may be modeled as random variables. Examples of random variables are: the cross section of a beam, the moment of inertia, the flange depth and thickness, the material strength, the elastic properties, and the applied loads on the structure such as vehicle loading, wind loading, or earthquake loading. In order to better understand the mentioned random variables, samples should be gathered and analyzed in order to calculate the mean and coefficient of variation (or standard deviation) in order to incorporate them into their respective distribution functions.

The analysis of safety of a failure mode (or limit state) of a structural system can be categorized as follows:

I. Safety is met when demand (or load effect) ≤ supply (resistance).

II. Failure will occur when demand (or load effect) > supply (resistance).

Based on above, the capacity and demand function of a structure is referred to as the performance function of the structure, in the form of:

$$Z = R - L \tag{1}$$

Where R = the supply and L = demand. It should be mentioned that when:

$Z = R - L > 0$, is referred as the survival state

$Z = R - L < 0$, is referred as the failure state

$Z = R - L = 0$, is referred as the limit state

Structures' components can have many limit state functions depending on the failure modes associated with given components. There are three types of limit state functions related to bridges and structures:

1. Ultimate limit state functions such as exceeding the moment carrying capacity, buckling of steel beam plates, shear failure, or combination of biaxial bending and axial loading.
2. Serviceability limit state functions, such as cracking, and deflection
3. Fatigue limit state functions, defined as the accumulation of damage and eventual failure under repeated loads. This limit state function applies to steel structures, and in particular to members in tension or reversal forces.

In 1996, The U.S National Science Foundation Workshop on Structural Reliability in Bridge engineering, held in Boulder, Colorado, discussed the importance of reliability techniques in Bridge design, management, and maintenance. A better understanding of the effect of local damage on the overall behavior of the bridge system was one of the issues raised. The representatives discussed also the importance of material non-linearity and the effect of secondary members, in some cases, to produce reserve strength for the overall bridge system. The workshop concluded that a better understanding of the reliability of the bridge structural system is needed.

The proposed study presents a new approach for evaluating redundancy of structural systems in general, and bridges in particular, based on reliability based system approach. Although, most of the researchers have mentioned that their proposed method for redundancy analysis is valid for all types of bridges, there still remain many unanswered questions, which the research is trying to answer. This presented methodology is also a powerful tool for depicting the true structural health condition of an existing building with localized failure.

The questions that may be raised are:

1. How significant modeling the bridge as a structural system in the pre-failure and post-failure phases is in the analysis of the global redundancy of the system in the pre-failure and post-failure phases?
2. How important is the presence of secondary members in the system, and their role in redistributing load, on their respective subsystems redundancy, and the global redundancy of the bridge structural system?
3. What is the effect of local failure on its related subsystem, and on the different subsystems of the global system?
4. How reliable the bridge system is, in the post-failure phase compared with a pre-set target reliability?
5. How can a bridge be identified as a redundant bridge, and what are the members that could cause the bridge to collapse?

In order to answer these questions, a new methodology is proposed, based on the analysis of a bridge structural system accounting for the component interaction and their integration in their respective subsystem, and the global system. In order to set a solid methodology for the redundancy analysis, the following two criteria are needed:

1. A better understanding of the structural topology of the structural system or bridge.
2. A better understanding of the effect of a failure of a component based on a certain limit state, on other failure modes of other components.

# 2. GENERAL METHODOLOGY

## 2.1. Structural System Analysis

A general methodology for the analysis of structural systems, based on reliability techniques, and can be applied to all types of structural systems is presented.

Prior to any type of structural reliability analysis, the following must be carried out:

1. Define all different failure modes for each component in the system. The limit state functions will be based on design codes to be followed
2. Define all the related random variables along with their means and standard deviation, and their respective distribution functions.
3. Analyze the structural system using a non-linear finite element software program, to determine the stresses caused by the dead load, live load, and lateral loadings.
4. Define the target reliability index $\beta_{target}$ for the system, subsystems and components. The assigned reliability index corresponds to the maximum probability of failure acceptable by the engineering community.
5. Evaluate the probability of failure of each component related to all existing failure modes. For the task mentioned, First Order Reliability Method (FORM), Advanced Second Moment (ASM), or Monte-Carlo simulation techniques can be used. Advanced second moment is generally used to evaluate the probability of failure of non-normal non-linear limit state functions.
6. Fit all failure modes related to each component into a series system, known also as the weakest link. If any limit state function reliability fall below the target level, the component fails. In figure 1, component $i$ is composed of multiple failure modes, such as buckling, shear, or torsion.

Figure 1: Component $i$

7. Define the main subsystems to be modeled as series systems, as failure of any subsystem will lead to the complete failure of the system.
8. Define the global structural topology of the system by proposing a general Block diagram that includes all the subsystems with all relevant components along with all related failure modes (Fig.2).

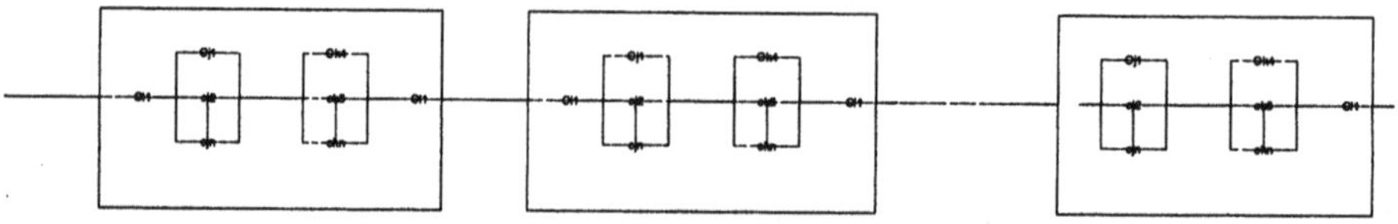


Figure 2: The Global System Layout

where $C_{ij}$ is the failure mode j of component $i$. Where $i = 1,2,3,...,n$, representing the number of components, and $j = 1,2,3,...,m$ representing the number of failure modes associated with the component. The general block diagram is unique for every system and the shown topology is for descriptive purposes only.
9. Calculate the reliability index along with the probability of failure of each subsystem and compare it with the target reliability index.
10. Calculate the reliability index along with the probability of failure of the global system.

## 2.2. Types of Uncertainties

In order to acquire a better understanding of random variables, types of uncertainties should be defined. "Uncertainties in civil engineering systems can be mainly attributed to vagueness in defining the variables and parameters of the systems and their relations" (Ayyub and McCuen, 1997).
There are three types of uncertainties and can be defined as follows:
1. Physical uncertainty: the variation of the loading applied on the structure. Examples are the wind loading, earthquake loading, and the live load loading on a bridge induced by trucks crossing the bridge.
2. Statistical Uncertainty: the uncertainty due to the limited sample size of the random variables.
3. Model Uncertainty: Uncertainty due to simplifying assumptions in modeling the physical problem from an engineering point of view. Unknown boundary conditions and unknown effects.
Based on the above, physical uncertainty requires a realistic understanding of the applied load on the structural system along with its random nature. For example, the live load on a structural system should be modeled based on data gathered after the system being monitored under this loading.

## 2.3. Types of Distribution Functions

There are many types of distribution functions used in the field of reliability engineering. The most common types used, related to bridge analysis under traffic loading, are normal distribution modeling the dead load randomness nature, the lognormal distribution that is commonly used to model the material strength as it can disregard samples with negative values. The extreme type I (largest), referred to as the Gumbel distribution, is best to model the live loading on a bridge (Nowak, 1995, and Moses and Ghosn, 1998).

**Probability of failure**

Failure occurs when the load effect (L) exceeds the resistance (R) of the structure, and can be derived by considering the probability density functions of R and L, along with their associated random variables. The main goal for the safety of the structure is to guarantee an R>L scenario throughout the design life of the structure. Assuming that R and L are normal, statistically independent, and positive random variables, it can be stated that:

$$P_f = \int_{-\infty}^{\infty} \int_{-\infty}^{L_j \geq R_i} f_R(R_i) f_L(L_J) dR_i dL_j$$

(2)

$f_R(R_i)$ is the probability density function for the structural resistance, and $f_L(L_j)$ is the probability density function of the external loading.

The reliability index is defined as:

$$\beta = -\Phi^{-1}(P_F)$$

(3)

**Performance Functions**

All limit states related to the structural system's components must be identified, prior to any analysis. In order to perform the redundancy analysis of structural system, based on reliability techniques, the limits states associated with each component must be identified.

# 3. REDUNDANCY ANALYSIS OF STRUCTURAL SYSTEMS

In the above section, a reliability-based analysis is presented of a structural system and its components. In this section, a new methodology for the redundancy analysis of structural systems is presented.

Modeling a structural system as a collection of structural elements in series and structural system is very essential for evaluation of its true redundancy and reliability.

In order to build the structural system, the following need to be carried out:

1. Analyze the structural system using a non-linear finite element package.
2. Fail each element separately, and re-analyze the structural system. New reliability indices for each component in the system are then generated.
3. Define the failure of the structural system. For the purpose of this study, the total collapse of the structural system is considered as the failure criteria.
4. If failure of element $i$ does not fail the system, then element $i$ is a redundant element.
5. If failure of element $i$ fails the system, then element $i$ is a non-redundant element.
6. Incorporate the elements in the system, according to their nature. If the element is redundant, then it is in parallel with the rest of the system. If the element is non-redundant, then it is in series with the rest of the system, thus failure of this element will result in the failure of the system.
7. Define the degree of redundancy of the system (D.O.R), $k$, based on the number of members, which can be cut simultaneously without resulting in the failure of the system, thus $k = 1,...,s$.
8. Number the non-redundant members from $i = 1,2,...,n$.
9. Number the redundant members from $j = n+1, n+2,...,n+m$.
10. Build a block diagram for the structural system according to steps 1 through 9, for the pre-failure phase, and also for the post-failure phase.

## 3.1. System Analysis in the Pre-Failure Phase

The following presentation is for redundant structural systems in the pre-failure phase. The pre-failure phase is defined as the intact system, prior to any redundant member failure or modification. Consider the following four examples structural systems:

1. Example system one, as illustrated in figure 3: A structural system composed of ten elements. If the degree of redundancy is $k = 1$, this implies that any failure of any one redundant member will not cause the structural system to fail. The number of non-redundant members equals to five, $i = 1,2,3,4,5$, and the number of redundant members equal to five, $j = 6,7,8,9,10$. Assuming that the components are fully correlated, ρ= 1, the probability of failure of the pre-failure phase $P_{f_{system\_pre\_failure}}$ is:

$$P_{f_{system\_pre\_failure}} = \max_i(P_{f_i}, \min(P_{f_6}, P_{f_7}), \min(P_{f_6}, P_{f_8}), \min(P_{f_6}, P_{f_9}), \min(P_{f_6}, P_{f_{10}}),$$
$$, \min(P_{f_7}, P_{f_8}), \min(P_{f_7}, P_{f_9}), \min(P_{f_7}, P_{f_{10}}), \min(P_{f_8}, P_{f_9}),$$
$$, \min(P_{f_8}, P_{f_{10}}), \min(P_{f_9}, P_{f_{10}}))$$

(4)

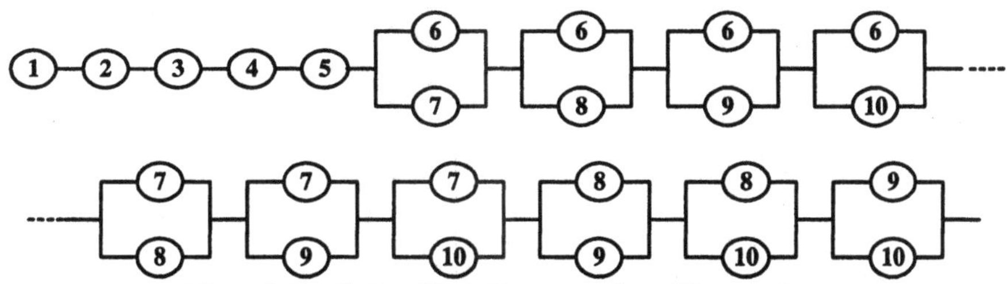


Figure 3: Pre-Failure Phase Representation of System One

2. Example system two, as illustrated in figure 4: A structural system composed of ten elements. If the degree of redundancy is $k = 2$, this implies that failure of any two redundant components will not cause the structural system to fail. The number of non-redundant members equals to five, $i = 1,2,3,4,5$, the number of redundant members equal to five, $j = 6,7,8,9,10$. Assuming that the components are fully correlated, □ = 1, the probability of failure of the pre-failure phase $P_{f_{system\_pre\_failure}}$ is:

$$P_{f_{system\_pre\_failure}} = \max_i(P_{f_i}, \min(P_{f_6}, P_{f_7}, P_{f_8}), \min(P_{f_6}, P_{f_7}, P_{f_9}),$$

$$, \min(P_{f_6}, P_{f_7}, P_{f_{10}}), \min(P_{f_6}, P_{f_8}, P_{f_9}), \min(P_{f_6}, P_{f_8}, P_{f_{10}}),$$
$$, \min(P_{f_6}, P_{f_9}, P_{f_{10}}), \min(P_{f_7}, P_{f_8}, P_{f_9}), \ldots$$
$$, \min(P_{f_7}, P_{f_8}, P_{f_{10}}), \min(P_{f_8}, P_{f_9}, P_{f_{10}}))$$

(5)

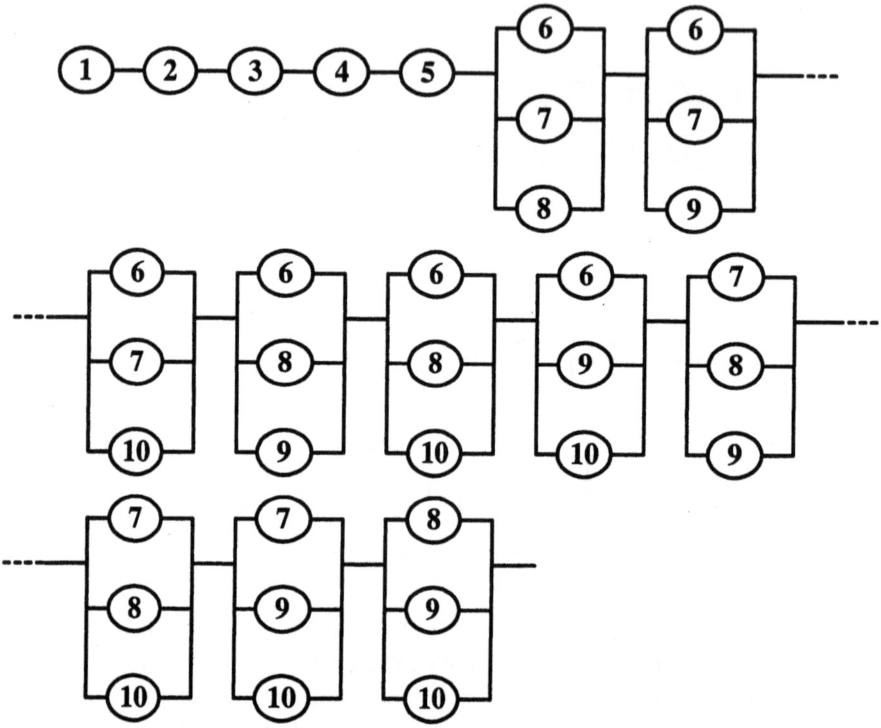

Figure 4: Pre-Failure Phase Representation of System Two

3. Example system three, as illustrated in figure 5: A structural system composed of ten elements. If the degree of redundancy is, $k = 3$, this implies that the failure of any three redundant components will not cause the structural system to fail. The number of non-redundant members equals to five, $i = 1,2,3,4,5$, and the number of redundant members equal to five, $j = 6,7,8,9,10$. Assuming that the components are fully correlated, $\rho = 1$, the probability of failure for the pre-failure phase $P_{f_{system\_pre\_failure}}$ is:

$$P_{f_{system\_pre\_failure}} = \max_i(P_{f_i}, \min(P_{f_6}, P_{f_7}, P_{f_8}, P_{f_9}), \min(P_{f_6}, P_{f_7}, P_{f_8}, P_{f_{10}}), \\ , \min(P_{f_6}, P_{f_7}, P_{f_9}, P_{f_{10}}), \min(P_{f_6}, P_{f_8}, P_{f_9}, P_{f_{10}}), \\ , \min(P_{f_7}, P_{f_8}, P_{f_9}, P_{f_{10}}))$$

(6)

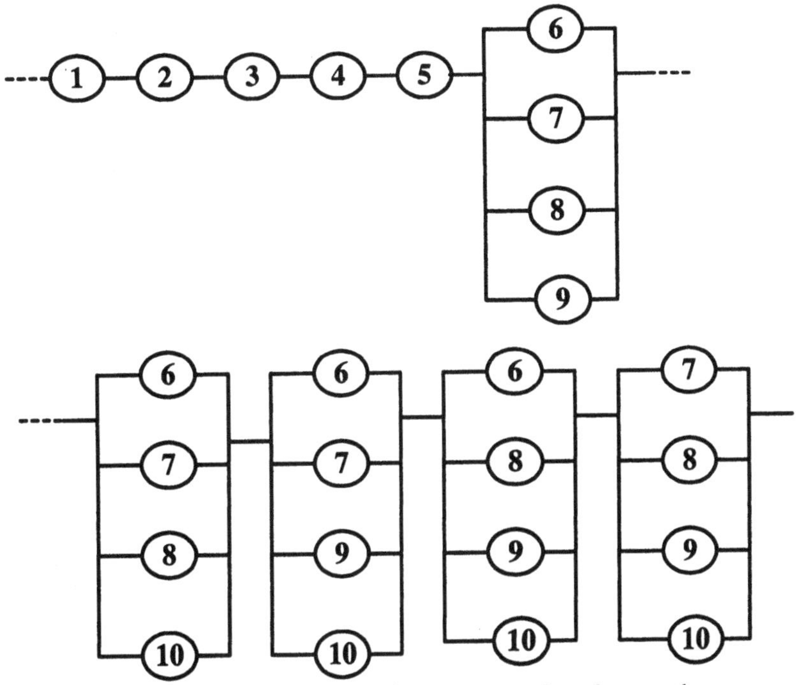


Figure 5: Pre-failure phase representation of system three

4. Example system four, as illustrated in figure 6: A structural system composed of ten elements. If the degree of redundancy is, $k = 4$, this implies that the failure of any four redundant members will not cause the structural system to fail. The number of non-redundant members equals to five, $i = 1,2,3,4,5$, and the number of redundant members equal to five, $j = 6,7,8,9,10$. Assuming that the components are fully correlated, $\rho = 1$, the probability of failure of the pre-failure phase $P_{f_{system\_pre\_failure}}$ is:

$$P_{f_{system\_pre\_failure}} = \max_i(P_{f_i}, \min(P_{f_6}, P_{f_7}, P_{f_8}, P_{f_9}, P_{f_{10}})) \tag{7}$$

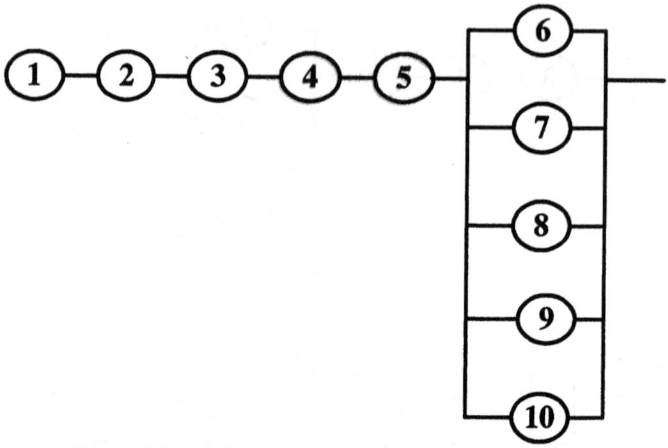


Figure 6: Pre-failure phase representation of system four

The above four example systems presented can be generalized for $k$ degrees of freedom and $m$ numbers of redundant members in a system. The probability of failure of the system in the pre-failure phase, $P_{f_{system\_pre\_failure}}$ can be presented as follows:

$$P_{f_{system\_pre\_failure}} = \max_{i,j}\left(P_{f_i}, P(F_j)\right) \tag{8}$$

where $P_{f_i}$ = probability of failure of the non-redundant members, $P(F_j)$ = the minimum probability of the occurrence of the failure event, $F_j$, of each possible arrangement of the redundant members, according to the system degree of redundancy, $k$, for fully correlated redundant members, $\rho = 1$. In order to identify the amount of failure event arrangements, let $r = k + 1$ be the combination of the failure events of the redundant elements from a set of $m$ amount of redundant elements, thus the amount of combination can be found as follows:

$$C_{r|m} = \frac{m!}{r!(m-r)!} \tag{9}$$

Let $C_{J\ r|m} = F_j$ = the series of the multiple failure events arrangements of the redundant members in the system. Thus, for a degree of redundancy, $k = 2$, and $r = k + 1 = 3$, and $j = n + 1,..., n + m$ is the amount of redundant members, the following series represent the possible arrangements of the failure events of redundant members:

$$\underset{\substack{j=n+1\\k=2}}{\overset{n+m}{F}} = F_{n+1}\,\mathrm{I}\,F_{n+2}\,\mathrm{I}\,F_{n+3}, F_{n+1}\,\mathrm{I}\,F_{n+2}\,\mathrm{I}\,F_{n+4},\ldots, F_{n+1}\,\mathrm{I}\,F_{n+2}\,\mathrm{I}\,F_{n+m},$$

$$, F_{n+2}\,\mathrm{I}\,F_{n+3}\,\mathrm{I}\,F_{n+4}, F_{n+2}\,\mathrm{I}\,F_{n+3}\,\mathrm{I}\,F_{n+5},\ldots, F_{n+2}\,\mathrm{I}\,F_{n+3}\,\mathrm{I}\,F_{n+m},$$

$$\ldots, F_{n+m-2}\,\mathrm{I}\,F_{n+m-1}\,\mathrm{I}\,F_{n+m} \qquad (10)$$

For a degree of redundancy $k = 3, r = k + 1 = 4$, and $j = n + 1,\ldots, n + m$ is the amount of redundant members, the following series represent the possible arrangements of the failure events of the redundant members:

$$\underset{\substack{j=n+1\\k=3}}{\overset{n+m}{F}} = F_{n+1}\,\mathrm{I}\,F_{n+2}\,\mathrm{I}\,F_{n+3}\,\mathrm{I}\,F_{n+4},\ldots, F_{n+1}\,\mathrm{I}\,F_{n+2}\,\mathrm{I}\,F_{n+3}\,\mathrm{I}\,F_{n+m},$$

$$, F_{n+2}\,\mathrm{I}\,F_{n+3}\,\mathrm{I}\,F_{n+4}\,\mathrm{I}\,F_{n+5},\ldots, F_{n+2}\,\mathrm{I}\,F_{n+3}\,\mathrm{I}\,F_{n+4}\,\mathrm{I}\,F_{n+m},$$

$$,\ldots, F_{n+m-3}\,\mathrm{I}\,F_{n+m-2}\,\mathrm{I}\,F_{n+m-1}\,\mathrm{I}\,F_{n+m} \qquad (11)$$

Similar equations such as equations (10) and (11) can be generated for $k = 1,\ldots s$, where $s < m$. Also, The above explanation of $P(F_j)$ is only applicable for fully correlated members, ρ = 1. When $P(F_j)$ is incorporated in equation (8), the maximum of the minimum of each combination is compared with all the probability of failure of the non-redundant members, and the maximum probability of failure would be the probability of failure for the pre-failure phase of the system $P_{f_{system\_pre\_failure}}$. In order to get a mathematical expression when the members are not fully correlated or in some cases no correlation exists between members, the definition of $P(F_j)$ should be modified to assess for the correlation between the random variables. The first order bound or the second order bounds can be used in order to narrow the gap between the upper and lower bounds. When the redundant members are not correlated, ρ = 0, the component $P(F_j) =$ the product of the probability of the occurrence of the failure event, $F_j$, of each possible arrangement of the redundant members, according to the system degree of redundancy, $k$. When $P(F_j)$ is incorporated in equation (8), the maximum of the product of each combination is compared with all the probability of failure of the non-redundant members, and the maximum probability of failure would be the probability of failure for the pre-failure phase of the system $P_{f_{system\_pre\_failure}}$.

## 3.2. System Analysis in the Post-Failure Phase

The following examples present the probability of failure for the post-failure phase of any redundant system. The post-failure phase is defined as the phase when the system looses a redundant member or members. Let $l = 1,\dots,v$ represents the member or members failed in the system. Consider the previously presented four systems:

1. Example system one, as illustrated in figure 7: for this system, consider for example the failure of member 6. The degree of redundancy $k = 1$ and $l = 6$, this implies that the redundant member 6 has failed. The probability of failure in the post-failure phase of the system $P_{f_{system\_post\_failure}}$ is:

$$P_{f_{system\_post\_failure}} = \max_i (P_{f_i}, P_{f_{\substack{j=n+1 \\ j\neq l}}^{j=n+m}}, \min(P_{f_7}, P_{f_8}), \min(P_{f_7}, P_{f_9}), \min(P_{f_7}, P_{f_{10}}),$$

$$, \min(P_{f_7}, P_{f_9}), \min(P_{f_7}, P_{f_{10}}), \min(P_{f_8}, P_{f_9}),$$
$$, \min(P_{f_8}, P_{f_{10}}), \min(P_{f_9}, P_{f_{10}})) \quad (12)$$

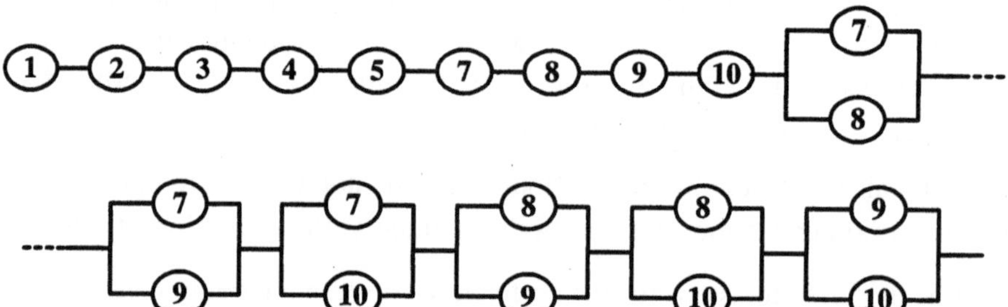

Figure 7: Post-failure Phase representation of System One

2. Example system two, as illustrated in figure 8: for this system, consider the failure of members 6 and 7. The degree of redundancy $k = 2$, $l = 6,7$, this implies that the two redundant members 6 and 7 have failed. The post-failure system is presented in figure 15. From figure 10, it can be noted that the combination of 8-9, 8-10, 9-10, do repeat, for the calculation of the probability of failure of the system, only once, the combination will be counted. The probability of failure for the post-failure phase of the system is:

$$P_{f_{system}} = \max_i (P_{f_i}, P_{f_{\substack{j=n+1 \\ j\neq l}}^{j=n+m}}, \min(P_{f_8}, P_{f_9}), \min(P_{f_8}, P_{f_{10}}),$$

$$, \min(P_{f_9}, P_{f_{10}}), \min(P_{f_8}, P_{f_9}, P_{f_{10}})) \quad (13)$$

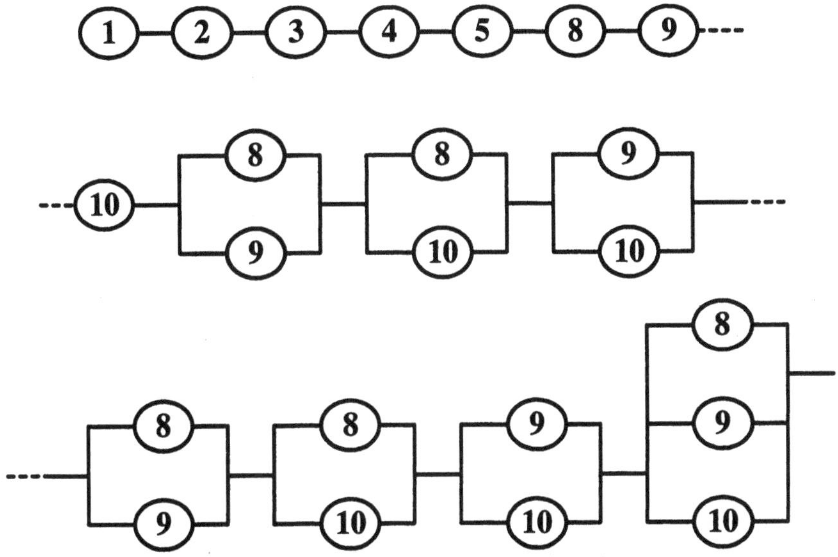

Figure 8: Post-failure phase representation of system two

3. Example system three, as illustrated in figure 9: for this system, consider for example the failure of members 6, 7, and 8. The degree of redundancy $k = 3, l = 6,7,8$, this implies that the three redundant members 6, 7, and 8, have failed. From figure 4.5.1.2.3, it can be noted that the combination of 9-10, do repeat, for the calculation of the probability of failure of the system, only once, the combination will be counted The probability of failure for the post-failure phase of the system is:

$$P_{f_{system}} = \max_{i}(P_{f_i}, P_{f\,\substack{j=n+m \\ j=n+1 \\ j \neq l}}, \min(P_{f_9}, P_{f_{10}})) \quad (14)$$

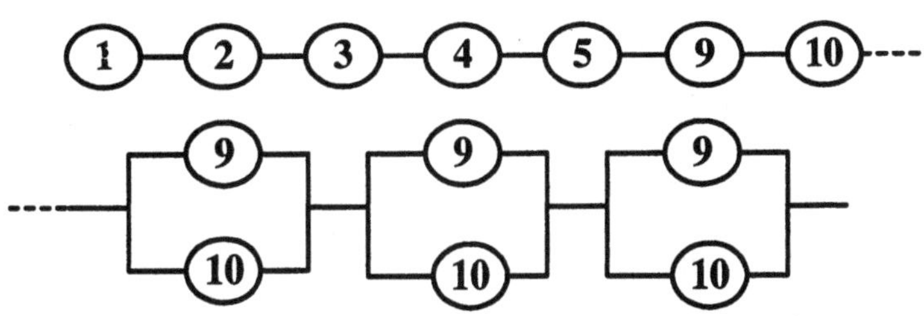

Figure 9: Post-failure phase representation of system three

4. Example system four, as illustrated in figure 10: for this system, consider for example the failure of members 6, 7, 8, and 9. The degree of redundancy $k = 4, l = 6,7,8,9$, this implies that the four redundant members 6, 7, and 8, have failed. The probability of failure for the post-failure phase of the system is:

$$P_{f_{system}} = \max_{i}(P_{f_i}, P_{f\,\substack{j=n+1\\ j\neq l}}^{\;n+m}) \tag{15}$$

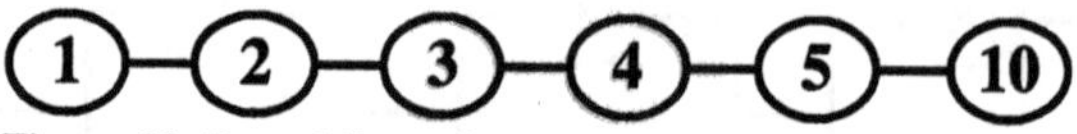


Figure 10: Post-failure phase representation of system four

From the above presented systems, it can be concluded that for redundant systems, in the post-failure phase, having $i = 1,\ldots,n$, $j = n+1, n+2,\ldots, n+m$, and $k = 1,\ldots, s$, $k$ is the degree of redundancy of the system, $l =$ the element or elements failed from the system. The probability of failure for the system of fully correlated components is:

$$P_{f_{systemp}} = \max_{i}(P_{f_i}, P_{f\,\substack{j=n+1\\ j\neq l}}^{\;j=n+m}) \tag{16}$$

It can be proven that when the members are fully correlated, the above presented methodology for redundant structural systems can also be used to evaluate effect of a failure of a two components when $k = 3$, for example, when used in its general form.

## 4. RELIABILITY MEASURES FOR REDUNDANT SYSTEMS

Based on the above two sections, the following equations can be derived for the pre-failure and post-failure phases from equations (8), and (16):
The probability of success of the system in the pre-failure phase is:

$$P_{s_{system\_pre\_failure}} = 1 - \max_{i,j}\left(P_{f_i}, P(F_j)\right) \tag{17}$$

The probability of success of the system in the post-failure phase is:

$$P_{s_{system\_post\_failure}} = 1 - \max_{i}(P_{f_i}, P_{f\,\substack{j=n+1\\ j\neq l}}^{\;j=n+m}) \tag{18}$$

The reliability index of the system in the pre-failure phase is:

$$\beta_{system\_pre\_failure} = -\Phi^{-1}\left(\max_{i,j}\left(P_{f_i}, P(F_j)\right)\right) \tag{19}$$

The reliability index of the system in the post-failure phase is:

$$\beta_{system\_post\_failure} = -\Phi^{-1}\left(\max_{i}(P_{f_i}, P_{f\,\substack{j=n+1\\ j\neq l}}^{\;j=n+m})\right) \tag{20}$$

# 5. RELIABILITY LIMITS FOR REDUNDANT STRUCTURAL SYSTEMS

In order to set limits for equations (17) through (20), for the redundancy evaluation of structural systems, a target reliability index should be chosen for the mentioned equations to be checked against, such as a target reliability index of $\beta_{t\arg et} = 2.0$.

The following limitations will depict if the system is redundant, in the pre-failure and post-failure phases, with respect to the pre-set target reliability index:

If: $\beta_{pre\_failure} > \beta_{post\_failure} > \beta_{t\arg et}$ (21)

The component, the subsystem, or the global system, is highly redundant.

If: $\beta_{pre\_failure} > \beta_{post\_failure} = \beta_{t\arg et}$ (22)

The component, the subsystem, or the global system, is redundant.

If: $\beta_{pre\_failure} > \beta_{post\_failure} < \beta_{t\arg et}$ (23)

The component, the subsystem, or the global system, is non-redundant.

## 5.1. Comments on the Presented Methodology

In this chapter, a new methodology for the analysis of redundant structural systems is presented along with twelve general equations to evaluate the probability of failure and the reliability index of the system in the pre-failure phase post-failure phases. The mentioned methodology is a reliable tool for evaluating the true redundancy of existing structural systems. Based on the system's degree of redundancy, the proposed methodology accounts for the possible amount of redundant members' combinations to be integrated in the system's block diagram. The presented methodology can also be applied on a family of structural systems, in order to generate safety factors coefficients to be incorporated in future codes.

## 5.2. Application of the Proposed Method on an Existing Concrete Structure

Based on the above presented methodology for structural system analysis, a complete risk assessment is being carried out. The reliability indices based on ultimate limit states functions (such as moment and shear capacities) and serviceability limit state functions (Such as deflection) for each component are calculated. The lowest reliability index related to a specific limit state function, of each component leading to the highest probability of failure, is considered. Table 1 shows the results of the reliability index and related probability of failure for the beams. All failure modes related to each structural component are laid in series configuration.

The structure's components along with their related failure modes are then laid in parallel and series configuration based on the previously mentioned methodology, in order to evaluate the reliability of the structural system. The structural reliability index of the system is found to be, $\beta = 1.22$. This leads to a probability of failure of the structural system of 11%.

An independent floor analysis is run, with the removal of beam 10 due to major cracks observed during site investigation (Figure 11). Accordingly, the floor is reanalyzed. It was noted that beams P6, P9, and P11 were overloaded. The probability of failure of the structural system is then recalculated and it is found to be 13% ($\beta = 1.13$).

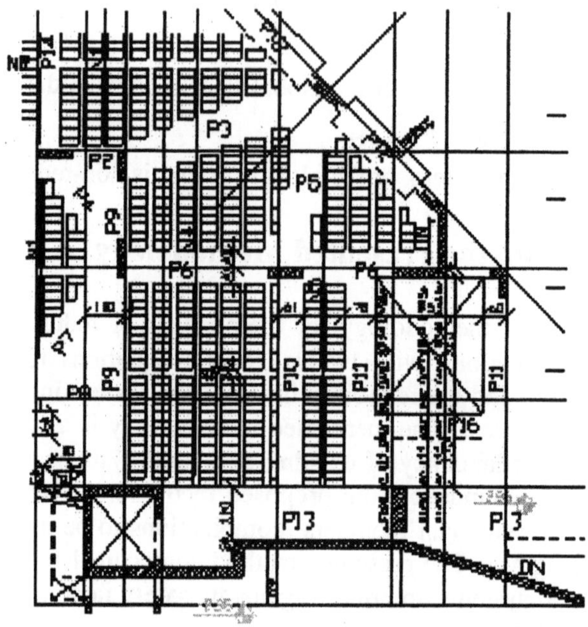


Figure 11: Part of structural drawing of the building

Table 1: Typical Reliability analysis of beams

| Beam Name | β for M+ | βfor M- | β for V | $P_f$ of Beam (%) |
|---|---|---|---|---|
| Beam 1 | 4.93 | 1.65 | 6.22 | 4.947145 |
| Beam 2 | 4.44 | 0.98 | 1.8 | 16.35430 |
| Beam 2b | 10.86 | 2.5 | 8.01 | 0.620968 |
| Beam 3 | 2.32 | 2.28 | 2.21 | 1.355253 |
| Beam 4 | 8.22 | 3.54 | 8.51 | 0.020010 |
| Beam 5 | 5.7 | 5.64 | 0.49 | 31.20669 |
| Beam 6 | 2.54 | 1.65 | 1.84 | 4.947145 |
| Beam 6b | 3.64 | 2.53 | 5.6 | 0.570314 |
| Beam 7 | 4.51 | 2.72 | 6.67 | 0.326414 |
| Beam 8 | 2.63 | 1.28 | 2.033 | 10.02726 |
| Beam 8b | 9.25 | 1.81 | 8.02 | 3.514783 |
| Beam 9 | 5.09 | 3.74 | ___ | 0.009203 |
| Beam 10 | 7.75 | 2.63 | ___ | 0.426928 |
| Beam 11 | 7.35 | 3.38 | ___ | 0.036248 |
| Beam 12 | 3.94 | 0.42 | 5.12 | 33.72427 |
| Beam 13 | 2.66 | 2.38 | 6.65 | 0.86563075 |
| Beam 14 | 5.37 | 7.89 | ___ | 3.9455E- |
| Beam 15 | 6.06 | 2.65 | ___ | 0.402463 |

## 6. CONCLUSION AND FUTURE RESEARCH

System analysis is used to determine the probability of failure of the structure based on the presented methodology. The probability of failure is determined to be relatively high (11%). The system probability of failure increased to 13% when beam 10 was removed from the floor analysis.

Through the complete analysis, the contribution of each component to the whole structure is clearly noted. Moreover, through system analysis it is noted that the foundation reliability affected drastically the reliability of the structural system. This is due to the fact that the footings are in series configuration with the whole structure. Hence a high probability of failure of the footing led to a high probability of failure of the structure. It may be noted also that an improvement in the structural design and construction of some of the footings will help increase the reliability of the structure.

As a result, the presented methodology helped in detecting the weak components of the structure. This may lead to a possible improvement that could be done to increase the reliability of the structure. However, it is important to be able to detect the structural topology in order to safely evaluate the reliability of the system. Also, incorporating a smart deterioration model into the assessment will help in truly finding the reliability of existing structures.

# REFERENCES

Ang, A.H., and Tang, W.H., 1975. "Probability Concepts in Engineering Planning and Design," Volume I, John Willey and Sons, New York.

Ang, A.H., and Tang, W.H., 1984. "Probability Concepts in Engineering Planning and Design," Volume II, John Willey and Sons, New York.

Ayyub, B.M, and McCuen R.H., 1997. "Probability, Statistics, & Reliability for Engineers" CRC Press, Florida.

Borri, A., and Speranzini E., 1997. "Structural Reliability Analysis Using a Standard Deterministic Finite Element Code," Structural Safety, Elsevier Science Ltd., 19(4), 361-382.

Ellingwood, B. R., 1996. "Reliability-based Condition Assessment and LRFD for Existing Structures," Structural Safety, Elsevier Science Ltd., 18(2/3), 67-80.

Elms, D. G., 1999. "Achieving Structural Safety: Theoretical Considerations," Structural Safety, Elsevier Science Ltd., 21(1999), 311-333.

Frangopol, D.M., and Curley, J.P., 1987. "Damage States, Redundancy and System Strength," In Effect of Damage and Redundancy on Structural Performance, (edited by D.M. Frangopol), ASCE, New York.

Frangopol, D.M., and Nakib, R., 1991. "Redundancy in Highway Bridges." Engineering Journal, AISC, 28(1), 45-50.

Freudenthal, A. M., 1947. "The Safety of Structures," Transactions of ASCE, 112 (2296), 125-180.

Ghosn, M., and Moses, F. (1998). "Redundancy in Highway Bridges." NCHRP report 406, Department of Civil Engineering, The City College of The City University of New York, New York, NY.

Kudsi, T.N. (2001) "Redundancy Analysis of Existing Truss Bridges: A System Reliability-Based Approach," in partial fulfillment of the requirement for the degree of Doctor of Philosophy, Department of Civil and Environmental Engineering, University of Maryland, College Park, MD 20742, (2001).

Melchers, R.E., 1987. "Structural Reliability: Analysis and Predictions," Ellis Horwood Ltd., Chichester.

Nowak, A.S., 1995. "Reliability of Structures," Class Notes, Department of Civil and Environmental Engineering, University of Michigan, Ann Arbor, Michigan.

Tang, J.P., and Yao, J.T.P., 1987. "Evaluation of Structural Damage and Redundancy," In Effect of Damage and Redundancy on Structural Performance, (edited by D.M. Frangopol), ASCE, New York.

Thoft-Christensen, P., and Murotsu, Y., 1986. "Application of structural system reliability theory," Springer-Verlag, Berlin.

# INDEX

MIX
Papier aus verantwortungsvollen Quellen
Paper from responsible sources
FSC® C105338

If you have any concerns about our products,
you can contact us on
**ProductSafety@springernature.com**

In case Publisher is established outside the EU,
the EU authorized representative is:
**Springer Nature Customer Service Center GmbH**
**Europaplatz 3, 69115 Heidelberg, Germany**

Printed by Libri Plureos GmbH
in Hamburg, Germany